T0178334

Modeling and Simulation in Science, Engineering and Technology

More information about this series at http://www.springer.com/series/4960

Vincenzo Capasso • David Bakstein

An Introduction to Continuous-Time Stochastic Processes

Theory, Models, and Applications to Finance, Biology, and Medicine

Third Edition

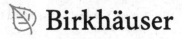

Vincenzo Capasso
ADAMSS (Interdisciplinary Centre for
  Advanced Applied Mathematical
  and Statistical Sciences)
Università degli Studi di Milano
Milan, Italy

David Bakstein
ADAMSS (Interdisciplinary Centre for
  Advanced Applied Mathematical
  and Statistical Sciences)
Università degli Studi di Milano
Milan, Italy

ISSN 2164-3679   ISSN 2164-3725 (electronic)
Modeling and Simulation in Science, Engineering and Technology
ISBN 978-1-4939-3836-0   ISBN 978-1-4939-2757-9 (eBook)
DOI 10.1007/978-1-4939-2757-9

Mathematics Subject Classification (2010): 60-01, 60FXX, 60GXX, 60G07, 60G10, 60G15, 60G22, 60G44, 60G51, 60G52, 60G57, 60H05, 60H10, 60H30, 60J25, 60J35, 60J60, 60J65, 60K35, 91GXX, 92BXX, 93E05, 93E15

Springer New York Heidelberg Dordrecht London
© Springer Science+Business Media New York 2005, 2012, 2015
Softcover reprint of the hardcover 3rd edition 2015

Printed on acid-free paper

Springer Science+Business Media LLC New York is part of Springer Science+Business Media (www.springer.com)

Preface to the Third Edition

In this third edition, we have included additional material for use in modern applications of stochastic calculus in finance and biology; in particular, Chap. 5 on stability and ergodicity is completely new. We have thought that this is an important addition for all those who use stochastic models in their applications.

The sections on infinitely divisible distributions and stable laws in Chap. 1, random measures, and Lévy processes in Chap. 2, Itô–Lévy calculus in Chap. 3, and Chap. 4, have been completely revisited.

Fractional calculus has gained a significant additional room, as requested by various applications.

The Karhunen-Loève expansion has been added in Chap. 2, as a useful mathematical tool for dealing with stochastic processes in statistics and in numerical analysis.

Various new examples and exercises have been added throughout the volume in order to guide the reader in the applications of the theory. The bibliography has been updated and significantly extended.

We have also made an effort to improve the presentation of parts already included in the previous editions, and we have corrected various misprints and errors made aware of by colleagues and students during class use of the book in the intervening years.

We are very grateful to all those who helped us in detecting them and suggested possible improvements. We are very grateful to Giacomo Aletti, Enea Bongiorno, Daniela Morale, and the many students for checking the final proofs and suggesting valuable changes. Among these, the PhD students Stefano Belloni and Sven Stodtmann at the University of Heidelberg deserve particular credit. Kevin Payne and (as usual) Livio Pizzocchero have been precious for bibliographical references, and advice.

Enea Bongiorno deserves once again special mention for his accurate final editing of the book.

We wish to pay our gratitude to Avner Friedman for having allowed us to grasp many concepts and ideas, if not pieces, from his vast volume of publications.

Allen Mann from Birkhäuser in New York deserves acknowledgement for encouraging the preparation of this third edition.

Last but not least, we acknowledge the precious editorial work of the many (without specific names) at Birkhäuser, who have participated in the preparation of the book.

Most of the preparation of this third edition has been carried out during the stays of VC at the Heidelberg University (which he wishes to acknowledge for support by BIOMS, IWR, and the local HGS), and at the "Carlos III" University in Madrid (which he wishes to thank for having offered him a Chair of Excellence there).

Milan, Italy   Vincenzo Capasso
Milan, Italy   David Bakstein

Preface to the Second Edition

In this second edition, we have included additional material for use in modern applications of stochastic calculus in finance and biology; in particular, the section on infinitely divisible distributions and stable laws in Chap. 1, Lévy processes in Chap. 2, the Itô–Lévy calculus in Chap. 3, and Chap. 4. Finally, a new appendix has been added that includes basic facts about semigroups of linear operators.

We have also made an effort to improve the presentation of parts already included in the first edition, and we have corrected the misprints and errors we have been made aware of by colleagues and students during class use of the book in the intervening years. We are very grateful to all those who helped us in detecting them and suggested possible improvements. We are very grateful to Giacomo Aletti, Enea Bongiorno, Daniela Morale, Stefania Ugolini, and Elena Villa for checking the final proofs and suggesting valuable changes.

Enea Bongiorno deserves special mention for his accurate editing of the book as you now see it.

Tom Grasso from Birkhäuser deserves acknowledgement for encouraging the preparation of a second, updated edition.

Milan, Italy   Vincenzo Capasso
Milan, Italy   David Bakstein

Preface to the First Edition

This book is a systematic, rigorous, and self-contained introduction to the theory of continuous-time stochastic processes. But it is neither a tract nor a recipe book as such; rather, it is an account of fundamental concepts as they appear in relevant modern applications and the literature. We make no pretense of its being complete. Indeed, we have omitted many results that we feel are not directly related to the main theme or that are available in easily accessible sources. Readers interested in the historical development of the subject cannot ignore the volume edited by Wax (1954).

Proofs are often omitted as technicalities might distract the reader from a conceptual approach. They are produced whenever they might serve as a guide to the introduction of new concepts and methods to the applications; otherwise, explicit references to standard literature are provided. A mathematically oriented student may find it interesting to consider proofs as exercises.

The scope of the book is profoundly educational, related to modeling real-world problems with stochastic methods. The reader becomes critically aware of the concepts involved in current applied literature and is, moreover, provided with a firm foundation of mathematical techniques. Intuition is always supported by mathematical rigor.

Our book addresses three main groups of readers: first, mathematicians working in a different field; second, other scientists and professionals from a business or academic background; third, graduate or advanced undergraduate students of a quantitative subject related to stochastic theory or applications.

As stochastic processes (compared to other branches of mathematics) are relatively new, yet increasingly popular in terms of current research output and applications, many pure as well as applied deterministic mathematicians have become interested in learning about the fundamentals of stochastic theory and modern applications. This book is written in a language that both groups will understand, and its content and structure will allow them to learn the essentials profoundly and in a time-efficient manner. Other scientist-practitioners and academics from fields like finance, biology, and medicine might be very familiar with a less mathematical approach to their specific fields and thus be interested in learning the mathematical techniques of modeling their applications.

Furthermore, this book would be suitable as a textbook accompanying a graduate or advanced undergraduate course or as secondary reading for students of mathematical or computational sciences. The book has evolved from course material that has already been tested for many years in various courses in engineering, biomathematics, industrial mathematics, and mathematical finance.

Last, but certainly not least, this book should also appeal to anyone who would like to learn about the mathematics of stochastic processes. The reader will see that previous exposure to probability, though helpful, is not essential and that the fundamentals of measure and integration are provided in a self-contained way. Only familiarity with calculus and some analysis is required.

The book is divided into three main parts. In Part I, comprising Chaps. 1–4, we introduce the foundations of the mathematical theory of stochastic processes and stochastic calculus, thereby providing the tools and methods needed in Part II (Chaps. 6 and 7), which is dedicated to major scientific areas of application. The third part consists of appendices, each of which gives a basic introduction to a particular field of fundamental mathematics (e.g., measure, integration, metric spaces) and explains certain problems in greater depth (e.g., stability of ODEs) than would be appropriate in the main part of the text.

In Chap. 1 the fundamentals of probability are provided following a standard approach based on Lebesgue measure theory due to Kolmogorov. Here the guiding textbook on the subject is the excellent monograph by Métivier (1968). Basic concepts from Lebesgue measure theory are also provided in Appendix A.

Chapter 2 gives an introduction to the mathematical theory of stochastic processes in continuous time, including basic definitions and theorems on processes with independent increments, martingales, and Markov processes. The two fundamental classes of processes, Poisson and Wiener, are introduced as well as the larger, more general, class of Lévy processes. Further, a significant introduction to marked point processes is also given as a support for the analysis of relevant applications.

Chapter 3 is based on Itô theory. We define the Itô integral, some fundamental results of Itô calculus, and stochastic differentials including Itô's formula, as well as related results like the martingale representation theorem.

Chapter 4 is devoted to the analysis of stochastic differential equations driven by Wiener processes and Itô diffusions and demonstrates the connections with partial differential equations of second order, via Dynkin and Feynman–Kac formulas.

Chapter 6 is dedicated to financial applications. It covers the core economic concept of arbitrage-free markets and shows the connection with martingales and Girsanov's theorem. It explains the standard Black–Scholes theory and relates it to Kolmogorov's partial differential equations and the Feynman–Kac formula. Furthermore, extensions and variations of the standard theory are discussed as are interest rate models and insurance mathematics.

Chapter 7 presents fundamental models of population dynamics such as birth and death processes. Furthermore, it deals with an area of important modern research—the fundamentals of self-organizing systems, in particular focusing on the social behavior of multiagent systems, with some applications to economics ("price herding"). It also includes a particular application to the neurosciences, illustrating the importance of stochastic differential equations driven by both Poisson and Wiener processes.

Problems and additions are proposed at the end of the volume, listed by chapter. In addition to exercises presented in a classical way, problems are proposed as a stimulus for discussing further concepts that might be of interest to the reader. Various sources have been used, including a selection of problems submitted to our students over the years. This is why we can provide only selected references.

The core of this monograph, on Itô calculus, was developed during a series of courses that one of the authors, VC, has been offering at various levels in many universities. That author wishes to acknowledge that the first drafts of the relevant chapters were the outcome of a joint effort by many participating students: Maria Chiarolla, Luigi De Cesare, Marcello De Giosa, Lucia Maddalena, and Rosamaria Mininni, among others. Professor Antonio Fasano is due our thanks for his continuous support, including producing such material as lecture notes within a series that he coordinated.

It was the success of these lecture notes, and the particular enthusiasm of coauthor DB, who produced the first English version (indeed, an unexpected Christmas gift), that has led to an extension of the material up to the present status, including, in particular, a set of relevant and updated applications that reflect the interests of the two authors.

VC would also like to thank his first advisor and teacher, Professor Grace Yang, who gave him the first rigorous presentation of stochastic processes and mathematical statistics at the University of Maryland at College Park, always referring to real-world applications. DB would like to thank the Meregalli and Silvestri families for their kind logistical help while he was in Milan. He would also like to acknowledge research funding from the EPSRC, ESF, Socrates–Erasmus, and Charterhouse and thank all the people he worked with at OCIAM, University of Oxford, over the years, as this is where he was based when embarking on this project.

The draft of the final volume was carefully read by Giacomo Aletti, Daniela Morale, Alessandra Micheletti, Matteo Ortisi, and Enea Bongiorno (who also took care of the problems and additions) whom we gratefully acknowledge.

Still, we are sure that some odd typos and other, hopefully noncrucial, mistakes remain, for which the authors take full responsibility.

We also wish to thank Professor Nicola Bellomo, editor of the "Modeling and Simulation in Science, Engineering and Technology" series, and Tom Grasso from Birkhäuser for supporting the project. Last but not least, we cannot neglect to thank Rossana (VC) and Casilda (DB) for their patience and great tolerance while coping with their "solitude" during the preparation of this monograph.

Milan, Italy   Vincenzo Capasso
Milan, Italy   David Bakstein

Contents

Part II   Applications of Stochastic Processes

Part I

Theory of Stochastic Processes

1

Fundamentals of Probability

We assume that the reader is already familiar with the basic motivations and notions of probability theory. In this chapter we recall the main mathematical concepts, methods, and theorems according to the Kolmogorov approach Kolmogorov (1956) by using as main references the books by Métivier (1968) and Neveu (1965). An interesting introduction can be found in Gnedenko (1963). We shall refer to Appendix A of this book for the required theory on measure and integration.

1.1 Probability and Conditional Probability

Definition 1.1. A *probability space* is an ordered triple $(\Omega, \mathcal{F}, P)$, where Ω is any set, $\mathcal{F}$ a σ-algebra of subsets of Ω, and $P : \mathcal{F} \to [0,1]$ a probability measure on $\mathcal{F}$ such that

1. $P(\Omega) = 1$ (and $P(\emptyset) = 0$).
2. For all $A_1, \ldots, A_n, \ldots \in \mathcal{F}$ with $A_i \cap A_j = \emptyset$, $i \neq j$:

$$P\left(\bigcup_i A_i\right) = \sum_i P(A_i).$$

The set Ω is called the *sample space*, $\emptyset$ the *empty set*, the elements of $\mathcal{F}$ are called *events*, and every element of Ω is called an *elementary event*.

Definition 1.2. A probability space $(\Omega, \mathcal{F}, P)$ is *finite* if Ω has finitely many elementary events.

Remark 1.3. If Ω is finite, then it suffices to only consider the σ-algebra of all subsets of Ω, i.e., $\mathcal{F} = \mathfrak{P}(\Omega)$.

Definition 1.4. Every finite probability space $(\Omega, \mathcal{F}, P)$ with $\mathcal{F} = \mathfrak{P}(\Omega)$ is an *equiprobable* or *uniform* space if

© Springer Science+Business Media New York 2015
V. Capasso, D. Bakstein, *An Introduction to Continuous-Time Stochastic Processes*, Modeling and Simulation in Science, Engineering and Technology, DOI 10.1007/978-1-4939-2757-9_1

$$\forall \omega \in \Omega : \qquad P(\{\omega\}) = k \text{ (constant)},$$

i.e., its elementary events are equiprobable.

Remark 1.5. Following the axioms of a probability space and the definition of a uniform space, if $(\Omega, \mathcal{F}, P)$ is equiprobable, then

$$\forall \omega \in \Omega : \qquad P(\{\omega\}) = \frac{1}{|\Omega|},$$

where $|\cdot|$ denotes the cardinal number of elementary events in Ω, and

$$\forall A \in \mathcal{F} \equiv \mathfrak{P}(\Omega) : \qquad P(A) = \frac{|A|}{|\Omega|}.$$

Intuitively, in this case we may say that $P(A)$ is the ratio of the number of favorable outcomes divided by the number of all possible outcomes.

Example 1.6. Consider an urn that contains 100 balls, of which 80 are red and 20 are black but which are otherwise identical, from which a player draws a ball. Define the event

$$R : \text{ The first drawn ball is red.}$$

Then

$$P(R) = \frac{|R|}{|\Omega|} = \frac{80}{100} = 0.8.$$

Definition 1.7. We shall call any event $F \in \mathcal{F}$ such that $P(F) = 0$, a *null event*.

Conditional Probability

Let $(\Omega, \mathcal{F}, P)$ be a probability space and $A, B \in \mathcal{F}$, with $P(B) > 0$. Under these assumptions the following equality is trivial:

$$P(A \cap B) = P(B) \frac{P(A \cap B)}{P(B)}.$$

In general one cannot expect that

$$P(A \cap B) = P(A) P(B),$$

in which case we would have

$$\frac{P(A \cap B)}{P(B)} = P(A).$$

This special case will be analyzed later. In general it makes sense to consider the following definition.

Definition 1.8. Let $(\Omega, \mathcal{F}, P)$ be a probability space and $A, B \in \mathcal{F}$, $P(B) > 0$. Then *the probability of A conditional on B*, denoted by $P(A|B)$, is any real number in $[0, 1]$ such that

$$P(A|B) = \frac{P(A \cap B)}{P(B)}.$$

This number is left unspecified whenever $P(B) = 0$.

We must at any rate notice that conditioning events of zero probability cannot be ignored. See later a more detailed account of this case in connection with the definition of conditional distributions.

Remark 1.9. Suppose that $P(B) > 0$. Then the mapping

$$P_B : A \in \mathcal{F} \mapsto P_B(A) = \frac{P(A \cap B)}{P(B)} \in [0, 1]$$

defines a probability measure P_B on $\mathcal{F}$. In fact, $0 \leq P_B(A) \leq 1$ and $P_B(\Omega) = \frac{P(B)}{P(B)} = 1$. Moreover, if $A_1, \ldots, A_n, \ldots \in \mathcal{F}$, $A_i \cap A_j = \emptyset$, $i \neq j$, then

$$P_B \left(\bigcup_{n \in \mathbb{N}} A_n \right) = \frac{P(\bigcup_n A_n \cap B)}{P(B)} = \frac{\sum_n P(A_n \cap B)}{P(B)} = \sum_n P_B(A_n).$$

From the preceding construction it follows that the probability measure P_B is an additional probability measure on $\mathcal{F}$ that in particular satisfies

$$P_B(B) = 1,$$

and for any event $C \in \mathcal{F}$ such that $B \subset C$ we have

$$P_B(C) = 1,$$

while if $C \cap B = \emptyset$, then

$$P_B(C) = 0.$$

It makes sense, then, to introduce the following definition.

Definition 1.10. Let $(\Omega, \mathcal{F}, P)$ be a probability space and $A, B \in \mathcal{F}$, $P(B) > 0$. Then the probability measure $P_B : \mathcal{F} \to [0, 1]$, such that

$$\forall A \in \mathcal{F}: \qquad P_B(A) := \frac{P(A \cap B)}{P(B)},$$

is called the *the conditional probability on $\mathcal{F}$ given B*.

Proposition 1.11. *If $A, B \in \mathcal{F}$, then*

1. $P(A \cap B) = P(A|B)P(B) = P(B|A)P(A)$.
2. If $A_1, \ldots, A_n \in \mathcal{F}$, then

$$P(A_1 \cap \cdots \cap A_n) = P(A_1)P(A_2|A_1)P(A_3|A_1 \cap A_2) \cdots P(A_n|A_1 \cap \cdots \cap A_{n-1}).$$

Proof. Statement 1 is obvious. Statement 2 is proved by induction. The proposition holds for $n = 2$. Assuming it holds for $n - 1$, $(n \geq 3)$, we obtain

$$P(A_1 \cap \cdots \cap A_n)$$
$$= P(A_1 \cap \cdots \cap A_{n-1})P(A_n|A_1 \cap \cdots \cap A_{n-1})$$
$$= P(A_1) \cdots P(A_{n-1}|A_1 \cap \cdots \cap A_{n-2})P(A_n|A_1 \cap \cdots \cap A_{n-1});$$

thus it holds for n as well. Since n was arbitrary, the proof is complete. $\square$

Definition 1.12. Two events A and B are *independent* if

$$P(A \cap B) = P(A)P(B).$$

Thus A is independent of B if and only if B is independent of A, and vice versa.

Proposition 1.13. *Let A, B be events and $P(A) > 0$; then, the following two statements are equivalent:*

1. A and B are independent.
2. $P(B|A) = P(B)$.

If $P(B) > 0$, then the statements hold with interchanged A and B as well.

Example 1.14. Considering the same experiment as in Example 1.6, we define the additional events (x, y) with $x, y \in \{B, R\}$ as, e.g.,

$$BR : \text{The first drawn ball is black, the second red,}$$
$$\cdot R : \text{The second drawn ball is red.}$$

Now the probability $P(\cdot R|R)$ depends on the rules of the draw.

1. If the draw is with subsequent replacement of the ball, then, due to the independence of the draws,

$$P(\cdot R|R) = P(\cdot R) = P(R) = 0.8.$$

2. If the draw is without replacement, then the second draw is dependent on the outcome of the first draw, and we have

$$P(\cdot R|R) = \frac{P(\cdot R \cap R)}{P(R)} = \frac{P(RR)}{P(R)} = \frac{80 \cdot 79 \cdot 100}{100 \cdot 99 \cdot 80} = \frac{79}{99}.$$

Definition 1.15. Two events A and B are *mutually exclusive* if $A \cap B = \emptyset$.

Proposition 1.16.

1. *Two events cannot be both independent and mutually exclusive, unless one of the two is a null event.*
2. *If A and B are independent events, then so are A and $\bar{B}$, $\bar{A}$ and B, and $\bar{A}$ and $\bar{B}$, where $\bar{A} := \Omega \setminus A$ is the complementary event.*

Definition 1.17. The events A, B, C are *independent* if

1. $P(A \cap B) = P(A)P(B)$
2. $P(A \cap C) = P(A)P(C)$
3. $P(B \cap C) = P(B)P(C)$
4. $P(A \cap B \cap C) = P(A)P(B)P(C)$

This definition can be generalized to any number of events.

Remark 1.18. If A, B, C are events that satisfy point 4 of Definition 1.17, then it is not true in general that it satisfies points 1–3 and vice versa.

Example 1.19. Consider a throw of two distinguishable, fair six-sided dice, and the events

$\quad$ A: the roll of the first dice results in 1, 2, or 5,

$\quad$ B: the roll of the second dice results in 4, 5, or 6,

$\quad$ C: the sum of the results of the rolls of the dice is 9.

Then $P(A) = P(B) = 1/2$ and $P(A \cap B) = 1/6 \neq 1/4 = P(A)P(B)$. But since $P(C) = 1/9$ and $P(A \cap B \cap C) = 1/36$, we have that

$$P(A)P(B)P(C) = \frac{1}{36} = P(A \cap B \cap C).$$

On the other hand, consider a uniformly shaped tetrahedron that has the colors white, green, and red on its separate surfaces and all three colors on the fourth. If we randomly choose one side, the events

$\quad$ W: the surface contains white,

$\quad$ G: the surface contains green,

$\quad$ R: the surface contains red

have the probabilities $P(W) = P(G) = P(R) = 1/2$. Hence $P(W \cap G) = P(W)P(G) = 1/4$, etc., but $P(W)P(G)P(R) = 1/8 \neq 1/4 = P(W \cap G \cap R)$.

Definition 1.20. Let $\mathcal{C}_1, \ldots, \mathcal{C}_k$ be subfamilies of the σ-algebra $\mathcal{F}$. They constitute k *mutually independent* classes of $\mathcal{F}$ if

$$\forall A_1 \in \mathcal{C}_1, \dots, \forall A_k \in \mathcal{C}_k : \qquad P(A_1 \cap \dots \cap A_k) = \prod_{i=1}^{k} P(A_i).$$

Definition 1.21. A family of elements $(B_i)_{i \in I}$ of $\mathcal{F}$, with $I \subset \mathbb{N}$, is called a *(countable) partition* of Ω if

1. I is a countable set
2. $i \neq j \Rightarrow B_i \cap B_j = \emptyset$
3. $P(B_i) \neq 0$ for all $i \in I$
4. $\Omega = \bigcup_{i \in I} B_i$

Theorem 1.22(Total law of probability). *Let $(B_i)_{i \in I}$ be a partition of Ω and $A \in \mathcal{F}$; then*

$$P(A) = \sum_{i \in I} P(A|B_i)P(B_i).$$

Proof.

$$\sum_i P(A|B_i)P(B_i) = \sum_i \frac{P(A \cap B_i)}{P(B_i)} P(B_i) = \sum_i P(A \cap B_i)$$

$$= P\left(\bigcup_i (A \cap B_i)\right) = P\left(A \cap \bigcup_i B_i\right)$$

$$= P(A \cap \Omega) = P(A).$$

$\square$

The following fundamental Bayes theorem provides a formula for the exchange of conditioning between two events; this is why it is also known as the *theorem for probability of causes.*

Theorem 1.23 (Bayes). *Let $(B_i)_{i \in I}$ be a partition of Ω and $A \in \mathcal{F}$, with $P(A) = 0$; then*

$$\forall i \in I : \qquad P(B_i|A) = \frac{P(B_i)}{P(A)} P(A|B_i) = \frac{P(A|B_i)P(B_i)}{\sum_{j \in I} P(A|B_j)P(B_j)}.$$

Proof. Since $A = \bigcup_{j=1}^{k} (B_j \cap A)$, then

$$P(A) = \sum_{j=1}^{k} P(B_j)P(A|B_j).$$

Also, because

$$P(B_i \cap A) = P(A)P(B_i|A) = P(B_i)P(A|B_i)$$

and by the total law of probability, we obtain

$$P(B_i|A) = \frac{P(B_i)P(A|B_i)}{P(A)} = \frac{P(B_i)P(A|B_i)}{\sum_{j=1}^{k} P(A|B_j)P(B_j)}.$$

$\square$

Example 1.24. Continuing with the experiment of Example 1.6, we further assume that there is a second indistinguishable urn (U_2) containing 40 red balls and 40 black balls. Randomly drawing one ball from one of the two urns, we make a probability estimate about which urn we had chosen:

$$P(U_2|B) = \frac{P(U_2)P(B|U_2)}{\sum_{i=1}^{2} P(U_i)P(B|U_i)} = \frac{1/2 \cdot 1/2}{1/2 \cdot 1/5 + 1/2 \cdot 1/2} = \frac{5}{7};$$

thus, $P(U_1|B) = 2/7$.

1.2 Random Variables and Distributions

A random variable is the concept of assigning a numerical magnitude to elementary outcomes of a random experiment, measuring certain of the latter's characteristics. Mathematically, we define it as a function $X : \Omega \to \mathbb{R}$ on the probability space $(\Omega, \mathcal{F}, P)$ such that for every elementary $\omega \in \Omega$ it assigns a numerical value $X(\omega)$. In general, we are then interested in finding the probabilities of events of the type

$$[X \in B] := \{\omega \in \Omega | X(\omega) \in B\} \subset \Omega \tag{1.1}$$

for every $B \subset \mathbb{R}$, i.e., the probability that the random variable will assume values that will lie within a certain range $B \subset \mathbb{R}$. In its simplest case, B can be a possibly unbounded interval or union of intervals of $\mathbb{R}$. More generally, B can be any subset of the Borel σ-algebra $\mathcal{B}_\mathbb{R}$, which is generated by the intervals of $\mathbb{R}$. This will require, among other things, the results of measure theory and Lebesgue integration in $\mathbb{R}$. Moreover, we will require the events (1.1) to belong to $\mathcal{F}$, and so to be P-measurable. We will later extend the concept of random variables to generic measurable spaces.

Definition 1.25. Let $(\Omega, \mathcal{F}, P)$ be a probability space. A real-valued *random variable* is any Borel-measurable mapping $X : \Omega \to \mathbb{R}$ such that for any $B \in \mathcal{B}_\mathbb{R} : X^{-1}(B) \in \mathcal{F}$. It will be denoted by $X : (\Omega, \mathcal{F}) \to (\mathbb{R}, \mathcal{B}_\mathbb{R})$. If X takes values in $\bar{\mathbb{R}}$, then it is said to be *extended*.

Definition 1.26. If $X : (\Omega, \mathcal{F}) \to (\mathbb{R}, \mathcal{B}_{\mathbb{R}})$ is a random variable, then the mapping $P_X : \mathcal{B}_{\mathbb{R}} \to \mathbb{R}$, where

$$P_X(B) = P(X^{-1}(B)) = P([X \in B]), \qquad \forall B \in \mathcal{B}_{\mathbb{R}},$$

is a probability on $\mathbb{R}$. It is called the *probability law* of X.

If a random variable X has a probability law P_X, we will use the notation $X \sim P_X$.

The following proposition shows that a random variable can be defined in a canonical way in terms of a given probability law on $\mathbb{R}$.

Proposition 1.27. *If $P : \mathcal{B}_{\mathbb{R}} \to [0, 1]$ is a probability, then there exists a random variable $X : \mathbb{R} \to \mathbb{R}$ such that P is identical to the probability law P_X associated with X.*

Proof. We identify $(\mathbb{R}, \mathcal{B}_{\mathbb{R}}, P)$ as the underlying probability space so that the mapping $X : \mathbb{R} \to \mathbb{R}$, with $X(s) = s$, for all $s \in \mathbb{R}$, is a random variable, and furthermore, denoting its associated probability law by P_X, we obtain

$$P_X(B) = P(X^{-1}(B)) = P(B) \qquad \forall B \in \mathcal{B}_{\mathbb{R}}.$$

$\square$

Definition 1.28. Let $X : (\Omega, \mathcal{F}) \to (\mathbb{R}, \mathcal{B}_{\mathbb{R}})$ be a random variable; the σ-algebra $\mathcal{F}_X := X^{-1}(\mathcal{B}_{\mathbb{R}})$ is called the *σ-algebra generated by X*.

Lemma 1.29. (Doob–Dynkin). *If $X, Y : \Omega \to \mathbb{R}^d$, then Y is $\mathcal{F}_X$-measurable if and only if there exists a Borel measurable function $g : \mathbb{R}^d \to \mathbb{R}^d$ such that $Y = g(X)$.*

Definition 1.30. Let X be a random variable. Then the mapping

$$F_X : \mathbb{R} \to [0, 1],$$

with

$$F_X(t) = P_X(]-\infty, t]) = P([X \le t]) \qquad \forall t \in \mathbb{R},$$

is called the *partition function* or *cumulative distribution function* of X.

Proposition 1.31.

1. *For all $a, b \in \mathbb{R}$, $a < b$: $F_X(b) - F_X(a) = P_X(]a, b])$.*
2. *F_X is right-continuous and increasing.*
3. *$\lim_{t \to +\infty} F_X(t) = 1$, $\lim_{t \to -\infty} F_X(t) = 0$.*

Proof. Points 1 and 2 are obvious, given that P_X is a probability. Point 3 can be demonstrated by applying points 2 and 4 of Proposition A.24. In fact, by the former we obtain

$$\lim_{t \to +\infty} F_X(t) = \lim_{t \to +\infty} P_X(]-\infty, t]) = \lim_n P_X(]-\infty, n])$$

$$= P_X\left(\bigcup_n]-\infty, n]\right) = P_X(\mathbb{R}) = 1.$$

Analogously, by point 4 of Proposition A.24 we get $\lim_{t \to -\infty} F_X(t) = 0$. $\square$

Proposition 1.32. *Conversely, if we assign a function $F : \mathbb{R} \to [0, 1]$ that satisfies points 2 and 3 of Proposition 1.31, then by point 1 we can define a probability $P_X : \mathcal{B}_\mathbb{R} \to \mathbb{R}$ associated with a random variable X whose cumulative distribution function is identical to F.*

Definition 1.33. If the probability law $P_X : \mathcal{B}_\mathbb{R} \to [0, 1]$ associated with the random variable X is endowed with a density with respect to Lebesgue measure[1] μ on $\mathbb{R}$, then this density is called the *probability density* of X. If $f : \mathbb{R} \to \mathbb{R}_+$ is the probability density of X, then

$$\forall t \in \mathbb{R}: \qquad F_X(t) = \int_{-\infty}^{t} f d\mu \text{ and } \lim_{t \to +\infty} F_X(t) = \int_{-\infty}^{+\infty} f d\mu = 1,$$

as well as

$$P_X(B) = \int_B f d\mu \qquad \forall B \in \mathcal{B}_\mathbb{R}.$$

We may notice that the Lebesgue–Stieltjes measure, canonically associated with F_X as defined in Definition A.51, is identical to P_X.

Definition 1.34. A random variable X is *continuous* if its cumulative distribution function F_X is continuous.

Remark 1.35. X is continuous if and only if $P(X = x) = 0$ for every $x \in \mathbb{R}$.

Definition 1.36. A random variable X is *absolutely continuous* if F_X is absolutely continuous or, equivalently, if P_X is defined through its density.[2]

Proposition 1.37. *Every absolutely continuous random variable is continuous, but the converse is not true.*

Example 1.38. Let $F : \mathbb{R} \to [0, 1]$ be an extension to the Cantor function $f : [0, 1] \to [0, 1]$, given by

$$\forall x \in \mathbb{R}: \qquad F(x) = \begin{cases} 1 & \text{if } x > 1, \\ f(x) & \text{if } x \in [0, 1], \\ 0 & \text{if } x < 0, \end{cases}$$

---

[1] See Definition A.53.
[2] See Proposition A.57.

where f is endowed with the following properties:

1. f is continuous and increasing.
2. $f' = 0$ almost everywhere.
3. f is not absolutely continuous.

Hence X is a random variable with continuous but not absolutely continuous distribution function F.

Remark 1.39. Henceforth we will use "continuous" in the sense of "absolutely continuous."

Remark 1.40. If $f : \mathbb{R} \to \mathbb{R}_+$ is a function that is integrable with respect to Lebesgue measure μ on $\mathbb{R}$ and

$$\int_{\mathbb{R}} f d\mu = 1,$$

then there exists an absolutely continuous random variable with probability density f. Defining

$$F(x) = \int_{-\infty}^{x} f(t)dt \qquad \forall x \in \mathbb{R},$$

then F is a cumulative distribution function.

Example 1.41. *(Continuous probability densities).*

1. Uniform [its distribution denoted by $U(a, b)$]:

$$\forall x \in [a, b]: \qquad f(x) = \frac{1}{b - a}, \qquad a, b \in \mathbb{R}, a < b.$$

2. Standard normal or standard Gaussian [its distribution denoted by $N(0, 1)$ or $\Phi(x)$]:

$$\forall x \in \mathbb{R}: \qquad \varphi(x) = \frac{1}{\sqrt{2\pi}} \exp\left\{-\frac{1}{2}x^2\right\}. \qquad (1.2)$$

3. Normal or Gaussian [its distribution denoted by $N(m, \sigma^2)$]:

$$\forall x \in \mathbb{R}: \qquad f(x) = \frac{1}{\sigma\sqrt{2\pi}} \exp\left\{-\frac{1}{2}\left(\frac{x - m}{\sigma}\right)^2\right\}, \qquad \sigma > 0, m \in \mathbb{R}.$$

4. Log-normal:

$$\forall x \in \mathbb{R}_+^*: \qquad f(x) = \frac{1}{x\sigma\sqrt{2\pi}} \exp\left\{-\frac{1}{2}\left(\frac{\ln x - m}{\sigma}\right)^2\right\}, \qquad (1.3)$$

where $\sigma > 0, m \in \mathbb{R}$.

5. Exponential [its distribution denoted by $E(\lambda)$]:

$$\forall x \in \mathbb{R}_+: \qquad f(x) = \lambda e^{-\lambda x},$$

where $\lambda > 0$.

6. Gamma [its distribution denoted by $\Gamma(\lambda, \alpha)$]:

$$\forall x \in \mathbb{R}_+: \qquad f(x) = \frac{e^{-\lambda x}}{\Gamma(\alpha)} \lambda (\lambda x)^{\alpha-1},$$

where $\lambda, \alpha \in \mathbb{R}_+^*$. Here

$$\Gamma(\alpha) = \int_0^\infty y^{\alpha-1} e^{-y} dy$$

is the gamma function, which for $n \in \mathbb{N}^*$ is $(n-1)!$, i.e., a generalized factorial.

7. Standard Cauchy [its distribution denoted by $C(0,1)$]:

$$\forall x \in \mathbb{R}: \qquad f(x) = \frac{1}{\pi} \frac{1}{1+x^2}.$$

8. Cauchy [its distribution denoted by $C(a,h)$]:

$$\forall x \in \mathbb{R}: \qquad f(x) = \frac{1}{\pi} \frac{h}{h^2 + (x-a)^2}.$$

Definition 1.42. Let X be a random variable and let D denote an at most countable set of real numbers $D = \{x_1, \ldots, x_n, \ldots\}$. If there exists a function $p : \mathbb{R} \to [0,1]$, such that

1. For all $x \in D : p(x) > 0$
2. For all $x \in \mathbb{R} \setminus D : p(x) = 0$
3. For all $B \in \mathcal{B}_{\mathbb{R}}$: $\sum_{x \in B} p(x) < +\infty$
4. For all $B \in \mathcal{B}_{\mathbb{R}}$: $P_X(B) = \sum_{x \in B} p(x)$

then X is *discrete* and p is the (discrete) distribution function of X. The set D is called the *support* of function p.

Because of 1 and 2, in 3 and 4 we clearly mean $\sum_{x \in B} = \sum_{x \in B \cap D}$.

Remark 1.43. Let p denote the discrete distribution function of the random variable X, having support D. The following properties hold:

1. $\sum_{x \in D} p(x) = 1$.
2. For all $B \in \mathcal{B}_{\mathbb{R}}$ such that $D \cap B = \emptyset$, $P_X(B) = 0$.
3. For all $x \in \mathbb{R}$:

$$P_X(\{x\}) = \begin{cases} 0 & \text{if } x \notin D, \\ p(x) & \text{if } x \in D. \end{cases}$$

Hence P_X corresponds to the discrete measure associated with the "masses" $p(x)$, $x \in D$.

Example 1.44. (Discrete probability distributions).

1. Uniform: Given $n \in \mathbb{N}^*$, and a set of n real numbers $D = \{x_1, \ldots, x_n\}$,

$$p(x) = \frac{1}{n}, \qquad x \in D.$$

2. Poisson [denoted by $P(\lambda)$]: Given $\lambda \in \mathbb{R}_+^*$,

$$p(x) = \exp\{-\lambda\} \frac{\lambda^x}{x!}, \qquad x \in \mathbb{N},$$

(λ is called intensity).

3. Binomial [denoted by $B(n, p)$]: Given $n \in \mathbb{N}^*$, and $p \in [0, 1]$,

$$p(x) = \frac{n!}{(n-x)!x!} p^x (1-p)^{n-x}, \qquad x \in \{0, 1, \ldots, n\}.$$

Remark 1.45. The cumulative distribution function F_X of a discrete random variable X is a right-continuous with left limit (RCLL) function with an at most countable number of finite jumps. If p is the distribution function of X, then

$$p(x) = F_X(x) - F_X(x^-) \qquad \forall x \in D$$

or, more generally,

$$p(x) = F_X(x) - F_X(x^-) \qquad \forall x \in \mathbb{R}.$$

1.2.1 Random Vectors

The concept of random variable can be extended to any function defined on a probability space $(\Omega, \mathcal{F}, P)$ and valued in a measurable space $(E, \mathcal{B})$, i.e., a set E endowed with a σ-algebra $\mathcal{B}$ of its parts.

Definition 1.46. Every measurable function $X : \Omega \to E$, with $X^{-1}(B) \in \mathcal{F}$, for all $B \in \mathcal{B}$, assigned on the probability space $(\Omega, \mathcal{F}, P)$ and valued in $(E, \mathcal{B})$ is a random variable. The probability law P_X associated with X is defined by translating the probability P on $\mathcal{F}$ into a probability on $\mathcal{B}$, through the mapping $P_X : \mathcal{B} \to [0, 1]$, such that

$$\forall B \in \mathcal{B}: \qquad P_X(B) = P(X^{-1}(B)) \equiv P(X \in B).$$

Definition 1.47. Let $(\Omega, \mathcal{F}, P)$ be a probability space and $(E, \mathcal{B})$ a measurable space. Further, let E be a normed space of dimension n, and let $\mathcal{B}$ be its Borel σ-algebra. Every measurable map $\mathbf{X} : (\Omega, \mathcal{F}) \to (E, \mathcal{B})$ is called a *random vector*. In particular, we can take $(E, \mathcal{B}) = (\mathbb{R}^n, \mathcal{B}_{\mathbb{R}^n})$.

Remark 1.48. The Borel σ-algebra on $\mathbb{R}^n$ is identical to the product σ-algebra of the family of n Borel σ-algebras on $\mathbb{R}$: $\mathcal{B}_{\mathbb{R}^n} = \bigotimes_n \mathcal{B}_{\mathbb{R}}$.

Proposition 1.49. *Let $(\Omega, \mathcal{F}, P)$ be a probability space and $\mathbf{X} : \Omega \to \mathbb{R}^n$ a mapping. Moreover, let, for all $i = 1, \ldots, n$, $\pi_i : \mathbb{R}^n \to \mathbb{R}$ be the ith projection, and thus $X_i = \pi_i \circ \mathbf{X}$, $i = 1, \ldots, n$, be the ith component of $\mathbf{X}$. Then the following statements are equivalent:*

1. *$\mathbf{X}$ is a random vector of dimension n.*
2. *For all $i \in \{1, \ldots, n\}$, X_i is a random variable.*

Proof. The proposition is an obvious consequence of Proposition A.18. $\square$

Definition 1.50. Under the assumptions of the preceding proposition, the probability measure

$$B_i \in \mathcal{B}_{\mathbb{R}} \mapsto P_{X_i}(B_i) = P(X_i^{-1}(B_i)) \in [0, 1], \qquad 1 \leq i \leq n,$$

is called the *marginal law* of the random variable X_i. The probability $P_{\mathbf{X}}$ associated with the random vector $\mathbf{X}$ is called the *joint probability* of the family of random variables $(X_i)_{1 \leq i \leq n}$.

Remark 1.51. If $\mathbf{X} : (\Omega, \mathcal{F}) \to (\mathbb{R}^n, \mathcal{B}_{\mathbb{R}^n})$ is a random vector of dimension n and if $X_i = \pi_i \circ \mathbf{X} : (\Omega, \mathcal{F}) \to (\mathbb{R}, \mathcal{B}_{\mathbb{R}})$, $1 \leq i \leq n$, then, knowing the joint probability law $P_{\mathbf{X}}$, it is possible to determine the marginal probability P_{X_i}, for all $i \in \{1, \ldots, n\}$. In fact, if we consider the probability law of X_i, $i \in \{1, \ldots, n\}$, as well as the induced probability $\pi_i(P_{\mathbf{X}})$ for all $i \in \{1, \ldots, n\}$, then we have the relation

$$P_{X_i} = \pi_i(P_{\mathbf{X}}), \qquad 1 \leq i \leq n.$$

Therefore, for every $B_i \in \mathcal{B}_{\mathbb{R}}$, we obtain

$$\begin{aligned} P_{X_i}(B_i) = P_{\mathbf{X}}(\pi_i^{-1}(B_i)) &= P_{\mathbf{X}}(X_1 \in \mathbb{R}, \ldots, X_i \in B_i, \ldots, X_n \in \mathbb{R}) \\ &= P_{\mathbf{X}}(\mathcal{C}_{B_i}), \end{aligned} \tag{1.4}$$

where $\mathcal{C}_{B_i}$ is the cylinder of base B_i in $\mathbb{R}^n$. This can be further extended by considering, instead of the projection π_i, the projections π_S, where $S \subset \{1, \ldots, n\}$. Then, for every measurable set B_S, we obtain

$$P_{X_S}(B_S) = P_{\mathbf{X}}(\pi_S^{-1}(B_S)).$$

Notice that in general the converse is not true; knowledge of the marginals does not imply knowledge of the joint distribution of a random vector $\mathbf{X}$ unless further conditions are imposed (see Remark 1.61).

Definition 1.52. Let $\mathbf{X} : (\Omega, \mathcal{F}) \to (\mathbb{R}^n, \mathcal{B}_{\mathbb{R}^n})$ be a random vector of dimension n. The mapping $F_{\mathbf{X}} : \mathbb{R}^n \to [0, 1]$, with

$$\mathbf{t} = (t_1, \ldots, t_n) : \qquad F_{\mathbf{X}}(\mathbf{t}) := P(X_1 \leq t_1, \ldots, X_n \leq t_n) \qquad \forall \mathbf{t} \in \mathbb{R}^n,$$

is called the *joint cumulative distribution function* of the random vector $\mathbf{X}$.

Remark 1.53. Analogous to the case of random variables, $F_{\mathbf{X}}$ is increasing and right-continuous on $\mathbb{R}^n$. Further, it is such that

$$\lim_{x_i \to +\infty, \forall i} F(x_1, \ldots, x_n) = 1,$$

and for any $i = 1, \ldots, n$:

$$\lim_{x_i \to -\infty} F(x_1, \ldots, x_n) = 0.$$

Conversely, given a distribution function F satisfying all the preceding properties, there exists an n-dimensional random vector X with F as its cumulative distribution function. The underlying probability space can be constructed in a canonical way. In the bidimensional case, if $F : \mathbb{R}^2 \to [0, 1]$ satisfies the preceding conditions, then we can define a probability $P : \mathcal{B}_{\mathbb{R}^2} \to [0, 1]$ in the following way:

$$P(]\mathbf{a}, \mathbf{b}]) = F(b_1, b_2) - F(b_1, a_2) + F(a_1, a_2) - F(a_1, b_2)$$

for all $\mathbf{a}, \mathbf{b} \in \mathbb{R}^2$, $\mathbf{a} = (a_1, a_2)$, $\mathbf{b} = (b_1, b_2)$. Hence there exists a bidimensional random vector $\mathbf{X}$ with P as its probability.

Remark 1.54. Let $\mathbf{X} : (\Omega, \mathcal{F}) \to (\mathbb{R}^n, \mathcal{B}_{\mathbb{R}^n})$ be a random vector of dimension n, let $X_i = \pi_i \circ \mathbf{X}$, $1 \leq i \leq n$, be the nth component of $\mathbf{X}$, and let F_{X_i}, $1 \leq i \leq n$, and $F_{\mathbf{X}}$ be the respective cumulative distribution functions of X_i and $\mathbf{X}$. The knowledge of $F_{\mathbf{X}}$ allows one to infer F_{X_i}, $1 \leq i \leq n$, through the relation

$$F_{X_i}(t_i) = P(X_i \leq t_i) = F_{\mathbf{X}}(+\infty, \ldots, t_i, \ldots, +\infty),$$

for every $t_i \in \mathbb{R}$.

Definition 1.55. Let $\mathbf{X} : (\Omega, \mathcal{F}) \to (\mathbb{R}^n, \mathcal{B}_{\mathbb{R}^n})$ be a random vector of dimension n. If the probability law $P_{\mathbf{X}} : \mathcal{B}_{\mathbb{R}^n} \to [0, 1]$ with respect to $\mathbf{X}$ is endowed with a density with respect to the Lebesgue measure μ_n on $\mathbb{R}^n$ (or product measure of Lebesgue measures μ on $\mathbb{R}$), then this density is called the probability density of $\mathbf{X}$. If $f : \mathbb{R}^n \to \mathbb{R}_+$ is the probability density of $\mathbf{X}$, then

$$F_{\mathbf{X}}(\mathbf{t}) = \int_{-\infty}^{\mathbf{t}} f \, d\mu_n \qquad \forall \mathbf{t} \in \mathbb{R}^n,$$

and moreover,

$$P_{\mathbf{X}}(B) = \int_B f(x_1, \ldots, x_n) d\mu_n \qquad \forall B \in \mathcal{B}_{\mathbb{R}}.$$

Proposition 1.56. *Under the assumptions of the preceding definition, defining $X_i = \pi_i \circ \mathbf{X}$, $1 \le i \le n$, then P_{X_i} is endowed with density with respect to Lebesgue measure μ on $\mathbb{R}$ and its density function $f_i : \mathbb{R} \to \mathbb{R}_+$ is given by*

$$f_i(x_i) = \int^i f(x_1, \ldots, x_n) d\mu_{n-1},$$

where we have denoted by $\int^i$ the integration with respect to all variables but the ith one.

Proof. By (1.4) we have that for all $B_i \in \mathcal{B}_{\mathbb{R}}$

$$P_{X_i}(B_i) = P_{\mathbf{X}}(\mathcal{C}_{B_i}) = \int_{\mathcal{C}_{B_i}} f(x_1, \ldots, x_n) d\mu_n$$

$$= \int_{\mathbb{R}} dx_1 \cdots \int_{B_i} dx_i \cdots \int_{\mathbb{R}} f(x_1, \ldots, x_n) dx_n$$

$$= \int_{B_i} dx_i \int^i f(x_1, \ldots, x_n) d\mu_{n-1}.$$

By setting $f_i(x_i) = \int^i f(x_1, \ldots, x_n) d\mu_{n-1}$, we see that f_i is the density of P_{X_i}. $\square$

Remark 1.57. The definition of a discrete random vector is analogous to Definition 1.42.

1.3 Independence

Definition 1.58. The random variables $X_1, \ldots, X_n$, defined on the same probability space $(\Omega, \mathcal{F}, P)$, are independent if they generate independent classes of σ-algebras. Hence

$$P(A_1 \cap \cdots \cap A_n) = \prod_{i=1}^{n} P(A_i) \qquad \forall A_i \in X_i^{-1}(\mathcal{B}_{\mathbb{R}}).$$

What follows is an equivalent definition.

Definition 1.59. The components X_i, $1 \le i \le n$, of an n-dimensional random vector $\mathbf{X}$ defined on the probability space $(\Omega, \mathcal{F}, P)$ are independent if

$$P_{\mathbf{X}} = \bigotimes_{i=1}^{n} P_{X_i},$$

where $P_{\mathbf{X}}$ and P_{X_i} are the probability laws of $\mathbf{X}$ and X_i, $1 \leq i \leq n$, respectively (see Proposition A.44.)

To show that Definitions 1.59 and 1.58 are equivalent, we need to show that the following equivalence holds:

$$P(A_1 \cap \cdots \cap A_n) = \prod_{i=1}^{n} P(A_i) \Leftrightarrow P_{\mathbf{X}} = \bigotimes_{i=1}^{n} P_{X_i} \qquad \forall A_i \in X_i^{-1}(\mathcal{B}_{\mathbb{R}}).$$

We may recall first that $P_{\mathbf{X}} = \bigotimes_{i=1}^{n} P_{X_i}$ is the unique measure on $\mathcal{B}_{\mathbb{R}^n}$ that factorizes on rectangles, i.e., if $B = \prod_{i=1}^{n} B_i$, with $B_i \in \mathcal{B}_{\mathbb{R}}$, we have

$$P_{\mathbf{X}}(B) = \prod_{i=1}^{n} P_{X_i}(B_i).$$

To prove the implication from left to right, we observe that if B is a rectangle in $\mathcal{B}_{\mathbb{R}^n}$ as defined above, then

$$P_{\mathbf{X}}(B) = P(\mathbf{X}^{-1}(B)) = P\left(\mathbf{X}^{-1}\left(\prod_{i=1}^{n} B_i\right)\right) = P\left(\bigcap_{i=1}^{n} X_i^{-1}(B_i)\right)$$

$$= \prod_{i=1}^{n} P(X_i^{-1}(B_i)) = \prod_{i=1}^{n} P_{X_i}(B_i).$$

Conversely, for all $i = 1, \ldots, n$:

$$A_i \in X_i^{-1}(\mathcal{B}_{\mathbb{R}}) \Rightarrow \exists B_i \in \mathcal{B}_{\mathbb{R}}, \text{ so that } A_i = X_i^{-1}(B_i).$$

Thus, since $A_1 \cap \cdots \cap A_n = \bigcap_{i=1}^{n} X_i^{-1}(B_i)$, we have

$$P(A_1 \cap \cdots \cap A_n) = P\left(\bigcap_{i=1}^{n} X_i^{-1}(B_i)\right) = P(\mathbf{X}^{-1}(B)) = P_{\mathbf{X}}(B)$$

$$= \prod_{i=1}^{n} P_{X_i}(B_i) = \prod_{i=1}^{n} P(X_i^{-1}(B_i)) = \prod_{i=1}^{n} P(A_i).$$

Proposition 1.60.

1. *The real-valued random variables $X_1, \ldots, X_n$ are independent if and only if, for every $\mathbf{t} = (t_1, \ldots, t_n)' \in \mathbb{R}^n$,*

$$F_{\mathbf{X}}(\mathbf{t}) := P(X_1 \leq t_1 \cap \cdots \cap X_n \leq t_n) = P(X_1 \leq t_1) \cdots P(X_n \leq t_n)$$
$$= F_{X_1}(t_1) \cdots F_{X_n}(t_n).$$

2. Let $\mathbf{X} = (X_1, \ldots, X_n)'$ be a real-valued random vector with density f and probability $P_{\mathbf{X}}$ that is absolutely continuous with respect to the measure μ_n. The following two statements are equivalent:
   - $X_1, \ldots, X_n$ are independent.
   - $f = f_{X_1} \cdots f_{X_n}$ almost surely a.s.

Remark 1.61. From the previous definition it follows that if a random vector $\mathbf{X}$ has independent components, then their marginal distributions determine the joint distribution of $\mathbf{X}$.

Example 1.62. Let $\mathbf{X}$ be a bidimensional random vector with uniform density $f(\mathbf{x}) = c \in \mathbb{R}$ for all $\mathbf{x} = (x_1, x_2)' \in \mathcal{R}$. If $\mathcal{R}$ is, say, a semicircle, then X_1 and X_2 are not independent. But if $\mathcal{R}$ is a rectangle, then X_1 and X_2 are independent.

Proposition 1.63. *Let $X_1, \ldots, X_n$ be independent random variables defined on $(\Omega, \mathcal{F}, P)$ and valued in $(E_1, \mathcal{B}_1), \ldots, (E_n, \mathcal{B}_n)$. If the mappings*

$$g_i : (E_i, \mathcal{B}_i) \to (F_i, \mathcal{U}_i), \qquad 1 \leq i \leq n,$$

are measurable, then the random variables $g_1(X_1), \ldots, g_n(X_n)$ are independent.

Proof. Defining $h_i = g_i(X_i)$, $1 \leq i \leq n$, gives

$$h_i^{-1}(U_i) = X_i^{-1}(g_i^{-1}(U_i)) \in X_i^{-1}(\mathcal{B}_i)$$

for every $U_i \in \mathcal{U}_i$. The assertion then follows from Definition 1.58. $\qquad\square$

Sums of Two Random Variables

Let X and Y be two real-valued, independent, continuous random variables on $(\Omega, \mathcal{F}, P)$ with densities f and g, respectively. Defining $Z = X + Y$, then Z is a random variable, and let F_Z be its cumulative distribution. It follows that

$$F_Z(t) = P(Z \leq t) = P(X + Y \leq t) = P_{(X,Y)}(\mathcal{R}_t),$$

where $\mathcal{R}_t = \{(x, y) \in \mathbb{R}^2 | x + y \leq t\}$. By Proposition 1.60 (X, Y) is continuous and its density is $f_{(X,Y)} = f(x)g(y)$, for all $(x, y) \in \mathbb{R}^2$. Therefore, for all $t \in \mathbb{R}$:

$$F_Z(t) = P_{(X,Y)}(\mathcal{R}_t) = \int \int_{\mathcal{R}_t} f(x)g(y)dxdy$$

$$= \int_{-\infty}^{+\infty} dx \int_{-\infty}^{t-x} f(x)g(y)dy = \int_{-\infty}^{+\infty} f(x)dx \int_{-\infty}^{t} g(z-x)dz$$

$$= \int_{-\infty}^{t} dz \int_{-\infty}^{+\infty} f(x)g(z-x)dx \qquad \forall z \in \mathbb{R}.$$

Hence, the function

$$f_Z(z) = \int_{-\infty}^{+\infty} f(x)g(z-x)dx \qquad (1.5)$$

is the density of the random variable Z.

Definition 1.64. The function f_Z defined by (1.5) is the convolution of f and g, denoted by $f * g$. Analogously it can be shown that if f_1, f_2, f_3 are the densities of the independent random variables X_1, X_2, X_3, then the random variable $Z = X_1 + X_2 + X_3$ has density

$$f_1 * f_2 * f_3(z) = \int_{-\infty}^{+\infty} \int_{-\infty}^{+\infty} f_1(x)f_2(y-x)f_3(z-y)dxdy$$

for every $z \in \mathbb{R}$. This extends to n independent random variables in an analogous way.

1.4 Expectations

Definition 1.65. Let $(\Omega, \mathcal{F}, P)$ be a probability space and $X : (\Omega, \mathcal{F}) \to (\mathbb{R}, \mathcal{B}_\mathbb{R})$ a real-valued random variable. Assume that X is P-integrable, i.e., $X \in \mathcal{L}^1(\Omega, \mathcal{F}, P)$; then

$$E[X] = \int_\Omega X(\omega)dP(\omega)$$

is the *expected value* or *expectation* of the random variable X.

Remark 1.66. By Proposition A.29 it follows that if X is integrable with respect to P, then its expected value is given by

$$E(X) = \int_\mathbb{R} I_\mathbb{R}(x)dP_X(x) := \int xdP_X.$$

Remark 1.67. If X is a continuous real-valued random variable with density function f of P_X, then

$$E[X] = \int xf(x)d\mu.$$

On the other hand, if f is discrete with probability function p, then

$$E[X] = \sum xp(x).$$

Proposition 1.68. *If $X : (\Omega, \mathcal{F}) \to (E, \mathcal{B})$ is a random variable with probability law P_X and $H : (E, \mathcal{B}) \to (F, \mathcal{U})$ a measurable function, then, defining $Y = H \circ X = H(X)$, Y is a random variable. Furthermore, if $H : (E, \mathcal{B}) \to (\mathbb{R}, \mathcal{B}_\mathbb{R})$, then $Y \in \mathcal{L}^1(P)$ is equivalent to $H \in \mathcal{L}^1(P_X)$ and*

$$E[Y] = \int H(x)P_X(dx).$$

Corollary 1.69 *Let* $\mathbf{X} = (X_1, \ldots, X_n)'$ *be a random vector defined on* $(\Omega, \mathcal{F}, P)$ *whose components are valued in* $(E_1, \mathcal{B}_1), \ldots, (E_n, \mathcal{B}_n)$, *respectively. If* $h : (E_1 \times \cdots \times E_n, \mathcal{B}_1 \otimes \cdots \otimes \mathcal{B}_n) \to (\mathbb{R}, \mathcal{B}_{\mathbb{R}})$, *then* $Y = h(\mathbf{X}) \equiv h \circ \mathbf{X}$ *is a real-valued random variable. Moreover,*

$$E[Y] = \int h(x_1, \ldots, x_n) dP_{\mathbf{X}}(x_1, \ldots, x_n),$$

where $P_{\mathbf{X}}$ *is the joint probability of the vector* $\mathbf{X}$.

Proposition 1.70. *Let* X *be a real, P-integrable random variable on the space* $(\Omega, \mathcal{F}, P)$. *For every* $\alpha, \beta \in \mathbb{R}$, $E[\alpha X + \beta] = \alpha E[X] + \beta$.

Definition 1.71. A real-valued P-integrable random variable X is *centered* if it has an expectation zero.

Remark 1.72. If X is a real, P-integrable random variable, then $X - E[X]$ is a centered random variable. This follows directly from the previous proposition.

Definition 1.73. Given a real P-integrable random variable X, if $E[(X - E[X])^n] < +\infty$, $n \in \mathbb{N}$, then it is the nth centered moment. The second centered moment is the *variance*, and its square root, the *standard deviation* of a random variable X, denoted by $Var[X]$ and $\sigma = \sqrt{Var[X]}$, respectively.

Proposition 1.74. *Let* $(\Omega, \mathcal{F})$ *be a probability space and* $X : (\Omega, \mathcal{F}) \to (\mathbb{R}, \mathcal{B}_{\mathbb{R}})$ *a random variable. Then the following two statements are equivalent:*

1. X *is square-integrable with respect to* P *(Definition A.62).*
2. X *is P-integrable and* $Var[X] < +\infty$.

Moreover, under these conditions,

$$Var[X] = E[X^2] - (E[X])^2. \tag{1.6}$$

Proof.

$1 \Rightarrow 2$: Because $\mathcal{L}^2(P) \subset \mathcal{L}^1(P)$, $X \in \mathcal{L}^1(P)$. Obviously, the constant $E[X]$ is P-integrable; thus, $X - E[X] \in \mathcal{L}^2(P)$ and $Var[X] < +\infty$.

$2 \Rightarrow 1$: By assumption, $E[X]$ exists and $X - E[X] \in \mathcal{L}^2(P)$; thus, $X = X - E[X] + E[X] \in \mathcal{L}^2(P)$. Finally, due to the linearity of expectations,

$$Var[X] = E[(X - E[X])^2] = E[X^2 - 2X E[X] + (E[X])^2]$$
$$= E[X^2] - 2(E[X])^2 + (E[X])^2 = E[X^2] - (E[X])^2.$$

$\square$

Proposition 1.75. *If* X *is a real-valued P-integrable random variable and* $Var[X] = 0$, *then* $X = E[X]$ *almost surely with respect to the measure* P.

Proof. $Var[X] = 0 \Rightarrow \int (X - E[X])^2 dP = 0$. With $(X - E[X])^2$ nonnegative, $X - E[X] = 0$ almost everywhere with respect to P; thus, $X = E[X]$ almost surely with respect to P. This is equivalent to

$$P(X \neq E[X]) = P(\{\omega \in \Omega | X(\omega) \neq E[X]\}) = 0.$$

$\square$

Proposition 1.76 (Markov's inequality). *Let X be a nonnegative real P-integrable random variable on a probability space $(\Omega, \mathcal{F}, P)$; then*

$$P(X \geq \lambda) \leq \frac{E[X]}{\lambda} \qquad \forall \lambda \in \mathbb{R}_+^*.$$

Proof. The cases $E[X] = 0$ and $\lambda \leq 1$ are trivial. So let $E[X] > 0$ and $\lambda > 1$; then setting $m = E[X]$ results in

$$m = \int_0^{+\infty} x dP_X \geq \int_{\lambda m}^{+\infty} x dP_X \geq \lambda m P(X \geq \lambda m),$$

thus $P(X \geq \lambda m) \leq 1/\lambda$.

$\square$

Proposition 1.77 (Chebyshev's inequality). *If X is a real-valued and P-integrable random variable with variance $Var[X]$ (possibly infinite), then*

$$P(|X - E[X]| \geq \epsilon) \leq \frac{Var[X]}{\epsilon^2}.$$

Proof. Apply Markov's inequality to the random variable $(X - E[X])^2$. $\square$

More in general, the following proposition holds.

Proposition 1.78. *Let X be a real-valued random variable on a probability space $(\Omega, \mathcal{F}, P)$, and let $h : \mathbb{R} \to \mathbb{R}_+$; then*

$$P(h(X) \geq \lambda) \leq \frac{E[h(X)]}{\lambda} \qquad \forall \lambda \in \mathbb{R}_+^*.$$

Proof. See, e.g., Jacod and Protter (2000, p. 22).

$\square$

Example 1.79.

1. If X is a P-integrable continuous random variable with density f, where the latter is symmetric around the axis $x = a$, $a \in \mathbb{R}$, then $E[X] = a$.
2. If X is a Gaussian variable, then $E[X] = m$ and $Var[X] = \sigma^2$.
3. If X is a discrete, Poisson-distributed random variable, then $E[X] = \lambda$, $Var[X] = \lambda$.
4. If X is binomially distributed, then $E[X] = np$, $Var[X] = np(1 - p)$.

5. If X is continuous and uniform with density $f(x) = I_{[a,b]}(x)\frac{1}{b-a}$, $a, b \in \mathbb{R}$, then $E[X] = \frac{a+b}{2}$, $Var[X] = \frac{(b-a)^2}{12}$.
6. If X is a Cauchy variable, then it does not admit an expected value.

Definition 1.80. Let $\mathbf{X} : (\Omega, \mathcal{F}) \to (\mathbb{R}^n, \mathcal{B}_{\mathbb{R}^n})$ be a vector of random variables with P-integrable components X_i, $1 \leq i \leq n$. The expected value of the vector $\mathbf{X}$ is

$$E[\mathbf{X}] = (E[X_1], \ldots, E[X_2])'.$$

Proposition 1.81. *Let $(X_i)_{1 \leq i \leq n}$ be a real, P-integrable family of random variables on the same space $(\Omega, \mathcal{F}, P)$. Then*

$$E[X_1 + \cdots + X_n] = \sum_{i=1}^{n} E[X_i].$$

Further, if α_i, $i = 1, \ldots, n$, is a family of real numbers, then

$$E[\alpha_1 X_1 + \cdots + \alpha_n X_n] = \sum_{i=1}^{n} \alpha_i E[X_i].$$

Definition 1.82. If X_1, X_2, and $X_1 X_2$ are P-integrable random variables, then
$$Cov[X_1, X_2] = E[(X_1 - E[X_1])(X_2 - E[X_2])]$$
is the *covariance* of X_1 and X_2.

Remark 1.83. Due to the linearity of the $E[\cdot]$ operator, if $E[X_1 X_2] < +\infty$, then

$$\begin{aligned}
Cov[X_1, X_2] &= E[(X_1 - E[X_1])(X_2 - E[X_2])] \\
&= E[X_1 X_2 - X_1 E[X_2] - E[X_1]X_2 + E[X_1]E[X_2]] \\
&= E[X_1 X_2] - E[X_1]E[X_2].
\end{aligned}$$

Proposition 1.84.

1. *If X is a square-integrable random variable with respect to P, and $a, b \in \mathbb{R}$, then*
$$Var[aX + b] = a^2 Var[X].$$

2. *If both X_1 and X_2 are in $\mathcal{L}^2(\Omega, \mathcal{F}, P)$, then*

$$Var[X_1 + X_2] = Var[X_1] + Var[X_2] + 2Cov[X_1, X_2].$$

Proof.

1. Since $Var[X] = E[X^2] - (E[X])^2$, then

$$Var[aX + b] = E[(aX + b)^2] - (E[aX + b])^2$$
$$= a^2 E[X^2] + 2abE[X] + b^2 - a^2(E[X])^2 - b^2 - 2abE[X]$$
$$= a^2(E[X^2] - (E[X])^2) = a^2 Var[X].$$

2.

$$Var[X_1] + Var[X_2] + 2Cov[X_1, X_2]$$
$$= E[X_1^2] - (E[X_1])^2 + E[X_2^2] - (E[X_2])^2 + 2(E[X_1 X_2] - E[X_1]E[X_2])$$
$$= E[(X_1 + X_2)^2] - 2E[X_1]E[X_2] - (E[X_1])^2 - (E[X_2])^2$$
$$= E[(X_1 + X_2)^2] - (E[X_1 + X_2])^2 = Var[X_1 + X_2].$$

$\square$

Definition 1.85. If X_1 and X_2 are square-integrable random variables with respect to P, having the respective standard deviations $\sigma_1 > 0$ and $\sigma_2 > 0$, then

$$\rho(X_1, X_2) = \frac{Cov[X_1, X_2]}{\sigma_1 \sigma_2}$$

is the *correlation coefficient* of X_1 and X_2.

Remark 1.86. If X_1 and X_2 are $\mathcal{L}^2(\Omega, \mathcal{F}, P)$ random variables, then, by the Cauchy–Schwarz inequality (1.20),

$$|\rho(X_1, X_2)| \leq 1;$$

moreover,

$$|\rho(X_1, X_2)| = 1 \Leftrightarrow \exists a, b \in \mathbb{R} \text{ so that } X_2 = aX_1 + b, \qquad \text{a.s.}$$

Proposition 1.87. *If X_1 and X_2 are real-valued independent random variables on $(\Omega, \mathcal{F}, P)$ and endowed with finite expectations, then their product $X_1 X_2 \in \mathcal{L}^1(\Omega, \mathcal{F}, P)$ and*

$$E[X_1 X_2] = E[X_1]E[X_2].$$

Proof. Given the assumption of independence of X_1 and X_2, it is a tedious though trivial exercise to show that $X_1 X_2 \in \mathcal{L}^1(\Omega, \mathcal{F}, P)$. For the second part, by Corollary 1.69:

$$E[X_1 X_2] = \int X_1 X_2 dP_{(X_1 X_2)} = \int X_1 X_2 d(P_{X_1} \otimes P_{X_2})$$
$$= \int X_1 dP_{X_1} \int X_2 dP_{X_2} = E[X_1]E[X_2].$$

$\square$

Remark 1.88.   From Definition 1.82 and Remark 1.83 it follows that the covariance of two independent variables is zero.

Proposition 1.89. *If two random variables X_1 and X_2 are independent, then the variance operator $Var[\cdot]$ is additive, but not homogeneous. This follows from Proposition 1.84 and Remark 1.88.*

Proposition 1.90. *Suppose X and Y are independent real-valued random variables such that $X + Y \in \mathcal{L}^2(P)$; then both X and Y are in $\mathcal{L}^2(P)$.*

Proof. We know that

$$X^2 + Y^2 \le (X+Y)^2 + 2|XY|;$$

because of independence we may state that

$$E[|XY|] \le E[|X|]E[|Y|].$$

It will then be sufficient to prove that both $X, Y \in \mathcal{L}^1(P)$.

Since $|Y| \le |x| + |x + Y|$, if by absurd $E[|Y|] = +\infty$, this would imply $E(|x + Y|) = +\infty$, for any $x \in \mathbb{R}$, hence $E[|X + Y|] = +\infty$, against the assumption that $X + Y \in \mathcal{L}^2(P)$.   □

Characteristic Functions

Let X be a real-valued random variable defined on the probability space $(\Omega, \mathcal{F}, P)$, and let P_X be its probability law. For any $t \in \mathbb{R}$, the random variables $\cos tX$ and $\sin tX$ surely belong to $\mathcal{L}^1$; hence, their expected values are well defined:

$$\mathbb{E}[e^{itX}] = \mathbb{E}[\cos tX + i \sin tX] \in \mathbb{C}.$$

Definition 1.91. The *characteristic function* associated with the random variable X is defined as the function

$$t \in \mathbb{R} \mapsto \phi_X(t) = \mathbb{E}[e^{itX}] = \int_{\mathbb{R}} e^{itx} P_X(dx) \in \mathbb{C}.$$

Example 1.92. The characteristic function of a standard normal random variable X is

$$\phi_X(s) = E\left[e^{isX}\right] = \frac{1}{\sqrt{2\pi}} \int_{-\infty}^{\infty} e^{isx} e^{-\frac{1}{2}x^2} dx$$

$$= e^{-\frac{s^2}{2}} \frac{1}{\sqrt{2\pi}} \int_{-\infty}^{\infty} e^{-\frac{1}{2}(x-is)^2} dx = e^{-\frac{s^2}{2}}.$$

Proposition 1.93 (Properties of a characteristic function).

1. $\phi_X(0) = 1$.
2. $|\phi_X(t)| \leq 1$ for all $t \in \mathbb{R}$.
3. ϕ_X is uniformly continuous in $\mathbb{R}$.
4. For $a, b \in \mathbb{R}$, let $X = aY + b$. Then $\phi_X(t) = e^{ibt}\phi_Y(at), t \in \mathbb{R}$.

Proof. See, e.g., Métivier (1968). □

Theorem 1.94 (Inversion theorem). *Let $\phi : \mathbb{R} \to \mathbb{C}$ be the characteristic function of a probability law P on $\mathcal{B}_{\mathbb{R}}$; then for any points of continuity $a, b \in \mathbb{R}$ of the cumulative distribution function associated with P, with $a < b$, the following holds:*

$$P((a, b]) = \lim_{c \to +\infty} \int_{-c}^{c} \frac{e^{-ita} - e^{-itb}}{it}\phi(t)dt.$$

Further, if $\phi \in \mathcal{L}^1(\nu^1)$ (where ν^1 is the usual Lebesgue measure on $\mathcal{B}_{\mathbb{R}}$), then P is absolutely continuous with respect to ν^1, and the function

$$f(x) = \int_{-\infty}^{+\infty} e^{-itx}\phi(t)dt \quad x \in \mathbb{R}$$

is a probability density function for P. The density function f is continuous.

Proof. See, e.g., Lukacs (1970, pp. 31–33). □

As a direct consequence of the foregoing result, the following theorem holds, according to which a probability law is uniquely identified by its characteristic function.

Theorem 1.95. *Let P_1 and P_2 be two probability measures on $\mathcal{B}_{\mathbb{R}}$, and let ϕ_{P_1} and ϕ_{P_2} be the corresponding characteristic functions. Then*

$$P_1 = P_2 \Leftrightarrow \phi_{P_1}(t) = \phi_{P_2}(t), \quad t \in \mathbb{R}.$$

Proof. See, e.g., Ash (1972). □

Theorem 1.96. *Let $\phi : \mathbb{R} \to \mathbb{C}$ be the characteristic function of a probability law on $\mathcal{B}_{\mathbb{R}}$. For any $x \in \mathbb{R}$ the limit*

$$p(x) = \lim_{T \to \infty} \int_{-T}^{T} e^{-itx}\phi(t)dt$$

exists and equals the amount of the jump of the cumulative distribution function corresponding to ϕ at point x.

Corollary 1.97 *Let $\phi : \mathbb{R} \to \mathbb{C}$ be the characteristic function of a continuous probability law on $\mathcal{B}_{\mathbb{R}}$. Then for any $x \in \mathbb{R}$,*

$$\lim_{T \to \infty} \int_{-T}^{T} e^{-itx} \phi(t)dt = 0.$$

Corollary 1.98 *A probability law on $\mathbb{R}$ is purely discrete if and only if its characteristic function is almost periodic (Bohr (1947), p. 60). In particular, if $\phi : \mathbb{R} \to \mathbb{C}$ is the characteristic function of a $\mathbb{Z}$-valued random variable X, then for any $x \in \mathbb{Z}$,*

$$P(X = x) = \frac{1}{2\pi} \int_{-\pi}^{\pi} e^{-itx} \phi(t)dt.$$

Proof. See, e.g., Lukacs (1970, pp. 35–36) and Fristedt and Gray (1997, p. 227). □

Characteristic Functions of Random Vectors

Let $\mathbf{X} = (X_1, \ldots, X_k)$ be a random vector defined on the probability space $(\Omega, \mathcal{F}, P)$ and valued in $\mathbb{R}^k$, for $k \in \mathbb{N}, k \geq 2$, and let $P_{\mathbf{X}}$ be its joint probability law on $\mathcal{B}_{\mathbb{R}^k}$.

Given $\mathbf{t}, \mathbf{x} \in \mathbb{R}^k$, let $\mathbf{t} \cdot \mathbf{x} := t_1 x_1 + \cdots t_k x_k \in \mathbb{R}$ be their scalar product.

Definition 1.99. The *characteristic function* associated with the random vector $\mathbf{X}$ is defined as the function

$$\phi_{\mathbf{X}}(\mathbf{t}) := \mathbb{E}[e^{i\mathbf{t} \cdot \mathbf{X}}] \in \mathbb{C}, \quad \mathbf{t} \in \mathbb{R}^k. \tag{1.7}$$

A uniqueness theorem holds in this case too.

Theorem 1.100. *Let $P_{\mathbf{X}}$ and $P_{\mathbf{Y}}$ be two probability laws on $\mathcal{B}_{\mathbb{R}^k}$ having the same characteristic function, i.e., for all $\mathbf{t} \in \mathbb{R}^k$,*

$$\phi_{\mathbf{X}}(\mathbf{t}) = \int \cdots \int_{\mathbb{R}^k} e^{i\mathbf{t} \cdot \mathbf{x}} P_{\mathbf{X}}(d\mathbf{x}) = \int \cdots \int_{\mathbb{R}^k} e^{i\mathbf{t} \cdot \mathbf{y}} P_{\mathbf{Y}}(d\mathbf{y}) = \phi_{\mathbf{Y}}(\mathbf{t}).$$

Then $\mathbf{X} \sim \mathbf{Y}$, i.e., $P_{\mathbf{X}} \equiv P_{\mathbf{Y}}$.

The characteristic function of a random vector satisfies the following properties, the proof of which is left as an exercise.

Proposition 1.101. *Let $\phi_{\mathbf{X}} : \mathbb{R}^k \to \mathbb{C}$ be the characteristic function of a random vector $\mathbf{X} : (\Omega, \mathcal{F}) \to (\mathbb{R}^k, \mathcal{B}_{\mathbb{R}^k})$; then*

1. *$\phi_{\mathbf{X}}(\mathbf{0}) = \phi_{\mathbf{X}}((0, \cdots, 0))) = 1$.*
2. *$|\phi_{\mathbf{X}}(\mathbf{t})| \leq 1$, for any $\quad \mathbf{t} \in \mathbb{R}^k$.*
3. *$\phi_{\mathbf{X}}$ is uniformly continuous in $\mathbb{R}^k$.*

4. *Let $\mathbf{Y}$ be a random vector of dimension k such that for any $i \in \{1, \ldots, k\}$, $Y_i = a_i X_i + b_i$, with $\mathbf{a}, \mathbf{b} \in \mathbb{R}^k$. Then, for any $\mathbf{t} = (t_1, \ldots, t_k) \in \mathbb{R}^k$,*

$$\phi_{\mathbf{Y}}(\mathbf{t}) = e^{i\mathbf{b}\cdot\mathbf{t}} \, \phi_{\mathbf{X}}(a_1 t_1 + \cdots a_k t_k).$$

An interesting consequence of the foregoing results is the following one.

Corollary 1.102 *Let $\phi_{\mathbf{X}} : \mathbb{R}^k \to \mathbb{C}$ be the characteristic function associated with a random vector $\mathbf{X} = (X_1, \ldots, X_k) : (\Omega, \mathcal{F}) \to (\mathbb{R}^k, \mathcal{B}_{\mathbb{R}^k})$; then, the characteristic function ϕ_{X_i} associated with the ith component $X_i : (\Omega, \mathcal{F}) \to (\mathbb{R}, \mathcal{B}_{\mathbb{R}})$ for $i \in \{1, \ldots, k\}$ is such that*

$$\phi_{X_i}(t) = \phi_{\mathbf{X}}(\mathbf{t}^{(i)}), \qquad t \in \mathbb{R},$$

where $\mathbf{t}^{(i)} = (t_j^{(i)})_{1 \le j \le k} \in \mathbb{R}^k$ is such that, for any $j = 1, \ldots, k$,

$$t_j^{(i)} = \begin{cases} 0, & se \ j \ne i; \\ t, & se \ j = i. \end{cases}$$

The following theorem extends to characteristic functions the factorization property of the joint distribution of independent random variables.

Theorem 1.103. *Let $\phi_{\mathbf{X}} : \mathbb{R}^k \to \mathbb{C}$ be the characteristic function of the random vector $\mathbf{X} = (X_1, \ldots, X_k) : (\Omega, \mathcal{F}) \to (\mathbb{R}^k, \mathcal{B}_{\mathbb{R}^k})$, and let $\phi_{X_i} : \mathbb{R} \to \mathbb{C}$ be the characteristic function of the component $X_i : (\Omega, \mathcal{F}) \to (\mathbb{R}, \mathcal{B}_{\mathbb{R}}), i \in \{1, \ldots, k\}$. A necessary and sufficient condition for the independence of the random variables $X_1, \ldots, X_k$ is*

$$\phi_{\mathbf{X}}(\mathbf{t}) = \prod_{i=1}^{k} \phi_{X_i}(t_i)$$

for any $\mathbf{t} = (t_1, \ldots, t_k) \in \mathbb{R}^k$.

Proof. We will limit ourselves to proving that the condition is necessary. Let us then assume the independence of the components X_i, per $i = 1, \ldots, k$. From (1.7) we obtain

$$\phi_{\mathbf{X}}(\mathbf{t}) = \mathbb{E}\left[e^{i\mathbf{t}\cdot\mathbf{X}}\right] = \mathbb{E}\left[e^{i\sum_i t_i X_i}\right]$$

$$= \mathbb{E}\left[\prod_{i=1}^{k} e^{it_i X_i}\right] = \prod_{i=1}^{k} \mathbb{E}\left[e^{it_i X_i}\right]$$

$$= \prod_{i=1}^{k} \phi_{X_i}(t_i).$$

$\square$

Corollary 1.104 *Let* $\mathbf{X} = (X_1, \ldots, X_k)$ *be a random vector with independent components. Let* $\phi_{\mathbf{X}}$ *and* ϕ_{X_i}*, for* $i = 1, \ldots, k$*, be the characteristic functions associated with* $\mathbf{X}$ *and its components, respectively. Consider the random variable sum of the components*

$$S := \sum_{i=1}^{k} X_i : (\Omega, \mathcal{F}) \to (\mathbb{R}, \mathcal{B}_{\mathbb{R}});$$

then the characteristic function ϕ_S *associated with it is such that*

$$\phi_S(t) = \prod_{i=1}^{k} \phi_{X_i}(t) = \phi_{\mathbf{X}}((t, t, \ldots, t)), \quad t \in \mathbb{R}.$$

In the case of identically distributed random variables we may further state the following corollary.

Corollary 1.105 *Let* $X_i, i = 1, \ldots, n$ *be a family of independent and identically distributed (i.i.d.) random variables.*

a. If $X = \sum_{i=1}^{n} X_i$*, then*

$$\phi_X(t) = (\phi_{X_1}(t))^n.$$

b. If $\bar{X} = \dfrac{1}{n} \sum_{i=1}^{n} X_i$*, then*

$$\phi_{\bar{X}}(t) = \left(\phi_{X_1} \left(\frac{t}{n} \right) \right)^n.$$

Example 1.106. An easy way, whenever applicable, to identify the probability law of a random variable is based on the uniqueness theorem of characteristic functions associated with probability laws.

An interesting application regards the distribution of the sum of independent random variables.

1. The sum of two independent binomial random variables distributed as $B(r_1, p)$ and $B(r_2, p)$ is distributed as $B(r_1 + r_2, p)$ for any $r_1, r_2 \in \mathbb{N}^*$ and any $p \in [0, 1]$.
2. The sum of two independent Poisson variables distributed as $P(\lambda_1)$ and $P(\lambda_2)$ is distributed as $P(\lambda_1 + \lambda_2)$ for any $\lambda_1, \lambda_2 \in \mathbb{R}_+^*$.
3. The sum of two independent Gaussian random variables distributed as $N(m_1, \sigma_1^2)$ and $N(m_2, \sigma_2^2)$ is distributed as $N(m_1 + m_2, \sigma_1^2 + \sigma_2^2)$ for any $m_1, m_2 \in \mathbb{R}$ and any $\sigma_1^2, \sigma_2^2 \in \mathbb{R}_+^*$. Note that

$$aN(m_1, \sigma_1^2) + b = N(am_1 + b, a^2 \sigma_1^2).$$

4. The sum of two independent Gamma random variables distributed as $\Gamma(\alpha_1, \lambda)$ and $\Gamma(\alpha_2, \lambda)$ is distributed as $\Gamma(\alpha_1 + \alpha_2, \lambda)$.

Definition 1.107. A family of random variables is said to be *reproducible* if it is closed with respect to the sum of independent random variables. Correspondingly, their probability distributions are called reproducible.

We may then state that binomial, Poisson, Gaussian, and Gamma distributions are reproducible.

Remark 1.108. Exponential distributions are not reproducible. In fact the sum of two independent exponential random variables $X_1 \sim \exp\{\lambda\}$ and $X_2 \sim \exp\{\lambda\}$ is not exponentially distributed, though it is Gamma distributed:

$$X = X_1 + X_2 \sim \Gamma(2, \lambda).$$

The following theorem, though an easy consequence of the previous results, is of great relevance.

Theorem 1.109 (Cramér–Wold theorem). *Consider the random vector $\mathbf{X} = (X_1, \ldots, X_k)$, valued in $\mathbb{R}^k$, and the vector of real numbers $\mathbf{c} = (c_1, \ldots, c_k) \in \mathbb{R}^k$. Let $Y_{\mathbf{c}}$ be the random variable defined by*

$$Y_{\mathbf{c}} := \mathbf{c} \cdot \mathbf{X} = \sum_{i=1}^{k} c_i X_i,$$

and let $\phi_{\mathbf{X}}$ and $\phi_{Y_{\mathbf{c}}}$ be the characteristic functions associated with the random vector $\mathbf{X}$ and the random variable $Y_{\mathbf{c}}$, respectively. Then

i. $\phi_{Y_{\mathbf{c}}}(t) = \phi_{\mathbf{X}}(t\mathbf{c})$ for any $t \in \mathbb{R}$.
ii. $\phi_{\mathbf{X}}(\mathbf{t}) = \phi_{Y_{\mathbf{t}}}(1)$ for any $\mathbf{t} \in \mathbb{R}^k$.

As a consequence the distribution of $Y_{\mathbf{c}}$ is determined by the joint distribution of the vector $\mathbf{X}$ and, conversely, the joint distribution of vector $\mathbf{X}$ is determined by the distribution of $Y_{\mathbf{c}}$ by varying $\mathbf{c} \in \mathbb{R}^k$.

1.5 Gaussian Random Vectors

The Cramér–Wold theorem suggests the following definition of Gaussian random vectors, also known as multivariate normal vectors.

Definition 1.110. A random vector $\mathbf{X} = (X_1, \ldots, X_k)'$, valued in $\mathbb{R}^k$, is said to be *multivariate normal* or a *Gaussian vector* if and only if the scalar random variable, valued in $\mathbb{R}$, defined by

$$Y_{\mathbf{c}} := \mathbf{c} \cdot \mathbf{X} = \sum_{i=1}^{k} c_i X_i,$$

has a normal distribution for any choice of the vector $\mathbf{c} = (c_1, \ldots, c_k)^T \in \mathbb{R}^k$.

Given a random vector $\mathbf{X} = (X_1, \ldots, X_k)'$, valued in $\mathbb{R}^k$, and such that $X_i \in \mathcal{L}^2$, $i \in \{1, \ldots, k\}$, it makes sense to define the vectors of the means

$$\mu_{\mathbf{X}} = \mathbb{E}(\mathbf{X}) := (\mathbb{E}(X_1), \ldots, \mathbb{E}(X_k))'$$

and the variance–covariance matrix

$$\Sigma_{\mathbf{X}} := cov(\mathbf{X}) := \mathbb{E}[(\mathbf{X} - \mu_{\mathbf{X}})(\mathbf{X} - \mu_{\mathbf{X}})'].$$

It is trivial to recognize that $\Sigma_{\mathbf{X}}$ is a symmetric and positive semidefinite square matrix; indeed, in the nontrivial cases it is positive definite.

Recall that a square matrix $A = (a_{ij}) \in \mathbb{R}^{k \times k}$ is said to be positive semidefinite on $\mathbb{R}^k$ if, for any vector $\mathbf{x} = (x_1, \ldots, x_k)^T \in \mathbb{R}^k$, $\mathbf{x} \neq \mathbf{0}$, it results in

$$\mathbf{x} \cdot A\mathbf{x} = \sum_{i=1}^{k} \sum_{j=1}^{k} x_i a_{ij} x_j \geq 0.$$

The same matrix is said to be positive definite if the last inequality is strict ($>$).

From the theory of matrices we know that a positive definite square matrix is nonsingular, hence invertible, and its determinant is positive; in this case, its inverse matrix is positive definite too. We will denote by A^{-1} the inverse matrix of A.

Let $\mathbf{X}$ be a multivariate normal vector valued in $\mathbb{R}^k$ for $k \in \mathbb{N}^*$ such that $\mathbf{X} \in \mathcal{L}^2$. If $\mu_{\mathbf{X}} \in \mathbb{R}^k$ is its mean vector, and $\Sigma_{\mathbf{X}} \in \mathbb{R}^{k \times k}$ is its variance–covariance matrix, then we will write

$$\mathbf{X} \sim N(\mu_{\mathbf{X}}, \Sigma_{\mathbf{X}}).$$

Theorem 1.111. *Let $\mathbf{X}$ be a multivariate normal vector valued in $\mathbb{R}^k$ for $k \in \mathbb{N}^*$, and let $\mathbf{X} \in \mathcal{L}^2$. If $\mu_{\mathbf{X}} \in \mathbb{R}^k$, and $\Sigma_{\mathbf{X}} \in \mathbb{R}^{k \times k}$ is a positive definite matrix, then the characteristic function of $\mathbf{X}$ is as follows:*

$$\phi_{\mathbf{X}}(\mathbf{t}) = e^{i\,\mathbf{t}'\mu_{\mathbf{X}} - \frac{1}{2}\mathbf{t}'\Sigma_{\mathbf{X}}\mathbf{t}}, \quad \mathbf{t} \in \mathbb{R}^k.$$

Further, $\mathbf{X}$ admits a joint probability density given by

$$f_{\mathbf{X}}(\mathbf{x}) = \left(\frac{1}{(2\pi)^k \det \Sigma_{\mathbf{X}}} \right)^{\frac{1}{2}} e^{-\frac{1}{2}(\mathbf{x} - \mu_{\mathbf{X}})' \Sigma_{\mathbf{X}}^{-1} (\mathbf{x} - \mu_{\mathbf{X}})}$$

for $\mathbf{x} \in \mathbb{R}^k$.

Proof. See, e.g., Billingsley (1986). $\qquad\qquad\qquad\qquad\qquad\qquad\qquad$ $\square$

The following propositions are a consequence of the foregoing results

Proposition 1.112. *If* $\mathbf{X}$ *is a multivariate normal vector valued in* $\mathbb{R}^k$ *for* $k \in \mathbb{N}^*$, *then its components* X_i, *per* $i = 1, \ldots, k$, *are themselves Gaussian (scalar random variables).*

The components are independent normal random variables if and only if the variance–covariance matrix of the random vector $\mathbf{X}$ *is diagonal.*

Proposition 1.113. *Let* $\mathbf{X}$ *be a multivariate normal vector valued in* $\mathbb{R}^k$ *for* $k \in \mathbb{N}^*$ *such that* $\mathbf{X} \sim N(\mu_{\mathbf{X}}, \Sigma_{\mathbf{X}})$. *Given a matrix* $D \in \mathbb{R}^{p \times k}$, *with* $p \in \mathbb{N}^*$, *and a vector* $\mathbf{b} \in \mathbb{R}^p$, *the random vector* $\mathbf{Y} = D\mathbf{X} + \mathbf{b}$ *is itself a Gaussian random vector:*

$$\mathbf{Y} \sim N(D\mu_{\mathbf{X}} + \mathbf{b}, D\,\Sigma_{\mathbf{X}}D^T).$$

Proof. The proof is not difficult and is left as an exercise. We may at any rate notice that, for well-known properties of expected values and covariances,

$$\mathbb{E}(\mathbf{Y}) = D\mu_{\mathbf{X}} + \mathbf{b},$$

whereas

$$\Sigma_{\mathbf{Y}} = D\,\Sigma_{\mathbf{X}}\,D^T.$$

We may now notice that, if Σ is a positive-definite square matrix, from the theory of matrices it is well known that there exists a nonsingular square matrix $P \in \mathbb{R}^{k \times k}$ such that

$$\Sigma = PP^T.$$

We may then consider the linear transformation

$$\mathbf{Z} = P^{-1}(\mathbf{X} - \mu_{\mathbf{X}}),$$

which leads to

$$\mathbb{E}(\mathbf{Z}) = P^{-1}\mathbb{E}(\mathbf{X} - \mu_{\mathbf{X}}) = \mathbf{0},$$

while

$$\begin{aligned}
\Sigma_{\mathbf{Z}} &= P^{-1}\,\Sigma_{\mathbf{X}}\,(P^{-1})^T \\
&= P^{-1}PP^T(P^{-1}) \\
&= (P^{-1}P)(P^{-1}P)^T \\
&= I_k,
\end{aligned}$$

having denoted by I_k the identity matrix of dimension k. From Theorem 1.111 it follows that $\mathbf{Z} \sim N(\mathbf{0}, I_k)$, so that its joint density is given by

$$f_{\mathbf{Z}}(\mathbf{z}) = \left(\frac{1}{2\pi}\right)^{\frac{k}{2}} e^{-\frac{1}{2}\mathbf{z}'\mathbf{z}}$$

for $\mathbf{z} \in \mathbb{R}^k$. It is thus proven that the random vector $\mathbf{Z}$ has all its components i.i.d. with distribution $N(0, 1)$.   □

A final consequence of the foregoing results is the following proposition.

Proposition 1.114. *Let* $\mathbf{X} = (X_1, \ldots, X_n)'$ *for* $n \in \mathbb{N}^*$ *be a multivariate normal random vector such that all its components are i.i.d. normal random variables,* $X_j \sim N(\mu, \sigma^2)$ *for any* $j \in \{1, \ldots, n\}$; *then any random vector that is obtained by applying to it a linear transformation is still a multivariate normal random vector, but its components may not necessarily be independent.*

1.6 Conditional Expectations

Let $X, Y : (\Omega, \mathcal{F}, P) \to (\mathbb{R}, \mathcal{B}_\mathbb{R})$ be two discrete random variables with joint discrete probability distribution p. There exists an, at most countable, subset $D \subset \mathbb{R}^2$ such that

$$p(x, y) \neq 0 \qquad \forall (x, y) \in D,$$

where $p(x, y) = P(X = x \cap Y = y)$. If, furthermore, D_1 and D_2 are the projections of D along its axes, then the marginal distributions of X and Y are given by

$$p_1(x) \doteq P(X = x) = \sum_y p(x, y) \neq 0 \qquad \forall x \in D_1,$$

$$p_2(y) = P(Y = y) = \sum_x p(x, y) \neq 0 \qquad \forall y \in D_2.$$

Definition 1.115. Given the preceding assumptions and fixing $y \in \mathbb{R}$, then the probability of y *conditional* on $X = x \in D_1$ is

$$p_2(y|x) = \frac{p(x, y)}{p_1(x)} = \frac{P(X = x \cap Y = y)}{P(X = x)} = P(Y = y | X = x).$$

Furthermore,

$$y \to p_2(y | X = x) \in [0, 1] \qquad \forall x \in D_1$$

is called the probability function of y conditional on $X = x$.

Definition 1.116. Analogous to the definition of expectation of a discrete random variable, the expectation of Y, conditional on $X = x$, is $\forall x \in D_1$,

$$E[Y | X = x] = \sum_y y p_2(y|x)$$

$$= \frac{1}{p_1(x)} \sum_y y p(x, y) = \frac{1}{p_1(x)} \sum_{y \in \mathbb{R}} y p(x, y)$$

$$= \frac{1}{p_1(x)} \int \int_{\mathcal{R}_x} y dP_{(X,Y)}(x, y)$$

$$= \frac{1}{P(X = x)} \int_{[X=x]} Y(\omega) dP(\omega),$$

with $\mathcal{R}_x = \{x\} \times \mathbb{R}$.

Definition 1.117. Let $X : (\Omega, \mathcal{F}) \to (E, \mathcal{B})$ be a discrete random variable and $Y : (\Omega, \mathcal{F}) \to (\mathbb{R}, \mathcal{B}_{\mathbb{R}})$ P-integrable. Then the mapping

$$x \to E[Y|X = x] = \frac{1}{P(X = x)} \int_{[X=x]} Y(\omega) dP(\omega) \qquad (1.8)$$

is the expected value of Y conditional on X, defined on the set $x \in E$ with $P_X(x) \neq 0$.

Remark 1.118. It is standard to extend the mapping (1.8) to the entire set E by fixing its value arbitrarily at the points $x \in E$ where $P([X = x]) = 0$. Hence there exists an entire equivalence class of functions f defined on E such that

$$f(x) = E[Y|X = x] \qquad \forall x \in E \text{ such that } P_X(x) \neq 0.$$

An element f of this class is said to be defined on E almost surely with respect to P_X. A generic element of this class is denoted by $E[Y|X = \cdot]$, $E[Y|\cdot]$, or $E^X[Y]$. Furthermore, its value at $x \in E$ is denoted by $E[Y|X = x]$, $E[Y|x]$, or $E^{X=x}[Y]$.

Definition 1.119. Let $X : (\Omega, \mathcal{F}) \to (E, \mathcal{B})$ be a discrete random variable and $x \in E$ so that $P_X(x) \neq 0$, and let $F \in \mathcal{F}$. The indicator of F, denoted by $I_F : \Omega \to \mathbb{R}$, is a real-valued, P-integrable random variable. The expression

$$P(F|X = x) = E[I_F|X = x] = \frac{P(F \cap [X = x])}{P(X = x)}$$

is the probability of F conditional upon $X = x$.

Remark 1.120. Let $X : (\Omega, \mathcal{F}) \to (E, \mathcal{B})$ be a discrete random variable. If we define $E_X = \{x \in E | P_X(x) \neq 0\}$, then for every $x \in E_X$ the mapping

$$P(\cdot|X = x) : \mathcal{F} \to [0, 1],$$

so that

$$P(F|X = x) = \frac{P(F \cap [X = x])}{P(X = x)} \qquad \forall F \in \mathcal{F}$$

is a probability measure on $\mathcal{F}$, conditional on $X = x$. Further, if we arbitrarily fix the value of $P(F|X = x)$ at the points $x \in E$ where P_X is zero, then we can extend the mapping

$$x \in E_X \to P(F|X = x)$$

to the whole of E, so that $P(\cdot|X = x) : \mathcal{F} \to [0, 1]$ is again a probability measure on $\mathcal{F}$, defined almost surely with respect to P_X.

Definition 1.121. The family of functions $(P(\cdot|X = x))_{x \in E}$ is called a *regular version of the conditional probability with respect to X.*

Proposition 1.122. *Let $(P(\cdot|X = x))_{x \in E}$ be a regular version of the conditional probability with respect to X. Then, for any $Y \in \mathcal{L}^1(\Omega, \mathcal{F}, P)$:*

$$\int Y(\omega) dP(\omega|X = x) = E[Y|X = x], \qquad P_X\text{-a.s.}$$

Proof. First, we observe that Y, being a random variable, is measurable.[3] Now from (1.8) it follows that

$$E[I_F|X = x] = P(F|X = x) = \int I_F(\omega) P(d\omega|X = x)$$

for every $x \in E$, $P_X(x) \neq 0$. Now let Y be an elementary function so that

$$Y = \sum_{i=1}^{n} \lambda_i I_{F_i}.$$

Then, for every $x \in E_X$:

$$E[Y|X = x] = \sum_{i=1}^{n} \lambda_i E[I_{F_i}|X = x] = \sum_{i=1}^{n} \lambda_i \int I_{F_i}(\omega) P(d\omega|X = x)$$

$$= \int \left(\sum_{i=1}^{n} \lambda_i I_{F_i} \right)(\omega) P(d\omega|X = x) = \int Y(\omega) dP(\omega|X = x).$$

If Y is a positive real-valued random variable, then, by Theorem A.14, there exists an increasing sequence $(Y_n)_{n \in \mathbb{N}}$ of elementary random variables so that

$$Y = \lim_{n \to \infty} Y_n = \sup_{n \in \mathbb{N}} Y_n.$$

Therefore, for every $x \in E$:

$$E[Y|X = x] = \sup_{n \in \mathbb{N}} E[Y_n|X = x] = \sup_{n \in \mathbb{N}} \int Y_n(\omega) dP(\omega|X = x)$$

$$= \int \left(\sup_{n \in \mathbb{N}} Y_n \right)(\omega) dP(\omega|X = x) = \int Y(\omega) dP(\omega|X = x),$$

where the first and third equalities are due to the property of Beppo–Levi (Proposition A.29). Lastly, if Y is a real-valued, P-integrable random variable, then it satisfies the assumptions, being the difference between two positive integrable functions. □

A notable extension of the preceding results and definitions is the subject of the following presentation.

---

[3] This only specifies its σ-algebras, not its measure.

Expectations Conditional on a σ-Algebra

Proposition 1.123. *Let $(\Omega, \mathcal{F}, P)$ be a probability space and $\mathcal{G}$ a σ-algebra contained in $\mathcal{F}$. For every real-valued random variable $Y \in \mathcal{L}^1(\Omega, \mathcal{F}, P)$, there exists a unique element $Z \in L^1(\Omega, \mathcal{G}, P)$ such that for all $G \in \mathcal{G}$:*

$$\int_G Y \, dP = \int_G Z \, dP.$$

Proof. First we consider Y nonnegative. The mapping $\nu : \mathcal{G} \to \mathbb{R}_+$ given by

$$\nu(G) = \int_G Y(\omega) dP(\omega) \qquad \forall G \in \mathcal{G}$$

is a bounded measure and absolutely continuous with respect to P on $\mathcal{G}$. In fact, for $G \in \mathcal{G}$

$$P(G) = 0 \Rightarrow \nu(G) = 0.$$

Since P is bounded, thus σ-finite, then, by the Radon–Nikodym Theorem A.54, there exists a unique $Z \in L^1(\Omega, \mathcal{G}, P)$ such that

$$\nu(G) = \int_G Z \, dP \qquad \forall G \in \mathcal{G}.$$

The case Y of arbitrary sign can be easily handled by the standard decomposition $Y = Y^+ - Y^-$. $\qquad\qquad\square$

Definition 1.124. Let $(\Omega, \mathcal{F}, P)$ be a probability space and $\mathcal{G}$ a σ-algebra contained in $\mathcal{F}$. Given a real-valued random variable $Y \in \mathcal{L}^1(\Omega, \mathcal{F}, P)$, any real-valued random variable $Z \in \mathcal{L}^1(\Omega, \mathcal{G}, P)$ that satisfies the condition

$$\int_G Y \, dP = \int_G Z \, dP, \qquad \forall G \in \mathcal{G} \tag{1.9}$$

will be called a version of the conditional expectation of Y given $\mathcal{G}$ and will be denoted by $E[Y|\mathcal{G}]$ or by $E^{\mathcal{G}}[Y]$.

Definition 1.125. Let now $X : (\Omega, \mathcal{F}) \to (\mathbb{R}^k, \mathcal{B}_{\mathbb{R}^k})$ be a random vector, and let $\mathcal{F}_X \subset \mathcal{F}$ be the σ-algebra generated by X. Given a real-valued random variable $Y \in \mathcal{L}^1(\Omega, \mathcal{F}, P)$, we define the conditional expectation of Y given X the real-valued random variable such that

$$E[Y|X] = E[Y|\mathcal{F}_X].$$

Again thanks to the Radon–Nikodym theorem, the following proposition can be shown directly.

Proposition 1.126. *Let $X : (\Omega, \mathcal{F}) \to (\mathbb{R}^k, \mathcal{B}_{\mathbb{R}^k})$ be a random vector and $Y : (\Omega, \mathcal{F}) \to (\mathbb{R}, \mathcal{B}_{\mathbb{R}})$ a P-integrable random variable. Then there exists a unique class of real-valued $h \in L^1(\mathbb{R}^k, \mathcal{B}_{\mathbb{R}^k}, P_X)$ such that*

$$\int_{X^{-1}(B)} Y(\omega)dP(\omega) = \int_B h \, dP_X, \qquad \forall B \in \mathcal{B}_{\mathbb{R}^k}. \tag{1.10}$$

By known results about integration with respect to image measures (change of integration variables), we may rewrite Equation (1.10) as follows:

$$\int_{X^{-1}(B)} Y(\omega)dP(\omega) = \int_{X^{-1}(B)} (h \circ X)(\omega)dP_X, \qquad \forall B \in \mathcal{B}_{\mathbb{R}^k}. \tag{1.11}$$

By direct comparison of (1.11) and (1.9), uniqueness implies that

$$E[Y|X] = h \circ X.$$

The usual interpretation of h is as follows. Given $x \in \mathbb{R}^k$,

$$h(x) = E[Y|X = x], \quad P_X\text{-a.s.}$$

We can then finally state that, for $\omega \in \Omega$,

$$E[Y|X](\omega) = E[Y|X = X(\omega)], \quad P\text{-a.s.}$$

Remark 1.127. We may obtain the preceding liaison $E[Y|X] = h \circ X$ as above by referring to the Doob–Dynkin Lemma 1.29. The quantity $E[Y|X] = E[Y|\mathcal{F}_X]$ surely is $\mathcal{F}_X$-measurable; hence, there exists a unique class of real-valued $h \in L^1(\mathbb{R}^k, \mathcal{B}_{\mathbb{R}^k}, P_X)$ such that $E[Y|X] = h(X)$ (Jeanblanc et al. (2009, p. 9)).

Proposition 1.128. *Let $\mathcal{G}$ be a sub-σ-algebra of $\mathcal{F}$. If Y is a real $\mathcal{G}$-measurable random variable in $\mathcal{L}^1(\Omega, \mathcal{G}, P)$, then*

$$E^{\mathcal{G}}[Y] = Y.$$

More generally, if Y is a real $\mathcal{G}$-measurable random variable and both Z and YZ are two real-valued random variables in $\mathcal{L}^1(\Omega, \mathcal{F}, P)$, then

$$E^{\mathcal{G}}[YZ] = YE^{\mathcal{G}}[Z].$$

Proof. The first statement follows from the fact that for all $G \in \mathcal{G} : \int_G Y dP = \int_G Y dP$, with Y $\mathcal{G}$-measurable and P-integrable.

For the second statement, see, e.g., Métivier (1968). □

Proposition 1.129 (tower law). *Let $Y \in \mathcal{L}^1(\Omega, \mathcal{F}, P)$. For any two subalgebras $\mathcal{G}$ and $\mathcal{B}$ of $\mathcal{F}$ such that $\mathcal{G} \subset \mathcal{B} \subset \mathcal{F}$, we have*

$$E[E[Y|\mathcal{B}]|\mathcal{G}] = E[Y|\mathcal{G}] = E[E[Y|\mathcal{G}]|\mathcal{B}].$$

Proof. For the first equality, by definition, we have

$$\int_G E[Y|\mathcal{G}]dP = \int_G YdP = \int_G E[Y|\mathcal{B}]dP = \int_G E[E[Y|\mathcal{B}]|\mathcal{G}]dP$$

for all $G \in \mathcal{G} \subset \mathcal{B}$, where comparing the first and last terms completes the proof. The second equality is proven along the same lines. $\square$

Definition 1.130. Let $(\Omega, \mathcal{F}, P)$ be a probability space, and let $\mathcal{G}$ be a sub-σ-algebra of $\mathcal{F}$. We say that a real random variable Y on $(\Omega, \mathcal{F}, P)$ is independent of $\mathcal{G}$ with respect to the probability measure P if

$$\forall B \in \mathcal{B}_{\mathbb{R}}, \ \forall G \in \mathcal{G}: \ P(G \cap Y^{-1}(B)) = P(G)P(Y^{-1}(B)).$$

Proposition 1.131. *Let $\mathcal{G}$ be a sub-σ-algebra of $\mathcal{F}$; if $Y \in \mathcal{L}^1(\Omega, \mathcal{F}, P)$ is independent of $\mathcal{G}$, then*

$$E[Y|\mathcal{G}] = E[Y], \text{ a.s.}$$

Proof. Let $G \in \mathcal{G}$; then, by independence,

$$\int_G YdP = \int I_G YdP = E[I_G Y] = E[I_G]E[Y] = P(G)E[Y] = \int_G E[Y]dP,$$

from which the proposition follows. $\square$

Proposition 1.132. *Let $(\Omega, \mathcal{F}, P)$ be a probability space and $\mathcal{F}'$ a sub-σ-algebra of $\mathcal{F}$. Furthermore, let Y and $(Y_n)_{n \in \mathbb{N}}$ be real-valued random variables, all belonging to $\mathcal{L}^1(\Omega, \mathcal{F}, P)$. The following properties hold:*

1. $E[E[Y|\mathcal{F}']] = E[Y]$;
2. $E[\alpha Y + \beta|\mathcal{F}'] = \alpha E[Y|\mathcal{F}'] + \beta$ *a.s.* $(\alpha, \beta \in \mathbb{R})$.
3. *(Extended monotone convergence theorem) Assume $|Y_n| \leq Z$ for all $n \in \mathbb{N}$, with $Z \in \mathcal{L}^1(\Omega, \mathcal{F}, P)$; if $Y_n \uparrow Y$ a.s., then $E[Y_n|\mathcal{F}'] \uparrow E[Y|\mathcal{F}']$ a.s.*
4. *(Fatou's lemma) Assume $|Y_n| \leq Z$ for all $n \in \mathbb{N}$, with $Z \in \mathcal{L}^1(\Omega, \mathcal{F}, P)$; $\limsup_{n \to \infty} E[Y_n|\mathcal{F}'] \leq E[\limsup_{n \to \infty} Y_n|\mathcal{F}']$ almost surely.*
5. *(Dominated convergence theorem) Assume $|Y_n| \leq Z$ for all $n \in \mathbb{N}$, with $Z \in \mathcal{L}^1(\Omega, \mathcal{F}, P)$; if $Y_n \to Y$ a.s., then $E[Y_n|\mathcal{F}'] \to E[Y|\mathcal{F}']$ almost surely.*
6. *If $\phi : \mathbb{R} \to \mathbb{R}$ is convex and $\phi(Y)$ P-integrable, then $\phi(E[Y|\mathcal{F}']) \leq E[\phi(Y)|\mathcal{F}']$ almost surely (Jensen's inequality).*

Proof.

1. This property follows from Proposition 1.123 with $B' = \Omega$.
2. This is obvious from the linearity of the integral.
3–5. These properties can be shown as the corresponding ones without conditioning as they derive from classical measure theory.

6. Here we use the fact that every convex function ϕ is of type $\phi(x) = \sup_n(a_n x + b_n)$. Therefore, defining $l_n(x) = a_n x + b_n$ for all n, we have that

$$l_n(E[Y|\mathcal{F}']) = E[l_n(Y)|\mathcal{F}'] \leq E[\phi(Y)|\mathcal{F}']$$

and thus

$$\phi(E[Y|\mathcal{F}']) = \sup_n l_n(E[Y|\mathcal{F}']) \leq E[\phi(Y)|\mathcal{F}'].$$

$\square$

Proposition 1.133. *If $Y \in L^p(\Omega, \mathcal{F}, P)$, then $E[Y|\mathcal{F}']$ is an element of $L^p(\Omega, \mathcal{F}', P)$ and*

$$\|E[Y|\mathcal{F}']\|_p \leq \|Y\|_p \qquad (1 \leq p < \infty). \tag{1.12}$$

Proof. With $\phi(x) = |x|^p$ being convex, we have that $|E[Y|\mathcal{F}']|^p \leq E[|Y|^p|\mathcal{F}']$ and thus $E[Y|\mathcal{F}'] \in L^p(\Omega, \mathcal{F}, P)$, and after integration we obtain (1.12). $\square$

Proposition 1.134. *The conditional expectation $E[Y|\mathcal{F}']$ is the unique $\mathcal{F}'$-measurable random variable Z such that for every $\mathcal{F}'$-measurable $X : \Omega \to \mathbb{R}$, for which the products XY and XZ are P-integrable, we have*

$$E[XY] = E[XZ]. \tag{1.13}$$

Proof. From the fact that $E[E[XY|\mathcal{F}']] = E[XY]$ (point 1 of Proposition 1.132) and because X is $\mathcal{F}'$-measurable, it follows from Proposition 1.128 that $E[E[XY|\mathcal{F}']] = E[XE[Y|\mathcal{F}']]$. On the other hand, if Z is an $\mathcal{F}'$-measurable random variable, so that for every $\mathcal{F}'$-measurable X, with $XY \in L^1(\Omega, \mathcal{F}, P)$ and $XZ \in L^1(\Omega, \mathcal{F}, P)$, it follows that $E[XY] = E[XZ]$. Taking $X = I_B$, $B \in \mathcal{F}'$ we obtain

$$\int_B Y dP = E[Y I_B] = E[Z I_B] = \int_B Z dP$$

and hence, by the uniqueness of $E[Y|\mathcal{F}']$, $Z = E[Y|\mathcal{F}']$ almost surely. $\square$

Theorem 1.135. *Let $(\Omega, \mathcal{F}, P)$ be a probability space, $\mathcal{F}'$ a sub-σ-algebra of $\mathcal{F}$, and Y a real-valued random variable on $(\Omega, \mathcal{F}, P)$. If $Y \in L^2(P)$, then $E[Y|\mathcal{F}']$ is the orthogonal projection of Y on $L^2(\Omega, \mathcal{F}', P)$, a closed subspace of the Hilbert space $L^2(\Omega, \mathcal{F}, P)$.*

Proof. By Proposition 1.133, from $Y \in L^2(\Omega, \mathcal{F}, P)$ it follows that

$$E[Y|\mathcal{F}'] \in L^2(\Omega, \mathcal{F}', P)$$

and, by equality (1.13), for all random variables $X \in L^2(\Omega, \mathcal{F}', P)$, it holds that

$$E[XY] = E[XE[Y|\mathcal{F}']],$$

completing the proof, by recalling that $(X, Y) \to E[XY]$ is the scalar product in L^2. $\qquad\qquad\qquad\qquad\qquad\qquad\qquad\qquad\qquad\qquad\qquad\qquad\qquad\square$

Remark 1.136. We may interpret the foregoing theorem by stating that $E[Y|\mathcal{F}']$ is the best mean square approximation of $Y \in L^2(\Omega, \mathcal{F}, P)$ in $L^2(\Omega, \mathcal{F}', P)$.

Definition 1.137. A family of random variables $(Y_n)_{n \in \mathbb{N}}$ is *uniformly integrable* if

$$\lim_{m \to \infty} \sup_n \int_{|Y_n| \geq m} |Y_n| dP = 0.$$

Proposition 1.138. *Let $(Y_n)_{n \in \mathbb{N}}$ be a family of random variables in $\mathcal{L}^1$. Then the following two statements are equivalent:*

1. *$(Y_n)_{n \subset \mathbb{N}}$ is uniformly integrable.*
2. *$\sup_{n \in \mathbb{N}} E[|Y_n|] < +\infty$, and for all ϵ there exists $\delta > 0$ such that $A \in \mathcal{F}$, $P(A) \leq \delta \Rightarrow E[|Y_n I_A|] < \epsilon$.*

Proposition 1.139. *Let $(Y_n)_{n \in \mathbb{N}}$ be a family of random variables dominated by a nonnegative $X \in \mathcal{L}^1$ on the same probability space $(\Omega, \mathcal{F}, P)$, so that $|Y_n(\omega)| \leq X(\omega)$ for all $n \in \mathbb{N}$. Then $(Y_n)_{n \in \mathbb{N}}$ is uniformly integrable.*

Theorem 1.140. *Let $Y \in \mathcal{L}^1$ be a random variable on $(\Omega, \mathcal{F}, P)$. Then the class $(E[Y|\mathcal{G}])_{\mathcal{G} \subset \mathcal{F}}$, where $\mathcal{G}$ are sub-σ-algebras, is uniformly integrable.*

Proof. See, e.g., Williams (1991). $\qquad\qquad\qquad\qquad\qquad\qquad\qquad\qquad\square$

Theorem 1.141. *Let $(Y_n)_{n \in \mathbb{N}}$ be a sequence of random variables in $\mathcal{L}^1$ and let $Y \in \mathcal{L}^1$. Then $Y_n \xrightarrow{\mathcal{L}^1} Y$ if and only if*

1. *$Y_n \xrightarrow[n]{P} Y$.*
2. *$(Y_n)_{n \in \mathbb{N}}$ is uniformly integrable.*

Proof. See, e.g., Williams (1991). $\qquad\qquad\qquad\qquad\qquad\qquad\qquad\qquad\square$

1.7 Conditional and Joint Distributions

Let $(\Omega, \mathcal{F}, P)$ be a probability space, $X : (\Omega, \mathcal{F}, P) \to (E, \mathcal{B})$ a random variable, and $F \in \mathcal{F}$. Following previous results, a unique element $E[I_F | X = x] \in L^1(E, \mathcal{B}, P_X)$ exists such that for any $B \in \mathcal{B}$

$$P(F \cap [X \in B]) = \int_{[X \in B]} I_F(\omega)dP(\omega) = \int_B E[I_F|X = x]dP_X(x). \quad (1.14)$$

We can write

$$P(F|X = \cdot) = E[I_F|X = \cdot].$$

Remark 1.142. By (1.14) the following properties hold:

1. For all $F \in \mathcal{F} : P(F|X = x) \geq 0$, almost surely with respect to P_X.
2. $P(\emptyset|X = x) = 0$, almost surely with respect to P_X.
3. $P(\Omega|X = x) = 1$, almost surely with respect to P_X.
4. For all $F \in \mathcal{F} : 0 \leq P(F|X = x) \leq 1$, almost surely with respect to P_X.
5. For all $(A_n)_{n \in \mathbb{N}} \in \mathcal{F}^{\mathbb{N}}$ collections of mutually exclusive sets:

$$P\left(\bigcup_{n \in \mathbb{N}} A_n|X = x\right) = \sum_{n \in \mathbb{N}} P(A_n|X = x), \qquad P_X\text{-a.s.}$$

If, for a fixed $x \in E$, points 3, 4, and 5 hold simultaneously, then $P(\cdot|X = x)$ is a probability, but in general they do not. For example, it is not in general the case that the set of points $x \in E$, $P_X(x) \neq 0$, for which 4 is satisfied, depends upon $F \in \mathcal{F}$. Even if the set of points for which 4 does not hold has zero measure, their union over $F \in \mathcal{F}$ will not necessarily have measure zero. This is also true for subsets $\mathcal{F}' \subset \mathcal{F}$. Hence, in general, given $x \in E$, $P(\cdot|X = x)$ is not a probability on $\mathcal{F}$, unless $\mathcal{F}$ is a countable family, or countably generated. If it happens that, apart from a set E_0 of P_X-measure zero, $P(\cdot|X = x)$ is a probability, then the collection $(P(\cdot|X = x))_{x \in E - E_0}$ is called a regular version of the conditional probability with respect to X on $\mathcal{F}$.

Definition 1.143. Let $X : (\Omega, \mathcal{F}) \to (E, \mathcal{B})$ and $Y : (\Omega, \mathcal{F}, P) \to (E_1, \mathcal{B}_1)$ be two random variables. We denote by $\mathcal{F}_Y$ the σ-algebra generated by Y, hence

$$\mathcal{F}_Y = Y^{-1}(\mathcal{B}_1) = \{Y^{-1}(B)|B \in \mathcal{B}_1\}.$$

If there exists a regular version $(P(\cdot|X = x))_{x \in E}$ of the probability conditional on X on the σ-algebra $\mathcal{F}_Y$, denoting by $P_Y(\cdot|X = x)$ the mapping defined on $\mathcal{B}_1$, then

$$P_Y(B|X = x) = P(Y \in B|X = x) \qquad \forall B \in \mathcal{B}_1, x \in E.$$

This mapping is a probability, called the distribution of Y conditional on X, with $X = x$.

Remark 1.144. From the properties of the induced measure it follows that

$$E[Y|X = x] = \int Y(\omega)dP(\omega|X = x) = \int Y dP_Y(Y|X = x).$$

Existence of Conditional Distributions

The following proposition shows the existence of a regular version of the conditional distribution of a random variable in a very special case.

Proposition 1.145. *Let $X : (\Omega, \mathcal{F}) \to (E, \mathcal{B})$ and $Y : (\Omega, \mathcal{F}) \to (E_1, \mathcal{B}_1)$ be two random variables. Then the necessary and sufficient condition for X and Y to be independent is*

$$\forall A \in \mathcal{B}_1: \qquad P(Y \in A|\cdot) = constant(A), \qquad P_X\text{-a.s.}$$

Therefore,

$$P(Y \in A|\cdot) = P(Y \in A), \qquad P_X\text{-a.s.},$$

and if Y is a real-valued integrable random variable, then

$$E[Y|\cdot] = E[Y], \qquad P_X\text{-a.s.}$$

Proof. The independence of X and Y is equivalent to

$$P([X \in B] \cap [Y \in A]) = P([X \in B])P([Y \in A]) \qquad \forall A \in \mathcal{B}_1, B \in \mathcal{B},$$

or

$$\int_{[X \in B]} I_{[Y \in A]}(\omega)P(d\omega) = P(Y \in A) \int I_B(x)dP_X(x)$$

$$= \int_B P(Y \in A)dP_X(x),$$

and this is equivalent to affirming that

$$P(Y \in A|\cdot) = P(Y \in A), \qquad P_X\text{-a.s.}, \tag{1.15}$$

which is a constant k for $x \in E$. If we can write

$$P(Y \in A|\cdot) = k(A), \qquad P_X\text{-a.s.},$$

then

$$\forall B \in \mathcal{B}: \int_{[X \in B]} I_{[Y \in A]}(\omega)dP(\omega) = \int_B k(A)dP_X(x) = k(A)P(X \in B),$$

from which it follows that

$$\forall B \in \mathcal{B}: \qquad P([X \in B] \cap [Y \in A]) = k(A)P(X \in B).$$

Therefore, for $B = E$ we have that

$$P(Y \in A) = k(A)P(X \in E) = k(A).$$

Now, we observe that (1.15) states that there exists a regular version of the probability conditional on X, relative to the σ-algebra $\mathcal{F}'$ generated by Y, where the latter is given by

$$P(Y \in A|\cdot) = P_Y(A) \qquad \forall x \in E.$$

Hence, by Remark 1.144, it can then be shown that $E[Y|\cdot] = E[Y]$.   □

We have already shown that if X is a discrete random variable, then the real random variable Y has a distribution conditional on X. The following theorem provides more general conditions under which this conditional distribution exists.

Theorem 1.146. *Let Y be a real-valued random variable on $(\Omega, \mathcal{F}, P)$, and let $\mathcal{G} \subset \mathcal{F}$ be a σ-algebra; there always exists a regular version $P_Y(\cdot \mid \mathcal{G})$ of the conditional distribution of Y given $\mathcal{G}$.*

Proof. See, e.g., Ash (1972, p. 263).   □

A further generalization to Polish spaces is possible, based on the following definition (Klenke (2008, p. 184)).

Definition 1.147. Two measurable spaces $(E, \mathcal{B}_E)$ and $(E_1, \mathcal{B}_{E_1})$ are called *isomorphic* if there exists a measurable bijection $\varphi : (E, \mathcal{B}_E) \to (E_1, \mathcal{B}_{E_1})$ such that its inverse $\widetilde{\varphi}$ is also measurable $\widetilde{\varphi} : (E_1, \mathcal{B}_{E_1}) \to (E, \mathcal{B}_E)$.

Definition 1.148. Two measure spaces $(E, \mathcal{B}_E, \mu)$ and $(E_1, \mathcal{B}_{E_1}, \mu_1)$ are called isomorphic if $(E, \mathcal{B}_E)$ and $(E_1, \mathcal{B}_{E_1})$ are isomorphic measurable spaces and $\mu_1 = \varphi(\mu)$.

In either case, φ is called an isomorphism.

Definition 1.149. A measurable space $(E, \mathcal{B}_E)$ is called a *Borel space* if there exists a Borel set $B \in \mathcal{B}_\mathbb{R}$ such that $(E, \mathcal{B}_E)$ and $(B, \mathcal{B}_B)$ are isomorphic measurable spaces.

The following theorem holds.

Theorem 1.150. *If E is a Polish space and $\mathcal{E}$ is its Borel σ-algebra, then $(E, \mathcal{E})$ is a Borel space.*

Proof. See, e.g., Ash (1972, Sect. 4.4, Problem 8).   □

Theorem 1.151. *Let $(\Omega, \mathcal{F}, P)$ be a probability space, and let $Y : (\Omega, \mathcal{F}) \to (\Omega', \mathcal{F}')$, where $(\Omega', \mathcal{F}')$ is a Borel space. Then there exists a regular version of the conditional distribution of Y with respect to any sub-σ-algebra $\mathcal{G} \subset \mathcal{F}$.*

Proof. Let $(\Omega', \mathcal{F}')$ be isomorphic to $(\mathbb{R}, \mathcal{B}_{\mathbb{R}})$, and let $\varphi : (\Omega', \mathcal{F}') \to (\mathbb{R}, \mathcal{B}_{\mathbb{R}})$ be the corresponding isomorphism. Consider the sub-σ-algebra $\mathcal{G} \subset \mathcal{F}$, and let

$$B \in \mathcal{B}_{\mathbb{R}} \mapsto Q_0(B) \equiv P(\varphi(Y) \in B \mid \mathcal{G})$$

be a regular version of the conditional probability of the real-valued random variable $\varphi(Y) : (\Omega, \mathcal{F}) \to (\mathbb{R}, \mathcal{B}_{\mathbb{R}})$ given $\mathcal{G}$.

Now let $A \in \mathcal{F}'$; from the foregoing discussion we obtain

$$P(Y \in A|\mathcal{G}) = P(\varphi(Y) \in \varphi(A)|\mathcal{G}) = Q_0(\varphi(A)) = Q(A)$$

if we denote $Q = \varphi^{-1}(Q_0)$. In fact,

$$Q(A) = Q_0((\varphi^{-1})^{-1}(A)) = Q_0(\varphi(A)).$$

Since Q_0 is a probability measure on $\mathcal{B}_{\mathbb{R}}$, the same will be Q on φ^{-1} $(\mathcal{B}_{\mathbb{R}}) = \mathcal{F}'$. $\qquad\square$

As a consequence of the preceding results, the following theorem holds.

Theorem 1.152 (Jirina). *Let X and Y be two random variables on $(\Omega, \mathcal{F}, P)$ with values in $(E, \mathcal{B})$ and $(E_1, \mathcal{B}_1)$, respectively. If E and E_1 are complete separable metric spaces with respective Borel σ-algebras $\mathcal{B}$ and $\mathcal{B}_1$, then there exists a regular version of the conditional distribution of Y given X.*

Definition 1.153. Given the assumptions of Definition 1.143, if $P_Y(\cdot|X = x)$ is defined by a density with respect to the measure μ_1 on $(E_1, \mathcal{B}_1)$, then this density is said to be conditional on X, written $X = x$, and denoted by $f_Y(\cdot|X = x)$.

Proposition 1.154. *Let $\mathbf{X} = (X_1, \ldots, X_n) : (\Omega, \mathcal{F}) \to (\mathbb{R}^n, \mathcal{B}_{\mathbb{R}^n})$ be a vector of random variables whose probability is defined through the density $f_{\mathbf{X}}(x_1, \ldots, x_n)$ with respect to Lebesgue measure μ_n on $\mathbb{R}^n$. Fixing $q = 1, \ldots, n$, we can consider the random vectors*

$$\mathbf{Y} = (X_1, \ldots, X_q) : (\Omega, \mathcal{F}) \to \mathbb{R}^q$$

and

$$\mathbf{Z} = (X_{q+1}, \ldots, X_n) : (\Omega, \mathcal{F}) \to \mathbb{R}^{n-q}.$$

Then $\mathbf{Z}$ admits a distribution conditional on $\mathbf{Y}$ for almost every $\mathbf{Y} \in \mathbb{R}$ defined through the function

$$f(x_{q+1}, \ldots, x_n | x_1, \ldots, x_q) = \frac{f_{\mathbf{X}}(x_1, \ldots, x_q, x_{q+1}, \ldots, x_n)}{f_{\mathbf{Y}}(x_1, \ldots, x_q)},$$

with respect to Lebesgue measure μ_{n-q} on $\mathbb{R}^{n-q}$. Hence, $f_{\mathbf{Y}}(x_1, \ldots, x_q)$ is the marginal density of $\mathbf{Y}$ at $(x_1, \ldots, x_q)$, given by

$$f_{\mathbf{Y}}(x_1, \ldots, x_q) = \int f_{\mathbf{X}}(x_1, \ldots, x_n) d\mu_{n-q}(x_{q+1}, \ldots, x_n).$$

Proof. Writing $\mathbf{y} = (x_1, \ldots, x_q)$ and $\mathbf{x} = (x_1, \ldots, x_n)$, let $B \in \mathcal{B}_{\mathbb{R}^q}$ and $B_1 \in \mathcal{B}_{\mathbb{R}^{n-q}}$. Then

$$P([\mathbf{Y} \in B] \cap [\mathbf{Z} \in B_1]) = P_{\mathbf{X}}((\mathbf{Y}, \mathbf{Z}) = \mathbf{X} \in B \times B_1) = \int_{B \times B_1} f_{\mathbf{X}}(\mathbf{x}) d\mu_n$$

$$= \int_B d\mu_q(x_1, \ldots, x_q) \int_{B_1} f_{\mathbf{X}}(\mathbf{x}) d\mu_{n-q}(x_{q+1}, \ldots, x_n)$$

$$= \int_B f_{\mathbf{Y}}(\mathbf{x}) d\mu_q \int_{B_1} \frac{f_{\mathbf{X}}(\mathbf{x})}{f_{\mathbf{Y}}(\mathbf{y})} d\mu_{n-q}$$

$$= \int_B dP_{\mathbf{Y}} \left(\int_{B_1} \frac{f_{\mathbf{X}}(\mathbf{x})}{f_{\mathbf{Y}(\mathbf{y})}} d\mu_{n-q} \right),$$

where the last equality holds for all points $\mathbf{y}$ for which $f_{\mathbf{Y}}(\mathbf{y}) \neq 0$. By the definition of density, the set of points $\mathbf{y}$ for which $f_{\mathbf{Y}}(\mathbf{y}) = 0$ has zero measure with respect to $P_{\mathbf{Y}}$, and therefore we can write in general

$$P([\mathbf{Y} \in B] \cap [\mathbf{Z} \in B_1]) = \int_B dP_{\mathbf{Y}}(\mathbf{y}) \int_{B_1} \frac{f_{\mathbf{X}}(\mathbf{x})}{f_{\mathbf{Y}}(\mathbf{y})} d\mu_{n-q}.$$

Thus the latter integral is an element of $P(\mathbf{Z} \in B_1 | \mathbf{Y} = \mathbf{y})$. Hence

$$\int_{B_1} \frac{f_{\mathbf{X}}(\mathbf{x})}{f_{\mathbf{Y}}(\mathbf{y})} d\mu_{n-q} = P(\mathbf{Z} \in B_1 | \mathbf{Y} = \mathbf{y}) = P_{\mathbf{Z}}(B_1 | \mathbf{Y} = \mathbf{y}),$$

from which it follows that $\frac{f_{\mathbf{X}}(\mathbf{x})}{f_{\mathbf{Y}}(\mathbf{y})}$ is the density of $P(\cdot | \mathbf{Y} = \mathbf{y})$. $\qquad\square$

Example 1.155. Let $f_{X,Y}(x, y)$ be the density of the bivariate Gaussian distribution. Then

$$f_{X,Y}(x, y) = k \exp \left\{ -\frac{1}{2} (a(x - m_1)^2 + 2b(x - m_1)(y - m_2) + c(y - m_2)^2) \right\},$$

where

$$k = \frac{1}{2\pi\sigma_x\sigma_y\sqrt{1 - \rho^2}}, \qquad a = \frac{1}{(1 - \rho^2)\sigma_x^2},$$

$$b = \frac{-\rho}{(1 - \rho^2)\sigma_x\sigma_y}, \qquad c = \frac{1}{(1 - \rho^2)\sigma_y^2}.$$

The distribution of Y conditional on X is defined through the density

$$f_Y(Y|X = x) = \frac{f_{X,Y}(x, y)}{f_X(x)}, \text{ where } f_X(x) = \frac{1}{\sigma_x\sqrt{2\pi}} \exp \left\{ -\frac{1}{2} \left(\frac{x - m}{\sigma_x} \right)^2 \right\}.$$

From this it follows that

$$f_Y(Y|X = x)$$

$$= \frac{1}{\sigma_y \sqrt{2\pi(1 - \rho^2)}} \exp\left\{ -\frac{1}{2(1 - \rho^2)} \left(\frac{y - m_2 - \frac{\sigma_y}{\sigma_x}(x - m_1)}{\sigma_y} \right)^2 \right\}.$$

Therefore, the conditional density is normal, but with mean

$$E[Y|X = x] = \int y dP_Y(y|X = x) = \int y f_Y(y|X = x) dy = m_2 + \rho \frac{\sigma_y}{\sigma_x}(x - m_1)$$

and variance $(1 - \rho^2)\sigma_y^2$. The conditional expectation in this case is also called the regression line of Y with respect to X.

Remark 1.156. Under the assumptions of Proposition 1.145, two generic random variables defined on the same probability space $(\Omega, \mathcal{F}, P)$ with values in $(E, \mathcal{B})$ and $(E, \mathcal{B}_1)$, respectively, are independent if and only if Y has a conditional distribution with respect to $X = x$, which is independent of x:

$$P_Y(A|X = x) = P_Y(A), \qquad P_X\text{-a.s.,} \qquad (1.16)$$

which can be rewritten to hold for every $x \in E$. If X and Y are independent, then their joint probability is given by

$$P_{(X,Y)} = P_X \otimes P_Y.$$

Integrating a function $f(x, y)$ with respect to $P(X, Y)$ by Fubini's theorem results in

$$\int f(x, y) P_{(X,Y)}(dx, dy) = \int dP_X(x) \int f(x, y) dP_Y(y). \qquad (1.17)$$

If we use (1.16), then (1.17) can be rewritten in the form

$$\int f(x, y) P_{(X,Y)}(dx, dy) = \int dP_X(x) \int f(x, y) dP_Y(y|X = x).$$

The following proposition asserts that this relation holds in general.

Proposition 1.157 (Generalization of Fubini's theorem). *Let X and Y be two generic random variables defined on the same probability space $(\Omega, \mathcal{F}, P)$ with values in $(E, \mathcal{B})$ and $(E, \mathcal{B}_1)$, respectively. Moreover, let P_X be the probability of X and $P_Y(\cdot|X = x)$ the probability of Y conditional on $X = x$ for every $x \in E$. Then, for all $M \in \mathcal{B} \otimes \mathcal{B}_1$, the function*

$$h : x \in E \rightarrow \int I_M(x, y) P_Y(dy|x)$$

is $\mathcal{B}$-measurable and positive, resulting in

$$P_{(X,Y)}(M) = \int P_X(dx) \left(\int I_M(x,y) P_Y(dy|x) \right). \qquad (1.18)$$

In general, if $f : E \times E_1 \to \mathbb{R}$ is $P_{(X,Y)}$-integrable, then the function

$$h' : x \in E \to \int f(x,y) P_Y(dy|x)$$

is defined almost surely with respect to P_X and is P_X-integrable. Thus we obtain

$$\int f(x,y) P_{(X,Y)}(dx,dy) = \int h'(x) P_X(dx). \qquad (1.19)$$

Proof. We observe that if $M = B \times B_1$, $B \in \mathcal{B}$, and $B_1 \in \mathcal{B}_1$, then

$$P_{(X,Y)}(B \times B_1) = P([X \in B] \cap [Y \in B_1]) = \int_B P(Y \in B_1 | X = x) dP_X(x),$$

and by the definition of conditional probability

$$P_{(X,Y)}(B \times B_1) = \int I_B(x) P_Y(B_1|x) dP_X(x)$$

$$= \int dP_X(x) \int P_Y(dy|x) I_B(x) I_{B_1}(y).$$

This shows that (1.18) holds for $M = B \times B_1$. It is then easy to show that (1.18) holds for every elementary function on $\mathcal{B} \otimes \mathcal{B}_1$. With the usual limiting procedure, we can show that for every $\mathcal{B} \otimes \mathcal{B}_1$-measurable positive f we obtain

$$\int^* f(x,y) dP_{(X,Y)}(x,y) = \int^* dP_X(x) \int^* f(x,y) P_Y(dy|x).$$

As usual, we have denoted by $\int^*$ the integral of a nonnegative measurable function, independently of its finiteness. If, then, f is measurable as well as both $P_{(X,Y)}$-integrable and positive, then

$$\int^* dP_X(x) \int^* f(x,y) P_Y(dy|x) < \infty,$$

where

$$\int^* f(x,y) P_Y(dy|x) < \infty, \qquad P_X\text{-a.s.,} x \in E.$$

Thus h' is defined almost surely with respect to P_X and (1.19) holds. Finally, if f is $P_{(X,Y)}$-integrable and of arbitrary sign, applying the preceding results to f^+ and f^-, we obtain that

$$\int f(x,y) P_Y(dy|x) = \int f^+(x,y) P_Y(dy|x) - \int f^-(x,y) P_Y(dy|x)$$

is defined almost surely with respect to P_X, and again (1.19) holds.   □

1.8 Convergence of Random Variables

Tail Events

Definition 1.158. Let $(A_n)_{n \in \mathbb{N}} \in \mathcal{F}^{\mathbb{N}}$ be a sequence of events and let

$$\sigma(A_n, A_{n+1}, \ldots), \qquad n \in \mathbb{N}$$

and

$$\mathcal{T} = \bigcap_{n=1}^{\infty} \sigma(A_n, A_{n+1}, \ldots)$$

be σ-algebras. Then $\mathcal{T}$ is the *tail σ-algebra* associated with the sequence $(A_n)_{n \in \mathbb{N}}$, and its elements are called *tail events*.

Example 1.159. The *essential supremum*

$$\limsup_n A_n = \bigcap_{n=1}^{\infty} \bigcup_{i=n}^{\infty} A_i$$

and *essential infimum*

$$\liminf_n A_n = \bigcup_{n=1}^{\infty} \bigcap_{i=n}^{\infty} A_i$$

are both tail events for the sequence $(A_n)_{n \in \mathbb{N}}$. If n is understood to be *time*, then we can write

$$\limsup A_n = \{A_n \text{ i.o.}\},$$

i.e., A_n occurs infinitely often (i.o.), thus, for infinitely many $n \in \mathbb{N}$. On the other hand, we may write

$$\liminf A_n = \{A_n \text{ a.a.}\},$$

i.e., A_n occurs almost always (a.a.), thus for all but finitely many $n \in \mathbb{N}$.

Theorem 1.160 (Kolmogorov's zero-one law). *Let $(A_n)_{n \in \mathbb{N}} \in \mathcal{F}^{\mathbb{N}}$ be a sequence of independent events. Then for any $A \in \mathcal{T}:, P(A) = 0$ or $P(A) = 1$.*

Lemma 1.161. (Borel–Cantelli).

1. *Let $(A_n)_{n \in \mathbb{N}} \in \mathcal{F}^{\mathbb{N}}$ be a sequence of events. If $\sum_n P(A_n) < +\infty$, then*

$$P\left(\limsup_n A_n\right) = 0.$$

2. *Let $(A_n)_{n \in \mathbb{N}} \in \mathcal{F}^{\mathbb{N}}$ be a sequence of independent events. If $\sum_n P(A_n) = +\infty$, then*

$$P\left(\limsup_n A_n\right) = 1.$$

Proof. See, e.g., Billingsley (1968). □

Almost Sure Convergence and Convergence in Probability

Definition 1.162. Let $(X_n)_{n\in\mathbb{N}}$ be a sequence of random variables on the probability space $(\Omega, \mathcal{F}, P)$ and X a further random variable defined on the same space. $(X_n)_{n\in\mathbb{N}}$ *converges almost surely* to X, denoted by $X_n \xrightarrow[n]{\text{a.s.}} X$ or, equivalently, $\lim_{n\to\infty} X_n = X$ almost surely if

$$\exists S_0 \in \mathcal{F} \text{ such that } P(S_0) = 0 \text{ and } \forall \omega \in \Omega \setminus S_0 : \lim_{n\to\infty} X_n(\omega) = X(\omega).$$

Definition 1.163. $(X_n)_{n\in\mathbb{N}}$ *converges in probability (or stochastically)* to X, denoted by $X_n \xrightarrow[n]{P} X$ or, equivalently, $P - \lim_{n\to\infty} X_n = X$ if

$$\forall \epsilon > 0 : \lim_{n\to\infty} P(|X_n - X| > \epsilon) = 0.$$

Theorem 1.164. *A sequence* $(X_n)_{n\in\mathbb{N}}$ *of random variables converges in probability to a random variable* X *if and only if*

$$\lim_{n\to\infty} E\left[\frac{|X_n - X|}{1 + |X_n - X|}\right] = 0.$$

Proof. See, e.g., Jacod and Protter (2000, p. 139). □

Theorem 1.165. *Consider a sequence* $(X_n)_{n\in\mathbb{N}}$ *of random variables and an additional random variable* X *on the same probability space, and let* $f : \mathbb{R} \to \mathbb{R}$ *be a continuous function. Then*

(a) $X_n \xrightarrow[n]{\text{a.s.}} X \Rightarrow f(X_n) \xrightarrow[n]{\text{a.s.}} f(X)$

(b) $X_n \xrightarrow[n]{P} X \Rightarrow f(X_n) \xrightarrow[n]{P} f(X)$

Proof. See, e.g., Jacod and Protter (2000, p. 142). □

Convergence in Mean of Order p

Definition 1.166. Let X be a real-valued random variable on the probability space $(\Omega, \mathcal{F}, P)$. X is *integrable to the pth exponent* $(p \geq 1)$ if the random variable $|X|^p$ is P-integrable; thus, $|X|^p \in \mathcal{L}^1(P)$. By $\mathcal{L}^p(P)$ we denote the whole of the real-valued random variables on $(\Omega, \mathcal{F}, P)$ that are integrable to the pth exponent. Then, by definition,

$$X \in \mathcal{L}^p(P) \Leftrightarrow |X|^p \in \mathcal{L}^1(P).$$

The following results are easy to show.

Theorem 1.167.

$$X, Y \in \mathcal{L}^p(P) \Rightarrow \begin{cases} \alpha X \in \mathcal{L}^p(P) \quad (\alpha \in \mathbb{R}), \\ X + Y \in \mathcal{L}^p(P), \\ \sup\{X, Y\} \in \mathcal{L}^p(P), \\ \inf\{X, Y\} \in \mathcal{L}^p(P). \end{cases}$$

Theorem 1.168. *If* $X \in \mathcal{L}^p(P)$, $Y \in \mathcal{L}^q(P)$ *with* $p, q > 1$ *and* $\frac{1}{p} + \frac{1}{q} = 1$, *then* $XY \in \mathcal{L}^1(P)$.

Corollary 1.169 *If* $1 \le p' \le p$, *then* $\mathcal{L}^p(P) \subset \mathcal{L}^{p'}(P)$.

Proposition 1.170. *Setting* $N_p(X) = (\int |X|^p dP)^{\frac{1}{p}}$ *for* $X \in \mathcal{L}^p(P)$ $(p \ge 1)$, *we obtain the following results.*

1. *Hölder's inequality: If* $X \in \mathcal{L}^p(P)$, $Y \in \mathcal{L}^q(P)$ *with* $p, q > 1$ *and* $\frac{1}{p} + \frac{1}{q} = 1$, *then* $N_1(XY) \le N_p(X) N_q(Y)$.
2. *Cauchy–Schwarz inequality:*

$$\left| \int XY dP \right| \le N_2(X) N_2(Y), \qquad X, Y \in \mathcal{L}^2(P). \tag{1.20}$$

3. *Minkowski's inequality:*

$$N_p(X + Y) \le N_p(X) + N_p(Y) \text{ for } X, Y \in \mathcal{L}^p(P), \ (p \ge 1).$$

Proposition 1.171. *The mapping* $N_p : \mathcal{L}^p(P) \to \mathbb{R}_+$ $(p \ge 1)$ *has the following properties:*

1. $N_p(\alpha X) = |\alpha| N_p(X)$ *for* $X \in \mathcal{L}^p(P)$, $\alpha \in \mathbb{R}$
2. $X = 0 \Rightarrow N_p(X) = 0$

By 1 and 2 of Proposition 1.171 as well as 3 of Proposition 1.170, we can assert that N_p is a *seminorm* on $\mathcal{L}^p(P)$, but not a norm. It is then defined the space $L^p(P)$ as the quotient space of $\mathcal{L}^p(P)$ with respect to the equivalence

$$X \sim Y \Leftrightarrow X = Y \ P - \text{a.s.}$$

Definition 1.172. Let $(X_n)_{n \in \mathbb{N}}$ be a sequence of elements of $L^p(P)$ and let X be another element of $L^p(P)$. Then the sequence $(X_n)_{n \in \mathbb{N}}$ *converges to* X *in mean of order* p (denoted by $X_n \xrightarrow[n]{L^p} X$) if $\lim_{n \to \infty} \|X_n - X\|_p = 0$.

Convergence in Distribution

Now we will define a different type of convergence of random variables that is associated with its partition function [see Loève (1963) for further references]. We consider a sequence of probabilities $(P_n)_{n \in \mathbb{N}}$ on $(\mathbb{R}, \mathcal{B}_{\mathbb{R}})$ and present the following definitions.

Definition 1.173. The sequence of probabilities $(P_n)_{n \in \mathbb{N}}$ *converges weakly* to a probability P if the following conditions are satisfied:

$$\text{for all } f : \mathbb{R} \to \mathbb{R} \text{ continuous and bounded: } \lim_{n \to \infty} \int f dP_n = \int f dP.$$

We write

$$P_n \xrightarrow[n \to \infty]{w} P.$$

Definition 1.174. Let $(X_n)_{n \in \mathbb{N}}$ be a sequence of random variables on the probability space $(\Omega, \mathcal{F}, P)$ and X a further random variable defined on the same space. $(X_n)_{n \in \mathbb{N}}$ *converges in distribution* to X if the sequence $(P_{X_n})_{n \in \mathbb{N}}$ converges weakly to P_X. We write

$$X_n \xrightarrow[n \to \infty]{d} X$$

or

$$X_n \underset{n \to \infty}{\Rightarrow} X.$$

Theorem 1.175. *Let $(X_n)_{n \in \mathbb{N}}$ be a sequence of random variables on the probability space $(\Omega, \mathcal{F}, P)$ and X another random variable defined on the same space. The following propositions are equivalent:*

(a) $(X_n)_{n \in \mathbb{N}}$ *converges in distribution to X*
(b) *For any continuous and bounded $f : \mathbb{R} \to \mathbb{R}$:*
    $\lim_{n \to \infty} E[f(X_n)] = E[f(X)]$
(c) *For any Lipschitz continuous $f : \mathbb{R} \to \mathbb{R}$:*
    $\lim_{n \to \infty} E[f(X_n)] = E[f(X)]$
(d) *For any uniformly continuous $f : \mathbb{R} \to \mathbb{R}$:*
    $\lim_{n \to \infty} E[f(X_n)] = E[f(X)]$

Theorem 1.176. *Denoting by F the partition function associated with X, and, for every $n \in \mathbb{N}$, by F_n the partition function associated with X_n, the following two conditions are equivalent:*

1. *For all $f : \mathbb{R} \to \mathbb{R}$ continuous and bounded: $\lim_{n \to \infty} \int f dP_{X_n} = \int f dP_X$.*
2. *For all $x \in \mathbb{R}$ such that F is continuous in x: $\lim_{n \to \infty} F_n(x) = F(x)$.*

Theorem 1.177.(Polya). *Under the assumptions of the previous theorem, if F is continuous and, for all $x \in \mathbb{R}$,:*

$$\lim_{n \to \infty} F_n(x) = F(x),$$

then the convergence is uniform on all bounded intervals of $\mathbb{R}$.

We will henceforth denote the characteristic functions associated with the random variables X and X_n by ϕ_X and ϕ_{X_n}, for all $n \in \mathbb{N}$, respectively.

Theorem 1.178 (Lévy's continuity theorem). *Let $(P_n)_{n \in \mathbb{N}}$ be a sequence of probability laws on $\mathbb{R}$ and $(\phi_n)_{n \in \mathbb{N}}$ the corresponding sequence of characteristic functions. If $(P_n)_{n \in \mathbb{N}}$ weakly converges to a probability law P having the characteristic function ϕ, then for all $t \in \mathbb{R}: \phi_n(t) \xrightarrow{n} \phi(t)$.*

If there exists $\phi : \mathbb{R} \to \mathbb{C}$ such that for all $t \in \mathbb{R} : \phi_n(t) \xrightarrow{n} \phi(t)$ and, moreover, ϕ is continuous in zero, then ϕ is the characteristic function of a probability P on $\mathcal{B}_\mathbb{R}$ such that $(P_n)_{n \in \mathbb{N}}$ converges weakly to P.

A trivial consequence of the foregoing theorem is the following result.

Corollary 1.179 *Let $(P_n)_{n \in \mathbb{N}}$ be a sequence of probability laws on $\mathbb{R}$ and $(\phi_n)_{n \in \mathbb{N}}$ the corresponding sequence of characteristic functions; let P be an additional probability law on $\mathbb{R}$ and ϕ the corresponding characteristic function.*

Then the following two statements are equivalent

(a) *$(P_n)_{n \in \mathbb{N}}$ weakly converges to P.*
(b) *For all $t \in \mathbb{R}: \phi_n(t) \xrightarrow{n} \phi(t)$.*

Relationships Between Different Types of Convergence

Theorem 1.180. *The following relationships hold:*

1. *Almost sure convergence $\Rightarrow$ convergence in probability $\Rightarrow$ convergence in distribution.*
2. *Convergence in mean $\Rightarrow$ convergence in probability.*
3. *If the limit is a degenerate random variable (i.e., a deterministic quantity), then convergence in probability $\Leftrightarrow$ convergence in distribution.*

The following theorems represent a kind of converses with respect to the preceding implications.

Theorem 1.181. *Consider a sequence $(X_n)_{n \in \mathbb{N}}$ of random variables and an additional random variable, X, on the same probability space; and suppose $X_n \xrightarrow{P}{n} X$; then there exists a subsequence $(X_{n_k})_{k \in \mathbb{N}}$ such that $X_{n_k} \xrightarrow{a.s.}{k} X$.*

Proof. See, e.g., Jacod and Protter (2000, p. 141).   □

Theorem 1.182 (Dominated convergence). *Consider a sequence* $(X_n)_{n \in \mathbb{N}}$ *of random variables and an additional random variable,* X, *on the same probability space; suppose* $X_n \xrightarrow[n]{P} X$ *and that there exists a random variable* $Y \in L^p$ *such that* $|X_n| \leq Y$ *for all* $n \in \mathbb{N}$; *then* $X_n, X \in L^p$ *and* $X_n \xrightarrow[n]{L^p} X$.

Proof. See, e.g., Jacod and Protter (2000, p. 142).   □

Theorem 1.183 (Skorohod representation theorem). *Consider a sequence* $(P_n)_{n \in \mathbb{N}}$ *of probability measures and a probability measure* P *on* $(\mathbb{R}^k, \mathcal{B}_{\mathbb{R}^k})$ *such that* $P_n \xrightarrow[n \to \infty]{\mathcal{W}} P$. *Then there exists a sequence of random variables* $(Y_n)_{n \in \mathbb{N}}$ *and a random variable* Y *defined on a common probability space* $(\Omega, \mathcal{F}, P)$, *with values in* $(\mathbb{R}^k, \mathcal{B}_{\mathbb{R}^k})$, *such that* Y_n *has probability law* P_n, Y *has probability law* P, *and*

$$Y_n \xrightarrow[n \to \infty]{\text{a.s.}} Y.$$

Proof. See, e.g., Billingsley (1968).   □

Laws of Large Numbers for Independent Random Variables

Consider a sequence $(X_n)_{n \in \mathbb{N} - \{0\}}$ of i.i.d. random variables on the same probability space $(\Omega, \mathcal{F}, P)$.

The sequence of cumulative sums of $(X_n)_{n \in \mathbb{N}}$ is

$$S_0 = 0, \quad S_n = X_1 + \cdots + X_n, \, n \in \mathbb{N} - \{0\},$$

so that the sequence of its arithmetic means is

$$\overline{X}_n = \frac{1}{n} S_n, \, n \in \mathbb{N} - \{0\}.$$

Theorem 1.184 [Weak law of large numbers (WLLN) for independent and identically distributed random variables] *Let* $(X_n)_{n \in \mathbb{N} - \{0\}}$ *be a sequence of independent and identically distributed (i.i.d.) random variables on the same probability space* $(\Omega, \mathcal{F}, P)$. *Suppose that they all belong to* $\mathcal{L}^2(\Omega, \mathcal{F}, P)$, *and denote* $m = E[X_1]$; *then*

$$\overline{X}_n \xrightarrow[n]{P} m.$$

Proof. This is a trivial consequence of Chebyshev's inequality.   □

Actually, the existence of the second moment is not a necessary condition for the WLLN; indeed, a stronger result holds.

Theorem 1.185 (Strong law of large numbers (SLLN) for i.i.d. random variables). *Let $(X_n)_{n \in \mathbb{N}-\{0\}}$ be a sequence of i.i.d. random variables on the same probability space $(\Omega, \mathcal{F}, P)$. Then*

$$\overline{X}_n \xrightarrow[n]{\text{a.s.}} a,$$

for some real constant $a \in \mathbb{R}$, if and only if all elements of the sequence of random variables belong to $\mathcal{L}^1(\Omega, \mathcal{F}, P)$.

Under this condition $a = m$.

Proof. See, e.g., Tucker (1967). □

Due to Theorem 1.180 it is now clear that for a WLLN the only requirement of existence of the first moment is sufficient.

A fundamental result for statistical applications is the well-known Glivenko–Cantelli theorem, sometimes called the *Fundamental Theorem of Statistics*.

Given a sequence $(X_n)_{n \in \mathbb{N}-\{0\}}$ of independent and identically distributed (i.i.d.) random variables on the same probability space $(\Omega, \mathcal{F}, P)$, its *empirical distribution function* $\widehat{F}_n$ is defined as

$$\widehat{F}_n(x) = \frac{1}{n} \sum_{j=1}^{n} I_{[X_j \leq x]}, \quad x \in \mathbb{R}.$$

Theorem 1.186 (Glivenko–Cantelli theorem). *Let $(X_n)_{n \in \mathbb{N}-\{0\}}$ be a sequence of i.i.d. random variables with arbitrary common distribution function F. Then*

$$\sup_{x \in \mathbb{R}} |\widehat{F}_n(x) - F(x)| \xrightarrow[n]{\text{a.s.}} 0.$$

Proof. See, e.g., Tucker (1967, P. 127). □

The Central Limit Theorem for Independent Random Variables

Theorem 1.187 (Central limit theorem for i.i.d. random variables). *Let $(X_n)_{n \in \mathbb{N}}$ be a sequence of i.i.d. random variables in $\mathcal{L}^2(\Omega, \mathcal{F}, P)$ with $m = E[X_i]$, $\sigma^2 = Var[X_i]$, for all i, and*

$$S_n = \frac{\frac{1}{n} \sum_{i=1}^{n} X_i - m}{\sigma/\sqrt{n}} = \frac{\sum_{i=1}^{n} X_i - nm}{\sigma\sqrt{n}}.$$

Then

$$S_n \xrightarrow[n \to \infty]{d} N(0, 1),$$

i.e., if we denote by $F_n = P(S_n \leq x)$ and the cumulative distribution function of S_n,

$$\Phi(x) = \int_{-\infty}^{x} \frac{1}{\sqrt{2\pi}} e^{-\frac{1}{2}y^2} dy, \qquad x \in \mathbb{R},$$

then $\lim_n F_n = \Phi$, uniformly in $\mathbb{R}$, and thus

$$\sup_{x \in \mathbb{R}} |F_n(x) - \Phi(x)| \underset{n}{\longrightarrow} 0.$$

A generalization of the central limit theorem that does not require the random variables to be identically distributed is possible.

Consider an independent array of centered random variables, i.e., for any $n \in \mathbb{N} - \{0\}$ consider a family $(X_{n1}, \ldots, X_{nn})$ of independent random variables in $\mathcal{L}^2(\Omega, \mathcal{F}, P)$, with $E[X_{nk}] = 0$, for all $k = 1, \ldots, n$. Let

$$\sigma_{nk}^2 := Var[X_{nk}] = E[X_{nk}^2] > 0, \quad k = 1, \ldots, n,$$

be such that

$$\sum_{k=1}^{n} \sigma_{nk}^2 = 1.$$

Take

$$S_n := \sum_{k=1}^{n} X_{nk}, \quad n \in \mathbb{N} - \{0\};$$

$$F_{nk} := P(X_{nk} \le x), \quad x \in \mathbb{R}, \quad n \in \mathbb{N} - \{0\}, \quad k = 1, \ldots, n.$$

Given the cumulative function of the standard normal distribution

$$\Phi(x) = \int_{-\infty}^{x} \frac{1}{\sqrt{2\pi}} e^{-\frac{1}{2}y^2} dy, \qquad x \in \mathbb{R},$$

denote

$$\Phi_{nk}(x) = \Phi\left(\frac{x}{\sigma_{nk}}\right), \qquad x \in \mathbb{R}.$$

We now introduce the following two conditions:

(L) [*Lindeberg*]   for all $\epsilon > 0$: $\quad \sum_{k=1}^{n} \int_{|x| > \epsilon} x^2 dF_{nk}(x) \underset{n \to \infty}{\longrightarrow} 0;$

(Λ) for all $\epsilon > 0$: $\quad \sum_{k=1}^{n} \int_{|x| > \epsilon} |x| \, |F_{nk}(x) - \Phi_{nk}(x)| \, dx \underset{n \to \infty}{\longrightarrow} 0.$

Theorem 1.188. *In general*

(i) $(L) \Rightarrow (\Lambda)$,
   but
(ii) if $\max\limits_{1 \leq k \leq n} E[X_{nk}^2] \xrightarrow[n \to \infty]{} 0$ (Feller condition),
   then $(\Lambda) \Rightarrow (L)$.

Proof. See, e.g., Shiryaev (1995).   □

Theorem 1.189. *Within the preceding framework,*

$$(\Lambda) \Longleftrightarrow S_n \xrightarrow[n \to \infty]{d} N(0, 1).$$

Proof. See, e.g., Shiryaev (1995).   □

   Thanks to the foregoing results, the Lindeberg theorem for noncentered random variables is a trivial corollary.

Corollary 1.190 (Lindeberg theorem). *Let $(X_n)_{n \in \mathbb{N}}$ be a sequence of independent random variables in $\mathcal{L}^2(\Omega, \mathcal{F}, P)$ with $m_n = E[X_n]$, $\sigma_n^2 = Var[X_n]$. Denote*

$$S_n := \sum_{k=1}^{n} X_k, \quad n \in \mathbb{N} - \{0\}$$

and

$$V_n^2 = Var[S_n] = \sum_{k=1}^{n} \sigma_k^2.$$

If for all $\epsilon > 0$

$$\lim_n \frac{1}{V_n^2} \sum_{k=1}^{n} \int_{|X_k - m_k| \geq \epsilon V_n} |X_k - m_k|^2 dP \xrightarrow[n \to \infty]{} 0,$$

then

$$\frac{S_n - E[S_n]}{\sqrt{Var S_n}} \xrightarrow[n \to \infty]{d} N(0, 1).$$

Theorem 1.191. *Let $(X_n)_{n \in \mathbb{N}}$ be a sequence of i.i.d. random variables, with $m = E[X_i]$ $\sigma^2 = Var[X_i]$ for all i, and let $(V_n)_{n \in \mathbb{N}}$ be a sequence of $\mathbb{N}$-valued random variables such that*

$$\frac{V_n}{n} \xrightarrow[n]{P} 1.$$

Then

$$\frac{1}{\sqrt{V_n}} \sum_{i=1}^{n} X_i \xrightarrow[n]{P} N\left(m, \sigma^2\right).$$

Proof. See, e.g., Chung (1974).   □

Note

For proofs of the various results, see also, e.g., Ash (1972), Bauer (1981), or Métivier (1968).

1.9 Infinitely Divisible Distributions

We will consider first the case of real-valued random variables; the treatment can be extended to the multidimensional case with suitable modifications.

Definition 1.192. Let X be a real-valued random variable on a probability space $(\Omega, \mathcal{F}, P)$, having cumulative distribution function (cdf) F and characteristic function ϕ. We say that X (or F or ϕ) is *infinitely divisible (i.d.)* iff, for any $n \in \mathbb{N} - \{0\}$, there exists a characteristic function ϕ_n such that

$$\phi(t) = [\phi_n(t)]^n, \qquad t \in \mathbb{R}.$$

In other words, for any $n \in \mathbb{N} - \{0\}$, X has the same distribution as the sum of n i.i.d. random variables.

Proposition 1.193. *The following three propositions are equivalent.*

(a) The random variable X is i.d.
(b) For any $n \in \mathbb{N} - \{0\}$, the probability law P_X of X is the convolution of n identical probability laws on $\mathcal{B}_{\mathbb{R}}$.
(c) For any $n \in \mathbb{N} - \{0\}$, the characteristic function ϕ_X of X is the nth power of a characteristic function of a real-valued random variable.

Proof. This is an easy consequence of the definition (e.g., Applebaum (2004, p. 23)). □

Proposition 1.194. *If ϕ is an i.d. characteristic function, it never vanishes.*

Proof. See, e.g., Ash (1972, p. 353), and Fristedt and Gray (1997, p. 294). □

Corollary 1.195 *The representation of an i.d. characteristic function in terms of the power of a characteristic function is unique.*

Corollary 1.196 *If P_X is the probability law of an i.d. random variable, then for any $n \in \mathbb{N} - \{0\}$, there exists a unique probability law P_Y such that*

$$P_X = (P_Y)^{*n}.$$

Example 1.197.

1. Poisson random variables.
   The characteristic function of a Poisson random variable X with parameter $\lambda > 0$ is

   $$\phi_X(t) = \exp\left\{\lambda(e^{it} - 1)\right\}, \quad t \in \mathbb{R},$$

   so that it can be rewritten as

   $$\phi_X(t) = \left(\exp\left\{\frac{\lambda}{n}(e^{it} - 1)\right\}\right)^n, \quad t \in \mathbb{R},$$

   for any $n \in \mathbb{N} - \{0\}$. Hence, for any $n \in \mathbb{N} - \{0\}$

   $$\phi_X(t) = (\phi_Y(t))^n, \quad t \in \mathbb{R},$$

   where ϕ_Y is the characteristic function of a Poisson random variable Y with parameter λ/n. So we may claim that a Poisson random variable is i.d.

2. Gaussian random variables.
   The characteristic function of a Gaussian random variable $X \sim N(m, \sigma^2)$ with parameters $m \in \mathbb{R}, \sigma^2 > 0$ is

   $$\phi_X(t) = \exp\left\{imt - \frac{1}{2}\sigma^2 t^2\right\}, \quad t \in \mathbb{R},$$

   so that it can be rewritten as

   $$\phi_X(t) = \left(\exp\left\{i\frac{m}{n}t - \frac{1}{2}\frac{\sigma^2}{n}t^2\right\}\right)^n, \quad t \in \mathbb{R},$$

   for any $n \in \mathbb{N} - \{0\}$. Hence, for any $n \in \mathbb{N} - \{0\}$

   $$\phi_X(t) = (\phi_Y(t))^n, \quad t \in \mathbb{R},$$

   where ϕ_Y is the characteristic function of a Gaussian random variable Y with parameters $m/n, \sigma^2/n$. So we may claim that a Gaussian random variable is i.d.

Theorem 1.198. *Let X be a real-valued random variable; the following two propositions are equivalent.*

(a) X is i.d.
(b) There exists a triangular array $(X_{n1}, \ldots, X_{nn})$, $n \in \mathbb{N} - \{0\}$ of i.i.d. random variables such that

$$\sum_{k=1}^{n} X_{nk} \xrightarrow[n \to \infty]{d} X.$$

Proof.

$(a) \Rightarrow (b)$: If X is i.d., then for any $n \in \mathbb{N} - \{0\}$, we may choose a family $(X_{n1}, \ldots, X_{nn})$, of i.i.d. random variables such that

$$\sum_{k=1}^{n} X_{nk} = X;$$

the consequence is obvious.

$(b) \Rightarrow (a)$: The proof of this part is a consequence of the Prohorov theorem on relative compactness B.87 (Ash (1972, p. 350)).

□

Theorem 1.199. *The weak limit of a sequence of i.d. probability measures is itself an i.d. probability measure.*

Proof. See Ash (1972, p. 352) or Lukacs (1970, p. 110).   □

Compound Poisson Random Variables

Definition 1.200. We say that X is a real-valued *compound Poisson* random variable if it can be expressed as

$$X = \sum_{k=0}^{N} Y_k,$$

where N is a Poisson random variable with some parameter $\lambda \in \mathbb{R}_+^*$, and $(Y_k)_{k \in \mathbb{N}^*}$ is a family of i.i.d. random variables, independent of N; it is assumed that $Y_0 = 0$, a.s., and that the common law of any Y_k has no atom at zero. If P_Y denotes the common law, then we write $X \sim P(\lambda, P_Y)$.

Proposition 1.201. *If X is a compound Poisson random variable, then it is i.d.*

Proof. Let P_Y denote the common law of the sequence of random variables $(Y_k)_{k \in \mathbb{N}^*}$ defining X, and let ϕ_Y denote the corresponding characteristic function. By conditioning and independence, the characteristic function of X will be, for any $t \in \mathbb{R}$,

$$\phi_X(t) = \sum_{n \in \mathbb{N}} E\left[\exp\left\{ it \sum_{k=0}^{N} Y_k \right\} | N = n \right] e^{-\lambda} \frac{\lambda^n}{n!},$$

$$= e^{-\lambda} \sum_{n \in \mathbb{N}} \frac{[\lambda \phi_Y(t)]^n}{n!}$$

$$= \exp\left\{ \lambda(\phi_Y(t) - 1) \right\}.$$

As a consequence,

$$\phi_X(t) = \exp\left\{\int_{\mathbb{R}} \lambda(e^{iyt} - 1)P_Y(dy)\right\}, \ t \in \mathbb{R}.$$

It is then clear that $X \sim P(\lambda, P_Y)$ is an i.d. random variable; for any $n \in \mathbb{N} - \{0\}$

$$\phi_X(t) = (\phi_{X_j^{(n)}}(t))^n, \quad t \in \mathbb{R}, \ j \in \{1, \ldots, n\},$$

where $\phi_{X_j^{(n)}}$ is the characteristic function of a compound Poisson random variable $X_j^{(n)} \sim P(\frac{\lambda}{n}, P_Y)$.   □

Theorem 1.202. *Any i.d. probability measure can be obtained as the weak limit of a sequence of compound Poisson probability laws.*

Proof. Let ϕ be the characteristic function of an i.d. law P_X on $\mathcal{B}_{\mathbb{R}}$, and let $P_X^{\frac{1}{n}}$ denote the probability law associated with the characteristic function $\phi^{\frac{1}{n}}$ for any $n \in \mathbb{N} - \{0\}$.
    Define

$$\phi_n(t) = \exp\left\{n(\phi^{\frac{1}{n}} - 1)\right\} = \exp\left\{\int_{\mathbb{R}} n(e^{iyt} - 1)P_X^{\frac{1}{n}}(dy)\right\},$$

so that ϕ_n is the characteristic function of a compound Poisson distribution.
    We may easily observe that

$$\phi_n(t) = \exp\left\{n(e^{\frac{1}{n}\ln\phi(t)} - 1)\right\} = \exp\left\{\ln\phi(t) + no(\frac{1}{n})\right\},$$

which converges to $\phi(t)$ as $n \to \infty$.
    The result follows from Levy's Continuity Theorem.   □

Theorem 1.203. *Let μ be a finite measure on $\mathcal{B}_{\mathbb{R}}$. Define*

$$\phi(t) = \exp\left\{\int_{\mathbb{R}} (e^{itx} - 1 - itx)\frac{1}{x^2}\mu(dx)\right\}, \ t \in \mathbb{R}. \tag{1.21}$$

Then ϕ is the characteristic function of an i.d. law on $\mathbb{R}$ with mean 0 and variance $\mu(\mathbb{R})$.

Proof. We recall that, for any $n \in \mathbb{N}$, and for any $x \in \mathbb{R}$,

$$\left|e^{ix} - \sum_{k=0}^{n} \frac{(ix)^k}{k!}\right| \le \min\left\{\frac{|x|^{n+1}}{(n+1)!}, 2\frac{|x|^n}{n!}\right\};$$

the first term on the right provides a sharp estimate for $|x|$ small, whereas the second one provides a sharp estimate for $|x|$ large.

As a consequence,

$$|e^{itx} - 1 - itx| \leq \min\left\{\frac{1}{2}t^2x^2, 2|tx|\right\},$$

so that, for $x \downarrow 0$ (using $n = 1$),

$$|(e^{itx} - 1 - itx)\frac{1}{x^2}| \leq \frac{1}{2}t^2,$$

and so the integrand in (1.21) is integrable.

Moreover, since (using $n = 2$)

$$|\frac{e^{itx} - 1 - itx}{x^2} + \frac{1}{2}t^2| \leq \frac{1}{6}|x|,$$

we may claim that the integrand tends to $-\frac{1}{2}t^2$ for $x \downarrow 0$; we may then assume for continuity that this is its value at 0.

We may further observe that if μ is purely atomic with a unique atom at 0, with mass $\mu(\{0\}) = \sigma^2$, then (1.21) is the characteristic function of a $N(0, \sigma^2)$ distribution.

On the other hand, if μ is purely atomic with a unique atom at $x_0 \neq 0$, having mass $\mu(\{x_0\}) = \lambda x_0^2$, for $\lambda \in \mathbb{R}^*$, then (1.21) is the characteristic function of the random variable $x_0(X - \lambda)$, where $X \sim P(\lambda)$.

Consequently, if μ is purely atomic with a finite number of atoms on the real line, then ϕ in (1.21) can be written as the product of a finite number of characteristic functions like those above, so that it is still a characteristic function.

We may now proceed with the general case. If $\mu \equiv 0$, then the result is trivial. If $\mu(\mathbb{R}) > 0$, then we may consider a discretization of μ by means of a sequence of atomic measures $\{\mu_k, \ k \in \mathbb{N}\}$, each of which has masses

$$\mu_k(j2^{-k}) = \mu((j2^{-k}, (j+1)2^{-k}]),$$

for $j = 0, \pm 1, \pm 2, \ldots, \pm 2^k$.

It can be shown that μ_k tends to μ for k tending to ∞, so that, for k sufficiently large, $0 < \mu_k(\mathbb{R}) < +\infty$. If we take μ_k instead of μ in (1.21), then we will obtain a sequence $\{\phi_k, \ k \in \mathbb{N}\}$ of characteristic functions such that

$$\phi_k(t) \to \phi(t), \quad \text{for any} \quad t \in \mathbb{R}.$$

By the Levy Continuity Theorem we may then claim that ϕ is itself a characteristic function.

As far as the infinite divisibility is concerned, let us take, for any $n \in \mathbb{N}^*$, ψ_n defined by (1.21), but with $\frac{1}{n}\mu$ instead of μ; then

$$\phi(t) = (\psi_n(t))^n, \quad \text{for any} \quad t \in \mathbb{R}.$$

The rest of the proof is a trivial consequence of the differentiability of ϕ at 0 up to the second order. □

An important characterization of i.d. probability laws is the following.

Theorem 1.204 (Lévy–Khintchine formula). *A function ϕ is an i.d. characteristic function if and only if there exist $a \in \mathbb{R}$, $\sigma \in \mathbb{R}$, and a measure λ_L on $\mathbb{R}$, concentrated on $\mathbb{R}^*$ satisfying*

$$\int_{\mathbb{R}^*} \min\left\{x^2, 1\right\} \lambda_L(dx) < +\infty,$$

such that

$$\ln \phi(s) = ias - \frac{\sigma^2 s^2}{2} + \int_{\mathbb{R}-\{0\}} (e^{isx} - 1 - is\chi(x))\lambda_L(dx), \text{ for any } s \in \mathbb{R}, \quad (1.22)$$

where

$$\chi(x) = -I_{]-\infty,1]}(x) + xI_{]-1,1[}(x) + I_{[1,+\infty[}.$$

The triplet (a, σ^2, λ_L) is called the generating triplet *of the i.d. characteristic function ϕ. Moreover the triplet (a, σ^2, λ_L) is unique. The function in Equation (1.22) is called the* characteristic exponent *of the i.d. law having ϕ as characteristic function.*

Proof.
See, e.g., Fristedt and Gray (1997, p. 295). □

Definition 1.205. A measure λ_L on $\mathbb{R}$, concentrated on $\mathbb{R}^*$, and satisfying

$$\int_{\mathbb{R}^*} \min\left\{x^2, 1\right\} \lambda_L(dx) < +\infty, \quad (1.23)$$

is called a *Lévy measure*.

Further examples of i.d. distributions are left as exercises (Sect. 1.11).

In what follows we will consider an independent triangular array $(X_{n1}, \ldots, X_{nn})$, $n \in \mathbb{N} - \{0\}$, of random variables, i.e., for any $n \in \mathbb{N} - \{0\}$, $(X_{n1}, \ldots, X_{nn})$ is a family of i.i.d. random variables.

We will further consider the following assumptions:

(H1) $E[X_{nk}] = 0$, $\sigma_{nk}^2 = E[X_{nk}^2] < +\infty$ for any $n \in \mathbb{N} - \{0\}$, $1 \le k \le n$

(H2) $\sup_n \sum_{k=1}^n \sigma_{nk}^2 < +\infty$

(H2) $\lim_n \max_{1 \le k \le n} \sigma_{nk}^2 = 0$

Theorem 1.206. *Let X be an i.d. real-valued random variable, having mean zero and finite variance. Then there exists an independent triangular array $\{(X_{nk})_{1 \le k \le n}, n \in \mathbb{N} - \{0\}\}$, satisfying conditions (H1)–(H3) such that*

$$\sum_{k=1}^{n} X_{nk} \underset{n\to\infty}{\Rightarrow} X.$$

Proof. If X is an i.d. real-valued random variable, then for each $n \in \mathbb{N} - \{0\}$ we may find a family $(X_{nk})_{1 \le k \le n}$ of i.i.d. random variables such that $X \sim \sum_{k=1}^{n} X_{nk}$. Then clearly

$$\sum_{k=1}^{n} X_{nk} \underset{n\to\infty}{\Rightarrow} X.$$

Moreover, for each $n \in \mathbb{N} - \{0\}$ and any $1 \le k \le n$ we have

$$E[X_{nk}] = 0; \qquad Var[X_{nk}] = E[X_{nk}^2] = \frac{\sigma^2}{n},$$

so that (H1)–(H3) are automatically satisfied. $\qquad\square$

Theorem 1.207. *Let $\{(X_{nk})_{1 \le k \le n}, \ n \in \mathbb{N} - \{0\}\}$ be an independent triangular array, satisfy conditions (H1)–(H3), and denote by F_{nk} the cumulative distribution function of the random variable X_{nk}. Consider the sequence of finite measures $\{\mu_n, \ n \in \mathbb{N} - \{0\}\}$ such that*

$$\mu_n((-\infty, x]) = \sum_{k=1}^{n} \int_{y \le x} y^2 dF_{nk}(y), \ x \in \mathbb{R}. \tag{1.24}$$

Note that, by setting $s_n^2 = \sum_{k=1}^{n} \sigma_{nk}^2$, we have, because of (H2), $\sup_n \mu_n(\mathbb{R}) = \sup_n s_n^2 < +\infty$.

Under the foregoing circumstances, the following two propositions are equivalent.

(a) $S_n := \sum_{k=1}^{n} X_{nk}$ converges in distribution to a random variable having a characteristic function of the form (1.21), where μ is a finite measure on $\mathbb{R}$.

(b) The sequence of finite measures $\{\mu_n, n \in \mathbb{N} - \{0\}\}$ defined in (1.24) weakly converges to the measure μ.

Proof.

(a) $\Rightarrow$ (b) If we denote by ϕ_n the characteristic function of S_n, then under (a) we may state that

$$\phi_n(t) \underset{n\to\infty}{\to} \phi(t), \ \text{for any } t \in \mathbb{R}. \tag{1.25}$$

Since $\mu_n(\mathbb{R}) = s_n^2$ is uniformly bounded for $n \in \mathbb{N} - \{0\}$, by Helly's theorem we can state that from $\{\mu_n, \ n \in \mathbb{N} - \{0\}\}$ we may extract a subsequence $\{\mu_{n_m}, \ m \in \mathbb{N} - \{0\}\}$ weakly convergent to some finite measure ν on $\mathcal{B}_{\mathbb{R}}$.

Because of the convergence (1.25) it must then also be

$$\phi(t) = \psi(t) = \overline{\exp}\left\{\int_{\mathbb{R}} (e^{itx} - 1 - itx)\frac{1}{x^2}\nu(dx)\right\}, \ t \in \mathbb{R}.$$

The same should hold for the derivatives

$$\phi''(t) = \psi''(t), \ t \in \mathbb{R},$$

i.e.,

$$\int_{\mathbb{R}} e^{itx}\mu(dx) = \int_{\mathbb{R}} e^{itx}\nu(dx), \ t \in \mathbb{R}.$$

By the uniqueness theorem for characteristic functions we may finally state that

$$\mu = \nu.$$

$(b) \Rightarrow (a)$ If

$$\mu_n \underset{n\to\infty}{\Rightarrow} \mu,$$

then, by known results,

$$\phi_n(t) \underset{n\to\infty}{\to} \phi(t), \ t \in \mathbb{R},$$

which implies (a).

$\square$

1.9.1 Examples

1. The Central Limit Theorem

The case

$$S_n \underset{n\to\infty}{\Rightarrow} N(0,1)$$

corresponds, by Theorem 1.207, to the Dirac measure at 0 in (1.21):

$$\mu = \varepsilon_0.$$

In fact, let us recall that we have taken as the value of the integrand at 0 in (1.21) the quantity $-\frac{t^2}{2}$, so that

$$\phi(t) = \exp\left\{\int_{\mathbb{R}} (e^{itx} - 1 - itx)\frac{1}{x^2}\delta_0(x)dx\right\} = \exp\left\{-\frac{t^2}{2}\right\}, \ t \in \mathbb{R}.$$

If we suppose $s_n^2 = 1$ for any $n \in \mathbb{N} - \{0\}$, condition (b) in Theorem 1.207 becomes the Lindeberg condition

$$(L) \ [Lindeberg] \qquad \text{for all } \epsilon > 0: \qquad \sum_{k=1}^{n} \int_{|x|>\epsilon} x^2 dF_{nk}(x) \underset{n\to\infty}{\longrightarrow} 0,$$

so that the result follows from Theorems 1.188 and 1.189.

2. The Poisson Limit

Let $\{(Z_{nk})_{1 \leq k \leq n},\ n \in \mathbb{N} - \{0\}\}$, be an independent triangular array in $\mathcal{L}^2$. If we denote $m_{nk} = E[Z_{nk}]$, then we take

$$X_{nk} := Z_{nk} - m_{nk},\ 1 \leq k \leq n,\ n \in \mathbb{N} - \{0\}.$$

According to Theorem 1.207,

$$\sum_{k=1}^{n} X_{nk} \underset{n \to \infty}{\Rightarrow} Z_\lambda - \lambda,$$

with $Z_\lambda \sim P(\lambda)$, if and only if

$$\mu_n \underset{n \to \infty}{\Rightarrow} \lambda \epsilon_1, \tag{1.26}$$

where ϵ_1 denotes the Dirac measure at 1.

If we assume that $\sigma_n^2 \underset{n \to \infty}{\to} \lambda$, then condition (1.26) is equivalent to

$$\mu_n([1 - \varepsilon, 1 + \varepsilon]) \underset{n \to \infty}{\to} \lambda \text{ for any } \varepsilon > 0$$

or to

$$\sum_{k=1}^{n} \int_{|Z_{nk} - m_{nk} - 1| > \varepsilon} (Z_{nk} - m_{nk})^2 dP \underset{n \to \infty}{\to} 0 \tag{1.27}$$

for any $\varepsilon > 0$.

Suppose that both

$$s_n^2 \underset{n \to \infty}{\to} \lambda \quad \text{and} \quad \sum_{k=1}^{n} m_{nk} \underset{n \to \infty}{\to} \lambda$$

hold; then (1.27) becomes a necessary and sufficient condition for

$$\sum_{k=1}^{n} Z_{nk} \underset{n \to \infty}{\Rightarrow} Z_\lambda \sim P(\lambda).$$

This case includes the circumstance that, for any $n \in \mathbb{N} - \{0\}$, and $1 \leq k \leq n$, $Z_{nk} \sim B(1, p_{nk})$, with

$$\max_{1 \leq k \leq n} p_{nk} \underset{n \to \infty}{\to} 0 \quad \text{and} \quad \sum_{k=1}^{n} p_{nk} \underset{n \to \infty}{\to} \lambda;$$

hence the well-known convergence of a sequence of binomial variables $B(n, p_n)$ to a Poisson variable $P(\lambda)$ is also included once $p_n \underset{n \to \infty}{\to} 0$ and $np_n \underset{n \to \infty}{\to} \lambda$.

1.10 Stable Laws

An important subclass of i.d. distributions is that of stable laws, which we will later relate to a corresponding subclass of Lévy processes. We will limit ourselves to the scalar case for simplicity; the interested reader may refer to excellent monographs such as Samorodnitsky and Taqqu (1994) and Sato (1999).

Definition 1.208. A real random variable X is defined as *stable* if for any two positive real numbers A and B there exist a real positive number C and a real number D such that

$$AX_1 + BX_2 \sim CX + D, \tag{1.28}$$

where X_1 and X_2 are two independent random variables having the same distribution as X (as usual the symbol $\sim$ means equality in distribution).

A stable random variable X is defined as *strictly stable* if $D = 0$ for any choice of A and B; it is defined as *symmetric* if its distribution is symmetric with respect to zero.

Remark 1.209. A stable symmetric random variable is strictly stable.

A second equivalent definition for the stability of real-valued random variables is as follows.

Definition 1.210. A real random variable X is defined as *stable* if for any $n \geq 2$ there exist a positive real number C_n and a real number D_n such that

$$X_1 + X_2 + \ldots + X_n \sim C_n X + D_n,$$

where $X_1, X_2, \ldots, X_n$ are a family of i.i.d. random variables having the same distribution as X. A stable random variable X is defined as *strictly stable* if $D_n = 0$ for all n.

The preceding definition implies the following result.

Proposition 1.211. *With reference to Definition 1.210 there exists a real number $\alpha \in (0, 2]$ such that*

$$C_n = n^{\frac{1}{\alpha}}.$$

The number α is called the *stability index* or *characteristic exponent*. A stable random variable X having index α is called α-*stable*.

Example 1.212. If X is a Gaussian random variable with mean $\mu \in \mathbb{R}$ and variance $\sigma^2 \in \mathbb{R}_+ - \{0\}$ ($X \sim N(\mu, \sigma^2)$), then X is stable with $\alpha = 2$ since

$$AX_1 + BX_2 \sim N((A + B)\mu, (A^2 + B^2)^{1/2}\sigma^2),$$

i.e., (1.28) is satisfied for $C = (A^2 + B^2)^{1/2}$ and $D = (A + B - C)\mu$.
It is trivial to recognize that a Gaussian random variable $X \sim N(\mu, \sigma^2)$ is symmetric if and only if $\mu = 0$.

The following theorems further characterize stable laws.

Theorem 1.213. *A random variable X is stable if and only if it admits a domain of attraction, i.e., there exist a sequence of i.i.d. random variables $(Y_n)_{n \in \mathbb{N} - \{0\}}$, a sequence of positive real numbers $(A_n)_{n \in \mathbb{N} - \{0\}}$, and a sequence of real numbers $(B_n)_{n \in \mathbb{N} - \{0\}}$ such that*

$$\frac{Y_1 + Y_2 + \ldots + Y_n}{A_n} - B_n \underset{n \to \infty}{\Rightarrow} X,$$

where $\Rightarrow$ denotes a convergence in law.

Theorem 1.214. *A random variable X is stable if and only if there exist parameters $0 < \alpha \leq 2$, $\sigma \geq 0$, $-1 \leq \beta \leq 1$ and a real number μ such that its characteristic function is of the following form:*

$$E\left[e^{isX}\right] = \begin{cases} exp\left\{-\sigma^\alpha |s|^\alpha \left(1 - i\beta \left(sign\, s\right) \tan\left(\frac{\pi\alpha}{2}\right)\right) + i\mu s\right\}, if \quad \alpha \neq 1, \\ exp\left\{-\sigma |s| \left(1 + i\beta\frac{2}{\pi} \left(sign\, s\right) \ln |s|\right) + i\mu s\right\}, \qquad if \quad \alpha = 1, \end{cases}$$

for $s \in \mathbb{R}$. The parameter α is the stability index of random variable X.

Here

$$sign\, s = \begin{cases} 1, & if \quad s > 0, \\ 0, & if \quad s = 0, \\ -1, & if \quad s < 0. \end{cases}$$

Proof. See, e.g., Chow and Teicher (1988, p. 449). $\qquad\qquad\square$

Theorem 1.214 shows that a stable random variable is characterized by the four parameters α, β, σ, and μ. This is why an α-stable random variable X is denoted by

$$X \sim S_\alpha \left(\sigma, \beta, \mu\right).$$

We have already stated that α is the stability index of the stable random variable X. As far as the other parameters are concerned, one can show the following results.

Proposition 1.215. *Let $X \sim S_\alpha \left(\sigma, \beta, \mu\right)$, and let a be a real constant. Then $X + a \sim S_\alpha \left(\sigma, \beta, \mu + a\right)$.*

We may then state that μ is a *parameter of location* of the distribution of X.

Proposition 1.216. *Let $X \sim S_\alpha \left(\sigma, \beta, \mu\right)$, and let $a \neq 0$ be a real number. Then*

$$aX \sim S_\alpha \left(\sigma |a|, sign(a)\beta, a\mu\right), \qquad\qquad for \quad \alpha \neq 1,$$

$$aX \sim S_1 \left(\sigma |a|, sign(a)\beta, a\mu - \frac{2}{\pi}a(\ln |a|)\sigma\beta\right), \quad for \quad \alpha = 1.$$

The preceding proposition characterizes σ as a *scaling parameter* of the distribution of X.

Proposition 1.217. *For any $\alpha \in (0, 2]$,*

$$X \sim S_\alpha(\sigma, \beta, 0) \Leftrightarrow -X \sim S_\alpha(\sigma, -\beta, 0).$$

$X \sim S_\alpha(\sigma, \beta, \mu)$ *is symmetric if and only if $\beta = 0$ and $\mu = 0$. It is symmetric with respect to μ if and only if $\beta = 0$.*

The preceding proposition characterizes β as a *parameter of asymmetry* of the distribution of X.

We usually write

$$X \sim S\alpha S$$

to denote that X is a symmetric α-stable random variable, i.e., when $\beta = \mu = 0$.

As a consequence of Theorem 1.214 we may recognize that the characteristic function of an α-stable law is such that, for some $c \in \mathbb{R}$,

$$|\phi(s)| = \exp\{-c|s|^\alpha\}, \quad s \in \mathbb{R}.$$

It is then easy to show that $\phi \in \mathcal{L}^1(\nu^1)$, so that, because of Theorem 1.94, we may finally state the following.

Proposition 1.218. *Any stable random variable is absolutely continuous.*

Unfortunately, the probability densities of stable random variables do not have in general a closed form, but for a few exceptions.

(*i*) Gaussian distributions: $X \sim N(\mu, 2\sigma^2) = S_2(\sigma, 0, \mu)$ with density

$$f(x) = \frac{1}{2\sigma\sqrt{\pi}} exp\left(-\frac{(x-\mu)^2}{4\sigma^2}\right).$$

(*ii*) Cauchy distributions: $X \sim Cauchy(\sigma, \mu) = S_1(\sigma, 0, \mu)$ with density

$$f(x) = \frac{\sigma}{\pi\left((x-\mu)^2 + \sigma^2\right)}.$$

(*iii*) Lèvy distributions: $X \sim \text{Lèvy}(\sigma, \mu) = S_{1/2}(\sigma, 1, \mu)$ with density

$$f(x) = \left(\frac{\sigma}{2\pi}\right)^{1/2} \frac{1}{(x-\mu)^{3/2}} exp\left\{-\frac{\sigma}{2(x-\mu)}\right\}.$$

Proposition 1.219. *A stable random variable is i.d.*

Remark 1.220. In general the converse of Proposition 1.219 does not hold. For example, a Poisson random variable is i.d. but not stable.

With reference to Definition 1.107 the following proposition holds.

Proposition 1.221. *A stable random variable is reproducible.*

Remark 1.222. The converse does not hold in general; in fact, we know that a Poisson distribution is not stable, though it is reproducible (Exercise 1.21).

1.10.1 Martingales

What follows extends the concept of sequences of random variables and introduces the concepts of (discrete-time) processes and martingales. The latter's continuous equivalents will be the subject of the following chapters.

Let $(\Omega, \mathcal{F}, P)$ be a probability space and $(\mathcal{F}_n)_{n \geq 0}$ a *filtration*, that is, an increasing family of sub-σ-algebras of $\mathcal{F}$:

$$\mathcal{F}_0 \subseteq \mathcal{F}_1 \subseteq \cdots \subseteq \mathcal{F}.$$

We define $\mathcal{F}_\infty := \sigma(\bigcup_n \mathcal{F}_n) \subseteq \mathcal{F}$. A process $X = (X_n)_{n \geq 0}$ is called *adapted* to the filtration $(\mathcal{F}_n)_{n \geq 0}$ if for each n, X_n is $\mathcal{F}_n$-measurable.

Definition 1.223. A sequence $X = (X_n)_{n \in \mathbb{N}}$ of real-valued random variables is called a *martingale* (relative to $(\mathcal{F}_n, P)$) if

- X is adapted
- $E[|X_n|] < \infty$ for all n $(\Leftrightarrow X_n \in \mathcal{L}^1)$
- $E[X_n | \mathcal{F}_{n-1}] = X_{n-1}$ almost surely $(n \geq 1)$

Proposition 1.224. *If $(X_n)_{n \in \mathbb{N}}$ is a martingale, then its expected value is constant, i.e., for all $n \in \mathbb{N}$, $E[X_n] = E[X_0]$.*

Example 1.225.

1. Show that if $(X_n)_{n \in \mathbb{N}}$ is a sequence of independent random variables with $E[X_n] = 0$ for all $n \in \mathbb{N}$, then $S_n = X_1 + X_2 + \cdots + X_n$ is a martingale with respect to $(\mathcal{F}_n = \sigma(X_1, \ldots, X_n), P)$ and $\mathcal{F}_0 = \{\emptyset, \Omega\}$.
2. Show that if $(X_n)_{n \in \mathbb{N}}$ is a sequence of independent random variables with $E[X_n] = 1$ for all $n \in \mathbb{N}$, then $M_n = X_1 \cdot X_2 \cdots \cdots X_n$ is a martingale with respect to $(\mathcal{F}_n = \sigma(X_1, \ldots, X_n), P)$ and $\mathcal{F}_0 = \{\emptyset, \Omega\}$.

Definition 1.226. A sequence $X = (X_n)_{n \in \mathbb{N}}$ of real-valued random variables is called a *submartingale* (respectively a *supermartingale*) (relative to $(\mathcal{F}_n, P)$) if

- X is adapted
- $E[|X_n|] < \infty$ for all n $(\Leftrightarrow X_n \in \mathcal{L}^1)$
- $E[X_n | \mathcal{F}_{n-1}] \geq X_{n-1}$ (respectively $E[X_n | \mathcal{F}_{n-1}] \leq X_{n-1}$) almost surely $(n \geq 1)$

Theorem 1.227 (Doob decomposition). *Let $(X_n)_{n \geq 0}$ be a submartingale. Then X admits a decomposition*

$$X = X_0 + M + A,$$

where M is a martingale null at $n = 0$ and A is a predictable increasing process null at $n = 0$. Moreover, such decomposition is a.s. unique, in the sense that if $X = X_0 + \widetilde{M} + \widetilde{A}$ is another such decomposition, then

$$P(M_n = \widetilde{M}_n, A_n = \widetilde{A}_n, \forall n) = 1.$$

Proof. See, e.g., Jacod and Protter (2000, p. 216).   □

Theorem 1.228. *Let $(X_n)_{n \in \mathbb{N}}$ be an adapted process with $X_n \in \mathcal{L}^1$ for all n. Prove that X admits an a.s. unique decomposition*

$$X = X_0 + M + A,$$

where M is a martingale null at $n = 0$ and A is a predictable process null at $n = 0$. $(X_n)_{n \geq 0}$ is a submartingale if and only if A is a predictable increasing process, in the sense that

$$P(A_n \leq A_{n+1}, \forall n) = 1.$$

Proof. See, e.g., Williams (1991, p. 120).   □

A discrete-time process $C = (C_n)_{n \geq 1}$ is called *predictable* if

$$C_n \text{ is } \mathcal{F}_{n-1}\text{-measurable } (n \geq 1).$$

We define

$$(C \bullet X)_n := \sum_{k=1}^{n} C_k(X_k - X_{k-1}).$$

Proposition 1.229 (Stochastic integration theorem). *If C is a bounded predictable process and X is a martingale, then $(C \bullet X)$ is a martingale null at $n = 0$.*

Definition 1.230. *Let $\overline{\mathbb{N}} = \mathbb{N} \cup \{+\infty\}$. A random variable $T : (\Omega, \mathcal{F}) \to (\overline{\mathbb{N}}, \mathcal{B}_{\overline{\mathbb{N}}})$ is a* stopping time *if and only if*

$$\forall n \in \overline{\mathbb{N}}: \ \{T \leq n\} \in \mathcal{F}_n.$$

Proposition 1.231. *Let X be a martingale with respect to the natural filtration $(\mathcal{F}_t)_{n \in \mathbb{R}_+}$ and let T be a stopping time with respect to the same filtration; then the stopped process $X_T; = (X_{n \wedge T(\omega)})_{n \geq 0}$ is a martingale with the same expected value of X.*

Proof. Hint: Consider the predictable process $C_n = I_{(T \geq n)}$ and apply the preceding results to the process $(X_T - X_0)_n = (C^T \bullet X)_n$.   □

Proposition 1.232(Martingale convergence theorem). *Let* $(X_n)_{n \in \mathbb{N}}$ *be a nonnegative submartingale, or a martingale bounded above or bounded below; then, the limit* $X_n \xrightarrow[n]{a.s.} X$ *exists, and* $X \in L^1$.

Proof. See, e.g., Jacod and Protter (2000, p. 226). *Warning:* we are not claiming that $X_n \xrightarrow[n]{L^1} X$, and indeed this is not true in general.   □

Theorem 1.233 (Martingale convergence theorem). *Let* $(X_n)_{n \in \mathbb{N}}$ *be a uniformly integrable martingale; then the limit* $X_n \xrightarrow[n]{a.s.} X$ *exists,* $X \in L^1$, *and, additionally,* $X_n \xrightarrow[n]{L^1} X$.

Moreover, $X_n = E[X \mid \mathcal{F}_n]$.

Proof. See, e.g., Jacod and Protter (2000, p. 232).   □

Theorem 1.234 (Martingale central limit theorem). *Let* $(X_n)_{n \in \mathbb{N}^*}$ *be a sequence of real-valued random variables on a given probability space* $(\Omega, \mathcal{F}, P)$ *endowed with a filtration* $(\mathcal{F}_n)_{n \in \mathbb{N}}$. *Assume that*

- $E[X_n | \mathcal{F}_{n-1}] = 0$ *almost surely* $(n \geq 1)$
- $E[X_n^2 | \mathcal{F}_{n-1}] = 1$ *almost surely* $(n \geq 1)$
- $E[X_n^3 | \mathcal{F}_{n-1}] \leq K < +\infty$ *almost surely* $(n \geq 1)$ *for a* $K > 0$

Consider

$$S_0 = 0; \qquad S_n = \sum_{i=1}^{n} X_i.$$

Then $\frac{1}{\sqrt{n}} S_n \xrightarrow[n]{d} N(0,1)$.

Proof. See, e.g., Jacod and Protter (2000, p. 229).   □

1.11 Exercises and Additions

1.1. Prove Proposition 1.16.

1.2. Prove all the points of Example 1.79.

1.3. Show that the statement of Example 1.62 is true.

1.4. Prove all points of Example 1.106 and, in addition, the following: Let X be a Cauchy distributed random variable, i.e., $X \sim C(0,1)$; then, $Y = a + hX \sim C(a, h)$.

1.5. Give an example of two random variables that are uncorrelated but not independent.

1.6. If X has an absolutely continuous distribution with pdf $f(x)$, its *entropy* is defined as (Khinchin (1957))

$$H(X) = -\int_D f(x)\ln f(x)dx,$$

where $D = \{x \in \mathbb{R} | f(x) > 0\}$.

1. Show that the maximal value of entropy within the set of nonnegative random variables with a given expected value μ is attained by the exponential $E(\mu^{-1})$.
2. Show that the maximal value of entropy within the set of real random variables with fixed mean μ and variance σ^2 is attained by the Gaussian $N(\mu, \sigma^2)$.

1.7. Show that an i.d. characteristic function never vanishes.

1.8. Let ϕ be an i.d. characteristic function. Show that $(\phi)^\alpha$ is an i.d. characteristic function for any real positive α. The converse is also true.

1.9. Let ψ be an arbitrary characteristic function, and suppose that λ is a positive real number. Then

$$\phi(t) = \exp\{\lambda[\psi(t) - 1]\}, \qquad t \in \mathbb{R},$$

is an i.d. characteristic function.

1.10.   1. Show that the negative binomial distribution is i.d.
2. Show that the exponential distribution $E(\lambda)$ is i.d.
3. Show that the characteristic function of a Gamma random variable $X \sim \Gamma(\alpha, \beta)$ is i.d.
4. Show that the characteristic function of a Cauchy random variable is i.d.
5. Show that the characteristic function of a uniform random variable $X \sim U(0, 1)$ is not i.d.

1.11. Show that any linear combination of independent i.d. random variables is itself an i.d. random variable (the reader may refer to Fristedt and Gray (1997, p. 294)).

1.12 (Kolmogorov). Show that a function ϕ is an i.d. characteristic function with finite variance if and only if

$$\ln \phi(s) = ias + \int_{\mathbb{R}} \frac{e^{isx} - 1 - isx}{x^2} G(dx) \text{ for any } s \in \mathbb{R},$$

where $a \in \mathbb{R}$ and G is a nondecreasing and bounded function such that $G(-\infty) = 0$. The representation is unique (the reader may refer to Lukacs (1970, p. 119)).

1.13. With reference to the Lévy–Khintchine formula (Theorem 1.204), show that in the generating triplet (a, σ^2, λ_L)

- (i) for a Gaussian random variable $N(a, \sigma^2)$, a equals the mean, σ^2 equals the variance, and $\lambda_L = 0$;
- (ii) or a Poisson random variable $P(\lambda)$, $a = 0$, $\sigma^2 = 0$, and $\lambda_L = \lambda \epsilon_1$, where ϵ_1 is the Dirac measure concentrated in 1;
- (iii) For a compound Poisson random variable $P(\lambda, \mu)$, $a = 0$, $\sigma^2 = 0$, and $\lambda_L = \lambda \mu$.

1.14. Show that ϕ is the characteristic function of a stable law if and only if for any a_1 and a_2 in $\mathbb{R}_+^*$ there exist two constants $a \in \mathbb{R}_+^*$ and $b \in \mathbb{R}$ such that

$$\phi(a_1 s)\phi(a_2 s) = e^{ibs}\phi(as).$$

1.15. Show that if ϕ is the characteristic function of a stable law that is symmetric about the origin, then there exist $c \in \mathbb{R}_+^*$ and $\alpha \in\,]0, 2]$ such that

$$\phi(s) = e^{-c|s|^\alpha} \text{ for any } x \in \mathbb{R}.$$

1.16. A stable random variable is symmetric if and only if its characteristic function is real. From Theorem 1.214 it may happen if and only if $\beta = 0$ and $\mu = 0$.

1.17. A stable symmetric random variable is strictly stable, but the converse is not true.

1.18. Let X_1 and X_2 be independent stable random variables such that $X_i \sim S_\alpha(\sigma_i, \beta_i, \mu_i)$, for $i = 1, 2$. Then $X_1 + X_2 \sim S_\alpha(\sigma, \beta, \mu)$, where

$$\sigma = (\sigma_1^\alpha + \sigma_2^\alpha)^{1/\alpha},$$

$$\beta = \frac{\beta_1 \sigma_1^\alpha + \beta_2 \sigma_2^\alpha}{\sigma_1^\alpha + \sigma_2^\alpha},$$

$$\mu = \mu_1 + \mu_2.$$

1.19. Let $X_1, \ldots, X_n$ be a family of i.i.d. stable random variables $S_\alpha(\sigma, \beta, \mu)$; then

$$X_1 + \ldots + X_n \overset{d}{=} n^{1/\alpha} X_1 + \mu\left(n - n^{1/\alpha}\right), \quad \text{if } \alpha \neq 1,$$

and

$$X_1 + \ldots + X_n \overset{d}{=} n^{1/\alpha} X_1 + \tfrac{2}{\pi}\sigma\beta n \ln n, \quad \text{if } \alpha = 1.$$

1.20. Show that every stable law is i.d. What about the converse?

1.21. Show that a Poisson random variable is i.d. but not stable.

1.22. If $\phi_1(t) = \sin t$ and $\phi_2(t) = \cos t$ are characteristic functions, then give an example of random variables associated with ϕ_1 and ϕ_2, respectively.

Let $\phi(t)$ be a characteristic function, and describe a random variable with characteristic function $|\phi(t)|^2$.

1.23. Let $X_1, X_2, \ldots, X_n$ be i.i.d. random variables with common density f, and

$$Y_j = j\text{th smallest of the} X_1, X_2, \ldots, X_n, \qquad j = 1, \ldots, n.$$

It follows that $Y_1 \leq \cdots \leq Y_j \leq \cdots \leq Y_n$. Show that

$$f_{Y_1, \ldots, Y_n} = \begin{cases} n! \prod_{i=1}^n f(y_i), & \text{if } y_1 < y_2 < \cdots < y_n, \\ 0, & \text{otherwise.} \end{cases}$$

1.24. Let X and $(Y_n)_{n \in \mathbb{N}}$ be random variables such that

$$X \sim E(1), \qquad Y_n(\omega) = \begin{cases} n, & \text{if } X(\omega) \leq \frac{1}{n}, \\ 0, & \text{otherwise.} \end{cases}$$

Give, if it exists, the limit $\lim_{n \to \infty} Y_n$:

- In distribution
- In probability
- Almost surely
- In mean of order $p \geq 1$

1.25. Let $(X_n)_{n \in \mathbb{N}}$ be a sequence of uncorrelated random variables with common expected value $E[X_i] = \mu$ and such that $\sup Var[X_i] < +\infty$.

Show that $\sum_{i=1}^n \dfrac{X_i}{n}$ converges to μ in mean of order $p = 2$.

1.26. Give an example of random variables $X, X_1, X_2, \ldots$ such that $(X_n)_{n \in \mathbb{N}}$ converges to X

- In probability but not almost surely
- In probability but not in mean
- Almost surely but not in mean and vice versa
- In mean of order 1 but not in mean of order $p = 2$ (generally $p > 1$)

1.27. Let $(X_n)_{n \in \mathbb{N}}$ be a sequence of i.i.d. random variables such that $X_i \sim B(p)$ for all i. Let Y be uniformly distributed on $[0, 1]$ and independent of X_i for all i. If $S_n = \frac{1}{n} \sum_{k=1}^n (X_k - Y)^2$, show that $(S_n)_{n \in \mathbb{N}}$ converges almost surely, and determine its limit.

1.28. Let $(X_n)_{n \in \mathbb{N}}$ be a sequence of i.i.d. random variables; determine the limit almost surely of

$$\frac{1}{n} \sum_{k=1}^n \sin\left(\frac{X_i}{X_{i+1}}\right)$$

in the following case:

- $X_i = \pm 1$ with probability $1/2$.
- X_i is a continuous random variable and its density function f_{X_i} is an even function.

(*Hint:* Consider the sum on the natural even numbers.)

1.29 (Large deviations). Let $(X_n)_{n \in \mathbb{N}}$ be a sequence of i.i.d. random variables and suppose that their moment-generating function $M(t) = E[e^{tX_1}]$ exists and is finite in $[0, a]$, $a \in \mathbb{R}_+^*$. Prove that for any $t \in [0, a]$

$$P(\bar{X} > E[X_1] + \epsilon) \le (e^{-t(E[X_1]+\epsilon)} M(t))^n < 1,$$

where $\bar{X}$ denotes the arithmetic mean of $X_1, \ldots, X_n$, $n \in \mathbb{N}$.

Apply the preceding result to the cases $X_1 \sim B(1, p)$ and $X_1 \sim N(0, 1)$.

1.30 (Chernoff). Let $(X_n)_{n \in \mathbb{N}}$ be a sequence of i.i.d. simple (finite-range) random variables, satisfying $E[X_n] < 0$ and $P(X_n > 0) > 0$ for any $n \in \mathbb{N}$, and suppose that their moment-generating function $M(t) = E[e^{tX_1}]$ exists and is finite in $[0, a]$, $a \in \mathbb{R}_+^*$. Show that

$$\lim_{n \to \infty} \frac{1}{n} \ln P(X_1 + \cdots + X_n \ge 0) = \ln \inf_t M(t).$$

For an extended treatment of large deviations we refer to Dembo and Zeitouni (2010).

1.31 (Law of iterated logarithms). Let $(X_n)_{n \in \mathbb{N}}$ be a sequence of i.i.d. simple (finite-range) random variables with mean zero and variance 1. Show that

$$P\left(\limsup_n \frac{S_n}{\sqrt{2n \ln \ln n}} = 1\right) = 1.$$

1.32. Let X be a d-dimensional Gaussian vector. Prove that for every Lipschitz function f on $\mathbb{R}^d$, with $\|f\|_{Lip} \le 1$, the following inequality holds for any $\lambda \ge 0$:

$$P(f(X) - E[f(X)] \ge \lambda) \le e^{-\frac{\lambda^2}{2}}.$$

1.33. Let X be an n-dimensional centered Gaussian vector. Show that

$$\lim_{r \to +\infty} \frac{1}{r^2} \ln P\left(\max_{1 \le i \le n} X_i \ge r\right) = -\frac{1}{2\sigma^2}.$$

1.34. Let $(Y_n)_{n\in\mathbb{N}}$ be a family of random variables in $\mathcal{L}^1$; then the following two statements are equivalent:

1. $(Y_n)_{n\in\mathbb{N}}$ is uniformly integrable.
2. $\sup_{n\in\mathbb{N}} E[|Y_n|] < +\infty$, and for all ϵ there exists a $\delta > 0$ such that $A \in \mathcal{F}$, $P(A) \leq \delta \Rightarrow E[|Y_n I_A|] < \epsilon$.

(*Hint:* $\int_A |Y_n| \leq rP(A) + \int_{|Y_n|>r} Y_n$ for $r > 0$.)

1.35. Show that the random variables $(Y_n)_{n\in\mathbb{N}}$ are uniformly integrable if and only if $\sup_n E[f(|Y_n|)] < \infty$ for some increasing function $f : \mathbb{R}_+ \to \mathbb{R}_+$ with $f(x)/x \to \infty$ as $n \to \infty$.

1.36. Show that for any $Y \in \mathcal{L}^1$ the family of conditional expectations $\{E[Y|\mathcal{G}],\ \mathcal{G} \subset F\}$ is uniformly integrable.

1.37. Show that if $(X_n)_{n\in\mathbb{N}}$ is a sequence of independent random variables with $E[X_n] = 0$ for all $n \in \mathbb{N}$, then $S_n = X_1 + X_2 + \cdots + X_n$ is a martingale with respect to $(\mathcal{F}_n = \sigma(X_1,\ldots,X_n), P)$ and $\mathcal{F}_0 = \{\emptyset, \Omega\}$.

1.38. Show that if $(X_n)_{n\in\mathbb{N}}$ is a sequence of independent random variables with $E[X_n] = 1$ for all $n \in \mathbb{N}$, then $M_n = X_1 \cdot X_2 \cdots\cdots X_n$ is a martingale with respect to $(\mathcal{F}_n = \sigma(X_1,\ldots,X_n), P)$ and $\mathcal{F}_0 = \{\emptyset, \Omega\}$.

1.39. Show that if $\{\mathcal{F}_n : n \geq 0\}$ is a filtration in $\mathcal{F}$ and $\xi \in \mathcal{L}^1(\Omega, \mathcal{F}, P)$, then $M_n \equiv E[\xi|\mathcal{F}_n]$ is a martingale.

1.40. An urn contains white and black balls; we draw a ball and replace it with two balls of the same color; the process is repeated many times. Let X_n be the proportion of white balls in the urn before the nth draw. Show that the process $(X_n)_{n\geq0}$ is a martingale.

1.41. Consider the model

$$\Delta X_n = X_{n+1} - X_n = pX_n + \Delta M_n,$$

where M_n is a zero-mean martingale. Prove that

$$\hat{p} = \frac{1}{n} \sum_{k=1}^{n} \frac{1}{X_j} \Delta X_j$$

is an unbiased estimator of p (i.e., $E[\hat{p}] = p$). (*Hint:* Use the stochastic integration theorem.)

2

Stochastic Processes

2.1 Definition

We commence along the lines of the founding work of Kolmogorov by
regarding stochastic processes as a family of random variables defined on
a probability space and thereby define a probability law on the set of trajectories of the process. More specifically, stochastic processes generalize the
notion of (finite-dimensional) vectors of random variables to the case of any
family of random variables indexed in a general set T. Typically, the latter
represents "time" and is an interval of $\mathbb{R}$ (in the continuous case) or $\mathbb{N}$ (in
the discrete case). For a nice and elementary introduction to this topic, the
reader may refer to Parzen (1962).

Definition 2.1. Let $(\Omega, \mathcal{F}, P)$ be a probability space, T an index set, and
$(E, \mathcal{B})$ a measurable space. An $(E, \mathcal{B})$-valued *stochastic process* on $(\Omega, \mathcal{F}, P)$
is a family $(X_t)_{t \in T}$ of random variables $X_t : (\Omega, \mathcal{F}) \to (E, \mathcal{B})$ for $t \in T$.

$(\Omega, \mathcal{F}, P)$ is called the underlying *probability space* of the process $(X_t)_{t \in T}$,
while $(E, \mathcal{B})$ is the *state space* or *phase space*. Fixing $t \in T$, the random
variable X_t is the *state of the process at "time"* t. Moreover, for all $\omega \in \Omega$,
the mapping $X(\cdot, \omega) : t \in T \to X_t(\omega) \in E$ is called the *trajectory* or *path of
the process* corresponding to ω. Any trajectory $X(\cdot, \omega)$ of the process belongs
to the space E^T of functions defined in T and valued in E. Our aim is to
introduce a suitable σ-algebra $\mathcal{B}^T$ on E^T that makes the family of trajectories
of our stochastic process a random function $X : (\Omega, \mathcal{F}) \to (E^T, \mathcal{B}^T)$.

More generally, let us consider the family of measurable spaces $(E_t, \mathcal{B}_t)_{t \in T}$
(as a special case, all E_t may coincide with a unique E) and define $W^T = \prod_{t \in T} E_t$. If $S \in \mathcal{S}$, where $\mathcal{S} = \{S \subset T | \text{ S is finite}\}$, then the product σ-algebra
$\mathcal{B}^S = \bigotimes_{t \in S} \mathcal{B}_t$ is well defined as the σ-algebra generated by the family of
rectangles with sides in $\mathcal{B}_t$, $t \in S$.

© Springer Science+Business Media New York 2015
V. Capasso, D. Bakstein, *An Introduction to Continuous-Time
Stochastic Processes*, Modeling and Simulation in Science, Engineering
and Technology, DOI 10.1007/978-1-4939-2757-9_2

Definition 2.2. If $A \in \mathcal{B}^S$, $S \in \mathcal{S}$, then the subset $\pi_{ST}^{-1}(A)$ is a *cylinder* in W^T with base A, where π_{ST} is the canonical projection of W^T on W^S.

It is easy to show that if C_A and $C_{A'}$ are cylinders with bases $A \in \mathcal{B}^S$ and $A' \in \mathcal{B}^{S'}$, $S, S' \in \mathcal{S}$, respectively, then $C_A \cap C_{A'}$, $C_A \cup C_{A'}$, and $C_A \setminus C_{A'}$ are cylinders with base in $W^{S \cup S'}$. From this it follows that the set of cylinders with a finite-dimensional base is a *ring* of subsets of W^T (or, better, an *algebra*). We denote by $\mathcal{B}^T$ the σ-algebra generated by it (see, e.g., Métivier (1968)).

Definition 2.3. The measurable space $(W^T, \mathcal{B}^T)$ is called the *product space of the measurable spaces* $(E_t, \mathcal{B}_t)_{t \in T}$.

From the definition of $\mathcal{B}^T$ we have the following result.

Theorem 2.4. $\mathcal{B}^T$ *is the smallest σ-algebra of the subsets of W^T that makes all canonical projections π_{ST} measurable.*

Furthermore, the following lemma is true.

Lemma 2.5. *The canonical projections π_{ST} are measurable if and only if $\pi_{\{t\}T}$ for all $t \in T$ are measurable as well.*

Moreover, from a well-known result of measure theory, we have the following proposition.

Proposition 2.6. *A function $f : (\Omega, \mathcal{F}) \to (W^T, \mathcal{B}^T)$ is measurable if and only if for all $t \in T$ the composite mapping $\pi_{\{t\}T} \circ f : (\Omega, \mathcal{F}) \to (E_t, \mathcal{B}_t)$ is measurable.*

For proofs of Theorem 2.4, Lemma 2.5, and Proposition 2.6, see, e.g., Métivier (1968).

Remark 2.7. Let $(\Omega, \mathcal{F}, P, (X_t)_{t \in T})$ be a stochastic process with state space $(E, \mathcal{B})$. Since the function space $E^T = \prod_{t \in T} E$, the mapping $f : \Omega \to E^T$, which associates every $\omega \in \Omega$ with its corresponding trajectory of the process, is $(\mathcal{F} - \mathcal{B}^T)$-measurable, and in fact we have that

$$\forall t \in T: \qquad \pi_{\{t\}T} \circ f(\omega) = \pi_{\{t\}T}(X(\cdot, \omega)) = X_t(\omega),$$

where $\pi_{\{t\}T} \circ f = X_t$, which is a random variable, is obviously measurable.

Definition 2.8. A function $f : \Omega \to E^T$ defined on a probability space $(\Omega, \mathcal{F}, P)$ and valued in a measurable space $(E^T, \mathcal{G})$ is called a *random function* if it is $(\mathcal{F}\text{-}\mathcal{G})$-measurable.

How can we define a probability law P^T on $(E^T, \mathcal{B}^T)$ for the stochastic process $(X_t)_{t \in T}$ defined on the probability space $(\Omega, \mathcal{F}, P)$ in a coherent way? We may observe that from a *physical* point of view, it is natural to assume that in principle we are able, from experiments, to define all possible *finite-dimensional* joint probabilities

$$P(X_{t_1} \in B_1, \ldots, X_{t_n} \in B_n)$$

for any $n \in \mathbb{N}$, for any $\{t_1, \ldots, t_n\} \subset T$, and for any $B_1, \ldots, B_n \in \mathcal{B}$, i.e., the joint probability laws P^S of all finite-dimensional random vectors $(X_{t_1}, \ldots, X_{t_n})$, for any choice of $S = \{t_1, \ldots, t_n\} \subset \mathcal{S}$, such that

$$P^S(B_1 \times \cdots \times B_n) = P(X_{t_1} \in B_1, \ldots, X_{t_n} \in B_n).$$

Accordingly, we require that, for any $S \subset \mathcal{S}$,

$$P^T(\pi_{ST}^{-1}(B_1 \times \cdots \times B_n)) = P^S(B_1 \times \cdots \times B_n) = P(X_{t_1} \in B_1, \ldots, X_{t_n} \in B_n).$$

A general answer comes from the following theorem. Having constructed the σ-algebra $\mathcal{B}^T$ on E^T, we now define a measure μ_T on $(W^T, \mathcal{B}^T)$, supposing that, for all $S \in \mathcal{S}$, a measure μ_S is assigned on $(W^S, \mathcal{B}^S)$. If $S \in \mathcal{S}$, $S' \in \mathcal{S}$, and $S \subset S'$, we denote the canonical projection of $W^{S'}$ on W^S by $\pi_{SS'}$, which is certainly $(\mathcal{B}^{S'}\text{-}\mathcal{B}^S)$-measurable.

Definition 2.9. If, for all $(S, S') \in \mathcal{S} \times \mathcal{S}'$, with $S \subset S'$, we have that $\pi_{SS'}(\mu_{S'}) = \mu_S$, then

$$(W^S, \mathcal{B}^S, \mu_S, \pi_{SS'})_{S, S' \in \mathcal{S}; S \subset S'}$$

is called a *projective system* of measurable spaces and $(\mu_S)_{S \in \mathcal{S}}$ is called a *compatible system* of measures on the finite products $(W^S, \mathcal{B}^S)_{S \in \mathcal{S}}$.

Theorem 2.10 (Kolmogorov–Bochner). *Let $(E_t, \mathcal{B}_t)_{t \in T}$ be a family of Polish spaces (i.e., metric, complete, separable) endowed with their respective Borel σ-algebras, and let $\mathcal{S}$ be the collection of finite subsets of T and, for all $S \in \mathcal{S}$ with $W^S = \prod_{t \in S} E_t$ and $\mathcal{B}^S = \bigotimes_{t \in S} \mathcal{B}_t$, let μ_S be a finite measure on $(W^S, \mathcal{B}^S)$. Under these assumptions the following two statements are equivalent:*

1. *There exists a μ_T measure on $(W^T, \mathcal{B}^T)$ such that for all $S \in \mathcal{S}$: $\mu_S = \pi_{ST}(\mu_T)$.*
2. *The system $(W^S, \mathcal{B}^S, \mu_S, \pi_{SS'})_{S, S' \in \mathcal{S}; S \subset S'}$ is projective.*

Moreover, in both cases, μ_T, as defined in 1, is unique.

Proof. See, e.g., Métivier (1968). □

Definition 2.11. The unique measure μ_T of Theorem 2.10 is called the *projective limit* of the projective system $(W^S, \mathcal{B}^S, \mu_S, \pi_{SS'})_{S, S' \in \mathcal{S}; S \subset S'}$.

As a special case consider a family of probability spaces $(E_t, \mathcal{B}_t, P_t)_{t \in T}$. If, for all $S \in \mathcal{S}$, we define $P_S = \bigotimes_{t \in S} P_t$, then $(W^S, \mathcal{B}^S, P_S, \pi_{SS'})_{S, S' \in \mathcal{S}; S \subset S'}$ is a projective system and the projective probability limit $\bigotimes_{t \in T} P_t$ is called the *probability product* of the family of probabilities $(P_t)_{t \in T}$.

With respect to the projective system of finite-dimensional probability laws $P_S = \bigotimes_{t \in S} P_{X_t}$ of a stochastic process $(X_t)_{t \in \mathbb{R}_+}$, the projective limit will be the required *probability law of the process*.

Theorem 2.12. *Two stochastic processes $(X_t)_{t\in\mathbb{R}_+}$ and $(Y_t)_{t\in\mathbb{R}_+}$ that have the same finite-dimensional probability laws have the same probability law.*

Definition 2.13. Two stochastic processes are *equivalent* if and only if they have the same projective system of finite-dimensional joint distributions.

A more stringent notion follows.

Definition 2.14. Two real-valued stochastic processes $(X_t)_{t\in\mathbb{R}_+}$ and $(Y_t)_{t\in\mathbb{R}_+}$ on the probability space $(\Omega, \mathcal{F}, P)$ are called *modifications* or *versions* of one another if,
$$\text{for any } t \in T, \ P(X_t = Y_t) = 1.$$

Remark 2.15. It is obvious that two processes that are modifications of one another are also equivalent.

An even more stringent requirement comes from the following definition.

Definition 2.16. Two processes are *indistinguishable* if
$$P(X_t = Y_t, \forall t \in \mathbb{R}_+) = 1.$$

Remark 2.17. It is obvious that two indistinguishable processes are modifications of each other.

Example 2.18. Let $(X_t)_{t\in T}$ be a family of independent random variables defined on $(\Omega, \mathcal{F}, P)$ and valued in $(E, \mathcal{B})$. [In fact, in this case, it is sufficient to assume that only finite families of $(X_t)_{t\in T}$ are independent.] We know that for all $t \in T$ the probability $P_t = X_t(P)$ is defined on $(E, \mathcal{B})$. Then
$$\forall S = \{t_1, \ldots, t_r\} \in \mathcal{S}: \qquad P_S = \bigotimes_{k=1}^{r} P_{t_k} \text{ for some } r \in \mathbb{N}^*,$$

and the system $(P_S)_{S\in\mathcal{S}}$ is compatible with its finite products $(E^S, \mathcal{B}^S)_{S\in\mathcal{S}}$. In fact, let $S, S' \in \mathcal{S}$, with $S = \{t_1, \ldots, t_r\} \subset S' = \{t_1, \ldots, t_{r'}\}$; if B is a rectangle of $\mathcal{B}^S$, i.e., $B = B_{t_1} \times \cdots \times B_{t_r}$, then

$$\begin{aligned}
P_S(B) &= P_S(B_{t_1} \times \cdots \times B_{t_r}) = P_{t_1}(B_{t_1}) \cdot \cdots \cdot P_{t_r}(B_{t_r}) \\
&= P_{t_1}(B_{t_1}) \cdot \cdots \cdot P_{t_r}(B_{t_r}) P_{t_{r+1}}(E) \cdot \cdots \cdot P_{t_{r'}}(E) \\
&= P_{S'}(\pi_{SS'}^{-1}(B)).
\end{aligned}$$

By the extension theorem we obtain that $P_S = \pi_{SS'}(P_{S'})$. As anticipated above, in this case we will write $P_T = \bigotimes_{t\in T} P_t$.

Remark 2.19. The compatibility condition $P_S = \pi_{SS'}(P_{S'})$, for all $S, S' \in \mathcal{S}$ and $S \subset S'$, can be expressed in an equivalent way by either the distribution function F_S of the probability P_S or its density f_S. For $E = \mathbb{R}$ we obtain, respectively,

1. For $S, S' \in \mathcal{S}$, with $S = \{t_1, \ldots, t_r\} \subset S' = \{t_1, \ldots, t_{r'}\}$; and for $(x_{t_1}, \ldots, x_{t_r}) \in \mathbb{R}^S$:
$$F_S(x_{t_1}, \ldots, x_{t_r}) = F_{S'}(x_{t_1}, \ldots, x_{t_r}, +\infty, \ldots, +\infty).$$

2. For $S, S' \in \mathcal{S}$, with $S = \{t_1, \ldots, t_r\} \subset S' = \{t_1, \ldots, t_{r'}\}$; and for $(x_{t_1}, \ldots, x_{t_r}) \in \mathbb{R}^S$:
$$f_S(x_{t_1}, \ldots, x_{t_r}) = \int \cdots \int dx_{t_{r+1}} \cdots dx_{t_{r'}} f_{S'}(x_{t_1}, \ldots, x_{t_r}, x_{t_{r+1}}, \ldots, x_{t_{r'}}).$$

Definition 2.20. A real-valued stochastic process $(X_t)_{t\in\mathbb{R}_+}$ is *continuous in probability* if
$$P - \lim_{s \to t} X_s = X_t, \qquad s, t \in \mathbb{R}_+.$$

Definition 2.21. A function $f : \mathbb{R}_+ \to \mathbb{R}$ is *right-continuous* if for any $t \in \mathbb{R}_+$, with $s > t$,
$$\lim_{s \downarrow t} f(s) = f(t).$$

Instead, the function is *left-continuous* if for any $t \in \mathbb{R}_+$, with $s < t$,
$$\lim_{s \uparrow t} f(s) = f(t).$$

Definition 2.22. A stochastic process $(X_t)_{t\in\mathbb{R}_+}$ is *right-(left-)continuous* if its trajectories are right-(left-)continuous almost surely. A stochastic process is continuous if its trajectories are continuous almost surely

Proposition 2.23. *A stochastic process that is continuous a.s. is continuous in probability. A stochastic process that is L^2-continuous is continuous in probability.*

Definition 2.24. A stochastic process $(X_t)_{t\in\mathbb{R}_+}$ is said to be *right-continuous with left limits* (RCLL) or *continu à droite avec limite à gauche* (càdlàg) if, almost surely, it has trajectories that are RCLL. The latter is denoted $X_{t-} = \lim_{s\uparrow t} X_s$.

Theorem 2.25. *Let $(X_t)_{t\in\mathbb{R}_+}$ and $(Y_t)_{t\in\mathbb{R}_+}$ be two RCLL processes. X_t and Y_t are modifications of each other if and only if they are indistinguishable.*

As discussed in Doob (1953) and in Billingsley (1986), the finite-dimensional distributions, which determine the existence of the probability law of a stochastic process according to the Kolmogorov–Bochner theorem, are not sufficient to determine the properties of the sample paths of the process. On the other hand, it is possible, under rather general conditions, to ensure the property of separability of a process, and from this property various other desirable properties of the sample paths follow, such as continuity for the Brownian paths.

Definition 2.26. A real-valued stochastic process $(X_t)_{t\in\mathbb{R}_+}$ on the probability space $(\Omega, \mathcal{F}, P)$ is called *separable* if

- There exists a $T_0 \subset \mathbb{R}_+$, countable and dense everywhere in $\mathbb{R}_+$
- There exists an $A \in \mathcal{F}$, $P(A) = 0$ (negligible)

such that

- For all $t \in \mathbb{R}_+$: there exists $(t_n)_{n \in \mathbb{N}} \in T_0^{\mathbb{N}}$ such that $\lim_{n \to \infty} t_n = t$.
- For all $\omega \in \Omega \setminus A$: $\lim_{n \to \infty} X_{t_n}(\omega) = X_t(\omega)$.

The subset T_0 of $\mathbb{R}_+$, as defined previously, is called the *separating set*.

Theorem 2.27. *Let $(X_t)_{t \in \mathbb{R}_+}$ be a separable process, having T_0 and A as its separating and negligible sets, respectively. If $\omega \notin A$, $t_0 \in \mathbb{R}_+$, and $\lim_{t \to t_0} X_t(\omega)$ for $t \in T_0$ exists, then so does the limit $\lim_{t \to t_0} X_t(\omega)$ for $t \in \mathbb{R}_+$, and they coincide.*

Proof. See, e.g., Ash and Gardner (1975).   □

Theorem 2.28. *Every real stochastic process $(X_t)_{t \in \mathbb{R}_+}$ admits a separable modification, almost surely finite, for any $t \in \mathbb{R}_+$.*

Proof. See, e.g., Ash and Gardner (1975).   □

Remark 2.29. By virtue of Theorem 2.28, we may henceforth only consider separable processes.

In general, it is not true that a function $f(\omega_1, \omega_2)$ is jointly measurable in both variables, even if it is separately measurable in each of them. It is therefore required to impose conditions that guarantee the joint measurability of f in both variables. Evidently, if $(X_t)_{t \in \mathbb{R}_+}$ is a stochastic process, then for all $t \in \mathbb{R}_+$: $X(t, \cdot)$ is measurable.

Definition 2.30. Let $(X_t)_{t \in \mathbb{R}_+}$ be a stochastic process defined on the probability space $(\Omega, \mathcal{F}, P)$ and valued in $(E, \mathcal{B}_E)$. The process $(X_t)_{t \in \mathbb{R}_+}$ is said to be *measurable* if it is measurable as a function defined on $\mathbb{R}_+ \times \Omega$ (with the σ-algebra $\mathcal{B}_{\mathbb{R}_+} \otimes \mathcal{F}$) and valued in E.

Proposition 2.31. *If the process $(X_t)_{t \in \mathbb{R}_+}$ is measurable, then the trajectory $X(\cdot, \omega) : \mathbb{R}_+ \to E$ is measurable for all $\omega \in \Omega$.*

Proof. Let $\omega \in \Omega$ and $B \in \mathcal{B}_E$. We want to show that $(X(\cdot, \omega))^{-1}(B)$ is an element of $\mathcal{B}_{\mathbb{R}_+}$. In fact,

$$(X(\cdot, \omega))^{-1}(B) = \{t \in \mathbb{R}_+ | X(t, \omega) \in B\} = \{t \in \mathbb{R}_+ | (t, \omega) \in X^{-1}(B)\},$$

meaning that $(X(\cdot, \omega))^{-1}(B)$ is the path ω of X^{-1}, which is certainly measurable, because $X^{-1}(B) \in \mathcal{B}_{\mathbb{R}_+} \otimes \mathcal{F}$ (as follows from the properties of the product σ-algebra).   □

If the process is measurable, then it makes sense to consider the integral $\int_a^b X(t, \omega) dt$ along a trajectory. By Fubini's theorem, we have

$$\int_\Omega P(d\omega) \int_a^b dt X(t,\omega) = \int_a^b dt \int_\Omega P(d\omega) X(t,\omega).$$

Definition 2.32. The process $(X_t)_{t\in\mathbb{R}_+}$ is said to be *progressively measurable* with respect to the filtration $(\mathcal{F}_t)_{t\in\mathbb{R}_+}$, which is an increasing family of subalgebras of $\mathcal{F}$, if, for all $t \in \mathbb{R}_+$, the mapping $(s,\omega) \in [0,t] \times \Omega \to X(s,\omega) \in E$ is $(\mathcal{B}_{[0,t]} \otimes \mathcal{F}_t)$-measurable. Furthermore, we henceforth suppose that $\mathcal{F}_t = \sigma(X(s), 0 \le s \le t)$, $t \in \mathbb{R}_+$, which is called the *generated* or *natural* filtration of the process X_t.

Proposition 2.33. *If the process $(X_t)_{t\in\mathbb{R}_+}$ is progressively measurable, then it is also measurable.*

Proof. Let $B \in \mathcal{B}_E$. Then

$$X^{-1}(B) = \{(s,\omega) \in \mathbb{R}_+ \times \Omega | X(s,\omega) \in B\}$$

$$= \bigcup_{n=0}^\infty \{(s,\omega) \in [0,n] \times \Omega | X(s,\omega) \in B\}.$$

Since

$$\forall n: \quad \{(s,\omega) \in [0,n] \times \Omega | X(s,\omega) \in B\} \in \mathcal{B}_{[0,n]} \otimes \mathcal{F}_n,$$

we obtain that $X^{-1}(B) \in \mathcal{B}_{\mathbb{R}_+} \otimes \mathcal{F}$. □

Theorem 2.34. *Let $(X_t)_{t\in\mathbb{R}_+}$ be a real stochastic process continuous in probability; then it admits a separable and progressively measurable modification.*

Proof. See, e.g., Ash and Gardner (1975). □

Definition 2.35. A filtered complete probability space $(\Omega, \mathcal{F}, P, (\mathcal{F}_t)_{t\in\mathbb{R}_+})$ is said to satisfy the *usual hypotheses* if

1. $\mathcal{F}_0$ contains all the P-null sets of $\mathcal{F}$.
2. $\mathcal{F}_t = \bigcap_{s>t} \mathcal{F}_s$, for all $t \in \mathbb{R}_+$, i.e., the filtration $(\mathcal{F}_t)_{t\in\mathbb{R}_+}$ is right-continuous.

Henceforth we will always assume that the usual hypotheses hold, unless specified otherwise.

Definition 2.36. Let $(\Omega, \mathcal{F}, P, (\mathcal{F}_t)_{t\in\mathbb{R}_+})$ be a filtered probability space. The σ-algebra on $\mathbb{R}_+ \times \Omega$ generated by all sets of the form $\{0\} \times A$, $A \in \mathcal{F}_0$, and $]a,b] \times A$, $0 \le a < b < +\infty$, $A \in \mathcal{F}_a$, is said to be the *predictable σ-algebra* for the filtration $(\mathcal{F}_t)_{t\in\mathbb{R}_+}$.

Definition 2.37. A real-valued process $(X_t)_{t\in\mathbb{R}_+}$ is called *predictable* with respect to a filtration $(\mathcal{F}_t)_{t\in\mathbb{R}_+}$, or $\mathcal{F}_t$-*predictable*, if as a mapping from $\mathbb{R}_+ \times \Omega \to \mathbb{R}$ it is measurable with respect to the predictable σ-algebra generated by this filtration.

Definition 2.38. A *simple predictable process* is of the form

$$X = k_0 I_{\{0\} \times A} + \sum_{i=1}^{n} k_i I_{]a_i, b_i] \times A_i},$$

where $A_0 \in \mathcal{F}_0$, $A_i \in \mathcal{F}_{a_i}$, $i = 1, \ldots, n$, and $k_0, \ldots, k_n$ are real constants.

Proposition 2.39. *Let $(X_t)_{t \in \mathbb{R}_+}$ be a process that is $\mathcal{F}_t$-predictable. Then, for any $t > 0$, X_t is $\mathcal{F}_{t-}$-measurable.*

Lemma 2.40. *Let $(X_t)_{t \in \mathbb{R}_+}$ be a left-continuous real-valued process adapted to $(\mathcal{F}_t)_{t \in \mathbb{R}_+}$. Then X_t is predictable.*

Lemma 2.41. *A process is predictable if and only if it is measurable with respect to the smallest σ-algebra on $\mathbb{R}_+ \times \Omega$ generated by the adapted left-continuous processes.*

Proposition 2.42. *Every predictable process is progressively measurable.*

Proposition 2.43. *If the process $(X_t)_{t \in \mathbb{R}_+}$ is right-(left-)continuous, then it is progressively measurable.*

Proof. See, e.g., Métivier (1968).   □

Let $(X_t)_{t \in \mathbb{R}_+}$ be an $\mathbb{R}^d$-valued stochastic process defined on the probability space $(\Omega, \mathcal{F}, P)$.

We say it is *continuous* (resp. *right-continuous*, *left-continuous*) if, for almost all $\omega \in \Omega$, the trajectory $(X_t(\omega))_{t \in \mathbb{R}_+}$ is *continuous* (resp. *right-continuous*, *left-continuous*) with respect to t.

We say it is *$\mathcal{F}$-adapted* (or simply *adapted*) if, for every $t \in \mathbb{R}_+$, X_t is $\mathcal{F}$-measurable.

Let $\mathcal{O}$ (resp. $\mathcal{P}$) be the smallest σ-algebra on $\mathbb{R}_+ \times \Omega$ with respect to which every càdlàg-adapted process (resp. left-continuous process) is a measurable function of (t, ω). We say that the stochastic process $(X_t)_{t \in \mathbb{R}_+}$ is *optional* (resp. *predictable*) if the process regarded as a function of (t, ω) is $\mathcal{O}$-measurable (resp. $\mathcal{P}$-measurable).

We say that a real-valued stochastic process $(X_t)_{t \in \mathbb{R}_+}$ is an *increasing* process if, for almost all $\omega \in \Omega$, $X_t(\omega)$ is nonnegative nondecreasing right-continuous with respect to $t \in \mathbb{R}_+$.

We say it is a *process of finite variation* if it can be decomposed as $X_t = \overline{X}_t - \widehat{X}_t$, with both $\overline{X}_t$ and $\widehat{X}_t$ increasing processes.

It is obvious that processes of finite variation are cadlag. Hence adapted processes of finite variation are optional.

The relations among various properties of stochastic properties are summarized below.

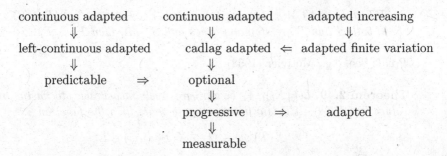

2.2 Stopping Times

In what follows we are given a probability space $(\Omega, \mathcal{F}, P)$ and a filtration $(\mathcal{F}_t)_{t \in \mathbb{R}_+}$ on $\mathcal{F}$.

Definition 2.44. A random variable T defined on Ω (endowed with the σ-algebra $\mathcal{F}$) and valued in $\bar{\mathbb{R}}_+$ is called a *stopping time* (or *Markov time*) with respect to the filtration $(\mathcal{F}_t)_{t \in \mathbb{R}_+}$, or simply an $\mathcal{F}_t$-*stopping time*, if

$$\forall t \in \mathbb{R}_+: \qquad \{\omega | T(\omega) \leq t\} \in \mathcal{F}_t.$$

The stopping time is said to be finite if $P(T = \infty) = 0$.

Remark 2.45. If $T(\omega) \equiv k$ (constant), then T is always a stopping time. If T is a stopping time with respect to the filtration $(\mathcal{F}_t)_{t \in \mathbb{R}_+}$ generated by the stochastic process $(X_t)_{t \in \mathbb{R}_+}$, $t \in \mathbb{R}_+$, then T is called the *stopping time of the process*.

Definition 2.46. Let T be an $\mathcal{F}_t$-stopping time. $A \in \mathcal{F}$ is said to *precede* T if, for all $t \in \mathbb{R}_+$: $A \cap \{T \leq t\} \in \mathcal{F}_t$.

Proposition 2.47. *Let T be an $\mathcal{F}_t$-stopping time, and let*

$$\mathcal{F}_T = \{A \in \mathcal{F} | A \text{ precedes } T\};$$

then $\mathcal{F}_T$ is a σ-algebra of the subsets of Ω. It is called the σ-algebra of T-preceding events.

Proof. See, e.g., Métivier (1968). $\qquad\qquad\qquad\qquad\qquad\qquad\qquad\qquad$ □

Theorem 2.48. *The following relationships hold:*

1. *If both S and T are stopping times, then so are $S \wedge T = \inf\{S, T\}$ and $S \vee T = \sup\{S, T\}$.*
2. *If T is a stopping time and $a \in [0, +\infty[$, then $T \wedge a$ is a stopping time.*
3. *If T is a finite stopping time, then it is $\mathcal{F}_T$-measurable.*

4. If both S and T are stopping times and $A \in \mathcal{F}_S$, then $A \cap \{S \leq T\} \in \mathcal{F}_T$.
5. If both S and T are stopping times and $S \leq T$, then $\mathcal{F}_S \subset \mathcal{F}_T$.

Proof. See, e.g., Métivier (1968).   □

Theorem 2.49. *Let $(X_t)_{t \in \mathbb{R}_+}$ be a progressively measurable stochastic process valued in $(S, \mathcal{B}_S)$. If T is a finite stopping time, then the function*

$$X(T) : \omega \in \Omega \to X(T(\omega), \omega) \in E$$

is $\mathcal{F}_T$-measurable (and hence a random variable).

Proof. We need to show that

$$\forall B \in \mathcal{B}_E: \qquad \{\omega | X(T(\omega)) \in B\} \in \mathcal{F}_T,$$

hence

$$\forall B \in \mathcal{B}_E, \forall t \in \mathbb{R}_+: \qquad \{\omega | X(T(\omega)) \in B\} \cap \{T \leq t\} \in \mathcal{F}_t.$$

Fixing $B \in \mathcal{B}_E$ we have

$$\forall t \in \mathbb{R}_+: \qquad \{\omega | X(T(\omega)) \in B\} \cap \{T \leq t\} = \{X(T \wedge t) \in B\} \cap \{T \leq t\},$$

where $\{T \leq t\} \in \mathcal{F}_t$, since T is a stopping time. We now show that

$$\{X(T \wedge t) \in B\} \in \mathcal{F}_t.$$

In fact, $T \wedge t$ is a stopping time (by point 2 of Theorem 2.48) and is $\mathcal{F}_{T \wedge t}$- measurable (by point 3 of Theorem 2.48). But $\mathcal{F}_{T \wedge t} \subset \mathcal{F}_t$ and thus $T \wedge t$ is $\mathcal{F}_t$-measurable. Now $X(T \wedge t)$ is obtained as a composite of the mapping

$$\omega \in \Omega \to (T \wedge t(\omega), \omega) \in [0, t] \times \Omega, \tag{2.1}$$

with

$$(s, \omega) \in [0, t] \times \Omega \to X(s, \omega) \in E. \tag{2.2}$$

The mapping (2.1) is $(\mathcal{F}_t - \mathcal{B}_{[0,t]} \otimes \mathcal{F}_t)$-measurable (because $T \wedge t$ is $\mathcal{F}_t$-measurable) and the mapping (2.2) is $(\mathcal{B}_{[0,t]} \otimes \mathcal{F}_t - \mathcal{B}_E)$-measurable since X is progressively measurable. Therefore, $X(T \wedge t)$ is $\mathcal{F}_t$-measurable, completing the proof.   □

2.3 Canonical Form of a Process

Let $(\Omega, \mathcal{F}, P, (X_t)_{t \in T})$ be a stochastic process valued in $(E, \mathcal{B})$ and, for every $S \in \mathcal{S}$, let P_S be the joint probability law for the random variables $(X_t)_{t \in S}$ that is the probability on $(E^S, \mathcal{B}^S)$ induced by P through the function

$$X^S : \omega \in \Omega \to (X_t(\omega))_{t \in S} \in E^S = \prod_{t \in S} E.$$

Evidently, if

$$S \subset S'(S, S' \in \mathcal{S}), \qquad X^S = \pi_{SS'} \circ X^{S'},$$

then it follows that

$$P_S = X^S(P) = (\pi_{SS'} \circ X^{S'})(P) = \pi_{SS'}(P_{S'}),$$

and therefore $(E^S, \mathcal{B}^S, P_S, \pi_{SS'})_{S,S' \in \mathcal{S}, S \subset S'}$ is a projective system of probabilities.

On the other hand, the random function $f : \Omega \to E^T$ that associates every $\omega \in \Omega$ with a trajectory of the process in ω is measurable (following Proposition 2.6). Hence we can consider the induced probability P_T on $\mathcal{B}^T$, $P_T = f(P)$; P_T is the projective limit of $(P_S)_{S \in \mathcal{S}}$. From this it follows that $(E^T, \mathcal{B}^T, P_T, (\pi_t)_{t \in T})$ is a stochastic process with the property that, for all $S \in \mathcal{S}$, the random vectors $(\pi_t)_{t \in S}$ and $(X_t)_{t \in S}$ have the same joint distribution.

Definition 2.50. The stochastic process $(E^T, \mathcal{B}^T, P_T, (\pi_t)_{t \in T})$ is called the *canonical form* of the process $(\Omega, \mathcal{F}, P, (X_t)_{t \in T})$.

Remark 2.51. From this it follows that two stochastic processes are equivalent if they admit the same canonical process.

2.4 L^2 Processes

Consider a real-valued stochastic process $X \equiv (X_t)_{t \in T}$ on a probability space $(\Omega, \mathcal{F}, P)$, which admits finite means and finite variances, i.e. $X_t \in L^2(\Omega)$, for any $t \in T$. For our treatment we may take T as an interval of $\mathbb{R}_+$. We shall denote by

$$m(t) = E[X_t], \ t \in T,$$

the mean value function of the process. The covariance function is well defined too, as

$$K(s,t) := Cov[X_s, X_t], \ s, t \in T.$$

It is then easily shown that

(i) K is a *symmetric* function, i.e.

$$K(s,t) = K(t,s), \ s, t \in T;$$

(ii) K is *nonnegative definite*, i.e. for all $t_1, \ldots, t_n \in T$, and all real numbers $a_1, \ldots, a_n$ $(n \in \mathbb{N}, n > 1)$,

$$\sum_{i,j=1}^{n} a_i a_j K(t_i, t_j) \geq 0.$$

2.4.1 Gaussian Processes

Definition 2.52. A real-valued stochastic process $(\Omega, \mathcal{F}, P, (X_t)_{t \in \mathbb{R}_+})$ is called a *Gaussian process* if, for all $n \in \mathbb{N}^*$ and for all $(t_1, \ldots, t_n) \in \mathbb{R}_+^n$, the n-dimensional random vector $\mathbf{X} = (X_{t_1}, \ldots, X_{t_n})'$ has a multivariate Gaussian distribution, with probability density

$$f_{t_1, \ldots, t_n}(\mathbf{x}) = \frac{1}{(2\pi)^{n/2} \sqrt{\det K}} \exp \left\{ -\frac{1}{2} (\mathbf{x} - \mathbf{m})' K^{-1} (\mathbf{x} - \mathbf{m}) \right\}, \qquad (2.3)$$

where

$$\begin{cases} m_i = E[X_{t_i}] \in \mathbb{R}, & i = 1, \ldots, n, \\ K_{ij} = Cov[X_{t_i}, X_{t_j}] \in \mathbb{R}, \; i, j = 1, \ldots, n. \end{cases}$$

The covariance matrix $K = (\sigma_{ij})$ is taken as positive-definite, i.e., for all $\mathbf{a} = (a_1, \ldots, a_n) \in \mathbb{R}^n \colon \sum_{i,j=1}^n a_i K_{ij} a_j > 0)$.

The existence of Gaussian processes is guaranteed by the following remarks. By assigning a real-valued function

$$m : \mathbb{R}_+ \to \mathbb{R},$$

and a positive-definite function

$$K : \mathbb{R}_+ \times \mathbb{R}_+ \to \mathbb{R},$$

thanks to well-known properties of multivariate Gaussian distributions, we may introduce a projective system of Gaussian laws $(P_S)_{S \in \mathcal{S}}$ (where $\mathcal{S}$ is the set of all finite subsets of $\mathbb{R}_+$) of the form (2.3) such that, for $S = \{t_1, \ldots, t_n\}$,

$$m_i = m(t_i), \quad i = 1, \ldots, n,$$

$$K_{ij} = K(t_i, t_j), \quad i, j = 1, \ldots, n.$$

Since $\mathbb{R}$ is a Polish space, by the Kolmogorov–Bochner Theorem 2.10, we can now assert that there exists a Gaussian process $(X_t)_{t \in \mathbb{R}_+}$ having the preceding $(P_S)_{S \in \mathcal{S}}$ as its projective system of finite-dimensional distributions.

Example 2.53. The *standard Brownian Bridge* (see later Proposition 2.187) is a centered Gaussian process $(X_t)_{t \in [0,1]}$ on $\mathbb{R}$ such that

$$\begin{cases} \forall t \in [0,1] \colon E[X_t] = 0; \\ \forall (s,t) \in [0,1] \times [0,1], s \le t \colon Cov[X_s, X_t] = s(1-t). \end{cases}$$

The above result concerning the existence of a Gaussian process, can be extended to any L^2 process.

Theorem 2.54 *Let* $K = K(s,t)$, $s,t \in T$, *be a symmetric, and nonnegative definite real valued function on* $T \times T$. *Then there exists an* L^2 *process* $(X_t)_{t \in T}$, *whose covariance function is precisely* K.

Remark 2.55. It is clear that Theorem 2.54 does not include uniqueness. In fact for any L^2 process one can always build a Gaussian process with the same mean and covariance functions.

Definition 2.56. Let $T \subset \mathbb{R}_+$; then we say that an L^2 process $(X_t)_{t \in T}$, is *weakly stationary* if and only if

(i) its mean function is independent of t, i.e.

$$m(t) = E[X_t] = const;$$

(ii) $K(s,t)$ depends only on the difference $t - s$, i.e. for any s, t, h such that $s, t, s+h, t+h \in T$,

$$K(s,t) = K(s+h, t+h).$$

Proposition 2.57 *Let $T \subset \mathbb{R}_+$. If the covariance function of an L^2 process is continuous at (t,t) for all $t \in T$, then it is continuous at any $(s,t) \in T \times T$.*

As usual, let us denote by m the mean value function of an L^2 process $(X_t)_{t \in T}$, and by K its covariance function.

Proposition 2.58 *Let $T \subset \mathbb{R}_+$. Assume that m is continuous on T. Then an L^2 process $(X_t)_{t \in T}$ is L^2−continuous at a point $t \in T$ if and only if K is continuous at (t,t).*

2.4.2 Karhunen-Loève Expansion

Let K be a real valued continuous covariance function, hence continuous, symmetric, and nonnegative function defined on $[a,b] \times [a,b]$, with $a, b \in \mathbb{R}_+, a < b$. Consider the operator

$$A : L^2([a,b]) \to L^2([a,b])$$

$$(Af)(s) = \int_a^b K(s,t)f(t)dt, \quad s \in [a,b]. \tag{2.4}$$

Due to the assumptions on K, the eigenfunctions of the operator A span the Hilbert space $L^2([a,b])$. Moreover A admits an at most countable number of nonnegative real eigenvalues which have 0 as the only possible limit point. We can build an orthonormal basis $e_n, n \in \mathbb{N}^*$ for the Hilbert subspace spanned by the eigenvectors corresponding to the nontrivial eigenvalues, in such a way that e_n is an eigenvector corresponding to the eigenvalue λ_n (see e.g. Renardy and Rogers (2004, p. 235)). For this specific case we may then apply Mercer's Theorem (Mercer (1909); see e.g. Parzen (1959), Courant and Hilbert (1966, p. 138), Bosq (2000, p. 24)) to state that

$$K(s,t) = \sum_{n=1}^{\infty} \lambda_n e_n(s) e_n(t), \quad s,t \in [a,b], \tag{2.5}$$

where the series converges absolutely and uniformly in both variables on $[a,b] \times [a,b]$.

By subtracting its mean function, an L^2 stochastic process will maintain its covariance function, so that we may reduce our analysis to zero mean stochastic processes.

Theorem 2.59 (Karhunen-Loève Theorem) *Let $(X_t)_{t \in [a,b]}$ be a zero mean L^2 process with a continuous covariance function K. Let $e_n, n \in \mathbb{N}^*$ be an orthonormal basis for the Hilbert subspace spanned by the eigenvectors corresponding to the nontrivial eigenvalues of the operator A associated with the covariance function K, as in (2.4), in such a way that e_n is an eigenvector corresponding to the eigenvalue λ_n. Then*

$$X_t = \sum_{n=1}^{\infty} Z_n e_n(t), \quad t \in [a,b], \tag{2.6}$$

where the Z_n, $n \in \mathbb{N}^*$ are orthogonal random variables such that, for any $n \in \mathbb{N}^*$,

(i) $Z_n = \int_a^b X_t e_n(t)$;
(ii) $E[Z_n] = 0$;
(iii) $E[Z_n^2] = \lambda_n$;
(iv) $E[Z_j Z_k] = 0$, for $j \neq k$.

The series in (2.6) converges in L^2 uniformly in t.

Corollary 2.60 *In the Karhunen-Loève expansion (2.6) for a Gaussian process, the random variables Z_n, $n \in \mathbb{N}^*$, form a Gaussian sequence of independent random variables.*

For example, it can be shown that the standard Wiener process for $t \in [0,1]$ admits the following Karhunen-Loève expansion

$$W_t = \sqrt{2} \sum_{n=0}^{\infty} Z_n \sin\left[(2n+1)\frac{\pi}{2}t\right], \; t \in [0,1],$$

where $(Z_n)_{n \in \mathbb{N}}$ is a sequence of independent zero-mean Gaussian random variables, with

$$e[Z_n^2] = \frac{4}{(2n+1)^2 \pi^2}, \; n \in \mathbb{N}.$$

For the proofs of the statements in this section, we refer, e.g., to Parzen (1959), Ash and Gardner (1975), and Bosq (2000, p. 24).

This theorem has gained a great relevance in functional principal component analysis (see Bosq (2000), Ramsey and Silverman (2005)).

In numerical analysis, a point of great practical importance is that the Karhunen-Loève expansion can be used in a numerical simulation scheme to obtain numerical realizations of the relevant process. The expansion is used in the fields of pattern recognition and image analysis as an efficient tool to store random processes (see, e.g., Devijver and Kittler (1982)).

Remark 2.61. It is worth remarking that the resulting structure in the Karhunen-Loève expansion splits the randomness (ω) and the time dependence (t) of the process $X_t(\omega)$, $t \in [a, b], \omega \in \Omega$. From the statistical and numerical point of view, it will be appreciated the fact that the basis of the expansion is now deterministic.

Remark 2.62. It may seem that any basis of the Hilbert space L^2 might be used instead of the Karhunen-Loève expansion. Actually it can be shown that this expansion possesses some desirable properties that make it a preferable choice in numerical analysis (see Ghanen and Spanos (2003, p. 21))

(i) The choice of the basis in the Karhunen-Loève expansion is optimal in the sense that the mean-square error resulting from a finite representation of a stochastic process is minimized.

(ii) The random variables appearing in an expansion of the kind (2.6) are orthogonal if and only if the vectors $e_n, n \in \mathbb{N}^*$, respectively, the constants $\lambda_n, n \in \mathbb{N}^*$, are the eigenvectors, respectively, the eigenvalues of the operator A associated with the covariance function K, as from (2.4).

2.5 Processes with Independent Increments

Definition 2.63. The stochastic process $(\Omega, \mathcal{F}, P, (X_t)_{t \in \mathbb{R}_+})$, with state space $(E, \mathcal{B})$, is called a *process with independent increments* if, for all $n \in \mathbb{N}$ and for all $(t_1, \ldots, t_n) \in \mathbb{R}_+^n$, where $t_1 < \cdots < t_n$, the random variables $X_{t_1}, X_{t_2} - X_{t_1}, \ldots, X_{t_n} - X_{t_{n-1}}$ are independent.

Theorem 2.64. *If $(\Omega, \mathcal{F}, P, (X_t)_{t \in \mathbb{R}_t})$ is a process with independent increments, then it is possible to construct a compatible system of probability laws $(P_S)_{S \in \mathcal{S}}$, where again $\mathcal{S}$ is a collection of finite subsets of the index set.*

Proof. To do this, we need to assign a joint distribution to every random vector $(X_{t_1}, \ldots, X_{t_n})$ for all $(t_1, \ldots, t_n)$ in $\mathbb{R}_+^n$ with $t_1 < \cdots < t_n$. Thus, let $(t_1, \ldots, t_n) \in \mathbb{R}_+^n$, with $t_1 < \cdots < t_n$, and $\mu_0, \mu_{s,t}$ be the distributions of X_0 and $X_t - X_s$, for every $(s, t) \in \mathbb{R}_+ \times \mathbb{R}_+$, with $s < t$, respectively. We define

$$Y_0 = X_0,$$
$$Y_1 = X_{t_1} - X_0,$$
$$\ldots$$

$$Y_n = X_{t_n} - X_{t_{n-1}},$$

where $Y_0, Y_1, \ldots, Y_n$ have the distributions $\mu_0, \mu_{0,t_1}, \ldots, \mu_{t_{n-1},t_n}$, respectively. Moreover, since the Y_i are independent, $(Y_0, \ldots, Y_n)$ have joint distribution $\mu_0 \otimes \mu_{0,t_1} \otimes \cdots \otimes \mu_{t_n,t_{n-1}}$. Let f be a real-valued, $\bigotimes^n \mathcal{B}$-measurable function, and consider the random variable $f(X_{t_1}, \ldots, X_{t_n})$. Then

$$E[f(X_{t_1}, \ldots, X_{t_n})]$$
$$\doteq E[f(Y_0 + Y_1, \ldots, Y_0 + \cdots + Y_n)]$$
$$= \int f(y_0 + y_1, \ldots, y_0 + \cdots + y_n) d(\mu_0 \otimes \mu_{0,t_1} \otimes \cdots \otimes \mu_{t_{n-1},t_n})(y_0, \ldots, y_n).$$

In particular, if $f = I_B$, with $B \in \bigotimes^n \mathcal{B}$, we obtain the joint distribution of $X_{t_1}, \ldots, X_{t_n}$:

$$P((X_{t_1}, \ldots, X_{t_n}) \in B) = E[I_B(X_{t_1}, \ldots, X_{t_n})]$$
$$= \int I_B(y_0 + y_1, \ldots, y_0 + \cdots + y_n) d(\mu_0 \otimes \mu_{0,t_1} \otimes \cdots \otimes \mu_{t_{n-1},t_n})(y_0, \ldots, y_n).$$

$$(2.7)$$

Having obtained P_S, where $S = \{t_1, \ldots, t_n\}$, with $t_1 < \cdots < t_n$, we show that $(P_S)_{S \in \mathcal{S}}$ is a compatible system. Let $S, S' \in \mathcal{S}; S \subset S', S = \{t_1, \ldots, t_n\}$, with $t_1 < \cdots < t_n$ and $S' = \{t_1, \ldots, t_j, s, t_{j+1}, \ldots, t_n\}$, with $t_1 < \cdots < t_j < s < t_{j+1} < \cdots < t_n$. For $B \in \mathcal{B}^S$ and $B' = \pi_{SS'}^{-1}(B)$, we will show that $P_S(B) = P_{S'}(B')$.

We can observe, by the definition of B', that

$$I_{B'}(x_{t_1}, \ldots, x_{t_j}, x_s, x_{t_{j+1}}, \ldots, x_{t_n})$$

does not depend on x_s and is therefore identical to $I_B(x_{t_1}, \ldots, x_{t_n})$. Thus putting $U = X_s - X_{t_j}$ and $V = X_{t_{j+1}} - X_s$, we obtain

$$P_{S'}(B') = \int I_{B'}(y_0 + y_1, \ldots, y_0 + \cdots + y_j, y_0 + \cdots + y_j + u, y_0 + \cdots$$
$$+ y_j + u + v, \ldots, y_0 + \cdots + y_n) d(\mu_0 \otimes \mu_{0,t_1} \otimes \cdots$$
$$\otimes \mu_{t_j,s} \otimes \mu_{s,t_{j+1}} \otimes \cdots \otimes \mu_{t_{n-1},t_n})(y_0, \ldots, y_j, u, v, y_{j+2}, \ldots, y_n)$$
$$= \int I_B(y_0 + y_1, \ldots, y_0 + \cdots + y_j, y_0 + \cdots + y_j + u + v, y_0 + \cdots$$
$$+ u + v + y_{j+2}, \ldots, y_0 + \cdots + y_n) d(\mu_0 \otimes \mu_{0,t_1} \otimes \cdots$$
$$\otimes \mu_{t_j,s} \otimes \mu_{s,t_{j+1}} \otimes \cdots \otimes \mu_{t_{n-1},t_n})(y_0, \ldots, y_j, u, v, y_{j+2}, \ldots, y_n).$$

Integrating with respect to all the variables except u and v, after applying Fubini's theorem, we obtain

$$P_{S'}(B') = \int h(u + v) d(\mu_{t_j,s} \otimes \mu_{s,t_{j+1}})(u, v).$$

Letting $y_{j+1} = u + v$ we have

$$P_{S'}(B') = \int h(y_{j+1})d(\mu_{t_j,s} * \mu_{s,t_{j+1}})(y_{j+1}).$$

Moreover, we observe that the definition of $y_{j+1} = u + v$ is compatible with the preceding notation $Y_{j+1} = X_{t_{j+1}} - X_{t_j}$. In fact, we have

$$u + v = x_s - x_{t_j} + x_{t_{j+1}} - x_s = x_{t_{j+1}} - x_{t_j}.$$

Furthermore, for the independence of $(X_{t_{j+1}} - X_s)$ and $(X_s - X_{t_j})$ the sum of random variables

$$X_{t_{j+1}} - X_s + X_s - X_{t_j} = X_{t_{j+1}} - X_{t_j}$$

must have the distribution $\mu_{t_j,s} * \mu_{s,t_{j+1}}$, where $*$ denotes the convolution product. Therefore, having denoted the distribution of $X_{t_{j+1}} - X_{t_j}$ by $\mu_{t_j,t_{j+1}}$, we obtain

$$\mu_{t_j,s} * \mu_{s,t_{j+1}} = \mu_{t_j,t_{j+1}}.$$

As a consequence we have

$$P_{S'}(B') = \int h(y_{j+1})d\mu_{t_j,t_{j+1}}(y_{j+1}).$$

This integral coincides with the one in (2.7), and thus

$$P_S(B') = P((X_{t_1},\ldots,X_{t_n}) \in B) = P_S(B).$$

If now $S' = S \cup \{s_1,\ldots,s_k\}$, the proof is completed by induction. $\qquad \square$

Definition 2.65. A process with independent increments is called *time-homogeneous* if

$$\mu_{s,t} = \mu_{s+h,t+h} \qquad \forall s,t,h \in \mathbb{R}_+, s < t.$$

If $(\Omega, \mathcal{F}, P, (X_t)_{t \in \mathbb{R}_+})$ is a homogeneous process with independent increments, then as a particular case we have

$$\mu_{s,t} = \mu_{0,t-s} \qquad \forall s,t \in \mathbb{R}_+, s < t.$$

Definition 2.66. A family of measures $(\mu_t)_{t \in \mathbb{R}_+}$ that satisfy the condition

$$\mu_{t_1+t_2} = \mu_{t_1} * \mu_{t_2}$$

is called a *convolution semigroup*.

Remark 2.67. A time-homogeneous process with independent increments is completely defined by assigning it a convolution semigroup.

2.6 Markov Processes

Definition 2.68. Let $(X_t)_{t \in \mathbb{R}_+}$ be a stochastic process on a probability space, valued in $(E, \mathcal{B})$ and adapted to the increasing family $(\mathcal{F}_t)_{t \in \mathbb{R}_+}$ of σ-algebras of subsets of $\mathcal{F}$. $(X_t)_{t \in \mathbb{R}_+}$ is a *Markov process* with respect to $(\mathcal{F}_t)_{t \in \mathbb{R}_+}$ if the following condition is satisfied:

$$\forall B \in \mathcal{B}, \forall (s,t) \in \mathbb{R}_+ \times \mathbb{R}_+, s < t: \qquad P(X_t \in B | \mathcal{F}_s) = P(X_t \in B | X_s) \text{ a.s.}$$
(2.8)

Remark 2.69. If, for all $t \in \mathbb{R}_+$, $\mathcal{F}_t = \sigma(X_r, 0 \leq r \leq t)$, then condition (2.8) becomes

$$P(X_t \in B | X_r, 0 \leq r \leq s) = P(X_t \in B | X_s) \text{ a.s.}$$

for all $B \in \mathcal{B}$, for all $(s,t) \in \mathbb{R}_+ \times \mathbb{R}_+$, and $s < t$.

Proposition 2.70. *Under the assumptions of Definition 2.68, the following two statements are equivalent:*

1. *For all $B \in \mathcal{B}$ and all $(s,t) \in \mathbb{R}_+ \times \mathbb{R}_+, s < t: P(X_t \in B | \mathcal{F}_s) = P(X_t \in B | X_s)$ almost surely.*
2. *For all $g: E \to \mathbb{R}, \mathcal{B}\text{-}\mathcal{B}_{\mathbb{R}}$-measurable such that $g(X_t) \in L^1(P)$ for all t, for all $(s,t) \in \mathbb{R}_+^2, s < t: E[g(X_t) | \mathcal{F}_s] = E[g(X_t) | X_s]$ almost surely.*

Proof. The proof is left to the reader as an exercise. $\qquad\square$

Lemma 2.71. *If $(Y_k)_{k \in \mathbb{N}^*}$ is a sequence of real, independent random variables, then, putting*

$$X_n = \sum_{k=1}^{n} Y_k \qquad \forall n \in \mathbb{N}^*,$$

the new sequence $(X_n)_{n \in \mathbb{N}^*}$ is Markovian with respect to the family of σ-algebras $(\sigma(Y_1, \ldots, Y_n))_{n \in \mathbb{N}^*}$.

Proof. From the definition of X_k it is obvious that

$$\sigma(Y_1, \ldots, Y_n) = \sigma(X_1, \ldots, X_n) \qquad \forall n \in \mathbb{N}^*.$$

We thus first prove that, for all $C, D \in \mathcal{B}_{\mathbb{R}}$, for all $n \in \mathbb{N}^*$:

$$P(X_{n-1} \in C, Y_n \in D | Y_1, \ldots, Y_{n-1})$$
$$= P(X_{n-1} \in C, Y_n \in D | X_{n-1}) \qquad \text{a.s.}$$
(2.9)

To do this we fix $C, D \in \mathcal{B}_{\mathbb{R}}$ and $n \in \mathbb{N}^*$ and separately look at the left- and right-hand sides of (2.9). We get

$$P(X_{n-1} \in C, Y_n \in D | Y_1, \ldots, Y_{n-1}) = E[I_C(X_{n-1}) I_D(Y_n) | Y_1, \ldots, Y_{n-1}]$$
$$= I_C(X_{n-1}) E[I_D(Y_n) | Y_1, \ldots, Y_{n-1}] = I_C(X_{n-1}) E[I_D(Y_n)] \text{ a.s.,}$$

$$(2.10)$$

where the second equality of (2.10) holds because $I_C(X_{n-1})$ is $\sigma(Y_1, \ldots, Y_{n-1})$-measurable, and for the last one we use the fact that $I_D(Y_n)$ is independent of $Y_1, \ldots, Y_{n-1}$. On the other hand, we obtain that

$$P(X_{n-1} \in C, Y_n \in D | X_{n-1}) = E[I_C(X_{n-1}) I_D(Y_n) | X_{n-1}]$$
$$= I_C(X_{n-1}) E[I_D(Y_n)] \text{ a.s.} \qquad (2.11)$$

In fact, $I_C(X_{n-1})$ is $\sigma(X_{n-1})$-measurable and $I_D(Y_n)$ is independent of $X_{n-1} = \sum_{k=1}^{n-1} Y_k$. For (2.10) and (2.11), (2.9) follows and hence

$$P((X_{n-1}, Y_n) \in C \times D | Y_1, \ldots, Y_{n-1})$$
$$= P((X_{n-1}, Y_n) \in C \times D | X_{n-1}) \qquad \text{a.s.} \qquad (2.12)$$

As (2.12) holds for the rectangles of $\mathcal{B}_{\mathbb{R}^2}$ ($= \mathcal{B}_{\mathbb{R}} \otimes \mathcal{B}_{\mathbb{R}}$), by the measure extension theorem (e.g., Bauer 1981), it follows that (2.12) is also true for every $B \in \mathcal{B}_{\mathbb{R}^2}$. If now $A \in \mathcal{B}_{\mathbb{R}}$, then the two events

$$\{X_{n-1} + Y_n \in A\} = \{(X_{n-1}, Y_n) \in B\},$$

where $B \in \mathcal{B}_{\mathbb{R}^2}$ is the inverse image of A for a generic mapping $+ : \mathbb{R}^2 \to \mathbb{R}$ (which is continuous and hence measurable), are identical. Applying (2.12) to B, we obtain

$$P(X_{n-1} + Y_n \in A | Y_1, \ldots, Y_{n-1}) = P(X_{n-1} + Y_n \in A | X_{n-1}) \text{ a.s.,}$$

and thus

$$P(X_{n-1} + Y_n \in A | X_1, \ldots, X_{n-1}) = P(X_{n-1} + Y_n \in A | X_{n-1}) \text{ a.s.,}$$

and then

$$P(X_n \in A | X_1, \ldots, X_{n-1}) = P(X_n \in A | X_{n-1}) \text{ a.s.}$$

Therefore, $(X_n)_{n \in \mathbb{N}^*}$ is Markovian with respect to $(\sigma(X_1, \ldots, X_n))_{n \in \mathbb{N}^*}$ or, equivalently, with respect to $(\sigma(Y_1, \ldots, Y_n))_{n \in \mathbb{N}^*}$. $\qquad \square$

Proposition 2.72. *Let $(X_t)_{t \in \mathbb{R}_+}$ be a real stochastic process defined on the probability space $(\Omega, \mathcal{F}, P)$. The following two statements are true:*

1. *If $(X_t)_{t \in \mathbb{R}_+}$ is a Markov process, then so is $(X_t)_{t \in J}$ for all $J \subset \mathbb{R}_+$.*
2. *If for all $J \subset \mathbb{R}_+$, J finite: $(X_t)_{t \in J}$ is a Markov process, then so is $(X_t)_{t \in \mathbb{R}_+}$.*

Proof. See, e.g., Ash and Gardner (1975).   □

Theorem 2.73. *Every real stochastic process $(X_t)_{t \in \mathbb{R}_+}$ with independent increments is a Markov process.*

Proof. We define $(t_1, \ldots, t_n) \in \mathbb{R}_+^n$ such that $0 < t_1 < \cdots < t_n$ and $t_0 = 0$. If, for simplicity, we further suppose that $X_0 = 0$, then $X_{t_n} = \sum_{k=1}^{n} (X_{t_k} - X_{t_{k-1}})$. Putting $Y_k = X_{t_k} - X_{t_{k-1}}$, then, for all $k = 1, \ldots, n$, the Y_k are independent (because the process $(X_t)_{t \in \mathbb{R}_+}$ has independent increments) and we have that

$$X_{t_n} = \sum_{k=1}^{n} Y_k.$$

From Lemma 2.71 we can assert that

$$\forall B \in \mathcal{B}_{\mathbb{R}}: \qquad P(X_{t_n} \in B | X_{t_1}, \ldots, X_{t_{n-1}}) = P(X_{t_n} \in B | X_{t_{n-1}}) \text{ a.s.}$$

Thus $\forall J \subset \mathbb{R}_+, J$ finite, $(X_t)_{t \in J}$ is Markovian. The theorem then follows by point 2 of Proposition 2.72.   □

Proposition 2.74. *Let $(E, \mathcal{B}_E)$ be a Polish space endowed with the σ-algebra $\mathcal{B}_E$ of its Borel sets. For $t_0, T \in \mathbb{R}$, with $t_0 < T$, let $(X_t)_{t \in [t_0, T]}$ be an E-valued Markov process, with respect to its natural filtration.*
The function

$$(s, t) \in [t_0, T] \times [t_0, T], \ s \leq t; \ x \in E; \ A \in \mathcal{B}_E \mapsto$$
$$p(s, x; t, A) := P(X_t \in A | X_s = x) \in [0, 1] \qquad (2.13)$$

satisfies the following properties:

(i) *For all $(s, t) \in [t_0, T] \times [t_0, T]$, $s \leq t$, and for all $A \in \mathcal{B}_E$, the function $x \in E \mapsto p(s, x, t, A)$ is $\mathcal{B}_E - \mathcal{B}_{\mathbb{R}}$-measurable.*

(ii) *For all $(s, t) \in [t_0, T] \times [t_0, T]$, $s \leq t$, and for all $x \in E$, the function $A \in \mathcal{B}_E \mapsto p(s, x, t, A)$ is a probability measure on $\mathcal{B}_E$ such that*

$$p(s, x, s, A) = \begin{cases} 1, \text{ if } x \in A, \\ 0, \text{ if } x \notin A. \end{cases}$$

(iii) *The function p defined in (2.13) satisfies the so-called Chapman–Kolmogorov equation, i.e., for all $x \in E$, for all $(s, r, t) \in [t_0, T] \times [t_0, T] \times [t_0, T]$, $s \leq r \leq t$, and for all $A \in \mathcal{B}_E$*

$$p(s, x, t, A) = \int_{\mathbb{R}} p(s, x, r, dy) p(r, y, t, A) \text{ a.s.} \qquad (2.14)$$

Proof. The proofs of properties (i) and (ii) are trivial consequences of the definitions (e.g., Ash and Gardner 1975). On the other hand, the proof of (iii) is left to later analysis, after the introduction of the semigroup associated with the Markov process.   □

Definition 2.75. Any nonnegative function $p(s, x, t, A)$ defined for $t_0 \leq s \leq t \leq T, x \in E, A \in \mathcal{B}_E$ that satisfies conditions (i), (ii), and (iii) is called a *Markov transition probability function.*

Definition 2.76. If $(X_t)_{t \in [t_0, T]}$ is a Markov process, then the distribution P_0 of $X(t_0)$ is the *initial distribution* of the process.

Theorem 2.77. *An $(E, \mathcal{B}_E)$-valued process $(X_t)_{t \in [t_0, T]}$ is a Markov process, with transition probability function $p(r, x, s, A), t_0 \leq r < s \leq T, x \in E, A \in \mathcal{B}_E$ and initial distribution P_0, if and only if, for any $t_0 < t_1 < \cdots < t_k, k \in \mathbb{N}^*,$ and for any family $f_i, i = 0, 1, \ldots, k$ of nonnegative Borel measurable real-valued functions*

$$E \left[\prod_{i=0}^{k} f_i(X_{t_i}) \right] = \int_E P_0(dx_0) f_0(x_0) \int_E p(0, x_0, t_1, dx_1) f_1(x_1) \cdots$$

$$\cdots \int_E p(t_{k-1}, x_{k-1}, t_k, dx_k) f_k(x_k).$$

Proof. See, e.g., Revuz and Yor (1991, p. 76).   □

Theorem 2.78. *Let E be a Polish space endowed with the σ-algebra $\mathcal{B}_E$ of its Borel sets, P_0 a probability measure on $\mathcal{B}_E$, and $p(r, x, s, A), t_0 \leq r < s \leq T, x \in E, A \in \mathcal{B}_E$ a Markov transition probability function. Then there exists a unique (in the sense of equivalence) Markov process $(X_t)_{t \in [t_0, T]}$ valued in E, with P_0 as its initial distribution and p as its transition probability.*

Proof. See, e.g., Ash and Gardner (1975), Dynkin (1965), Applebaum (2004, p. 124), and Kallenberg (1997, p. 120).   □

Remark 2.79. From Theorem 2.78 we can deduce that

$$p(s, x, t, A) = P(X_t \in A | X_s = x), \text{a.s.} \qquad t_0 \leq s \leq t \leq T, x \in E, A \in \mathcal{B}_E.$$

Remark 2.80. It is of great importance to remark here that for Markov processes, the *compatibility condition* required by the Kolmogorov-Bochner Theorem 2.10 for the well posedness of a process, is due to the Chapman-Kolmogorov equation (2.14).

Semigroups Associated with Markov Transition Probability Functions

In this section we will consider the case $E = \mathbb{R}$ as a technical simplification.

Let $BC(\mathbb{R})$ be the space of all continuous and bounded functions on $\mathbb{R}$, endowed with the norm $\|f\| = \sup_{x \in \mathbb{R}} |f(x)| (< \infty)$, and let $p(s, x, t, A)$ be a transition probability function ($t_0 \leq s < t \leq T, x \in \mathbb{R}, A \in \mathcal{B}_{\mathbb{R}}$). We consider the operator

$$T_{s,t} : BC(\mathbb{R}) \to BC(\mathbb{R}), \qquad t_0 \leq s < t \leq T,$$

defined by assigning, for all $f \in BC(\mathbb{R})$,

$$(T_{s,t}f)(x) = \int_{\mathbb{R}} f(y) p(s, x, t, dy) = E[f(X(t))|X(s) = x].$$

Proposition 2.81. *The family $\{T_{s,t}\}_{t_0 \leq s \leq t \leq T}$ associated with the transition probability function $p(s, x, t, A)$ (or with its corresponding Markov process) is a semigroup of linear operators on $BC(\mathbb{R})$, i.e., it satisfies the following properties.*

1. *For any $t_0 \leq s \leq t \leq T$, $T_{s,t}$ is a linear operator on $BC(\mathbb{R})$.*
2. *For any $t_0 \leq s \leq T$, $T_{s,s} = I$ (the identity operator).*
3. *For any $t_0 \leq s \leq t \leq T$, $T_{s,t} 1 = 1$.*
4. *For any $t_0 \leq s \leq t \leq T$, $\|T_{s,t}\| \leq 1$ (contraction semigroup).*
5. *For any $t_0 \leq s \leq t \leq T$, and $f \in BC(\mathbb{R})$, $f \geq 0$ implies $T_{s,t}f \geq 0$.*
6. *For any $t_0 \leq r \leq s \leq t \leq T$, $T_{r,s}T_{s,t} = T_{r,t}$ (Chapman–Kolmogorov).*

Proof. All the preceding statements, apart from 4 and 6, are a direct consequence of the definitions that we are going to prove.

Proof of 4: Let $t_0 \leq s \leq t \leq T$, and $f \in BC(\mathbb{R})$;

$$\begin{aligned}
\|T_{s,t}f\| &= \sup_{x \in \mathbb{R}} |E(f(X(t))|X(s) = x)| \\
&\leq \sup_{x \in \mathbb{R}} E(|f(X(t))| \, |X(s) = x) \\
&\leq \sup_{x \in \mathbb{R}} |f(x)| \sup_{x \in \mathbb{R}} E(1|X(s) = x) \\
&= \|f\| 1 = \|f\|,
\end{aligned}$$

as stated. This fact lets us claim in particular that indeed $T_{s,t} : BC(\mathbb{R}) \to BC(\mathbb{R})$, for $t_0 \leq s < t \leq T$.

Proof of 6: Let $t_0 \leq r \leq s \leq t \leq T$, and $f \in BC(\mathbb{R})$; for any $x \in \mathbb{R}$

$$(T_{r,t}f)(x) = E[f(X(t))|X(r) = x]$$

$$\text{(by the tower property)} = E[E[f(X(t))|\mathcal{F}_s]|X(r) = x]$$

$$(\text{since } \mathcal{F}_r \subset \mathcal{F}_s) = E[E[f(X(t))|X(s)]|X(r) = x]$$
$$= E[(T_{s,t}f)(X(s))|X(r) = x]$$
$$= (T_{r,s}(T_{s,t}f))(x),$$

as stated. □

As the transition probability function $p(s, x, t, A)$ defines the semigroup $\{T_{s,t}\}_{t_0 \leq s \leq t \leq T}$ associated with it, conversely we may obtain the transition probability function from the semigroup, since we may easily recognize that

$$p(s, x, t, A) = P(X_t \in A|X_s = x) = (T_{s,t}I_A)(x) \text{ a.s.}$$

for $t_0 \leq s \leq t \leq T, x \in \mathbb{R}, A \in \mathcal{B}_{\mathbb{R}}$.

We may now finally prove the following proposition.

Proposition 2.82. *Let X be a real-valued Markov process, indexed in $\mathbb{R}$; the function p defined in (2.13) satisfies the so-called Chapman–Kolmogorov equation, i.e., for all $x \in \mathbb{R}$, for all $(s, r, t) \in [t_0, T] \times [t_0, T] \times [t_0, T]$, $s \leq r \leq t$, and for all $A \in \mathcal{B}_{\mathbb{R}}$*

$$p(s, x, t, A) = \int_{\mathbb{R}} p(s, x, r, dy)p(r, y, t, A) \text{ a.s.}$$

Proof. From definitions and Proposition 2.81 we easily obtain

$$p(s, x, t, A) = (T_{s,t}I_A)(x) = (T_{s,r}(T_{r,t}I_A))(x)$$
$$= \int_{\mathbb{R}} (T_{r,t}I_A)(y)p(s, x, r, dy)$$
$$= \int_{\mathbb{R}} p(s, x, r, dy)p(r, y, t, A) \text{ a.s.}$$

□

Definition 2.83. Let $(X_t)_{t \in \mathbb{R}_+}$ be a Markov process with transition probability function $p(s, x, t, A)$, and let $\{T_{s,t}\}$ $(s, t \in \mathbb{R}_+, s \leq t)$ be its associated semigroup. If, for all $f \in BC(\mathbb{R})$, the function

$$(t, x) \in \mathbb{R}_+ \times \mathbb{R} \to (T_{t,t+\lambda}f)(x) = \int_{\mathbb{R}} p(t, x, t + \lambda, dy)f(y) \in \mathbb{R}$$

is continuous for all $\lambda > 0$, then we say that the process satisfies the *Feller property*.

Theorem 2.84. *If* $(X_t)_{t \in \mathbb{R}_+}$ *is a Markov process with right-continuous trajectories satisfying the Feller property, then, for all* $t \in \mathbb{R}_+$, $\mathcal{F}_t = \mathcal{F}_{t+}$, *where* $\mathcal{F}_{t+} = \bigcap_{t' > t} \sigma(X(s), 0 \leq s \leq t')$, *and the filtration* $(\mathcal{F}_t)_{t \in \mathbb{R}_+}$ *is right-continuous.*

Proof. See, e.g., Friedman (1975).   □

Remark 2.85. It can be shown that $\mathcal{F}_{t+}$ is a σ-algebra.

Example 2.86. Examples of processes with the Feller property, or simply *Feller processes*, include Wiener processes (Brownian motions), Poisson processes, and all Lévy processes (see later sections).

Definition 2.87. If $(X_t)_{t \in \mathbb{R}_+}$ is a Markov process with transition probability function p and associated semigroup $\{T_{s,t}\}$, then the operator

$$\mathcal{A}_s f = \lim_{h \downarrow 0} \frac{T_{s,s+h}f - f}{h}, \qquad s \geq 0, f \in BC(\mathbb{R})$$

is called the *infinitesimal generator of the Markov process* $(X_t)_{t \geq 0}$. Its domain $\mathcal{D}_{\mathcal{A}_s}$ consists of all $f \in BC(\mathbb{R})$ for which the preceding limit exists uniformly (and therefore in the norm of $BC(\mathbb{R})$) (see, e.g., Feller 1971).

Remark 2.88. From the preceding definition we observe that

$$(\mathcal{A}_s f)(x) = \lim_{h \downarrow 0} \frac{1}{h} \int_{\mathbb{R}} [f(y) - f(x)] p(s, x, s + h, dy).$$

Remark 2.89. Up to this point we have been referring to the space $BC(\mathbb{R}^d)$ of bounded and continuous functions on $\mathbb{R}^d$. Actually, a more accurate analysis would require us to refer to its subspace $C_0(\mathbb{R}^d)$ of continuous functions, which tend to zero at infinity. This one is still a Banach space with the sup norm. In such a space it can be shown that a Feller semigroup is completely characterized by its infinitesimal generator (e.g., Kallenberg 1997, p. 317).

Examples of Stopping Times

Let $(\mathbf{X}_t)_{t \in \mathbb{R}_+}$ be a continuous Markov process taking values in $\mathbb{R}^v$, and suppose that the filtration $(\mathcal{F}_t)_{t \in \mathbb{R}_+}$, generated by the process, is right-continuous. Let $B \in \mathcal{B}_{\mathbb{R}^v} \setminus \{\emptyset\}$, and we define $T : \Omega \to \bar{\mathbb{R}}_+$ as

$$\forall \omega \in \Omega, \qquad T(\omega) = \begin{cases} \inf \{t \geq 0 | \mathbf{X}(t, \omega) \in B\} & \text{if the set is } \neq \emptyset, \\ +\infty & \text{if the set is } = \emptyset. \end{cases}$$

This gives rise to the following theorem.

Theorem 2.90. *If* B *is an open or closed subset of* $\mathbb{R}^v$, *then* T *is a stopping time.*

Proof. For B open, let $t \in \mathbb{R}_+$. In this case it can be shown that

$$\{T < t\} = \bigcup_{r < t, r \in \mathbb{Q}^+} \{\omega | \mathbf{X}(r, \omega) \in B\}.$$

Since $\mathbf{X}(r)$ is $\mathcal{F}$-measurable,

$$\{\omega | \mathbf{X}(r, \omega) \in B\} \in \mathcal{F}_r \subset \mathcal{F}_t \quad \forall r < t, r \in \mathbb{Q}^+,$$

and therefore the (countable) union of such events will be an element of $\mathcal{F}_t$ as well, and thus $\{T < t\} \in \mathcal{F}_t$. Now, fixing $\delta > 0$ and $N \in \mathbb{N}$ such that $\delta > \frac{1}{N}$, we have that

$$\forall n \in \mathbb{N}, n \geq N: \qquad \left\{ T < t + \frac{1}{n} \right\} \in \mathcal{F}_{t+\delta}.$$

Hence

$$\{T \leq t\} = \bigcap_{n=N}^{\infty} \left\{ T < t + \frac{1}{n} \right\} \in \mathcal{F}_{t+\delta}$$

and, due to the arbitrary choice of δ, this results in

$$\{T \leq t\} \in \bigcap_{\delta > 0} \mathcal{F}_{t+\delta} = \mathcal{F}_t^+ = \mathcal{F}_t.$$

For B closed, for all $n \in \mathbb{N}$, we define $V_n = \left\{ \mathbf{x} \in \mathbb{R}^v | d(\mathbf{x}, B) < \frac{1}{n} \right\}$ and

$$T_n = \begin{cases} \inf \{t \geq 0 | \mathbf{X}(t, \omega) \in V_n\} & \text{if the set is} \neq \emptyset, \\ +\infty & \text{if the set is} = \emptyset. \end{cases}$$

It can be shown that $B = \bigcap_{n \in \mathbb{N}} V_n$ and $\{T \leq t\} = \bigcap_{n \in \mathbb{N}} \{T_n < t\}$, and, since (with V_n open) $\{T_n < t\} \in \mathcal{F}_{t^+}$, we finally get that $\{T \leq t\} \in \mathcal{F}_{t^+} = \mathcal{F}_t$. $\square$

Definition 2.91. The stopping time T is the *first hitting time* of B or, equivalently, the *first exit time* from $\mathbb{R}^v \setminus B$.

Definition 2.92. A Markov process $(X_t)_{t \in \mathbb{R}_+}$ with transition probability function $p(s, x, t, A)$ is said to have the *strong Markov property* if, for any stopping time T of the process and for all $A \in \mathcal{B}_\mathbb{R}$,

$$P(X(T+t) \in A | \mathcal{F}_T) = p(T, X(T), T+t, A) \qquad \text{a.s.} \qquad (2.15)$$

Remark 2.93. Equation (2.15) is formally analogous to the Markov property

$$P(X(t) \in A | \mathcal{F}_s) = p(s, X(s), t, A) \text{ for } s < t,$$

with which it coincides when $T = s$ (constant).

Proposition 2.94. *Equation (2.15) is equivalent to the assertion that for all* $f : \mathbb{R} \to \mathbb{R}$, *measurable, bounded,*

$$E[f(X(T+t))|\mathcal{F}_T] = E[f(X(T+t))|X(T)] \qquad a.s.$$

Proof. See, e.g., Ash and Gardner (1975). $\square$

Remark 2.95. By Proposition 2.43 and Theorem 2.49, if $(X_t)_{t \in \mathbb{R}_+}$ is right-continuous and if T is a finite stopping time of the process, then $X(T)$ is $\mathcal{F}_T$-measurable.

Lemma 2.96. *Every Markov process* $(X_t)_{t \in \mathbb{R}_+}$ *that satisfies the Feller property has the strong Markov property, at least for a discrete stopping time* T.

Proof. Let T be a discrete stopping time of the process $(X_t)_{t \in \mathbb{R}_+}$ and $\{t_j\}_{j \in \mathbb{N}}$ its codomain. Fixing a $j \in \mathbb{N}$ we have $\{T \leq t_j\} \in \mathcal{F}_{t_j}$ and $\{T < t_j\} = \bigcup_{t_l < t_j} \{T \leq t_l\} \in \mathcal{F}_{t_j}$. Therefore,

$$G_j \equiv \{T = t_j\} = \{T \leq t_j\} \setminus \{T < t_j\} \in \mathcal{F}_{t_j}$$

and

$$\forall t \in \mathbb{R}_+, \qquad G_j \cap \{T \leq t\} = \begin{cases} \emptyset & \text{for } t_j > t, \\ G_j & \text{for } t \geq t_j. \end{cases}$$

From this we obtain, for all $t \in \mathbb{R}_+$, $G_j \cap \{T \leq t\} \in \mathcal{F}_t$, that is, $G_j \in \mathcal{F}_T$. Proving (2.15) is equivalent to showing that if $t \in \mathbb{R}_+, A \in \mathcal{B}_{\mathbb{R}}$, then:

1. $p(T, X(T), T + t, A)$ is $\mathcal{F}_T$-measurable.
2. For all $E \in \mathcal{F}_T$, $P((X(T + t) \in A) \cap E) = \int_E p(T, X(T), T + t, A) dP$.

Before proving point 1, we will show that 2 holds. Let $E \in \mathcal{F}_T$; then, by $\Omega = \bigcup_{j \in \mathbb{N}} G_j$, it follows that

$$P((X(T+t) \in A) \cap E) = \sum_{j \in \mathbb{N}} P((X(T+t) \in A) \cap E \cap G_j)$$

$$= \sum_{j \in \mathbb{N}} P((X(T+t) \in A) \cap E \cap (T = t_j))$$

$$= \sum_{j \in \mathbb{N}} P((X(t+t_j) \in A) \cap E \cap (T = t_j))$$

$$= \sum_{j \in \mathbb{N}} P((X(t+t_j) \in A) \cap E \cap G_j). \qquad (2.16)$$

But

$$E \cap G_j = E \cap (\{T \leq t_j\} \setminus \{T < t_j\}) \in \mathcal{F}_{t_j}$$

(in fact, $E \cap \{T \leq t_j\} \in \mathcal{F}_{t_j}$ following point 4 of Theorem 2.48), and therefore

$$P((X(t+t_j) \in A) \cap E \cap G_j) = \int_{E \cap G_j} P(X(t+t_j) \in A | \mathcal{F}_{t_j}) dP.$$

Moreover, by the Markov property,

$$P(X(t+t_j) \in A | \mathcal{F}_{t_j}) = p(t_j, X(t_j), t_j + t, A) \text{ a.s.} \qquad (2.17)$$

Using (2.16) and (2.17), we obtain

$$P((X(T+t) \in A) \cap E) = \bigcup_{j \in \mathbb{N}} \int_{E \cap G_j} p(t_j, X(t_j), t_j + t, A) dP$$

$$= \bigcup_{j \in \mathbb{N}} \int_{E \cap \{T=t_j\}} p(t_j, X(t_j), t_j + t, A) dP$$

$$= \bigcup_{j \in \mathbb{N}} \int_{E \cap \{T=t_j\}} p(T, X(T), T + t, A) dP$$

$$= \int_E p(T, X(T), T + t, A) dP.$$

For the proof of 1, we now observe that, by the Feller property, the mapping

$$(r, z) \in \mathbb{R}_+ \times \mathbb{R} \to \int_{\mathbb{R}} p(r, z, r + t, dy) f(y) \in \mathbb{R}$$

is continuous [for $f \in BC(\mathbb{R})$]. Furthermore, T and $X(T)$ are $\mathcal{F}_T$-measurable, and therefore the mapping

$$\omega \in \Omega \to (T(\omega), X(T(\omega), \omega))$$

is $\mathcal{F}_T$-measurable. Hence the composite of the two mappings

$$\omega \in \Omega \to \int_{\mathbb{R}} p(T, X(T), T + t, dy) f(y) \in \mathbb{R}$$

is $\mathcal{F}_T$-measurable [for $f \in BC(\mathbb{R})$]. Now let $(f_m)_{m \in \mathbb{N}} \in (BC(\mathbb{R}))^{\mathbb{N}}$ be a sequence of uniformly bounded functions such that $\lim_{m \to \infty} f_m = I_A$. Then, from our previous observations,

$$\forall m \in \mathbb{N}, \qquad \int_{\mathbb{R}} p(T, X(T), T + t, dy) f_m(y)$$

is $\mathcal{F}_T$-measurable and, following Lebesgue's theorem on integral limits, we get

$$\int_{\mathbb{R}} p(T, X(T), T + t, dy) I_A(y) = \lim_{m \to \infty} \int_{\mathbb{R}} p(T, X(T), T + t, dy) f_m(y),$$

and thus

$$p(T, X(T), T + t, A) = \int_{\mathbb{R}} p(T, X(T), T + t, dy) I_A(y)$$

is $\mathcal{F}_T$-measurable.   □

Before generalizing Lemma 2.96, we assert the following lemma.

Lemma 2.97. *If T is a stopping time of the stochastic process $(X_t)_{t \in \mathbb{R}_+}$, then there exists a sequence of stopping times $(T_n)_{n \in \mathbb{N}}$ such that*

1. *For all $n \in \mathbb{N}$, T_n has a codomain that is at most countable.*
2. *For all $n \in \mathbb{N}$, $T_n \geq T$.*
3. *$T_n \downarrow T$ almost surely for $n \to \infty$.*

Moreover, $\{T_n = \infty\} = \{T = \infty\}$ for every n.

Proof. See, e.g., Friedman (1975).   □

Theorem 2.98. *If $(X_t)_{t \in \mathbb{R}_+}$ is a right-continuous Markov process that satisfies the Feller property, then it satisfies the strong Markov property.*

Proof. Let T be a finite stopping time of the process $(X_t)_{t \in \mathbb{R}_+}$ and $(T_n)_{n \in \mathbb{N}}$ a sequence of stopping times satisfying properties 1–3 of Lemma 2.97 with respect to T. We observe that, for all $n \in \mathbb{N}$, $\mathcal{F}_T \subset \mathcal{F}_{T_n}$. In fact, if $A \in \mathcal{F}_T$, then

$$\forall t \in \mathbb{R}_+, \qquad A \cap \{T_n \leq t\} = (A \cap \{T \leq t\}) \cap \{T_n \leq t\} \in \mathcal{F}_t,$$

provided that $A \cap \{T \leq t\} \in \mathcal{F}_t, \{T_n \leq t\} \in \mathcal{F}_t$. Just like for Lemma 2.96, we will need to show that points 1 and 2 of its proof hold in this present case. Following Proposition 2.94, point 2 is equivalent to asserting that for all $E \in \mathcal{F}_T$ and all $f \in BC(\mathbb{R})$:

$$\int_E f(X(T + t)) dP = \int_E dP \int_{\mathbb{R}} p(T, X(T), T + t, dy) f(y). \qquad (2.18)$$

Then, by Proposition 2.43, for all $n \in \mathbb{N}$, we have that for all $E \in \mathcal{F}_{T_n}$ and all $f \in BC(\mathbb{R})$

$$\int_E f(X(T_n + t)) dP = \int_E dP \int_{\mathbb{R}} p(T_n, X(T_n), T_n + t, dy) f(y).$$

Moreover, since $T_n \downarrow T$ for $n \to \infty$ and by the right-continuity of process X, it follows that

$$X(T_n) \to X(T) \text{ for } n \to \infty.$$

From the continuity[4] of the mapping

$$(\lambda, x) \in \mathbb{R}_+ \times \mathbb{R} \to \int_{\mathbb{R}} p(\lambda, x, \lambda + t, \lambda y) f(y) \text{ for } f \in BC(\mathbb{R}),$$

we have that, for $n \to \infty$,

$$\int_{\mathbb{R}} p(T_n, X(T_n), T_n + t, dy) f(y) \to \int_{\mathbb{R}} p(T, X(T), T + t, dy) f(y). \qquad (2.19)$$

On the other hand, if f is continuous, then we also get

$$f(X(T_n + t)) \to f(X(T + t)) \text{ for } n \to \infty. \qquad (2.20)$$

Therefore, if $E \in \mathcal{F}_T$ and $f \in BC(\mathbb{R})$, then $E \in \mathcal{F}_{T_n}$ for all n, and we have

$$\lim_{n \to \infty} \int_E f(X(T_n + t)) dP = \lim_{n \to \infty} \int_E dP \int_{\mathbb{R}} p(T_n, X(T_n), T_n + t, dy) f(y).$$

Since f and p are bounded, following Lebesgue's theorem, we can take the limit of the integral and then (2.18) follows from (2.19) and (2.20). The proof of point 1 is entirely analogous to the proof of Lemma 2.97. $\qquad \square$

The preceding results may be extended to more general, possibly uncountable, state spaces. In particular, we will assume that E is a subset of $\mathbb{R}^d$ for $d \in \mathbb{N}^*$.

Time-Homogeneous Markov Processes

An important class of Markov processes is the time-homogeneous case.

Definition 2.99. A Markov process $(X_t)_{t \in [t_0, T]}$ is said to be *time-homogeneous* if the transition probability functions $p(s, x, t, A)$ depend on t and s only through their difference $t - s$. Therefore, for all $(s, t) \in [t_0, T]^2$, $s < t$, for all $u \in [0, T - t]$, for all $A \in \mathcal{B}_{\mathbb{R}}$, and for all $x \in \mathbb{R}$:

$$p(s, x, t, A) = p(s + u, x, t + u, A) \qquad \text{a.s.}$$

Remark 2.100. If $(X_t)_{t \in [t_0, T]}$ is a homogeneous Markov process with transition probability function p, then, for all $(s, t) \in [t_0, T]^2$, $s < t$, for all $A \in \mathcal{B}_{\mathbb{R}}$, and for all $x \in \mathbb{R}$, we obtain

$$p(t_0, x, t_0 + t - s, A) = p(s, x, t, A) \text{ a.s.},$$

where $p(t_0, x, t_0 + t - s, A)$ is denoted by $p(\bar{t}, x, A)$, with $\bar{t} = (t - s) \in [0, T - t_0]$, $x \in \mathbb{R}, A \in \mathcal{B}_{\mathbb{R}}$.

---

[4] By the Feller property.

If we consider the time-homogeneous case, a Markov process $(X_t)_{t \in \mathbb{R}_+}$ on $(E, \mathcal{B}_E)$ will be defined in terms of a transition kernel $p(t, x, B)$ for $t \in \mathbb{R}_+$, $x \in E$, $B \in \mathcal{B}_E$, such that

$$p(h, X_t, B) = P(X_{t+h} \in B | \mathcal{F}_t) \qquad \forall t, h \in \mathbb{R}_+, B \in \mathcal{B}_E,$$

given that $(\mathcal{F}_t)_{t \in \mathbb{R}_+}$ is the natural filtration of the process. Equivalently, if we denote by $BC(E)$ the Banach space of all continuous and bounded functions on E, endowed with the *sup norm*, then

$$E[g(X_{t+h})|\mathcal{F}_t] = \int_E g(y) p(h, X_t, dy) \qquad \forall t, h \in \mathbb{R}_+, g \in BC(E).$$

In this case the transition semigroup of the process is such that

$$T_{s,t} = T_{0,t-s} =: T(t-s)$$

for any $s, t \in \mathbb{R}_+$, $s \leq t$, which defines a one-parameter contraction semigroup $(T(t), t \in \mathbb{R}_+)$ on $BC(E)$; it is then such that

$$T(t)g(x) := \int_E g(y) p(t, x, dy) = E[g(X_t)|X_0 = x], \qquad x \in E,$$

for any $g \in BC(E)$.

Hence in this case the semigroup property reduces to

$$T(s+t) = T(s)T(t) = T(t)T(s)$$

for any $s, t \in \mathbb{R}_+$.

Up to now we have referred to $BC(\mathbb{R})$, i.e., the family of bounded and continuous functions on $\mathbb{R}$. For various reasons, as the reader may see later, it is more convenient to refer to its Banach subspace $C_0(\mathbb{R})$, the family of continuous functions vanishing at infinity, since it has nicer analytical properties.

Definition 2.101. Let $(T(t))_{t \in \mathbb{R}_+}$ be the transition semigroup associated with a time-homogeneous Markov process $X = (X_t)_{t \in \mathbb{R}_+}$. We say that X is a *Feller process* if the following statements hold:

(i) $T(t)(C_0(\mathbb{R})) \subset C_0(\mathbb{R})$ for all $t \in \mathbb{R}_+$.
(ii) $\lim_{t \to 0} \|T(t)f - f\| = 0$ for all $f \in C_0(\mathbb{R})$.

In this case we say that the semigroup $(T(t))_{t \in \mathbb{R}_+}$ is a *Feller semigroup*.

Proposition 2.102. *For any Feller semigroup on $C_0(\mathbb{R})$ there exists a unique time-homogeneous transition probability measure $p(t, x, B)$ on $(\mathbb{R}, \mathcal{B}_\mathbb{R})$ such that, for all $f \in C_0(\mathbb{R})$,*

$$T(t)f(x) = \int_\mathbb{R} f(y) p(t, x, dy), \qquad x \in \mathbb{R}, \quad t \in \mathbb{R}_+.$$

Proof. See, e.g., Revuz-Yor (1991, p. 83).   □

Definition 2.103. A time-homogeneous transition probability measure associated with a Feller semigroup is called a *Feller transition function*.

For time-homogeneous Markov processes the infinitesimal generator will be time independent. It is defined as

$$\mathcal{A}g = \lim_{t \to 0+} \frac{1}{t}(T(t)g - g)$$

for $g \in \mathcal{D}(\mathcal{A})$, the subset of $BC(E)$ for which the preceding limit exists, in $BC(E)$, with respect to the sup norm. Given the preceding definitions, it is obvious that for all $g \in \mathcal{D}(\mathcal{A})$,

$$\mathcal{A}g(x) = \lim_{t \to 0+} \frac{1}{t}E[g(X_t) - g(X_0)|X_0 = x], \qquad x \in E.$$

If $(T(t), t \in \mathbb{R}_+)$ is the contraction semigroup associated with a Markov process, it is not difficult to show that the mapping $t \to T(t)g$ is right-continuous in $t \in \mathbb{R}_+$ provided that $g \in BC(E)$ is such that the mapping $t \to T(t)g$ is right-continuous in $t = 0$.

The following properties hold, by considering Riemann integrals and strong derivatives (Applebaum 2004, p. 129).

1. For any $t \geq 0$: $T(t)\mathcal{D}(\mathcal{A}) \subset \mathcal{D}(\mathcal{A})$.
2. For any $t \geq 0$ and for any $g \in \mathcal{D}(\mathcal{A})$: $T(t)\mathcal{A}g = \mathcal{A}T(t)g$.
3. For any $t \geq 0$ and for any $g \in \mathcal{D}(\mathcal{A})$: $\int_0^t T(s)gds \in \mathcal{D}(\mathcal{A})$.
4. For any $t \geq 0$ and for any $g \in \mathcal{D}(\mathcal{A})$:

$$T(t)g - g = \mathcal{A}\int_0^t T(s)gds = \int_0^t \mathcal{A}T(s)gds = \int_0^t T(s)\mathcal{A}gds.$$

5. For any $t \geq 0$ and for any $g \in \mathcal{D}(\mathcal{A})$:

$$\frac{d}{dt}[T(t)g] = \mathcal{A}[T(t)g] = T(t)[\mathcal{A}g].$$

6. For any $g \in \mathcal{D}(\mathcal{A})$, the function $t \in \mathbb{R}_+ \mapsto T(t)g \equiv u(t) \in \mathcal{D}(\mathcal{A})$ is a solution of the following initial value problem in the Banach space $BC(\mathbb{R}^d)$:

$$\begin{cases} \dfrac{d}{dt}u(t) = \mathcal{A}u(t), \\ u(0) \quad = g. \end{cases}$$

These results justify the notation $T(t) = e^{t\mathcal{A}}$.

Holding Times for a Markov Process

Suppose that a Markov process $(X_t)_{t \in \mathbb{R}_+}$ on $\mathbb{R}$ starts at a point x. We wish to evaluate the probability distribution of the *holding time* at x, i.e., the time it spends in that state before leaving it:

$$\tau_x := \inf \{t \in \mathbb{R}_+ \mid X_s = x, \ X_{t+s} \neq x\}$$

for any given $s \in \mathbb{R}_+$.

The following proposition holds.

Proposition 2.104. *For any right-continuous time-homogeneous Markov process, we have*

$$F_x(t) = P(\tau_x \leq t) = 1 - \exp\{-c_x t\}, \quad t \in \mathbb{R}_+,$$

for some $c_x \in [0, +\infty]$.

Proof. See, e.g., Lamperti (1977, p. 195).   □

Markov Diffusion Processes

Definition 2.105. A Markov process on $\mathbb{R}$ with transition probability function $p(s, x, t, A)$ is called a *diffusion process* if

1. It has a.s. continuous trajectories.
2. For all $\epsilon > 0$, for all $t \geq 0$, and for all $x \in \mathbb{R}$: $\lim_{h \downarrow 0} \frac{1}{h} \int_{|x-y|>\epsilon} p(t, x, t + h, dy) = 0$.
3. There exist $a(t, x)$ and $b(t, x)$ such that, for all $\epsilon > 0$, for all $t \geq 0$, and for all $x \in \mathbb{R}$,

$$\lim_{h \downarrow 0} \frac{1}{h} \int_{|x-y|<\epsilon} (y - x) p(t, x, t + h, dy) = a(t, x),$$

$$\lim_{h \downarrow 0} \frac{1}{h} \int_{|x-y|<\epsilon} (y - x)^2 p(t, x, t + h, dy) = b(t, x),$$

where $a(t, x)$ is the *drift coefficient* and $b(t, x)$ the *diffusion coefficient* of the process.

Lemma 2.106. *Conditions 1 and 2 of Definition 2.105 are satisfied if*

1.* *There exists a $\delta > 0$ such that, for all $t \geq 0$ and for all $x \in \mathbb{R}$,*
   $\lim_{h \downarrow 0} \frac{1}{h} \int_{\mathbb{R}} |x - y|^{2+\delta} p(t, x, t + h, dy) = 0$.
2.* *For all $t \geq 0$ and for all $x \in \mathbb{R}$,*

$$\lim_{h \downarrow 0} \frac{1}{h} \int_{\mathbb{R}} (y - x) p(t, x, t + h, dy) = a(t, x),$$

$$\lim_{h \downarrow 0} \frac{1}{h} \int_{\mathbb{R}} (y - x)^2 p(t, x, t + h, dy) = b(t, x).$$

Proof. We fix $\epsilon > 0$, $x \in \mathbb{R}$, $|x - y| > \epsilon \Rightarrow \dfrac{|y - x|^{2+\delta}}{\epsilon^{2+\delta}} \geq 1$, and hence

$$\frac{1}{h} \int_{|x-y|>\epsilon} p(t, x, t + h, dy) \leq \frac{1}{h\epsilon^{2+\delta}} \int_{|x-y|>\epsilon} |y - x|^{2+\delta} p(t, x, t + h, dy)$$

$$\leq \frac{1}{h\epsilon^{2+\delta}} \int_{\mathbb{R}} |y - x|^{2+\delta} p(t, x, t + h, dy).$$

From this, due to 1*, point 1 of Definition 2.105 follows. Analogously, for $j = 1, 2$,

$$\frac{1}{h} \int_{|x-y|>\epsilon} |y - x|^j p(t, x, t + h, dy) \leq \frac{1}{h\epsilon^{2+\delta-j}} \int_{\mathbb{R}} |y - x|^{2+\delta} p(t, x, t + h, dy),$$

and again from 1* we obtain

$$\lim_{h\downarrow 0} \frac{1}{h} \int_{|x-y|>\epsilon} |y - x|^j p(t, x, t + h, dy) = 0.$$

Moreover,

$$\lim_{h\downarrow 0} \frac{1}{h} \int_{\mathbb{R}} |y - x|^j p(t, x, t + h, dy) = \lim_{h\downarrow 0} \frac{1}{h} \left(\int_{|x-y|>\epsilon} |y - x|^j p(t, x, t + h, dy) \right.$$

$$\left. + \int_{|x-y|<\epsilon} |y - x|^j p(t, x, t + h, dy) \right),$$

which, along with 2*, gives point 2 of Definition 2.105. $\qquad\square$

Proposition 2.107. *If $(X_t)_{t\in\mathbb{R}_+}$ is a diffusion process with transition probability function p and drift and diffusion coefficients $a(x,t)$ and $b(x,t)$, respectively, and if $\mathcal{A}_s$ is the infinitesimal generator associated with p, then we have that*

$$(\mathcal{A}_s f)(x) = a(s, x)\frac{\partial f}{\partial x} + \frac{1}{2}b(s, x)\frac{\partial^2 f}{\partial x^2},$$

provided that f is bounded and twice continuously differentiable.

Proof. Let $f \in BC(\mathbb{R}) \cap C^2(\mathbb{R})$. From Taylor's formula we obtain

$$f(y) - f(x) = f'(x)(y - x) + \frac{1}{2}f''(x)(y - x)^2 + o(|y - x|^2) \qquad (2.21)$$

for $|y - x| < \delta$ (which is in a suitable neighborhood of x), and thus

$$(\mathcal{A}_s f)(x) = \lim_{h\downarrow 0} \frac{1}{h} \int_{\mathbb{R}} [f(y) - f(x)] p(s, x, s + h, dy)$$

$$= \lim_{h\downarrow 0} \frac{1}{h} \int_{|y-x|<\delta} f'(x)(y - x) p(s, x, s + h, dy)$$

$$+\frac{1}{2}\lim_{h\downarrow 0}\frac{1}{h}\int_{|y-x|<\delta} f''(x)(y-x)^2 p(s,x,s+h,dy)$$

$$+\lim_{h\downarrow 0}\frac{1}{h}\int_{|y-x|<\delta} o(|y-x|^2)p(s,x,s+h,dy)$$

$$+\lim_{h\downarrow 0}\frac{1}{h}\int_{|y-x|\geq\delta} [f(y)-f(x)]p(s,x,s+h,dy).$$

Because $f \in BC(\mathbb{R})$,

$$\lim_{h\downarrow 0}\frac{1}{h}\int_{|y-x|\geq\delta} [f(y)-f(x)]p(s,x,s+h,dy)$$

$$\leq \lim_{h\downarrow 0}\frac{1}{h}c\int_{|y-x|\geq\delta} p(s,x,s+h,dy) = 0,$$

by point 1 of Definition 2.105, where c is a constant. By point 2 of the same definition:

$$\lim_{h\downarrow 0}\frac{1}{h}\int_{|y-x|<\delta} f'(x)(y-x)p(s,x,s+h,dy)$$

$$= f'(x)\lim_{h\downarrow 0}\frac{1}{h}\int_{|y-x|<\delta} (y-x)p(s,x,s+h,dy)$$

$$= f'(x)a(t,x),$$

as well as

$$\frac{1}{2}\lim_{h\downarrow 0}\frac{1}{h}\int_{|y-x|<\delta} f''(x)(y-x)^2 p(s,x,s+h,dy) = \frac{1}{2}f''(x)b(x,t).$$

Fixing $\epsilon > 0$, we finally observe that if we choose δ such that Taylor's formula (2.21) holds, so that

$$|y-x| < \delta \Rightarrow \frac{o(|y-x|^2)}{|y-x|^2} < \epsilon,$$

we get

$$\lim_{h\downarrow 0}\frac{1}{h}\int_{|y-x|<\delta} o(|y-x|^2)p(s,x,s+h,dy)$$

$$\leq \lim_{h\downarrow 0}\frac{1}{h}\int_{|y-x|<\delta} \epsilon|y-x|^2 p(s,x,s+h,dy)$$

$$= \epsilon b(t,x)$$

and, from the fact that ϵ is arbitrary, we conclude that

$$\lim_{h\downarrow 0}\frac{1}{h}\int_{|y-x|<\delta} o(|y-x|^2)p(s,x,s+h,dy) = 0.$$

$\square$

A detailed account of conditions that ensure a.s. path continuity of the trajectories of Markov processes can be found in Lamperti (1977, p. 188).

Markov Jump Processes

Consider a Markov process $(X_t)_{t \in \mathbb{R}_+}$ valued in a countable set E (say, $\mathbb{N}$ or $\mathbb{Z}$). In such a case it is sufficient (with respect to Theorem 2.78) to provide the so-called *one-point* transition probability function

$$p_{ij}(s,t) := p(s,i,t,j) := P(X_t = j | X_s = i)$$

for $t_0 \leq s < t$, $i, j \in E$. It follows from the general structure of Markov processes that the one-point transition probabilities satisfy the following relations:

(a) $p_{ij}(s,t) \geq 0$
(b) $\sum_{j \in E} p_{ij}(s,t) = 1$
(c) $p_{ij}(s,t) = \sum_{k \in E} p_{ik}(s,r) p_{kj}(r,t)$

provided $t_0 \leq s \leq r \leq t$, in $\mathbb{R}_+$, and $i, j \in E$. To these three conditions we need to add

(d)

$$\lim_{t \to s+} p_{ij}(s,t) = p_{ij}(s,s) = \delta_{ij} = \begin{cases} 1 & \text{for } i = j, \\ 0 & \text{for } i \neq j. \end{cases}$$

The time-homogeneous case gives transition probabilities $(\tilde{p}_{ij}(t))_{t \in \mathbb{R}_+}$ such that

$$p_{ij}(s,t) = \tilde{p}_{ij}(t-s), \qquad s \leq t.$$

Henceforth we shall limit our analysis to the time-homogeneous case whose transition probabilities will be denoted by $(p_{ij}(t))_{t \in \mathbb{R}_+}$. The following theorems hold (Gihman and Skorohod 1974, pp. 304–306).

Theorem 2.108. *The transition probabilities* $(p_{ij}(t))_{t \in \mathbb{R}_+}$ *of a homogeneous Markov process on a countable state space E are uniformly continuous in $t \in \mathbb{R}_+$ for any fixed $i, j \in E$.*

Theorem 2.109. *The limit*

$$q_i = \lim_{h \to 0+} \frac{1 - p_{ii}(h)}{h} \leq +\infty$$

always exists (finite or not), and for arbitrary $t > 0$:

$$\frac{1 - p_{ii}(t)}{t} \leq q_i.$$

If $q_i < +\infty$, then for all $t > 0$ the derivatives $p'_{ij}(t)$ exist for any $i, j \in E$ and are continuous. They satisfy the following relations:

1. $p'_{ij}(t+s) = \sum_{k \in E} p'_{ik}(t) p_{kj}(s)$
2. $\sum_{j \in E} p'_{ij}(t) = 0$
3. $\sum_{j \in E} |p'_{ij}(t)| \leq 2q_i$

In the following theorem the condition $q_i < +\infty$ is not required.

Theorem 2.110. *The limits*

$$\lim_{t \to 0+} \frac{p_{ij}(t)}{t} = p'_{ij}(0) =: q_{ij} < +\infty$$

always exist (finite) for any $i \neq j$.

As a consequence of Theorems 2.109 and 2.110, provided $q_i < +\infty$, we obtain evolution equations for $p_{ij}(t)$:

$$p'_{ij}(t) = \sum_{k \in E} q_{ik} p_{kj}(t),$$

with $q_{ii} = -q_i$. These equations are known as *Kolmogorov backward equations*. Consider the family of matrices $(P(t))_{t \in \mathbb{R}_+}$, with entries $(p_{ij}(t))_{t \in \mathbb{R}_+}$, for $i, j \in E$. We may rewrite conditions (c) and (d) in matrix form as follows:

(c') $P(s + t) = P(s)P(t)$ for any $s, t \geq 0$
(d') $\lim_{h \to 0+} P(h) = P(0) = I$

A family of stochastic matrices fulfilling conditions (c') and (d') is called a *matrix transition function*. If the matrix Q satisfies the condition

$$\sum_{j \neq i} q_{ij} = -q_{ii} \equiv q_i < +\infty$$

for any $i \in E$, it is called *conservative*. The matrix $Q = (q_{ij})_{i,j \in E}$ is called the *intensity matrix*. The Kolmogorov backward equations can be rewritten in matrix form as

$$P'(t) = QP(t), \qquad t > 0,$$

subject to

$$P(0) = I.$$

If Q is a finite-dimensional matrix, then the function $exp\{tQ\}$ for $t > 0$ is well defined.

Theorem 2.111 (Karlin and Taylor 1975, p. 152). *If E is finite, then the matrix transition function can be represented in terms of its intensity matrix Q via*

$$P(t) = e^{tQ}, \qquad t \geq 0.$$

Given an intensity matrix Q of a conservative Markov jump process with stationary (time-homogeneous) transition probabilities, we have that (Doob 1953)

$$P(X_u = i \; \forall u \in]s, s + t] | X_s = i) = e^{-q_i t}$$

for every $s, t \in \mathbb{R}_+$, and state $i \in E$. This shows that the sojourn time in state i is exponentially distributed with parameter q_i. This is independent of the initial time $s \geq 0$.

Furthermore, let π_{ij}, $i \neq j$, be the conditional probability of a jump to state j, given that a jump from state i has occurred. It can be shown (Doob 1953) that

$$\pi_{ij} = \frac{q_{ij}}{q_i},$$

provided that $q_i > 0$. For $q_i = 0$, state i is *absorbing*, which obviously means that once state i is entered, the process remains there permanently. Indeed,

$$P(X_u = i, \text{ for all } u \in]s, s+t] \mid X_s = i) = e^{-q_i t} = 1$$

for all $t \geq 0$. A state i for which $q_i = +\infty$ is called an *instantaneous state*. The expected sojourn time in such a state is zero. A state i for which $0 \leq q_i < +\infty$ is called a *stable state*.

Example 2.112. If $(X_t)_{t \in \mathbb{R}_+}$ is a homogeneous Poisson process with intensity $\lambda > 0$, then

$$p_{ij}(t) = \begin{cases} e^{-\lambda t} \frac{(\lambda t)^{j-i}}{(j-i)!} & \text{for } j > i, \\ 0 & \text{otherwise.} \end{cases}$$

This implies that

$$q_{ij} = p'_{ij}(0) \begin{cases} \lambda & \text{for } j = i+1, \\ -\lambda & \text{for } j = i, \\ 0 & \text{otherwise.} \end{cases}$$

For the following result, we refer again to Doob (1953).

Theorem 2.113. *For any $x \in E$ there exists a unique RCLL Markov process associated with a given intensity matrix Q and such that $P(X(0) = x) = 1$.*

Further discussions on this topic may be found in Doob (1953) and Karlin and Taylor (1981) [an additional and updated source regarding discrete-space continuous-time Markov chains is Anderson (1991)]. For applications, see, for example, Robert (2003).

2.7 Martingales

Extension of the concept of continuous-time martingales is mainly due to P.A. Meyer and his coworkers (Meyer (1966)).

Definition 2.114. Let $(X_t)_{t \in \mathbb{R}_+}$ be a real-valued family of random variables defined on the probability space $(\Omega, \mathcal{F}, P)$, and let $(\mathcal{F}_t)_{t \in \mathbb{R}_+}$ be a filtration.

The stochastic process $(X_t)_{t \in \mathbb{R}_+}$ is said to be *adapted* to the family $(\mathcal{F}_t)_{t \in \mathbb{R}_+}$ if, for all $t \in \mathbb{R}_+$, X_t is $\mathcal{F}_t$-measurable.

Definition 2.115. The stochastic process $(X_t)_{t \in \mathbb{R}_+}$, adapted to the filtration $(\mathcal{F}_t)_{t \in \mathbb{R}_+}$, is a *martingale* with respect to this filtration, provided the following conditions hold:

1. X_t is P-integrable for all $t \in \mathbb{R}_+$.
2. For all $(s, t) \in \mathbb{R}_+ \times \mathbb{R}_+, s < t : E[X_t | \mathcal{F}_s] = X_s$ almost surely.

$(X_t)_{t \in \mathbb{R}_+}$ is said to be a *submartingale* (*supermartingale*) with respect to $(\mathcal{F}_t)_{t \in \mathbb{R}_+}$ if, in addition to condition 1 and instead of condition 2, we have:

2'. For all $(s, t) \in \mathbb{R}_+ \times \mathbb{R}_+, s < t : E[X_t | \mathcal{F}_s] \geq X_s$ $(E[X_t | \mathcal{F}_s] \leq X_s)$ almost surely.

Remark 2.116. When the filtration $(\mathcal{F}_t)_{t \in \mathbb{R}_+}$ is not specified, it is understood to be the increasing σ-algebra generated by the random variables of the process $(\sigma(X_s, 0 \leq s \leq t))_{t \in \mathbb{R}_+}$. In this case we can write $E[X_t | X_r, 0 \leq r \leq s]$, instead of $E[X_t | \mathcal{F}_s]$.

Example 2.117. The evolution of a gambler's wealth in a game of chance, the latter specified by the sequence of real-valued random variables $(X_n)_{n \in \mathbb{N}}$, will serve as a descriptive example of the preceding definitions. Suppose that two players flip a coin and the loser pays the winner (who guessed head or tail correctly) the amount α after every round. If $(X_n)_{n \in \mathbb{N}}$ represents the cumulative fortune of player 1, then after n throws he holds

$$X_n = \sum_{i=0}^{n} \Delta_i.$$

The random variables Δ_i (just like every flip of the coin) are independent and take values α and $-\alpha$ with probabilities p and q, respectively. Therefore, we see that

$$E[X_{n+1} | X_0, \dots, X_n] = E[\Delta_{n+1} + X_n | X_0, \dots, X_n]$$
$$= X_n + E[\Delta_{n+1} | X_0, \dots, X_n].$$

Since Δ_{n+1} is independent of every $\sum_{i=0}^{k} \Delta_i, k = 0, \dots, n$, we obtain

$$E[X_{n+1} | X_0, \dots, X_n] = X_n + E[\Delta_{n+1}] = X_n + \alpha(p - q).$$

- If the game is fair, then $p = q$ and $(X_n)_{n \in \mathbb{N}}$ is a martingale.
- If the game is in player 1's favor, then $p > q$ and $(X_n)_{n \in \mathbb{N}}$ is a submartingale.
- If the game is to the disadvantage of player 1, then $p < q$ and $(X_n)_{n \in \mathbb{N}}$ is a supermartingale.

Example 2.118. Let $(X_t)_{t\in\mathbb{R}_+}$ be (for all $t \in \mathbb{R}_+$) a P-integrable stochastic process on $(\Omega, \mathcal{F}, P)$ with independent increments. Then $(X_t - E[X_t])_{t\in\mathbb{R}_+}$ is a martingale. In fact[5]:

$$E[X_t|\mathcal{F}_s] = E[X_t - X_s|\mathcal{F}_s] + E[X_s|\mathcal{F}_s], \qquad s < t,$$

and recalling that X_s is $\mathcal{F}_s$-measurable and that $(X_t - X_s)$ is independent of $\mathcal{F}_s$, we obtain that

$$E[X_t|\mathcal{F}_s] = E[X_t - X_s] + X_s = X_s, \qquad s < t.$$

Proposition 2.119. *Let $(X_t)_{t\in\mathbb{R}_+}$ be a real-valued martingale. If the function $\phi : \mathbb{R} \to \mathbb{R}$ is both convex and measurable and such that*

$$\forall t \in \mathbb{R}_+: \qquad E[|\phi(X_t)|] < +\infty,$$

then $(\phi(X_t))_{t\in\mathbb{R}_+}$ is a submartingale.

Proof. Let $(s,t) \in \mathbb{R}_+ \times \mathbb{R}_+, s < t$. Following Jensen's inequality and the properties of the martingale $(X_t)_{t\in\mathbb{R}_+}$, we have that

$$\phi(X_s) = \phi(E[X_t|\mathcal{F}_s]) \le E[\phi(X_t)|\mathcal{F}_s].$$

Letting

$$\mathcal{V}_s = \sigma(\phi(X_r), 0 \le r \le s) \qquad \forall s \in \mathbb{R}_+$$

and with the measurability of ϕ, it is easy to verify that $\mathcal{V}_s \subset \mathcal{F}_s$ for all $s \in \mathbb{R}_+$, and therefore

$$\phi(X_s) = E[\phi(X_s)|\mathcal{V}_s] \le E[E[\phi(X_t)|\mathcal{F}_s]|\mathcal{V}_s] = E[\phi(X_t)|\mathcal{V}_s].$$

$\square$

Lemma 2.120. *Let X and Y be two positive real random variables defined on $(\Omega, \mathcal{F}, P)$. If $X \in L^p(P)$ $(p > 1)$ and if, for all $\alpha > 0$,*

$$\alpha P(Y \ge \alpha) \le \int_{\{Y \ge \alpha\}} X dP, \qquad (2.22)$$

then $Y \in L^p(P)$ and $\|Y\|_p \le q\|X\|_p$, where $\frac{1}{p} + \frac{1}{q} = 1$.

---

[5] For simplicity, but without loss of generality, we will assume that $E[X_t] = 0$, for all t. In the case where $E[X_t] \ne 0$, we can always define a variable $Y_t = X_t - E[X_t]$, so that $E[Y_t] = 0$. In that case $(Y_t)_{t\in\mathbb{R}_+}$ will again be a process with independent increments, so that the analysis is analogous.

Proof. We have

$$
E[Y^p] = \int_\Omega Y^p(\omega) dP(\omega) = \int_\Omega dP(\omega) p \int_0^{Y(\omega)} \lambda^{p-1} d\lambda
$$

$$
= p \int_\Omega dP(\omega) \int_0^\infty \lambda^{p-1} I_{\{\lambda \leq Y(\omega)\}}(\lambda) d\lambda
$$

$$
= p \int_0^\infty d\lambda \lambda^{p-1} \int_\Omega dP(\omega) I_{\{\lambda \leq Y(\omega)\}}(\lambda)
$$

$$
= p \int_0^\infty d\lambda \lambda^{p-1} P(\lambda \leq Y) = p \int_0^\infty d\lambda \lambda^{p-2} \lambda P(Y \geq \lambda)
$$

$$
\leq p \int_0^\infty d\lambda \lambda^{p-2} \int_{\{Y \geq \lambda\}} X dP
$$

$$
= p \int_\Omega dP(\omega) X(\omega) \int_0^\infty d\lambda \lambda^{p-2} I_{\{Y(\omega) \geq \lambda\}}(\lambda)
$$

$$
= p \int_\Omega dP(\omega) X(\omega) \int_0^{Y(\omega)} d\lambda \lambda^{p-2} = \frac{p}{p-1} \int_\Omega dP(\omega) X(\omega) Y^{p-1}(\omega)
$$

$$
= \frac{p}{p-1} E[Y^{p-1} X],
$$

where, throughout, λ denotes the Lebesgue measure, and when changing the order of integration we invoke Fubini's theorem. By Hölder's inequality, we obtain

$$
E[Y^p] \leq \frac{p}{p-1} E[Y^{p-1} X] \leq \frac{p}{p-1} E[X^p]^{\frac{1}{p}} E[Y^p]^{\frac{p-1}{p}},
$$

which, after substitution and rearrangement, gives

$$
E[Y^p]^{\frac{1}{p}} \leq q E[X^p]^{\frac{1}{p}},
$$

as long as $E[Y^p] < +\infty$ (in such a case we may, in fact, divide the left- and right-hand sides by $E[Y^p]^{\frac{p-1}{p}}$). But in any case we can consider the sequence of random variables $(Y \wedge n)_{n \in \mathbb{N}}$ ($Y \wedge n$ is the random variable defined letting, for all $\omega \in \Omega$, $Y \wedge n(\omega) = \inf\{Y(\omega), n\}$); since, for all n, $Y \wedge n$ satisfies condition (2.22), then we obtain

$$
\|Y \wedge n\|_p \leq q \|X\|_p,
$$

and in the limit

$$
\|Y\|_p = \lim_{n \to \infty} \|Y \wedge n\|_p \leq q \|X\|_p.
$$

$\square$

Proposition 2.121. *Let $(X_n)_{n \in \mathbb{N}^*}$ be a sequence of real random variables defined on the probability space $(\Omega, \mathcal{F}, P)$, and let X_n^+ be the positive part of X_n.*

1. *If $(X_n)_{n \in \mathbb{N}^*}$ is a submartingale, then*

$$P\left(\max_{1 \le k \le n} X_k > \lambda\right) \le \frac{1}{\lambda} E[X_n^+], \qquad \lambda > 0, n \in \mathbb{N}^*.$$

2. *If $(X_n)_{n \in \mathbb{N}^*}$ is a martingale and if, for all $n \in \mathbb{N}^*$, $X \in L^p(P)$, $p > 1$, then*

$$E\left[\left(\max_{1 \le k \le n} |X_k|\right)^p\right] \le \left(\frac{p}{p-1}\right)^p E[|X_n|^p], \qquad n \in \mathbb{N}^*.$$

(Points 1 and 2 are called Doob's inequalities.*)*

Proof.

1. For all $k \in \mathbb{N}^*$ we put $A_k = \bigcap_{j=1}^{k-1} \{X_j \le \lambda\} \cap \{X_k > \lambda\}$ $(\lambda > 0)$, where all A_k are pairwise disjoint and $A = \{\max_{1 \le k \le n} X_k > \lambda\}$. Thus it is obvious that $A = \bigcup_{k=1}^{n} A_k$. Because in A_k, X_k is greater than λ, we have

$$\int_{A_k} X_k dP \ge \lambda \int_{A_k} dP.$$

Therefore,

$$\forall k \in \mathbb{N}^*, \qquad \lambda P(A_k) \le \int_{A_k} X_k dP,$$

resulting in

$$\lambda P(A) = \lambda P\left(\bigcup_{k=1}^{n} A_k\right) = \lambda \sum_{k=1}^{n} P(A_k)$$

$$\le \sum_{k=1}^{n} \int_{A_k} X_k dP = \sum_{k=1}^{n} \int_{\Omega} X_k I_{A_k} dP = \sum_{k=1}^{n} E[X_k I_{A_k}]. \quad (2.23)$$

Now we have

$$E[X_n^+] = \int_{\Omega} X_n^+ dP$$

$$\ge \int_A X_n^+ dP = \sum_{k=1}^{n} \int_{A_k} X_n^+ dP = \sum_{k=1}^{n} \int_{\Omega} X_n^+ I_{A_k} dP$$

$$= \sum_{k=1}^{n} E[X_n^+ I_{A_k}] = \sum_{k=1}^{n} E[E[X_n^+ I_{A_k} | X_1, \ldots, X_k]]$$

$$= \sum_{k=1}^{n} E[I_{A_k} E[X_n^+ | X_1, \ldots, X_k]] \ge \sum_{k=1}^{n} E[I_{A_k} E[X_n | X_1, \ldots, X_k]],$$

where the last row follows from the fact that I_{A_k} is $\sigma(X_1, \ldots, X_k)$-measurable. Moreover, since $(X_n)_{n \in \mathbb{N}^*}$ is a submartingale, we have

$$E[X_n^+] \geq \sum_{k=1}^n E[I_{A_k} X_k]. \tag{2.24}$$

By (2.23) and (2.24), $E[X_n^+] \geq \lambda P(A)$, and this completes the proof of 1. We can also observe that

$$\sum_{k=1}^n E[I_{A_k} X_n^+] = \sum_{k=1}^n E[E[X_n^+ I_{A_k} | X_1, \ldots, X_k]]$$

$$\geq \sum_{k=1}^n E[I_{A_k} E[X_n | X_1, \ldots, X_k]] \geq \sum_{k=1}^n E[I_{A_k} X_k] \geq \lambda P(A),$$

and therefore

$$\lambda P \left(\max_{1 \leq k \leq n} X_k > \lambda \right) \leq \sum_{k=1}^n E[I_{A_k} X_n^+]. \tag{2.25}$$

2. Let $(X_n)_{n \in \mathbb{N}^*}$ be a martingale such that $X_n \in L^p(P)$ for all $n \in \mathbb{N}^*$. Since $\phi = |x|$ is a convex function, it follows from Proposition 2.119 that $(|X_n|)_{n \in \mathbb{N}^*}$ is a submartingale. Thus from (2.25) we have

$$\lambda P \left(\max_{1 \leq k \leq n} |X_k| > \lambda \right) \leq \sum_{k=1}^n E[I_{A_k} |X_n^+|] = \sum_{k=1}^n E[I_{A_k} |X_n|]$$

$$= \sum_{k=1}^n \int_{A_k} |X_n| dP = \int_A |X_n| dP \quad (\lambda > 0, n \in \mathbb{N}^*).$$

Putting $X = \max_{1 \leq k \leq n} |X_k|$ and $Y = |X_n|$, we obtain

$$\lambda P(X > \lambda) \leq \int_A Y dP = \int_{\{X > \lambda\}} Y dP,$$

and from Lemma 2.120 it follows that $\|X\|_p \leq q\|Y\|_p$. Thus $E[X^p] \leq q^p E[Y^p]$, proving 2.

$\square$

Remark 2.122. Because

$$\max_{1 \leq k \leq n} |X_k|^p = \left(\max_{1 \leq k \leq n} |X_k| \right)^p,$$

by point 2 of Proposition 2.121 it is also true that

$$E \left[\max_{1 \leq k \leq n} |X_k|^p \right] \leq \left(\frac{p}{p-1} \right)^p E[|X_n|^p].$$

Corollary 2.123. *If $(X_n)_{n \in \mathbb{N}^*}$ is a martingale such that $X_n \in L^p(P)$ for all $n \in \mathbb{N}^*$, then*

$$P\left(\max_{1\le k\le n}|X_k|>\lambda\right)\le\frac{1}{\lambda^p}E[|X_n|^p], \qquad \lambda>0.$$

Proof. From Proposition 2.119 we can assert that $(|X_n|^p)_{n\in\mathbb{N}^*}$ is a submartingale. In fact, $\phi(x)=|x|^p, p>1$, is convex. By point 1 of Proposition 2.121, it follows that

$$P\left(\max_{1\le k\le n}|X_k|^p>\lambda^p\right)\le\frac{1}{\lambda^p}E[|X_n|^p],$$

which is equivalent to

$$P\left(\max_{1\le k\le n}|X_k|>\lambda\right)\le\frac{1}{\lambda^p}E[|X_n|^p].$$

$\square$

Lemma 2.124. *The following statements are true:*

1. *If $(X_t)_{t\in\mathbb{R}_+}$ is a martingale, then so is $(X_t)_{t\in I}$ for all $I\subset\mathbb{R}_+$.*
2. *If, for all $I\subset\mathbb{R}_+$ and I finite, $(X_t)_{t\in I}$ is a (discrete) martingale, then so is $(X_t)_{t\in\mathbb{R}_+}$.*

Proof.

1. Let $I\subset\mathbb{R}_+, (s,t)\in I^2, s<t$. Because $(X_r)_{r\in\mathbb{R}_+}$ is a martingale,

$$X_s=E[X_t|X_r,0\le r\le s,r\in\mathbb{R}_+].$$

Observing that

$$\sigma(X_r,0\le r\le s,r\in I)\subset\sigma(X_r,0\le r\le s,r\in\mathbb{R}_+)$$

and remembering that in general

$$E[X|B_1]=E[E[X|B_2]|B_1], \qquad B_1\subset B_2\subset\mathcal{F},$$

we obtain

$$
\begin{aligned}
&E[X_t|X_r,0\le r\le s,r\in I]\\
&= E[E[X_t|X_r,0\le r\le s,r\in\mathbb{R}_+]|X_r,0\le r\le s,r\in I]\\
&= E[X_s|X_r,0\le r\le s,r\in I]\\
&= X_s.
\end{aligned}
$$

The last equality holds because X_s is measurable with respect to $\sigma(X_r,0\le r\le s,r\in I)$.
2. See, e.g., Doob (1953).

$\square$

Proposition 2.125. *Let $(X_t)_{t\in\mathbb{R}_+}$ be a stochastic process on $(\Omega,\mathcal{F},P)$ valued in $\mathbb{R}$.*

1. *If* $(X_t)_{t \in \mathbb{R}_+}$ *is a submartingale, then*

$$P\left(\sup_{0 \leq s \leq t} X_s > \lambda\right) \leq \frac{1}{\lambda} E[X_t^+], \qquad \lambda > 0, t \geq 0.$$

2. *If* $(X_t)_{t \in \mathbb{R}_+}$ *is a martingale such that, for all* $t \geq 0$, $X_t \in L^p(P)$, $p > 1$, *then*

$$E\left[\sup_{0 \leq s \leq t} |X_s|^p\right] \leq \left(\frac{p}{p-1}\right)^p E[|X_t|^p].$$

Proof. See, e.g., Doob (1953). □

Definition 2.126. A subset H of $L^1(\Omega, \mathcal{F}, P)$ is *uniformly integrable* if

$$\lim_{c \to \infty} \sup_{Y \in H} \int_{\{|Y| > c\}} |Y| dP = 0.$$

Theorem 2.127. *A martingale is uniformly integrable if and only if it is of the form* $M_n = E[Y|\mathcal{F}_n]$, *where* $Y \in L^1(\Omega, \mathcal{F}, P)$. *Under these conditions* $\{M_n\}_n$ *converges almost surely and in* L^1.

Proof. See, e.g., Baldi (1984). □

The subsequent proposition specifies the limit of a uniformly integrable martingale.

Proposition 2.128. *Let* $Y \in L^1(\Omega, \mathcal{F}, P)$, $\{\mathcal{F}_n\}_n$ *be a filtration and* $\mathcal{F}_\infty = \bigcup_n \mathcal{F}_n$ *the* σ-*algebra generated by* $\{\mathcal{F}_n\}_n$. *Then*

$$\lim_{n \to \infty} E[Y|\mathcal{F}_n] = E[Y|\mathcal{F}_\infty] \text{ almost surely and in } L^1.$$

Proof. See, e.g., Baldi (1984). □

Doob–Meyer Decomposition

In the sequel, whenever not explicitly specified we will refer to the natural filtration of a process, suitably completed.

Proposition 2.129. *Every martingale has a right-continuous version.*

Theorem 2.130. *Let* X_t *be a supermartingale. Then the mapping* $t \to E[X_t]$ *is right-continuous if and only if there exists an RCLL modification of* X_t. *This modification is unique.*

Proof. See, e.g., Protter (1990). □

Definition 2.131.   Consider the set $\mathcal{S}$ of stopping times T, with $P(T < \infty) = 1$, of the filtration $(\mathcal{F}_t)_{t \in \mathbb{R}_+}$. The right-continuous adapted process $(X_t)_{t \in \mathbb{R}_+}$ is said to be of *class D* if the family $(X_T)_{T \in \mathcal{S}}$ is uniformly integrable. Instead, if $\mathcal{S}_a$ is the set of stopping times with $P(T \leq a) = 1$, for a finite $a > 0$, and the family $(X_T)_{T \in \mathcal{S}_a}$ is uniformly integrable, then it is said to be of *class DL*.

Proposition 2.132. *Let $(X_t)_{t \in \mathbb{R}_+}$ be a right-continuous submartingale. Then X_t is of class DL under either of the following two conditions:*

1. *$X_t \geq 0$ almost surely for every $t \geq 0$.*
2. *X_t has the form*

$$X_t = M_t + A_t, \qquad t \in \mathbb{R}_+,$$

where $(M_t)_{t \in \mathbb{R}_+}$ is a martingale and $(A_t)_{t \in \mathbb{R}_+}$ an adapted increasing process.

Lemma 2.133. *If $(X_t)_{t \in \mathbb{R}_+}$ is a uniformly integrable martingale, then it is of class D.*

If $(X_t)_{t \in \mathbb{R}_+}$ is a martingale, or it is bounded from below, then it is of class DL.

Definition 2.134. Let $(X_t)_{t \in \mathbb{R}_+}$ be an adapted stochastic process with RCLL trajectories. It is said to be *decomposable* if it can be written as

$$X_t = X_0 + M_t + Z_t,$$

where $M_0 = Z_0 = 0$, M_t is a locally square-integrable martingale and Z_t has RCLL-adapted trajectories of bounded variation.

Theorem 2.135 (Doob–Meyer).   *Let $(X_t)_{t \in \mathbb{R}_+}$ be an adapted right-continuous process. It is a submartingale of class D, with $X_0 = 0$ almost surely if and only if it can be decomposed as*

$$\forall t \in \mathbb{R}_+, \qquad X_t = M_t + A_t \, a.s.,$$

where M_t is a uniformly integrable martingale with $M_0 = 0$ and $A_t \in L^1(P)$ is an increasing predictable process with $A_0 = 0$. The decomposition is unique and if, in addition, X_t is bounded, then M_t is uniformly integrable and A_t is integrable.

Proof. See, e.g., Ethier and Kurtz (1986).   □

Corollary 2.136. *Let $X = (X_t)_{t \in \mathbb{R}_+}$ be an adapted right-continuous submartingale of class DL; then, there exists a unique (up to indistinguishability) right-continuous increasing predictable process A adapted to the same filtration as X, with $A_0 = 0$ almost surely, such that*

$$M_t = X_t - A_t, \quad t \in \mathbb{R}_+$$

is a martingale, adapted to the same filtration as X.

Definition 2.137. Resorting to the notation of Theorem 2.135, the process $(A_t)_{t \in \mathbb{R}_+}$ is called the *compensator* of X_t.

Proposition 2.138. *Under the assumptions of Theorem 2.135, the compensator A_t of X_t is continuous if and only if X_t is regular in the sense that for every predictable finite stopping time T we have that $E[X_T] = E[X_{T-}]$.*

Definition 2.139. A stochastic process $(M_t)_{t \in \mathbb{R}_+}$ is a *local martingale* with respect to the filtration $(\mathcal{F}_t)_{t \in \mathbb{R}_+}$ if there exists a "localizing" sequence $(T_n)_{n \in \mathbb{N}}$ such that for each $n \in \mathbb{N}$, $(M_{t \wedge T_n})_{t \in \mathbb{R}_+}$ is an $\mathcal{F}_t$-martingale.

Definition 2.140. Let $(X_t)_{t \in \mathbb{R}_+}$ be a stochastic process. *Property $\mathcal{P}$* is said to hold locally if

1. There exists $(T_n)_{n \in \mathbb{N}}$, a sequence of stopping times, with $T_n < T_{n+1}$.
2. $\lim_n T_n = +\infty$ almost surely.

such that $X_{T_n} I_{\{T_n > 0\}}$ has property $\mathcal{P}$ for $n \in \mathbb{N}^*$.

Theorem 2.141. *Let $(M_t)_{t \in \mathbb{R}_+}$ be an adapted and RCLL stochastic process, and let $(T_n)_{n \in \mathbb{N}}$ be as in the preceding definition. If $M_{T_n} I_{\{T_n > 0\}}$ is a martingale for each $n \in \mathbb{N}^*$, then M_t is a local martingale.*

Lemma 2.142. *Any martingale is a local martingale.*

Proof. Simply take $T_n = n$ for all $n \in \mathbb{N}^*$. □

Theorem 2.143 (Local form Doob–Meyer). *Let $(X_t)_{t \in \mathbb{R}_+}$ be a nonnegative right-continuous $\mathcal{F}_t$-local submartingale with $(\mathcal{F}_t)_{t \in \mathbb{R}_+}$ a right-continuous filtration. Then there exists a unique increasing right-continuous predictable process $(A_t)_{t \in \mathbb{R}_+}$ such that $A_0 = 0$ almost surely and $P(A_t < \infty) = 1$ for all $t > 0$, so that $X_t - A_t$ is a right-continuous local martingale.*

Definition 2.144. A martingale $M = (M_t)_{t \in \mathbb{R}_+}$ is square-integrable if, for all $t \in \mathbb{R}_+$, $E[|M_t|^2] < +\infty$.

We will denote by $\mathcal{M}$ the family of all right-continuous square-integrable martingales.

Remark 2.145. If $M \in \mathcal{M}$, then M^2 satisfies the conditions of Corollary 2.136; let $\langle M \rangle$ be the increasing process given by the theorem with $X = M^2$. Then $\langle M_0 \rangle = 0$, and $M_t^2 - \langle M_t \rangle$ is a martingale.

Definition 2.146. For two martingales M and N, in $\mathcal{M}$ the process

$$\langle M, N \rangle = \frac{1}{4}(\langle M + N \rangle - \langle M - N \rangle)$$

is called the *predictable covariation* of M and N. Evidently $\langle M, M \rangle = \langle M \rangle$, and so it is called the *predictable variation* of M.

Remark 2.147. Hence $\langle M, N \rangle$ is the unique finite variation predictable RCLL process such that $\langle M, N \rangle_0 = 0$ and $MN - \langle M, N \rangle$ is a martingale. Furthermore, if $\langle M, N \rangle = 0$, then the two martingales are said to be *orthogonal*. Thus M and N are orthogonal if and only if MN is a martingale.

Definition 2.148. A martingale M is said to be a *purely discontinuous martingale* if and only if $M_0 = 0$ and it is orthogonal to any continuous martingale.

Definition 2.149. Two local martingales M and N are said to be *orthogonal* if and only if MN is a local martingale.

Definition 2.150. A local martingale M is said to be a *purely discontinuous local martingale* if and only if $M_0 = 0$ and it is orthogonal to any continuous local martingale.

Having denoted by $\mathcal{M}$ the family of all right-continuous square-integrable martingales, let $\mathcal{M}_c \subset \mathcal{M}$ denote the family of all continuous square integrable martingales, and let $\mathcal{M}_d \subset \mathcal{M}$ denote the family of all purely discontinuous square-integrable martingales.

Theorem 2.151. *Any local martingale M admits a unique (up to indistinguishability) decomposition*

$$M = M_0 + M^c + M^d,$$

where M_c is a continuous local martingale and M_d is a purely discontinuous local martingale, with $M_0^c = M_0^d = 0$.

Proof. See, e.g., Jacod and Shiryaev (1987, p. 43).   □

Remark 2.152. The reader has to be cautious about the meaning of the term "purely discontinuous"; it is indeed referring just to an orthogonality property with respect to the continuous case, but it does refer to the kind of discontinuities of its trajectories (e.g., Jacod and Shiryaev 1987, p. 40).

Proposition 2.153. *Let $X = (X_t)_{t \in \mathbb{R}_+}$ be a right-continuous martingale. Then there exists a right-continuous increasing process, denoted by $[X]$, such that for each $t \in \mathbb{R}_+$, and each sequence of partitions $(t_k^{(n)})_{n \in \mathbb{N}, 0 \le k \le n}$ of $[0, t]$, with $\max_k(t_{k+1}^{(n)} - t_k^{(n)}) \xrightarrow{n} \infty$:*

$$\sum_k (X(t_{k+1}^{(n)}) - X(t_k^{(n)}))^2 \xrightarrow[n\to\infty]{P} [X](t). \qquad (2.26)$$

If $X \in \mathcal{M}$, then the convergence in (2.26) is in L^1. If $X \in \mathcal{M}_c$, then $[X]$ can be taken to be continuous.

Proof. See, e.g., Ethier and Kurtz (1986).   □

Definition 2.154. The process $[X]$ introduced above is known as the *quadratic variation* process associated with X.

Proposition 2.155. *If $M \in \mathcal{M}$, then $M^2 - [M]$ is a martingale.*

Remark 2.156. If $M \in \mathcal{M}_c$, then, by Proposition 2.26, $[M]$ is continuous, and Proposition 2.155 implies, by uniqueness, that $[M] = \langle M \rangle$, up to indistinguishability.

Proposition 2.157. *Let $M \in \mathcal{M}_c$. Then $\langle M \rangle = 0$ if and only if M is constant, i.e., $M_t = M_0$, a.s., for any $t \in \mathbb{R}_+$.*

Proof. See, e.g., Revuz-Yor (1991, p. 119).   □

2.7.1 The martingale problem for Markov processes

The following so-called *Dynkin's formula* establishes a fundamental link between Markov processes and martingales (Rogers and Williams 1994, p. 253).

Given a process $(X_t)_{t\in\mathbb{R}_+}$, we will denote by $(\mathcal{F}_t)_{t\in\mathbb{R}_+}$ its natural filtration.

Theorem 2.158. *Assume $(X_t)_{t\in\mathbb{R}_+}$ is a Markov process on $(E, \mathcal{B}_E)$, with transition kernel $p(t, x, B)$, $t \in \mathbb{R}_+$, $x \in E$, $B \in \mathcal{B}_E$. Let $(T(t), t \in \mathbb{R}_+)$ denote its transition semigroup and $\mathcal{A}$ its infinitesimal generator. Then, for any $g \in \mathcal{D}(\mathcal{A})$, the stochastic process*

$$M(t) := g(X_t) - g(X_0) - \int_0^t \mathcal{A}g(X_s)ds$$

is an $\mathcal{F}_t$-martingale (indeed a zero-mean martingale).

Proof. The following equations hold:

$$E[M(t+h)|\mathcal{F}_t] + g(X_0)$$

$$= E\left[g(X_{t+h}) - \int_t^{t+h} \mathcal{A}g(X_s)ds \,\Big|\, \mathcal{F}_t\right] - \int_0^t \mathcal{A}g(X_s)ds.$$

Now, thanks to the Markov property,

$$E\left[g(X_{t+h})|\,\mathcal{F}_t\right] = E\left[g(X_{t+h})|\,X_t\right] = T(h)g(X_t),$$

and

$$E\left[\int_t^{t+h} \mathcal{A}g(X_s)ds|\mathcal{F}_t\right] = \int_t^{t+h} ds\, E[\mathcal{A}g(X_s)|\mathcal{F}_t]$$

$$= \int_t^{t+h} ds\, E[\mathcal{A}g(X_s)|X_t] = \int_t^{t+h} ds\, T(s-t)\mathcal{A}g(X_t)$$

$$= \int_0^h ds\, T(s)\mathcal{A}g(X_t) = \int_0^h ds\, \mathcal{A}T(s)g(X_t)$$

$$= \int_0^h d[T(s)g(X_t)] = T(h)g(X_t) - T(0)g(X_t)$$

$$= T(h)g(X_t) - g(X_t).$$

As a consequence

$$E[M(t+h)|\mathcal{F}_t] + g(X_0)$$

$$= T(h)g(X_t) - T(h)g(X_t) + g(X_t) - \int_0^t \mathcal{A}g(X_s)ds$$

$$= g(X_t) - \int_0^t \mathcal{A}g(X_s)ds = M(t) + g(X_0).$$

$\square$

The next proposition shows that a Markov process is indeed character-ized by its infinitesimal generator via a martingale problem (e.g., Rogers and Williams 1994, p. 253).

Theorem 2.159 (Martingale problem for Markov processes). *If an RCLL Markov process $(X_t)_{t\in\mathbb{R}_+}$ is such that*

$$g(X_t) - g(X_0) - \int_0^t \mathcal{A}g(X_s)ds$$

is an $\mathcal{F}_t$-martingale for any function $g \in \mathcal{D}(\mathcal{A})$, where $\mathcal{A}$ is the infinitesimal generator of a contraction semigroup on E, then X_t is equivalent to a Markov process having $\mathcal{A}$ as its infinitesimal generator.

Remark 2.160. Note that, from

$$M(t) := g(X_t) - g(X_0) - \int_0^t \mathcal{A}g(X_s)ds$$

one may derive

$$g(X_t) - g(X_0) = \int_0^t \mathcal{A}g(X_s)ds + M(t).$$

Formally, by a suitable definition of differential of a martingale, this may be rewritten as

$$dg(X_t) = \mathcal{A}g(X_t) + dM(t).$$

Hence, apart from the "noise" $M(t)$, the evolution of any function $g(X_t)$ of a Markov process $\{X_t, t \in \mathbb{R}_+\}$ is determined by its infinitesimal generator.

Theorem 2.161. *Let $\{X_t, t \in \mathbb{R}_+\}$ be a Feller process on $\mathbb{R}$ having infinitesimal generator $\mathcal{A}$ with domain $\mathcal{D}_{\mathcal{A}}$. If $g \in C_0(\mathbb{R})$ and there exists an $f \in C_0(\mathbb{R})$ such that*

$$M(t) := g(X_t) - g(X_0) - \int_0^t f(X_s)ds, \ t \in \mathbb{R}_+,$$

is an $\mathcal{F}_t$-martingale, then $g \in \mathcal{D}_{\mathcal{A}}$, and $f = \mathcal{A}g$.

Proof. See, e.g., Revuz-Yor (1991, p. 262) □

The preceding results lead to an extension of the concept of infinitesimal generator called an *extended infinitesimal generator* (Revuz-Yor 1991, p. 263).

In what follows we shall denote by $C_K^p(\mathbb{R})$, $p \in \overline{\mathbb{R}}_+^*$, the set of real functions with compact support, which are continuous with their derivatives up to the order p.

Theorem 2.162. *Let $\{X_t, t \in \mathbb{R}_+\}$ be a Feller process on $\mathbb{R}$, having infinitesimal generator $\mathcal{A}$ with domain $\mathcal{D}_{\mathcal{A}}$, such that $C_K^\infty(\mathbb{R}) \subset \mathcal{D}_{\mathcal{A}}$. Then*

(i) $C_K^2(\mathbb{R}) \subset \mathcal{D}_{\mathcal{A}}$.
(ii) For any relatively compact open set U there exist functions a, b, and $c \leq 0$ on U, and a kernel measure $N(x, B)$, $x \in U$, $B \in \mathbb{R} - \{0\}$, which is a Radon measure for any $x \in U$ such that, for $f \in C_K^2(\mathbb{R})$ and $x \in U$,

$$(\mathcal{A}f)(x) = \frac{1}{2}b^2(x)\frac{\partial^2 f}{\partial x^2} + a(x)\frac{\partial f}{\partial x} + c(x)$$

$$+ \int_{\mathbb{R} - \{0\}} \left[f(y) - f(x) - I_U(y)(y - x)\frac{\partial f}{\partial x} \right] N(x, dy).$$

If the process $\{X_t, t \in \mathbb{R}_+\}$ has continuous paths, then

$$(\mathcal{A}f)(x) = \frac{1}{2}b^2(x)\frac{\partial^2 f}{\partial x^2} + a(x)\frac{\partial f}{\partial x} + c(x).$$

Proof. See, e.g., Revuz-Yor (1991, p. 267) □

Example 2.163. A Poisson process (see the following section for more details) is an integer-valued Markov process $(N_t)_{t \in \mathbb{R}_+}$. If its intensity parameter is $\lambda > 0$, then the process $(X_t)_{t \in \mathbb{R}_+}$, defined by $X_t = N_t - \lambda t$, is a stationary Markov process with independent increments. The transition kernel of X_t is

$$p(h, x, B) = \sum_{k=0}^{\infty} \frac{(\lambda h)^k}{k!} e^{-\lambda h} I_{\{x + k - \lambda h \in B\}} \text{ for } x \in \mathbb{N}, h \in \mathbb{R}_+, B \subset \mathbb{N}.$$

Its transition semigroup is then

$$T(h)g(x) = \sum_{k=0}^{\infty} \frac{(\lambda h)^k}{k!} e^{-\lambda h} g(x + k - \lambda h) \quad \text{for } x \in \mathbb{N}, g \in BC(\mathbb{R}).$$

The infinitesimal generator is then

$$\mathcal{A}g(x) = \lambda(g(x+1) - g(x)) - \lambda g'(x+).$$

According to previous theorems,

$$M(t) = g(X_t) - \int_0^t ds(\lambda(g(X_s + 1) - g(X_s)) - \lambda g'(X_s+)$$

is a martingale for any $g \in BC(\mathbb{R})$ (where $g(0) = 0$).

The martingale problem for Markov jump processes

Consider a time-homogeneous Markov jump process $(X_t)_{t \in \mathbb{R}_+}$ on a countable state space E with a conservative intensity matrix $Q = (q_{ij})_{i,j \in E}$,

$$q_{ij} \geq 0, \quad i \neq j,$$

$$q_i := -q_{ii} = \sum_{j \neq i} q_{ij}.$$

The matrix Q can be seen as a functional $Q : f \to Q(f)$ on functions $f : E \to \mathbb{R}_+$, by setting

$$Q(f)(i) := \sum_{j \in E} q_{ij} f(j) = \sum_{j \neq i} q_{ij}(f(j) - f(i)), \quad \text{for any} \quad i \in E.$$

For f bounded in E we have, for any $s, t \in \mathbb{R}_+$,

$$E[f(X_{t+s})] - E[f(X_t)]$$
$$= E[E[f(X_{t+s}) - f(X_t)|X_t]]$$
$$= \sum_{i \in E} P(X_t = i) \sum_{j \neq i} (f(j) - f(i)) P(X_{t+s} = j|X_t = i).$$

Since the process is time homogeneous

$$E[f(X_{t+s})] - E[f(X_t)]$$
$$= \sum_{i \in E} P(X_t = i) \sum_{j \neq i} (f(j) - f(i)) P(X_s = j|X_0 = i)$$
$$= \sum_{i \in E} P(X_t = i) \sum_{j \neq i} (f(j) - f(i)) p_{ij}(s).$$

We may then compute the limit

$$\lim_{s \downarrow 0} \frac{1}{s} \{ E[f(X_{t+s})] - E[f(X_t)] \}$$

$$= \lim_{s \downarrow 0} \{ \sum_{i \in E} P(X_t = i) \sum_{j \neq i} \frac{1}{s} p_{ij}(s)(f(j) - f(i)) \},$$

Suppose we may interchange the limit with the sums of the series

$$\frac{d}{dt} E[f(X(t))] = \sum_{i \in E} P(X_t = i) \sum_{j \neq i} q_{ij}(f(j) - f(i))$$

which can be written as

$$\frac{d}{dt} E[f(X(t))] = E[Q(f)(X_t)].$$

By returning to the integral formulation, we gain Dynkin's formula for Markov jump processes in terms of the intensity matrix Q

$$E[f(X_t)] - E[f(X_0)] = \int_0^t E[Q(f)(X_s)] ds. \tag{2.27}$$

Indeed, from Rogers and Williams (1994, pp. 30–37) we obtain the following theorem.

Theorem 2.164. *For any function $g \in C^{1,0}(\mathbb{R}_+ \times E)$ such that the mapping*

$$t \to \frac{\partial}{\partial t} g(t, x)$$

is continuous for all $x \in E$, the process

$$\left(g(t, X_t) - g(0, X_0) - \int_0^t \left(\frac{\partial g(s, \cdot)}{\partial s} + Q(g(s, \cdot)) \right)(X_s) ds \right)_{t \in \mathbb{R}_+}$$

is a local martingale.

Corollary 2.165. *For any real function f defined on E, the process*

$$\left(f(X_t) - f(X_0) - \int_0^t Q(f)(X_s) ds \right)_{t \in \mathbb{R}_+} \tag{2.28}$$

is a local martingale. Whenever the local martingale is a martingale, we recover (2.27).

Proposition 2.166 (Martingale problem for Markov jump processes). *Given an intensity matrix Q, if an RCLL Markov process $X \equiv (X_t)_{t \in \mathbb{R}_+}$ on E is such that the process (2.28) is a local martingale, then Q is the intensity matrix of the Markov process X.*

2.8 Brownian Motion and the Wiener Process

A small particle (e.g., a pollen corn) suspended in a liquid is subject to infinitely many collisions with atoms, and therefore it is impossible to observe its exact trajectory. With the help of a microscope it is only possible to confirm that the movement of the particle is entirely chaotic. This type of movement, discovered under similar circumstances by the botanist Robert Brown, is called Brownian motion. As its mathematical inventor Einstein had already observed, it is necessary to make approximations in order to describe the process. The formalized mathematical model defined on the basis of these facts is called a Wiener process. Henceforth, we will limit ourselves to the study of the one-dimensional Wiener process in $\mathbb{R}$, under the assumption that the three components determining its motion in space are independent.

Definition 2.167. The real-valued process $(W_t)_{t \in \mathbb{R}_+}$ is a *Wiener process* if it satisfies the following conditions:

1. $W_0 = 0$ almost surely.
2. $(W_t)_{t \in \mathbb{R}_+}$ is a process with independent increments.
3. $W_t - W_s$ is normally distributed with $N(0, t - s)$, $(0 \le s < t)$.

Remark 2.168. From point 3 of Definition 2.167 it becomes obvious that every Wiener process is homogeneous.

Proposition 2.169. *If $(W_t)_{t \in \mathbb{R}_+}$ is a Wiener process, then*

1. $E[W_t] = 0 \qquad$ *for all $t \in \mathbb{R}_+$*
2. $K(s, t) = Cov[W_t, W_s] = \min\{s, t\}, \qquad s, t \in \mathbb{R}_+$

Proof.

1. By fixing $t \in \mathbb{R}$, we observe that $W_t = W_0 + (W_t - W_0)$ and, thus, $E[W_t] = E[W_0] + E[W_t - W_0] = 0$. The latter is given by the fact that $E[W_0] = 0$ (by point 1 of Definition 2.167) and $E[W_t - W_0] = 0$ (by point 3 of Definition 2.167).
2. Let $s, t \in \mathbb{R}_+$ and $Cov[W_t, W_s] = E[W_t W_s] - E[W_t]E[W_s]$, which (by point 1) gives $Cov[W_t, W_s] = E[W_t W_s]$. For simplicity, if we suppose that $s < t$, then

$$E[W_t W_s] = E[W_s(W_s + (W_t - W_s))] = E[W_s^2] + E[W_s(W_t - W_s)].$$

Since $(W_t)_{t \in \mathbb{R}_+}$ has independent increments, we obtain

$$E[W_s(W_t - W_s)] = E[W_s]E[W_t - W_s],$$

and by point 1 of Proposition 2.136 (or point 3 of Definition 2.167) it follows that this is equal to zero, thus

$$Cov[W_t, W_s] = E[W_s^2] = Var[W_s].$$

If we now observe that $W_s = W_0 + (W_s - W_0)$ and hence $Var[W_s] = Var[W_0 + (W_s - W_0)]$, then, by the independence of the increments of the process, we get

$$Var[W_0 + (W_s - W_0)] = Var[W_0] + Var[W_s - W_0].$$

Therefore, by points 1 and 3 of Definition 2.167 it follows that

$$Var[W_s] = s = \inf\{s, t\},$$

which completes the proof.

$\square$

Proposition 2.170. *The Wiener process is a Gaussian process.*

Proof. In fact, if $n \in \mathbb{N}^*, (t_1, \ldots, t_n) \in \mathbb{R}_+^n$ with $0 = t_0 < t_1 < \ldots < t_n$ and $(a_1, \ldots, a_n) \in \mathbb{R}^n, (b_1, \ldots, b_n) \in \mathbb{R}^n$, such that $a_i \leq b_i, i = 1, 2, \ldots, n$, then it can be shown that

$$P(a_1 \leq W_{t_1} \leq b_1, \ldots, a_n \leq W_{t_n} \leq b_n)$$
$$= \int_{a_1}^{b_1} \cdots \int_{a_n}^{b_n} g(0|x_1, t_1) g(x_1|x_2, t_2 - t_1) \cdots g(x_{n-1}|x_n, t_n - t_{n-1}) dx_n \cdots dx_1,$$

$$(2.29)$$

where

$$g(x|y, t) = \frac{e^{-\frac{|x-y|^2}{2t}}}{\sqrt{2\pi t}}.$$

In order to prove that the density of $(W_{t_1}, \ldots, W_{t_n})$ is given by the integrand of (2.29), by the uniqueness of the characteristic function, it is sufficient to show that the characteristic function ϕ' of the n-dimensional real-valued random vector, whose density is given by the integrand of (2.29), is identical to the characteristic function ϕ of $(W_{t_1}, \ldots, W_{t_n})$. Thus, let $\boldsymbol{\lambda} = (\lambda_1, \ldots, \lambda_n) \in \mathbb{R}^n$. Then

$$\phi(\boldsymbol{\lambda}) = E\left[e^{i(\lambda_1 W_{t_1} + \cdots + \lambda_n W_{t_n})}\right]$$
$$= E\left[e^{i(\lambda_n(W_{t_n} - W_{t_{n-1}}) + (\lambda_n + \lambda_{n-1})(W_{t_{n-1}} - W_{t_{n-2}}) + \cdots + (\lambda_1 + \cdots + \lambda_n)W_{t_1})}\right]$$
$$= E\left[e^{i\lambda_n(W_{t_n} - W_{t_{n-1}})}\right] E\left[e^{i(\lambda_n + \lambda_{n-1})(W_{t_{n-1}} - W_{t_{n-2}})}\right] \cdots$$
$$\cdots E\left[e^{i(\lambda_1 + \cdots + \lambda_n)W_{t_1}}\right],$$

where we exploit the independence of the random variables $W_{t_i} - W_{t_{i-1}}$, $i = 1, \ldots, n$. Furthermore, because $(W_{t_i} - W_{t_{i-1}})$ is $N(0, t_i - t_{i-1}), i = 1, \ldots, n$, we get

$$\phi(\boldsymbol{\lambda}) = e^{-\frac{\lambda_n^2}{2}(t_n - t_{n-1})} e^{\frac{-(\lambda_n + \lambda_{n-1})^2}{2}(t_{n-1} - t_{n-2})} \cdots e^{\frac{-(\lambda_1 + \cdots + \lambda_n)^2}{2} t_1}.$$

We continue by calculating the characteristic function ϕ':

$$\phi'(\boldsymbol{\lambda}) = \int_{-\infty}^{+\infty} \cdots \int_{-\infty}^{+\infty} e^{i\boldsymbol{\lambda} \cdot \mathbf{x}} g(0|x_1, t_1) \cdots g(x_{n-1}|x_n, t_n - t_{n-1}) dx_n \cdots dx_1$$

$$= \int_{-\infty}^{+\infty} \cdots \left(\int_{-\infty}^{+\infty} e^{i\lambda_n x_n} g(x_{n-1}|x_n, t_n - t_{n-1}) dx_n \right) \cdots dx_1.$$

Because

$$\int_{-\infty}^{+\infty} e^{i\lambda x} \frac{1}{\sigma\sqrt{2\pi}} e^{\frac{-|x-m|^2}{2\sigma^2}} dx = e^{im\lambda - \frac{\lambda^2 \sigma^2}{2}}, \tag{2.30}$$

we obtain

$$\phi'(\boldsymbol{\lambda}) = \int_{-\infty}^{+\infty} \cdots \left(e^{i\lambda_n x_{n-1} - \frac{\lambda_n^2}{2}(t_n - t_{n-1})} \right) \cdots dx_1$$

$$= e^{-\frac{\lambda_n^2}{2}(t_n - t_{n-1})} \int_{-\infty}^{+\infty} \cdots$$

$$\left(\int_{-\infty}^{+\infty} e^{i(\lambda_n + \lambda_{n-1})x_{n-1}} g(x_{n-2}|x_{n-1}, t_{n-1} - t_{n-2}) dx_{n-1} \right) \cdots dx_1.$$

Recalling (2.30) and applying it to each variable, we obtain

$$\phi'(\boldsymbol{\lambda}) = e^{\frac{-\lambda_n^2}{2}(t_n - t_{n-1})} e^{\frac{-(\lambda_n + \lambda_{n-1})^2}{2}(t_{n-1} - t_{n-2})} \cdots e^{\frac{-(\lambda_1 + \cdots + \lambda_n)^2}{2} t_1},$$

and hence $\phi'(\boldsymbol{\lambda}) = \phi(\boldsymbol{\lambda})$. We now show that $g(0|x_1, t_1) \cdots g(x_{n-1}|x_n, t_n - t_{n-1})$ is of the form

$$\frac{1}{(2\pi)^{\frac{n}{2}} \sqrt{\det K}} e^{-\frac{1}{2}(\mathbf{x}-\mathbf{m})' K^{-1}(\mathbf{x}-\mathbf{m})}.$$

We will only show it for the case where $n = 2$; then

$$g(0|x_1, t_1) g(x_1|x_2, t_2 - t_1) = \frac{1}{2\pi\sqrt{t_1(t_2 - t_1)}} e^{-\frac{1}{2}\left[\frac{x_1^2}{t_1} + \frac{(x_2 - x_1)^2}{t_2 - t_1} \right]}$$

$$= \frac{1}{2\pi\sqrt{t_1(t_2 - t_1)}} e^{-\frac{1}{2}\left[\frac{x_1^2(t_2 - t_1) + (x_2 - x_1)^2 t_1}{t_1(t_2 - t_1)} \right]}.$$

If we put

$$K = \begin{pmatrix} t_1 & t_1 \\ t_1 & t_2 \end{pmatrix} \qquad (\text{where } K_{ij} = Cov[W_{t_i}, W_{t_j}]; i, j = 1, 2),$$

then

$$K^{-1} = \begin{pmatrix} \frac{t_2}{t_1(t_2-t_1)} & -\frac{1}{t_2-t_1} \\ -\frac{1}{t_2-t_1} & \frac{1}{t_2-t_1} \end{pmatrix},$$

resulting in

$$g(0|x_1,t_1)g(x_1|x_2,t_2-t_1) = \frac{1}{2\pi\sqrt{\det K}}e^{-\frac{1}{2}(\mathbf{x}-\mathbf{m})'K^{-1}(\mathbf{x}-\mathbf{m})},$$

where $m_1 = E[W_{t_1}] = 0, m_2 = E[W_{t_2}] = 0$. Thus

$$g(0|x_1,t_1)g(x_1|x_2,t_2-t_1) = \frac{1}{2\pi\sqrt{\det K}}e^{-\frac{1}{2}\mathbf{x}'K^{-1}\mathbf{x}},$$

completing the proof. $\square$

Remark 2.171. By point 1 of Definition 2.167, it follows, for all $t \in \mathbb{R}_+$, that $W_t = W_t - W_0$ almost surely and, by point 3 of the same definition, that W_t is distributed as $N(0,t)$. Thus

$$P(a \le W_t \le b) = \frac{1}{\sqrt{2\pi t}}\int_a^b e^{-\frac{x^2}{2t}}\,dx, \qquad a \le b.$$

Proposition 2.172. *If $(W_t)_{t\in\mathbb{R}_+}$ is a Wiener process, then it is a martingale.*

Proof. The proposition follows from Example 2.118 because $(W_t)_{t\in\mathbb{R}_+}$ is a centered process with independent increments. $\square$

Theorem 2.173 (Kolmogorov's continuity theorem). *Let $(X_t)_{t\in\mathbb{R}_+}$ be a separable real-valued stochastic process. If there exist positive real numbers r, c, ϵ, δ such that*

$$\forall h < \delta, \forall t \in \mathbb{R}_+, \qquad E[|X_{t+h} - X_t|^r] \le ch^{1+\epsilon}, \tag{2.31}$$

then, for almost every $\omega \in \Omega$, the trajectories are continuous in $\mathbb{R}_+$.

Proof. For simplicity, we will only consider the interval $I =]0,1[$, instead of $\mathbb{R}_+$, so that $(X_t)_{t\in]0,1[}$. Let $t \in]0,1[$ and $0 < h < \delta$ such that $t + h \in]0,1[$.
    Then by the Markov inequality and by (2.31) we obtain

$$P(|X_{t+h} - X_t| > h^k) \le h^{-rk}E[|X_{t+h} - X_t|^r] \le ch^{1+\epsilon-rk} \tag{2.32}$$

for $k > 0$ and $\epsilon - rk > 0$. Therefore,

$$\lim_{h\to 0} P\left(|X_{t+h} - X_t| > h^k\right) = 0;$$

namely, the process is continuous in probability and, by hypothesis, separable. Under these two conditions, it can be shown that any arbitrary countable dense subset T_0 of $]0,1[$ can be regarded as a separating set. Thus we define

$$T_0 = \left\{ \frac{j}{2^n} \,\middle|\, j = 1, \ldots, 2^n - 1; n \in \mathbb{N}^* \right\}$$

and observe that, by (2.32),

$$P\left(\max_{1 \leq j \leq 2^n - 2} \left| X_{\frac{j+1}{2^n}} - X_{\frac{j}{2^n}} \right| \geq \frac{1}{2^{nk}} \right) \leq \sum_{j=1}^{2^n - 2} P\left(\left| X_{\frac{j+1}{2^n}} - X_{\frac{j}{2^n}} \right| \geq \frac{1}{2^{nk}} \right)$$

$$\leq 2^n c 2^{-n(1+\epsilon-rk)} = c 2^{-n(\epsilon-rk)}.$$

Because $(\epsilon - rk) > 0$ and $\sum_n 2^{-n(\epsilon-rk)} < \infty$, we can apply the Borel–Cantelli Lemma 1.161 to the sets

$$F_n = \left\{ \max_{0 \leq j \leq 2^n - 1} \left| X_{\frac{j+1}{2^n}} - X_{\frac{j}{2^n}} \right| \geq \frac{1}{2^{nk}} \right\},$$

yielding $P(B) = 0$, where $B = \limsup_n F_n = \bigcap_n \bigcup_{k \leq n} F_k$. As a consequence, if $\omega \notin B$, then $\omega \in \Omega \setminus (\bigcap_n \bigcup_{k \geq n} F_k)$, i.e., there exists an $N = N(\omega) \in \mathbb{N}^*$ such that, for all $n \geq N$,

$$\left| X_{\frac{j+1}{2^n}}(\omega) - X_{\frac{j}{2^n}}(\omega) \right| < \frac{1}{2^{nk}}, \qquad j = 0, \ldots, 2^n - 1. \tag{2.33}$$

Now, let $\omega \notin B$ and s be a rational number such that

$$s = j2^{-n} + a_1 2^{-(n+1)} + \cdots + a_m 2^{-(n+m)}, \qquad s \in [j2^{-n}, (j+1)2^{-n}[,$$

where either $a_j = 0$ or $a_j = 1$ and $m \in \mathbb{N}^*$. If we put

$$b_r = j2^{-n} + a_1 2^{-(n+1)} + \cdots + a_r 2^{-(n+r)},$$

with $b_0 = j2^{-n}$ and $b_m = s$ for $r = 0, \ldots, m$, then

$$|X_s(\omega) - X_{j2^{-n}}(\omega)| \leq \sum_{r=0}^{m-1} |X_{b_{r+1}}(\omega) - X_{b_r}(\omega)|.$$

If $a_{r+1} = 0$, then $[b_r, b_{r+1}[= \emptyset$; if $a_{r+1} = 1$, then $[b_r, b_{r+1}[$ is of the form $[l2^{-(n+r+1)}, (l+1)2^{-(n+r+1)}[$. Hence from (2.33) it follows that

$$|X_s(\omega) - X_{j2^{-n}}(\omega)| \leq \sum_{r=0}^{m-1} 2^{-(n+r+1)k} \leq 2^{-nk} \sum_{r=0}^{\infty} 2^{-(r+1)k} \leq M2^{-nk},$$

$$\tag{2.34}$$

with $M \geq 1$. Fixing $\epsilon > 0$, there exists an $N_1 > 0$ such that, for all $n \geq N_1$, $M2^{-nk} < \frac{\epsilon}{3}$, and from the fact that $M \geq 1$ it also follows that, for all $n \geq N_1$, $2^{-nk} < \frac{\epsilon}{3}$. Let t_1, t_2 be elements of T_0 (separating set) such that $|t_1 - t_2| < \min\{2^{-N_1}, 2^{-N(\omega)}\}$. If $n = \max\{N_1, N(\omega)\}$, then there is at most one rational number of the form $\frac{j+1}{2^n}$ $(j = 1, \ldots, 2^n - 1)$ between t_1 and t_2. Therefore, by (2.33) and (2.34), it follows that

$$|X_{t_1}(\omega) - X_{t_2}(\omega)|$$

$$\leq \left|X_{t_1}(\omega) - X_{\frac{j}{2^n}}(\omega)\right| + \left|X_{\frac{j+1}{2^n}}(\omega) - X_{\frac{j}{2^n}}(\omega)\right| + \left|X_{t_2}(\omega) - X_{\frac{j+1}{2^n}}(\omega)\right|$$

$$< \frac{\epsilon}{3} + \frac{\epsilon}{3} + \frac{\epsilon}{3} = \epsilon.$$

Hence the trajectory is uniformly continuous almost everywhere in T_0 and has a continuous extension in $[0, 1]$. By Theorem 2.27, the extension coincides with the original trajectory. Therefore, the trajectory is continuous almost everywhere in $]0, 1[$. $\qquad\square$

Theorem 2.174. *If $(W_t)_{t \in \mathbb{R}_+}$ is a real-valued Wiener process, then it has continuous trajectories almost surely.*

Proof. Let $t \in \mathbb{R}_+$ and $h > 0$. Because $W_{t+h} - W_t$ is normally distributed as $N(0, h)$, if we put $Z_{t,h} = \frac{W_{t+h} - W_t}{\sqrt{h}}$, then $Z_{t,h}$ has a standard normal distribution. Therefore, it is clear that there exists an $r > 2$ such that $E[|Z_{t,h}|^r] > 0$, and thus $E[|W_{t+h} - W_t|^r] = E[|Z_{t,h}|^r] h^{\frac{r}{2}}$. If we write $r = 2(1 + \epsilon)$, then we obtain $E[|W_{t+h} - W_t|^r] = ch^{1+\epsilon}$, with $c = E[|Z_{t,h}|^r]$. The assertion then follows by Kolmogorov's continuity theorem. $\qquad\square$

Remark 2.175. Since Brownian motion is continuous in probability, then by Theorem 2.34, it admits a separable and progressively measurable modification.

Theorem 2.176. *Every Wiener process $(W_t)_{t \in \mathbb{R}_+}$ is a Markov diffusion process. Its transition density is*

$$g(x, y; t) = \frac{e^{-\frac{(x-y)^2}{2t}}}{\sqrt{2\pi t}}, \quad \text{for} \quad x, y \in \mathbb{R}, \ t \in \mathbb{R}_+^*.$$

Its infinitesimal generator is

$$\mathcal{A} = \frac{1}{2} \frac{\partial^2}{\partial x^2}, \quad \text{with domain} \quad \mathcal{D}_{\mathcal{A}} = C^2(\mathbb{R}).$$

Proof. The theorem follows directly by Theorem 2.73. See also Lamperti (1977) and Revuz-Yor (1991, p. 264) $\qquad\square$

Theorem 2.177 (Lévy characterization of Brownian motion). *Let $(X_t)_{t \in \mathbb{R}_+}$ be a real-valued continuous random process on a probability space $(\Omega, \mathcal{F}, P)$. Then the following two statements are equivalent:*

1. $(X_t)_{t \in \mathbb{R}_+}$ is a P-Brownian motion.
2. $(X_t)_{t \in \mathbb{R}_+}$ and $(X_t^2 - t)_{t \in \mathbb{R}_+}$ are P-martingales (with respect to their respective natural filtrations).

Proof. (For example, Ikeda and Watanabe (1989)). Here we shall only prove that statement 1 implies statement 2.

The Wiener process $(W_t)_{t \in \mathbb{R}_+}$ is a continuous square-integrable martingale, with $W_t - W_s \sim N(0, t - s)$, for all $0 \le s < t$. To show that $W_t^2 - t$ is also a martingale, we need to show that either

$$E[W_t^2 - t | \mathcal{F}_s] = W_s^2 - s \qquad \forall 0 \le s < t$$

or, equivalently, that

$$E[W_t^2 - W_s^2 | \mathcal{F}_s] = t - s \qquad \forall 0 \le s < t.$$

In fact,

$$E[W_t^2 - W_s^2 | \mathcal{F}_s] = E\left[(W_t - W_s)^2 \middle| \mathcal{F}_s \right] = Var[W_t - W_s] = t - s.$$

Because of uniqueness, we can say that $\langle W_t \rangle = t$ for all $t \ge 0$ by indistinguishability. $\qquad \square$

Additional characterizations are offered by the following proposition.

Proposition 2.178. *Let $(X_t)_{t \in \mathbb{R}_+}$ be a real-valued continuous process starting at 0 at time 0, and let $\mathcal{F}^X$ denote its natural filtration. It is a Wiener process if and only if either of the following statements applies:*

(i) *For any real number λ, the process $\left(\exp\left\{ \lambda X_t - \dfrac{\lambda^2}{2} t \right\} \right)_{t \in \mathbb{R}_+}$ is an $\mathcal{F}^X$-local martingale.*

(ii) *For any real number λ, the process $\left(\exp\left\{ i\lambda X_t + \dfrac{\lambda^2}{2} t \right\} \right)_{t \in \mathbb{R}_+}$ is an $\mathcal{F}^X$-local martingale.*

Proof. See, e.g., Revuz-Yor (1991). $\qquad \square$

We may state the converse of Proposition 2.170 as follows.

Proposition 2.179. *Let $(X_t)_{t \in \mathbb{R}_+}$ be a real-valued continuous process starting at 0 at time 0. If the process is a Gaussian process satisfying*

1. $E[X_t] = 0 \qquad$ *for all $t \in \mathbb{R}_+$*
2. $K(s, t) = Cov[X_t, X_s] = \min\{s, t\}, \qquad s, t \in \mathbb{R}_+$

then it is a Wiener process.

Proof. See, e.g., Revuz-Yor (1991, p. 35). $\qquad \square$

Lemma 2.180. *Let* $(W_t)_{t \in \mathbb{R}_+}$ *be a real-valued Wiener process. If* $a > 0$, *then*

$$P\left(\max_{0 \leq s \leq t} W_s > a\right) = 2P(W_t > a).$$

Proof. We employ the *reflection principle* by defining the process $(\tilde{W}_t)_{t \in \mathbb{R}}$ as

$$\begin{cases} \tilde{W}_t = W_t & \text{if } W_s < a, \forall s < t, \\ \tilde{W}_t = 2a - W_t & \text{if } \exists s < t \text{ such that } W_s = a. \end{cases}$$

The name arises because once $W_s = a$, then $\tilde{W}_s$ becomes a reflection of W_s about the *barrier* a. It is obvious that $(\tilde{W}_t)_{t \in \mathbb{R}}$ is a Wiener process as well. Moreover, we can observe that

$$\max_{0 \leq s \leq t} W_s > a$$

if and only if either $W_t > a$ or $\tilde{W}_t > a$. These two events are mutually exclusive and thus their probabilities are additive. As they are both Wiener processes, it is obvious that the two events have the same probability, and thus

$$P\left(\max_{0 \leq s \leq t} W_s > a\right) = P(W_t > a) + P(\tilde{W}_t > a) = 2P(W_t > a),$$

completing the proof. For a more general case, see (B.8). $\square$

Theorem 2.181. *If* $(W_t)_{t \in \mathbb{R}_+}$ *is a real-valued Wiener process, then*

1. $P(\sup_{t \in \mathbb{R}_+} W_t = +\infty) = 1$
2. $P(\inf_{t \in \mathbb{R}_+} W_t = -\infty) = 1$

Proof. For $a > 0$,

$$P\left(\sup_{t \in \mathbb{R}_+} W_t > a\right) \geq P\left(\sup_{0 \leq s \leq t} W_s > a\right) = P\left(\max_{0 \leq s \leq t} W_s > a\right),$$

where the last equality follows by continuity of trajectories. By Lemma 2.180:

$$P\left(\sup_{t \in \mathbb{R}_+} W_t > a\right) \geq 2P(W_t > a) = 2P\left(\frac{W_t}{\sqrt{t}} > \frac{a}{\sqrt{t}}\right), \text{ for } t > 0.$$

Because W_t is normally distributed as $N(0, t)$, $\frac{W_t}{\sqrt{t}}$ is standard normal and, denoting by Φ its cumulative distribution, we get

$$2P\left(\frac{W_t}{\sqrt{t}} > \frac{a}{\sqrt{t}}\right) = 2\left(1 - \Phi\left(\frac{a}{\sqrt{t}}\right)\right).$$

By $\lim_{t\to\infty} \Phi(\frac{a}{\sqrt{t}}) = \frac{1}{2}$, it follows that

$$\lim_{t\to\infty} 2P\left(\frac{W_t}{\sqrt{t}} > \frac{a}{\sqrt{t}}\right) = 1,$$

and because

$$\left\{\sup_{t\in\mathbb{R}_+} W_t = +\infty\right\} = \bigcap_{a=1}^{\infty}\left\{\sup_{t\in\mathbb{R}_+} W_t > a\right\},$$

we obtain 1.

Point 2 follows directly from point 1, by the observation that if $(W_t)_{t\in\mathbb{R}_+}$ is a real-valued Wiener process, then so is $(-W_t)_{t\in\mathbb{R}_+}$. $\square$

Theorem 2.182. *If $(W_t)_{t\in\mathbb{R}_+}$ is a real-valued Wiener process, then*

$$\forall h > 0, \qquad P\left(\max_{0\leq s\leq h} W_s > 0\right) = P\left(\min_{0\leq s\leq h} W_s < 0\right) = 1.$$

Moreover, for almost every $\omega \in \Omega$ the process $(W_t)_{t\in\mathbb{R}_+}$ has a zero (i.e., crosses the spatial axis) in $]0, h]$ for all $h > 0$.

Proof. If $h > 0$ and $a > 0$, then it is obvious that

$$P\left(\max_{0\leq s\leq h} W_s > 0\right) \geq P\left(\max_{0\leq s\leq h} W_s > a\right).$$

Then, by Lemma 2.180,

$$P\left(\max_{0\leq s\leq h} W_s > a\right) = 2P(W_h > a) = 2P\left(\frac{W_h}{\sqrt{h}} > \frac{a}{\sqrt{h}}\right) = 2\left(1 - \Phi\left(\frac{a}{\sqrt{h}}\right)\right).$$

For $a \to 0$, $2(1 - \Phi(\frac{a}{\sqrt{h}})) \to 1$, and thus $P(\max_{0\leq s\leq h} W_s > 0) = 1$. Furthermore,

$$P\left(\min_{0\leq s\leq h} W_s < 0\right) = P\left(\max_{0\leq s\leq h} (-W_s) > 0\right) = 1.$$

Now we can observe that

$$P\left(\max_{0\leq s\leq h} W_s > 0, \forall h > 0\right) \geq P\left(\bigcap_{n=1}^{\infty}\left(\max_{0<s\leq \frac{1}{n}} W_s > 0\right)\right) = 1.$$

Hence

$$P\left(\max_{0\leq s\leq h} W_s > 0, \forall h > 0\right) = 1$$

and, analogously,

$$P\left(\min_{0 \le s \le h} W_s < 0, \forall h > 0\right) = 1.$$

From this it can be deduced that for almost every $\omega \in \Omega$ the process $(W_t)_{t \in \mathbb{R}_+}$ becomes zero in $]0, h]$ for all $h > 0$. On the other hand, since $(W_t)_{t \in \mathbb{R}_+}$ is a time-homogeneous Markov process with independent increments, it has the same behavior in $]h, 2h]$ as in $]0, h]$, and thus it has zeros in every interval. $\square$

Theorem 2.183. *Almost every trajectory of the Wiener process $(W_t)_{t \in \mathbb{R}_+}$ is nowhere differentiable.*

Proof. Let $D = \{\omega \in \Omega | W_t(\omega)$ be differentiable for at least one $t \in \mathbb{R}_+\}$. We will show that $D \subset G$, with $P(G) = 0$ (obviously, if P is complete, then $D \in \mathcal{F}$). Let $k > 0$ and

$$A_k = \left\{\omega \middle| \limsup_{h \downarrow 0} \frac{|W_{t+h}(\omega) - W_t(\omega)|}{h} < k \text{ for at least one } t \in [0, 1[\right\}.$$

Then, if $\omega \in A_k$, we can choose $m \in \mathbb{N}$ sufficiently large such that $\frac{j-1}{m} \le t < \frac{j}{m}$ for $j \in \{1, \ldots, m\}$, and for $t \le s \le \frac{j+3}{m}$, $W(s, \omega)$ is enveloped by a cone with slope k. Then, for an integer $j \in \{1, \ldots, m\}$, we get

$$\left|W_{\frac{j+1}{m}}(\omega) - W_{\frac{j}{m}}(\omega)\right| \le \left|W_{\frac{j+1}{m}}(\omega) - W_t(\omega)\right| + \left|-W_t(\omega) + W_{\frac{j}{m}}(\omega)\right|$$
$$< \left(\frac{j+1}{m} - \frac{j-1}{m}\right)k + \left(\frac{j}{m} - \frac{j-1}{m}\right)k$$
$$= \frac{2k}{m} + \frac{k}{m} = \frac{3k}{m}. \tag{2.35}$$

Analogously, we obtain that

$$\left|W_{\frac{j+2}{m}}(\omega) - W_{\frac{j+1}{m}}(\omega)\right| \le \frac{5k}{m} \tag{2.36}$$

and

$$\left|W_{\frac{j+3}{m}}(\omega) - W_{\frac{j+2}{m}}(\omega)\right| \le \frac{7k}{m}. \tag{2.37}$$

Because $\frac{W_{t+h}-W_t}{\sqrt{h}}$ is distributed as $N(0,1)$, it follows that

$$P(|W_{t+h}-W_t|<a) = P\left(\frac{|W_{t+h}-W_t|}{\sqrt{h}}<\frac{a}{\sqrt{h}}\right)$$

$$= \int_{-\frac{a}{\sqrt{h}}}^{\frac{a}{\sqrt{h}}} \frac{1}{\sqrt{2\pi}}\exp\left\{-\frac{x^2}{2}\right\}dx$$

$$\le \frac{1}{\sqrt{2\pi}}2\frac{a}{\sqrt{h}} = \frac{2a}{\sqrt{2\pi h}}.$$

Putting $A_{m,j} = \{\omega|(2.35),(2.36),(2.37)$ are true$\}$, because the process has independent increments, we obtain

$$P(A_{m,j}) = P(\{\omega|(2.35) \text{ is true}\})P(\{\omega|(2.36) \text{ is true}\})P(\{\omega|(2.37) \text{ is true}\})$$

$$\le 8\left(\frac{2\pi}{m}\right)^{-\frac{3}{2}}\frac{3k}{m}\frac{5k}{m}\frac{7k}{m},$$

and thus $P(A_{m,j}) \le cm^{-\frac{3}{2}}$, $j=1,\ldots,m$. Putting $A_m = \bigcup_{j=1}^m A_{m,j}$, then

$$P(A_m) \le \sum_{j=1}^m P(A_{m,j}) \le cm^{-\frac{1}{2}}.$$

Now let $m=n^4$ $(n\in\mathbb{N}^*)$; we obtain $P(A_{n^4}) \le cn^{-2} = \frac{c}{n^2}$ and thus

$$\sum_n P(A_{n^4}) \le c\sum_n \frac{1}{n^2} < \infty.$$

Therefore, by the Borel–Cantelli Lemma 1.161,

$$P\left(\limsup_n A_{n^4}\right) = 0.$$

It can now be shown that

$$A_k \subset \liminf_m A_m \equiv \bigcup_m \bigcap_{i\ge m} A_i \subset \liminf_n A_{n^4} \subset \limsup_n A_{n^4},$$

hence $A_k \subset A_{n^4}''$ and $P(A_{n^4}'') = 0$. Let

$$D_0 = \{\omega|W(\cdot,\omega) \text{ is differentiable in at least one } t\in[0,1[\}.$$

Then $D_0 \subset \bigcup_{k=1}^\infty A_k = G_0$, which means that D_0 is contained in a set of probability zero, namely, $D_0 \subset G_0$ and $P(G_0) = 0$. Decomposing $\mathbb{R}_+ = \bigcup_n [n,n+1[$, since the motion is Brownian and of independent increments,

$$D_n = \{\omega|W(\cdot,\omega) \text{ is differentiable in at least one } t\in[n,n+1[\},$$

analogously to D_0, will be contained in a set of probability zero, i.e., $D_n \subset G_n$ and $P(G_n) = 0$. But $D \subset \bigcup_n D_n \subset \bigcup_n G_n$, thus completing the proof. $\square$

A trivial consequence of the preceding theorem is the following corollary.

Corollary 2.184. *Almost every trajectory of a Wiener process* $(W_t)_{t \in \mathbb{R}_+}$ *is of unbounded variation on any finite interval.*

An important property of the trajectories of a Brownian motion is their Hölder continuity. We may recall that a real function f defined on a real line satisfies a Hölder condition, or is Hölder continuous, when there are nonnegative real constants C, α such that

$$|f(y) - f(x)| \leq C|y - x|^\alpha$$

for all x and y in the domain of f. The number α is called the exponent or order of the Hölder condition.

The following theorem holds.

Theorem 2.185. *Almost every trajectory of a Wiener process* $(W_t)_{t \in \mathbb{R}_+}$ *is Hölder continuous for any order* $\alpha < \dfrac{1}{2}$.

Almost every trajectory of a Wiener process $(W_t)_{t \in \mathbb{R}_+}$ *is not Hölder continuous for any order* $\alpha \geq \dfrac{1}{2}$.

Proof. See, e.g., $\square$

Proposition 2.186. *Let* $(W_t)_{t \in \mathbb{R}_+}$ *be a Wiener process; then the following properties hold:*

(i) *(Symmetry) The process* $(-W_t)_{t \in \mathbb{R}_+}$ *is a Wiener process.*

(ii) *(Time scaling) The time-scaled process* $(\tilde{W}_t)_{t \in \mathbb{R}_+}$ *defined by*

$$\tilde{W}_t = tW_{1/t}, \, t > 0, \quad \tilde{W}_0 = 0$$

is also a Wiener process.

(iii) *(Space scaling) For any* $c \neq 0$, *the space-scaled process* $(\tilde{W}_t)_{t \in \mathbb{R}_+}$ *defined by*

$$\tilde{W}_t = cW_{t/c^2}, \, t > 0, \quad \tilde{W}_0 = 0,$$

is also a Wiener process.

Proof. See, e.g., Karlin and Taylor (1975), Rogers and Williams (1994, p. 4), Klenke (2014, p. 465). $\square$

Proposition 2.187. *If $(W_t)_{t\in\mathbb{R}_+}$ is a Wiener process, then the process*

$$X_t = W_t - tW_1, \quad t \in [0,1]$$

is a Brownian bridge.

Proof. See, e.g., Revuz-Yor (1991, p. 35). □

We may observe that $X_0 = X_1 = 0$, from which the name follows.

Proposition 2.188 (Strong law of large numbers). *Let $(W_t)_{t\in\mathbb{R}_+}$ be a Wiener process. Then*

$$\frac{W_t}{t} \to 0, \qquad as \qquad t \to +\infty, \qquad a.s.$$

Proof. See, e.g., Karlin and Taylor (1975). □

Proposition 2.189 (Law of iterated logarithms). *Let $(W_t)_{t\in\mathbb{R}_+}$ be a Wiener process. Then*

$$\limsup_{t\to+\infty} \frac{W_t}{\sqrt{2t\ln\ln t}} = 1, \qquad a.s.,$$

$$\liminf_{t\to+\infty} \frac{W_t}{\sqrt{2t\ln\ln t}} = -1, \qquad a.s.$$

As a consequence, for any $\epsilon > 0$ there exists a $t_0 > 0$ such that for any $t > t_0$ we have

$$-(1+\epsilon)\sqrt{2t\ln\ln t} \le W_t \le (1+\epsilon)\sqrt{2t\ln\ln t}, \qquad a.s.$$

Moreover,

$$P(W_t \ge (1+\epsilon)\sqrt{2t\ln\ln t}, \; i.o.) = 0;$$

while

$$P(W_t \ge (1-\epsilon)\sqrt{2t\ln\ln t}, \; i.o.) = 1.$$

Proof. See, e.g., Breiman (1968). □

Proposition 2.190. *For almost every $\omega \in \Omega$ the trajectory $(W_t(\omega))_{t\in\mathbb{R}_+}$ of a Wiener process is locally Hölder continuous with exponent δ if $\delta \in (0,\frac{1}{2})$. But for almost every $\omega \in \Omega$ it is nowhere Hölder continuous with exponent δ if $\delta > \frac{1}{2}$.*

Wiener Process Started at x

Let $(W_t)_{t \in \mathbb{R}_+}$ be a Wiener process. For any $x \in \mathbb{R}$ the process $(W_t^x)_{t \in \mathbb{R}_+}$, defined by

$$W_t^x := x + W_t \quad t \in \mathbb{R}_+,$$

is called a *Wiener process started at x*. It is such that for any $t \in \mathbb{R}_+$ and any $B \in \mathcal{B}_\mathbb{R}$

$$P(W_t^x \in B) = \frac{1}{\sqrt{2\pi t}} \int_B e^{-\frac{(y-x)^2}{2t}} dy.$$

Reflected Brownian Motion

If $(W_t)_{t \in \mathbb{R}_+}$ is a Wiener process, the process $(|W_t|)_{t \in \mathbb{R}_+}$ is valued in $\mathbb{R}_+$; its transition density is

$$g(x, y; t) = \frac{1}{\sqrt{2\pi t}} \left[\exp\left\{ -\frac{(y-x)^2}{2t} \right\} + \exp\left\{ -\frac{(y+x)^2}{2t} \right\} \right],$$

for $x, y \in \mathbb{R}_+$, $t \in \mathbb{R}_+^*$.

It is known as *reflected Brownian motion*.

Its infinitesimal generator is

$$\mathcal{A} = \frac{1}{2} \frac{\partial^2}{\partial x^2}, \quad \text{with domain} \quad \mathcal{D}_\mathcal{A} = \{ f \in C^2(\mathbb{R}_+) \, | \, f'(0) = 0 \}$$

(Lamperti (1977), pp. 126 and 173).

Absorbed Brownian Motion

Let $(W_t)_{t \in \mathbb{R}_+}$ be a Wiener process; for a given $a \in \mathbb{R}$ let τ_a denote the first passage time of the process started at $W_0 = 0$. The stopped process $(X_t)_{t \in \mathbb{R}_+}$ defined by

$$X_t = W_t, \quad \text{for} \quad 0 \le t \le \tau_a$$
$$X_t = a, \quad \text{for} \quad t \ge \tau_a,$$

is called *absorbed Brownian motion*.

Its cumulative probability distribution is given by

$$P(X_t \le y) = \begin{cases} \frac{1}{\sqrt{2\pi t}} \int_{-\infty}^y e^{-\frac{z^2}{2t}} dz - \int_{2a-y}^{+\infty} e^{-\frac{z^2}{2t}} dz & \text{for} \quad y < a, \\ 1 & \text{for} \quad y \ge a, \end{cases}$$

for any $t \in \mathbb{R}_+$ and any $y \in \mathbb{R}$.

Its infinitesimal generator is

$$\mathcal{A} = \frac{1}{2} \frac{\partial^2}{\partial x^2},$$

with domain $\mathcal{D}_\mathcal{A} = \{ f \in C^2(\mathbb{R}_+) \, | \, f(x) = 0, \text{ for } x \ge a \}$ (Schuss (2010, p. 58)).

Brownian Motion After a Stopping Time

Let $(W(t))_{t \in \mathbb{R}_+}$ be a Wiener process with a finite stopping time T and $\mathcal{F}_T$ the σ-algebra of events preceding T. By Remark 2.175 and Theorem 2.49, $W(T)$ is $\mathcal{F}_T$-measurable and, hence, measurable.

Remark 2.191. Brownian motion is endowed with the Feller property and therefore also with the strong Markov property (This can be shown using the representation of the semigroup associated with $(W(t))_{t \in \mathbb{R}_+}$.).

Theorem 2.192. *Resorting to the previous notation, we have that*

1. *The process $y(t) = W(T + t) - W(T), t \geq 0$, is again a Brownian motion.*
2. *$\sigma(y(t), t \geq 0)$ is independent of $\mathcal{F}_T$.*

(Thus a Brownian motion remains a Brownian motion after a stopping time.)

Proof. If $T = s$ (s constant), then the assertion is obvious. We now suppose that T has a countable codomain $(s_j)_{j \in \mathbb{N}}$ and that $B \in \mathcal{F}_T$. If we consider further that $0 \leq t_1 < \cdots < t_n$ and that $A_1, \ldots, A_n$ are Borel sets of $\mathbb{R}$, then

$$P(y(t_1) \in A_1, \ldots, y(t_n) \in A_n, B)$$
$$= \sum_{j \in \mathbb{N}} P(y(t_1) \in A_1, \ldots, y(t_n) \in A_n, B, T = s_j)$$
$$= \sum_{j \in \mathbb{N}} P((W(t_1 + s_j) - W(s_j)) \in A_1, \ldots$$
$$\ldots, (W(t_n + s_j) - W(s_j)) \in A_n, B, T = s_j).$$

Moreover, $(T = s_j) \cap B = (B \cap (T \leq s_j)) \cap (T = s_j) \in \mathcal{F}_{s_j}$ (as observed in the proof of Theorem 2.49), and since a Wiener process has independent increments, the events $((W(t_1 + s_j) - W(s_j)) \in A_1, \ldots, (W(t_n + s_j) - W(s_j)) \in A_n)$ and $(B, T = s_j)$ are independent; therefore,

$$P(y(t_1) \in A_1, \ldots, y(t_n) \in A_n, B)$$
$$= \sum_{j \in \mathbb{N}} P((W(t_1 + s_j) - W(s_j)) \in A_1, \ldots$$
$$\ldots, (W(t_n + s_j) - W(s_j)) \in A_n) P(B, T = s_j)$$
$$= \sum_{j \in \mathbb{N}} P(W(t_1) \in A_1, \ldots, W(t_n) \in A_n) P(B, T = s_j)$$
$$= P(W(t_1) \in A_1, \ldots, W(t_n) \in A_n) P(B),$$

where we note that $W(t_k + s_j) - W(s_j)$ has the same distribution as $W(t_k)$. From these equations (having factorized) follows point 2. Furthermore, if we take $B = \Omega$, we obtain

$$P(y(t_1) \in A_1, \ldots, y(t_n) \in A_n) = P(W(t_1) \in A_1, \ldots, W(t_n) \in A_n).$$

This shows that the finite-dimensional distributions of the process $(y(t))_{t\geq 0}$ coincide with those of W. Therefore, by the Kolmogorov–Bochner theorem, the proof of 1 is complete.

Let T be a generic finite stopping time of the Wiener process $(W_t)_{t\geq 0}$ and (as in Lemma 2.97) $(T_n)_{n\in\mathbb{N}}$ a sequence of stopping times such that $T_n \geq T, T_n \downarrow T$ as $n \to \infty$ and T_n has an at most countable codomain. We put, for all $n \in \mathbb{N}$, $y_n(t) = W(T_n + t) - W(T_n)$ and let $B \in \mathcal{F}_T, 0 \leq t_1 \leq \cdots \leq t_k$. Then, because for all $n \in \mathbb{N}$, $\mathcal{F}_T \subset \mathcal{F}_{T_n}$ (see the proof of Theorem 2.98) and for all $n \in \mathbb{N}$, the theorem holds for T_n (as already shown above), we have

$$P(y_n(t_1) \leq x_1, \ldots, y_n(t_k) \leq x_k, B) = P(W(t_1) \leq x_1, \ldots, W(t_k) \leq x_k)P(B).$$

Moreover, since W is continuous, from $T_n \downarrow T$ as $n \to \infty$, it follows that $y_n(t) \to y(t)$ a.s. for all $t \geq 0$. Thus, if $(x_1, \ldots, x_k)$ is a point of continuity of the k-dimensional distribution F_k of $(W(t_1), \ldots, W(t_k))$, we get by Lévy's continuity Theorem 1.178

$$P(y(t_1) \leq x_1, \ldots, y(t_k) \leq x_k, B)$$
$$= P(W(t_1) \leq x_1, \ldots, W(t_k) \leq x_k)P(B). \tag{2.38}$$

Since F_k is continuous almost everywhere (given that Gaussian distributions are absolutely continuous with respect to the Lebesgue measure and thus have density), (2.38) holds for every $x_1, \ldots, x_k$. Therefore, for every Borel set $A_1, \ldots, A_k$ of $\mathbb{R}$, we have that

$$P(y(t_1) \in A_1, \ldots, y(t_k) \in A_k, B) = P(W(t_1) \in A_1, \ldots, W(t_n) \in A_k)P(B),$$

completing the proof.   □

Definition 2.193. The real-valued process $(W_1(t), \ldots, W_n(t))'_{t\geq 0}$ is said to be an n-*dimensional Wiener process (or Brownian motion)* if

1. For all $i \in \{1, \ldots, n\}$, $(W_i(t))_{t\geq 0}$ is a Wiener process
2. The processes $(W_i(t))_{t\geq 0}$, $i = 1, \ldots, n$, are independent

(thus, the σ-algebras $\sigma(W_i(t), t \geq 0)$, $i = 1, \ldots, n$, are independent).

Proposition 2.194. *If* $(W_1(t), \ldots, W_n(t))'_{t\geq 0}$ *is an* n-*dimensional Brownian motion, then it can be shown that*

1. $(W_1(0), \ldots, W_n(0)) = (0, \ldots, 0)$ *almost surely.*
2. $(W_1(t), \ldots, W_n(t))'_{t\geq 0}$ *has independent increments.*
3. $(W_1(t), \ldots, W_n(t))' - (W_1(s), \ldots, W_n(s))', 0 \leq s < t$, *has multivariate normal distribution* $N(\mathbf{0}, (t - s)I)$ *(where* $\mathbf{0}$ *is the null vector of order* n *and* I *is the* $n \times n$ *identity matrix).*

Proof. The proof follows from Definition 2.193.   □

2.9 Counting, and Poisson Processes

Whereas Brownian motion and the Wiener process are continuous in space and time, there exists a family of processes that are continuous in time, but discontinuous in space, admitting jumps. The simplest of these is a counting process, of which the Poisson process is a special case. The latter also allows many explicit results. The most general process admitting both continuous and discontinuous movements is the Lévy process, which contains both Brownian motion and the Poisson process. Finally, a stable process is a particular type of Lévy process, which reproduces itself under addition.

Though not necessary in general, here we refer to simple counting processes (see later for a definition).

Definition 2.195. Let $(\tau_i)_{i \in \mathbb{N}^*}$ be a strictly increasing sequence of positive random variables on the space $(\Omega, \mathcal{F}, P)$, with $\tau_0 \equiv 0$. Then the process $(N_t)_{t \in \mathbb{R}_+}$ given by

$$N_t = \sum_{i \in \mathbb{N}^*} I_{[\tau_i, +\infty]}(t), \qquad t \in \bar{\mathbb{R}}_+,$$

valued in $\bar{\mathbb{N}}$, is called a *counting process* associated with the sequence $(\tau_i)_{i \in \mathbb{N}^*}$. Moreover, the random variable $\tau = \sup_i \tau_i$ is the *explosion time* of the process. If $\tau = \infty$ almost surely, then N_t is *nonexplosive*.

We may easily notice that, due to the following equality, which holds for any $t_1, t_2, \ldots, t_n \in \mathbb{R}_+$,

$$P(\tau_1 \le t_1, \tau_2 \le t_2, \ldots, \tau_n \le t_n) = P(N(t_1) \ge 1, N(t_2) \ge 2, \ldots, N(t_n) \ge n),$$

we may claim that it is equivalent to knowledge of the probability law of $(N_t)_{t \in \mathbb{R}_+}$ and that of $(\tau_n)_{n \in \mathbb{N}^*}$.

Theorem 2.196. *Let $(\mathcal{F}_t)_{t \in \mathbb{R}_+}$ be a filtration that satisfies the usual hypotheses (Definition 2.35). A counting process $(N_t)_{t \in \mathbb{R}_+}$ is adapted to $(\mathcal{F}_t)_{t \in \mathbb{R}_+}$ if and only if its associated random variables $(\tau_i)_{i \in \mathbb{N}^*}$ are stopping times.*

Proof. See, e.g., Protter (1990, p. 13). □

The following proposition holds (Protter 1990, p. 16).

Theorem 2.197. *Let $(N_t)_{t \in \mathbb{R}_+}$ be a counting process. Then its natural filtration is right-continuous.*

Hence, by a suitable extension, we may consider as underlying filtered space the given probability space $(\Omega, \mathcal{F}, P)$ endowed with the natural filtration $\mathcal{F}_t = \sigma\{N_s | s \le t\}$.

With respect to the natural filtration, the jump times τ_n for $n \in \mathbb{N}^*$ are stopping times.

Remark 2.198. A nonexplosive counting process is RCLL. Its trajectories are right-continuous step functions with upward jumps of magnitude 1 and $N_0 = 0$ almost surely.

Proposition 2.199. *An RCLL process may admit at most jump discontinuities.*

Definition 2.200. We say that a process $(X_t)_{t \in \mathbb{R}_+}$ has a *fixed jump* at a time t if $P(X_t \neq X_{t-}) > 0$.

Poisson Process

Definition 2.201. A counting process $(N_t)_{t \in \mathbb{R}_+}$ is a *Poisson process* if it is a process with time-homogeneous independent increments.

Theorem 2.202(Çynlar (1975, p. 71); Protter 1990, p. 13). *Let $(N_t)_{t \in \mathbb{R}_+}$ be a Poisson process. Then a $\lambda > 0$ exists such that, for any $t \in \mathbb{R}_+$, N_t has a Poisson distribution with parameter λt, i.e.,*

$$P(N_t = n) = e^{-\lambda t} \frac{(\lambda t)^n}{n!}, \qquad n \in \mathbb{N}.$$

Moreover $(N_t)_{t \in \mathbb{R}_+}$ is continuous in probability and does not have explosions.

Proposition 2.203 (Chung (1974)). *Let $(N_t)_{t \in \mathbb{R}_+}$ be a Poisson process. Then*

$$P(\tau_\infty = \infty) = 1,$$

namely, almost all sample functions are step functions.

The following theorem specifies the distribution of the random variable N_t, $t \in \mathbb{R}_+$.

Theorem 2.204. *Let $(N_t)_{t \in \mathbb{R}_+}$ be a Poisson process of intensity $\lambda > 0$. Then for any $t \in \mathbb{R}_+$, $E[N_t] = \lambda t$, $Var[N_t] = \lambda t$, its characteristic function is*

$$\phi_{N_t}(u) = E\left[e^{iuN_t}\right] = e^{-\lambda t(1 - \exp\{iu\})},$$

and its probability-generating function is

$$g_{N_t}(u) = E\left[u^{N_t}\right] = e^{\lambda t(u-1)}, \qquad u \in \mathbb{R}_+^*.$$

Proof. All formulas are a consequence of the Poisson distribution of N_t for any $t \in \mathbb{R}_+$:

$$E[N_t] = \sum_{n=0}^{\infty} n \frac{(\lambda t)^n}{n!} e^{-\lambda t} = \lambda t \sum_{n=0}^{\infty} \frac{(\lambda t)^{n-1}}{(n-1)!} e^{-\lambda t} = \lambda t,$$

$$E\left[N_t^2\right] = \sum_{n=0}^{\infty} n^2 \frac{(\lambda t)^n}{n!} e^{-\lambda t} = \lambda t \sum_{n=0}^{\infty} ((n-1)+1) \frac{(\lambda t)^{n-1}}{(n-1)!} e^{-\lambda t}$$

$$= (\lambda t)^2 + \lambda t,$$

$$Var[N_t] = E\left[N_t^2\right] - (E[N_t])^2,$$

$$E\left[e^{iuN_t}\right] = \sum_{n=0}^{\infty} e^{iun} \frac{(\lambda t)^n}{n!} e^{-\lambda t} = e^{-\lambda t(1-\exp\{iu\})} \sum_{n=0}^{\infty} \frac{\left(\lambda t e^{iu}\right)^n}{n!} e^{-\lambda t \exp\{iu\}}$$

$$= e^{-\lambda t(1-\exp\{iu\})},$$

$$E\left[u^{N_t}\right] = \sum_{n=0}^{\infty} u^n \frac{(\lambda t)^n}{n!} e^{-\lambda t} = e^{\lambda t(u-1)} \sum_{n=0}^{\infty} \frac{(u\lambda t)^n}{n!} e^{-u\lambda t} = e^{\lambda t(u-1)}.$$

$\square$

Due to the independence of the increments, the following theorem holds.

Theorem 2.205. *A Poisson process* $(N_t)_{t\in\mathbb{R}_+}$ *is an RCLL Markov process.*

Proposition 2.206 (Rolski et al. (1999), p. 157; Billingsley (1986), p. 307). *Let* $(N_t)_{t\in\mathbb{R}_+}$ *be a counting process. From the definition,* $\tau_n = \inf\{t \in \mathbb{R}_+ : N_t \geq n\}$; *we denote by* $T_n = \tau_n - \tau_{n-1}$, *for* $n \in \mathbb{N} \setminus \{0\}$, *the interarrival times. The following statements are all equivalent:*

P^1 : $(N_t)_{t\in\mathbb{R}_+}$ *is a Poisson process with intensity parameter* $\lambda > 0$.

P^2 : T_n *are independent exponentially distributed random variables with parameter* λ.

P^3 : *For any* $t \in \mathbb{R}_+$, *and for any* $n \in \mathbb{N} - \{0\}$, *the joint conditional distribution of* $(T_1, \ldots, T_n)$, *given* $\{N_t = n\}$, *has density*

$$\frac{n!}{t^n} \mathbf{1}_{\{0 < t_1 < \cdots < t_n\}}$$

with respect to the Lebesgue measure, i.e., it has the same distribution of the order statistics of n independent real random variables having uniform law on $[0, t]$.

P^4 : *For any* $0 < t_1 < \cdots < t_k$ *the increments* $N_{t_2} - N_{t_1}, \ldots, N_{t_k} - N_{t_{k-1}}$ *are independent and each of them is Poisson distributed:*

$$N_{t_i} - N_{t_{i-1}} \sim P(\lambda(t_i - t_{i-1})).$$

P^5 : $(N_t)_{t\in\mathbb{R}_+}$ *has time-homogeneous independent increments and, as* $h \downarrow 0$,

$$P(N_h = 1) = \lambda h + o(h),$$
$$P(N_h \geq 2) = o(h);$$

moreover, $(N_t)_{t\in\mathbb{R}_+}$ *has no fixed jumps.*

Theorem 2.207. *A process $(N_t)_{t \in \mathbb{R}_+}$ with stationary increments has a version in which it is constant on all sample paths except for upward jumps of magnitude 1 if and only if there exists a parameter $\lambda > 0$ so that its characteristic function*

$$\phi_{N_t}(u) = E\left[e^{iuN_t}\right] = e^{-\lambda t(1 - \exp\{iu\})}$$

or, equivalently, $N_t \sim P(\lambda t)$.

Proof. See, e.g., Breiman (1968). □

Remark 2.208. Let us consider a Poisson process $(N_t)_{t \in \mathbb{R}_+}$ with intensity parameter $\lambda > 0$; for any $a, b \in \mathbb{R}_+$, $a < b$ we denote

$$N((a, b]) = N_b - N_a. \tag{2.39}$$

Due to the fact that $(N_t)_{t \in \mathbb{R}_+}$ is nondecreasing and *càdlàg*, with $N(0) = 0$, by means of (2.39) it will generate a random measure on $\mathcal{B}_{\mathbb{R}_+}$ in the usual way. It is such that, for any $B \in \mathcal{B}_{\mathbb{R}_+}$,

$$N(B) = \sharp\{n \in \mathbb{N}^* | \tau_n \in B\}.$$

It is not difficult to show that $N(B)$ is a Poisson random variable with parameter $\lambda \nu^1(B)$, where ν^1 denotes the usual Lebesgue measure on $\mathcal{B}_{\mathbb{R}_+}$.

A particular consequence of this is the fact that a Poisson process $(N_t)_{t \in \mathbb{R}_+}$ cannot have a fixed jump at any $t_0 \in \mathbb{R}_+$ since

$$P(N_{t_0} - N_{t_0-}) = P(N(\{t_0\} > 0) = 1 - e^{-\lambda \nu^1(\{t_0\})} = 1 - 1 = 0.$$

Theorem 2.209. *Let $(N_t)_{t \in \mathbb{R}_+}$ be a Poisson process of intensity λ. Then $(N_t - \lambda t)_{t \in \mathbb{R}_+}$ and $((N_t - \lambda t)^2 - \lambda t)_{t \in \mathbb{R}_+}$ are martingales.*

Remark 2.210. Because $M_t = (N_t - \lambda t)^2 - \lambda t$ is a martingale, by uniqueness, the process $(\lambda t)_{t \in \mathbb{R}_+}$ is the predictable compensator of $(N_t - \lambda t)^2$, i.e., $\langle (N_t - \lambda t)^2 \rangle = \lambda t$, for all $t \in \mathbb{R}_+$, as well as the compensator of the Poisson process $(N_t)_{t \in \mathbb{R}_+}$ directly.

The following theorem, known as the *Watanabe characterization*, provides the converse of Theorem 2.209 (e.g., Bremaud (1981), p. 25).

Theorem 2.211. *Let $(N_t)_{t \in \mathbb{R}_+}$ be a counting process. Suppose that a deterministic $\lambda \in \mathbb{R}_+^*$ exists such that $(N_t - \lambda t)_{t \in \mathbb{R}_+}$ is a martingale with respect to the natural filtration of the process; then, $(N_t)_{t \in \mathbb{R}_+}$ is a Poisson process.*

Corollary 2.212 (Çynlar (1975, p. 76)). *Let $(N_t)_{t \in \mathbb{R}_+}$ be an integer-valued stochastic process such that its trajectory almost surely satisfies the following statements:*

1. *It is nondecreasing.*
2. *It increases by jumps only.*
3. *It is right continuous.*
4. $N_0 = 0$.

Then $(N_t)_{t \in \mathbb{R}_+}$ is a Poisson process with (deterministic) intensity $\lambda \in \mathbb{R}_+^*$ if and only if

(a) *Almost surely each jump of the process is of unit magnitude.*
(b) *For any $s, t \in \mathbb{R}_+$*

$$E[N_{t+s} - N_t | \mathcal{F}_t] = \lambda s \quad \text{a.s.}$$

Theorem 2.211 can be extended to the nonhomogeneous case.

Theorem 2.213. *Let $(N_t)_{t \in \mathbb{R}_+}$ be a counting process, and let $\lambda : \mathbb{R}_+ \to \mathbb{R}_+$ be a locally integrable function such that $(N_t - \int_0^t \lambda(s)ds)_{t \in \mathbb{R}_+}$ is a martingale with respect to the natural filtration of the process; then, $(N_t)_{t \in \mathbb{R}_+}$ is a Poisson process, with nonhomogeneous intensity $\lambda(t)$, i.e., for all $0 \leq s \leq t$, $N_t - N_s$ is a Poisson random variable with parameter $\int_s^t \lambda(\tau)d\tau$, independent of $\mathcal{F}_s$.*

Definition 2.214. A counting process $(N_t)_{t \in \mathbb{R}_+}$ is *simple* if

$$P(N_t - N_{t^-} \in \{0, 1\} \quad \text{for any} \quad t \in \mathbb{R}_+) = 1.$$

Definition 2.215. A counting process $(N_t)_{t \in \mathbb{R}_+}$ is *orderly* if

$$\lim_{h \downarrow 0} \frac{1}{h} P(N_t \geq 2) = 0.$$

Proposition 2.216. *A Poisson process with intensity $\lambda > 0$ is orderly and simple.*

Proof. A Poisson process has time-homogeneous increments; since it is orderly, it is also simple by Proposition 3.3.VI in Daley and Vere-Jones (1988). Further, since λ is finite, simplicity implies orderliness by Dobrushin's lemma (e.g., Daley and Vere-Jones (1988), p. 48). $\qquad \square$

Theorem 2.217. *Let $(N_t)_{t \in \mathbb{R}_+}$ be a simple counting process on $\mathbb{R}_+$ adapted to $\mathcal{F}_t$. If the $\mathcal{F}_t$-compensator $(A_t)_{t \in \mathbb{R}_+}$ of $(N_t)_{t \in \mathbb{R}_+}$ is a continuous and $\mathcal{F}_0$-measurable random process, then $(N_t)_{t \in \mathbb{R}_+}$ is a doubly stochastic Poisson process (with stochastic intensity), directed by A_t, also known as a Cox process.*

Proof. For $u \in \mathbb{R}$ let

$$M_t(u) = e^{iuN_t - (\exp\{iu\} - 1)A_t}.$$

Then, using the properties of stochastic integrals, it can be shown that

$$E[M_t(u)|\mathcal{F}_0] = E\left[e^{iuN_t-(\exp\{iu\}-1)A_t}\,\middle|\,\mathcal{F}_0\right] = 1.$$

Because A_t is assumed to be $\mathcal{F}_0$-measurable,

$$E\left[e^{iuN_t}|\mathcal{F}_0\right] = e^{(\exp\{iu\}-1)A_t},$$

representing the characteristic function of a Poisson distribution with (stochastic) intensity A_t. □

Definition 2.218. We say that $(Z_t)_{t\in\mathbb{R}_+}$ is a *compound Poisson process* if it can be expressed as

$$Z_0 = 0$$

and

$$Z_t = \sum_{k=1}^{N_t} Y_k, \text{ for } t > 0,$$

where N_t is a Poisson process with intensity parameter $\lambda \in \mathbb{R}_+^*$ and $(Y_k)_{k\in\mathbb{N}^*}$ is a family of i.i.d. random variables, independent of N_t, whose common law P_Y has no atom at zero.

By proceeding as for the compound Poisson distribution, we may easily show that the characteristic function of a compound Poisson process, for any $t \in \mathbb{R}_+$, is given by

$$\phi_{X_t}(u) = \exp\left\{-t\int_{\mathbb{R}} \lambda(1-e^{iuy})P_Y(dy)\right\}$$
$$= \exp\left\{-t\int_{\mathbb{R}} (1-e^{iuy})\nu(dy)\right\}, \qquad u \in \mathbb{R}, \qquad (2.40)$$

if we set $\nu := \lambda P_Y$.

2.10 Random Measures

Consider a locally compact Polish space $(E, \mathcal{B}_E)$ (for example, $E = \mathbb{R}^d$, $d \in \mathbb{N}^*$); we denote by $\mathcal{N}$ the family of all σ-finite measures on $(E, \mathcal{B}_E)$; we define the measurable space $(\mathcal{N}, \mathcal{B}_\mathcal{N})$ by assigning $\mathcal{B}_\mathcal{N}$ as the smallest σ-algebra on $\mathcal{N}$ with respect to which all maps

$$\mu \in \mathcal{N} \mapsto \left\{\mu(B) \in \mathcal{B}_{\overline{\mathbb{R}}_+}, B \in \mathcal{B}_E\right\}$$

are measurable.

Definition 2.219. Given a probability space $(\Omega, \mathcal{F}, P)$, a *random measure* on $(E, \mathcal{B}_E)$ is any measurable function

$$N : (\Omega, \mathcal{F}) \to (\mathcal{N}, \mathcal{B}_{\mathcal{N}}).$$

We say that N is a *random point measure* on $(E, \mathcal{B}_E)$ if $\mathcal{N}$ is the family of all σ−finite integer valued measures on $(E, \mathcal{B}_E)$, i.e., for any $B \in \mathcal{B}_E$, $N(B) \in \overline{\mathbb{N}} := \mathbb{N} \cup \{\infty\}$.

2.10.1 Poisson Random Measures

Definition 2.220. Given a probability space $(\Omega, \mathcal{F}, P)$, a *Poisson random measure* with intensity measure Λ on the Polish space $(E, \mathcal{B}_E)$ is a random point measure N on $(E, \mathcal{B}_E)$ such that

(i) for any $B \in \mathcal{B}_E$, $N(B)$ is an integer valued random variable on $(\Omega, \mathcal{F}, P)$, admitting a Poisson distribution with parameter $\Lambda(B)$, i.e.

$$P(N(B) = k) = e^{-\Lambda(B)} \frac{(\Lambda(B))^k}{k!}, \quad k \in \mathbb{N},$$

where Λ is a deterministic σ−finite measure on $\mathcal{B}_E$, called the *intensity measure* of the process.
An obvious consequence is that

$$\Lambda(B) = E[N(B)], \quad B \in \mathcal{B}_E.$$

(ii) for any finite family of disjoint sets $B_1, B_2, \ldots, B_k, k \in \mathbb{N} \setminus \{0, 1\}$, of elements of $\mathcal{B}_E$, the random variables $N(B_1), N(B_2), \ldots, N(B_k)$ are independent.

In (i) we assume that whenever $\Lambda(B) = 0$, then $N(B) = 0$, a.s.; while whenever $\Lambda(B) = +\infty$, then $N(B) = \infty$, a.s.

Theorem 2.221 *Given a deterministic σ−finite measure Λ on a Polish space $(E, \mathcal{B}_E)$, there exists a Poisson random measure N on $(E, \mathcal{B}_E)$, such that for any $B \in \mathcal{B}_E$,*
$$\Lambda(B) = E[N(B)].$$

Proof. See, e.g., Ikeda and Watanabe (1989, p. 42). $\qquad\qquad\square$

Proposition 2.222 *Suppose that N is a Poisson random measure with intensity measure Λ on the Polish space $(E, \mathcal{B}_E)$; then the support of N is P−a.s. countable. If in addition Λ is a finite measure, then the support of N is P−a.s. finite.*

Proof. See, e.g., Kyprianou (2014, p. 42). $\qquad\qquad\square$

The following theorem somehow extends to Poisson random measures the known characterization of the Poisson process as from Section 2.9.

Theorem 2.223 *Let Λ be an atom-free measure on the Polish space $(E, \mathcal{B}_E)$; that is $\Lambda(\{x\}) = 0$, for any $x \in E$. Let N be a random point measure on $(E, \mathcal{B}_E)$. Then the following statements are equivalent:*

(i) N *is a Poisson random measure with intensity measure Λ.*
(ii) N *is a simple point measure; i.e.*

$$P(N(\{x\}) > 1, \quad \text{for some} \quad x \in E) = 0,$$

and

$$P(N(B) = 0) = e^{-\Lambda(B)}, \quad \text{for all bounded} \quad B \in \mathcal{B}_E. \tag{2.41}$$

Proof. See, e.g., Klenke (2008, p. 529). □

Theorem 2.224 *Consider*

a) a Poisson random measure N with intensity measure Λ on the Polish space $(E, \mathcal{B}_E)$;
b) a measurable function $f : (E, \mathcal{B}_E) \to (\mathbb{R}, \mathcal{B}_\mathbb{R})$.

Then

(i)

$$X(f) := \int_E f(x) N(dx)$$

is absolutely convergent if and only if

$$\int_E (1 \wedge |f(x)|) \Lambda(dx) < +\infty; \tag{2.42}$$

(ii)   *when Condition (2.42) holds, then, for any $\beta \in \mathbb{R}$,*

$$E\left[e^{i\beta X(f)}\right] = \exp\left\{-\int_E \left(1 - e^{i\beta f(x)}\right) \Lambda(dx)\right\}; \tag{2.43}$$

hence the characteristic functional of N is

$$\varphi_N(f) := E\left[e^{iX(f)}\right] = \exp\left\{-\int_E \left(1 - e^{if(x)}\right) \Lambda(dx)\right\}; \tag{2.44}$$

(iii)   *further, when*

$$\int_E |f(x)| \Lambda(dx) < +\infty, \tag{2.45}$$

then

$$E[X(f)] = E[\int_E f(x)N(dx)] = \int_E f(x)\Lambda(dx); \qquad (2.46)$$

and when both

$$\int_E (f(x))^2 \Lambda(dx) < +\infty, \quad \int_E |f(x)|\Lambda(dx) < +\infty, \qquad (2.47)$$

then

$$E[X(f)^2] = \int_E (f(x))^2 \Lambda(dx) + \left(\int_E f(x)\Lambda(dx)\right)^2; \qquad (2.48)$$

hence

$$Var[X(f)] = E[X(f)^2] - (E[X(f)])^2; \qquad (2.49)$$

i.e.

$$Var[\int_E f(x)N(dx)] = \int_E (f(x))^2 \Lambda(dx). \qquad (2.50)$$

Proof. See, e.g., Kyprianou (2014, p. 43), and Klenke (2008, p. 530). □

Remark 2.225. Equation (2.46) is known as the *first Campbell formula*.

2.11 Marked Counting Processes

We now extend our presentation to the class of counting processes at large, including the so-called marked counting processes.

For a more detailed updated account on random measures and point processes, the reader may refer to Daley and Vere-Jones (2008), Karr (1986).

2.11.1 Counting Processes

A counting process N, introduced in Definition 2.195, can be represented as a random point measure on $\mathbb{R}_+$, as follows:

$$N = \sum_{n \in \mathbb{N}^*} \epsilon_{\tau_n},$$

defined by the sequence of random times $(\tau_n)_{n \in \mathbb{N}^*}$ on the underlying probability space $(\Omega, \mathcal{F}, P)$. Here ϵ_t is the Dirac measure (also called point mass) on $\mathbb{R}_+$, i.e.,

$$\forall A \in \mathcal{B}_{\mathbb{R}_+}: \qquad \epsilon_t(A) = \begin{cases} 1 & \text{if } t \in A, \\ 0 & \text{if } t \notin A. \end{cases}$$

Definition 2.226. (A^*): Let $\mathcal{F}_t = \sigma(N_s, 0 \le s \le t)$, $t \in \mathbb{R}_+$, be the natural filtration of the counting process $(N_t)_{t \in \mathbb{R}_+}$. We assume that

1. The filtered probability space $(\Omega, \mathcal{F}, (\mathcal{F}_t)_{t \in \mathbb{R}_+}, P)$ satisfies the usual hypotheses (Definition 2.35).
2. $E[N_t] < \infty$ for all $t \in \mathbb{R}_+$, thus avoiding the problem of exploding martingales in the Doob–Meyer decomposition (Theorem 2.143).

Proposition 2.227. *Under assumption (A^*) of Definition 2.226, there exists a unique increasing right-continuous predictable process $(A_t)_{t \in \mathbb{R}_+}$ such that*

1. $A_0 = 0$.
2. $P(A_t < \infty) = 1$ for any $t > 0$.
3. *The process $(M_t)_{t \in \mathbb{R}_+}$ defined as $M_t = N_t - A_t$ is a right-continuous zero-mean martingale.*

The process $(A_t)_{t \in \mathbb{R}_+}$ is called the compensator of the process $(N_t)_{t \in \mathbb{R}_+}$.

Proposition 2.228 (Bremaud (1981); Karr (1986)). *For every nonnegative $\mathcal{F}_t$-predictable process $(C_t)_{t \in \mathbb{R}_+}$, by Proposition 2.227, we have that*

$$E\left[\int_0^\infty C_t dN_t\right] = E\left[\int_0^\infty C_t dA_t\right]. \tag{2.51}$$

Theorem 2.229. *Given a point (or counting) process $(N_t)_{t \in \mathbb{R}_+}$ satisfying assumption (A^*) of Definition 2.226 and a predictable random process $(A_t)_{t \in \mathbb{R}_+}$, the following two statements are equivalent:*

1. $(A_t)_{t \in \mathbb{R}_+}$ *is the compensator of $(N_t)_{t \in \mathbb{R}_+}$.*
2. *The process $M_t = N_t - A_t$ is a zero-mean martingale.*

Remark 2.230. In infinitesimal form, (2.51) provides the heuristic expression

$$dA_t = E[dN_t | \mathcal{F}_{t-}],$$

giving a dynamical interpretation to the compensator. In fact, the increment $dM_t = dN_t - dA_t$ is the unpredictable part of dN_t over $[0, t[$, therefore also known as the *innovation martingale* of $(N_t)_{t \in \mathbb{R}_+}$.

In the case where the innovation martingale M_t is bounded in L^2, we may apply Theorem 2.145 and introduce the predictable variation process $\langle M \rangle_t$, with $\langle M \rangle_0 = 0$ and $M_t^2 - \langle M \rangle_t$ being a uniformly integrable martingale. Then the variation process can be compensated in terms of A_t by the following theorem.

Theorem 2.231 (Karr (1986, p. 64)) *Let $(N_t)_{t \in \mathbb{R}_+}$ be a point process on $\mathbb{R}_+$ with compensator $(A_t)_{t \in \mathbb{R}_+}$, and let the innovation process $M_t = N_t - A_t$ be an L^2-martingale. Defining $\Delta A_t = A_t - A_{t-}$, then*

$$\langle M \rangle_t = \int_0^t (1 - \Delta A_s) dA_s.$$

Remark 2.232. In particular, if A_t is continuous in t, then $\Delta A_t = 0$, so that $\langle M \rangle_t = A_t$. Formally, in this case we have

$$E\left[(dN_t - E[dN_t|\mathcal{F}_{t-}])^2|\mathcal{F}_{t-}\right] = dA_t = E[dN_t|\mathcal{F}_{t-}],$$

so that the counting process has locally and conditionally the typical behavior of a Poisson process.

Let N be a simple point process on $\mathbb{R}_+$ with a compensator A, satisfying the assumptions of Proposition 2.227.

Definition 2.233. We say that N admits an $\mathcal{F}_t$-*stochastic intensity* if a (nontrivial) nonnegative, predictable process $\lambda = (\lambda_t)_{t \in \mathbb{R}_+}$ exists such that

$$A_t = \int_0^t \lambda_s ds, \qquad t \in \mathbb{R}_+.$$

Remark 2.234. Due to the uniqueness of the compensator, the stochastic intensity, whenever it exists, is unique.

Formally, from

$$dA_t = E[dN_t|\mathcal{F}_{t-}]$$

it follows that

$$\lambda_t dt = E[dN_t|\mathcal{F}_{t-}],$$

i.e.,

$$\lambda_t = \lim_{\Delta t \to 0+} \frac{1}{\Delta t} E[\Delta N_t|\mathcal{F}_{t-}],$$

and, because of the simplicity of the process, we also have

$$\lambda_t = \lim_{\Delta t \to 0+} \frac{1}{\Delta t} P(\Delta N_t = 1|\mathcal{F}_{t-}),$$

meaning that $\lambda_t dt$ is the conditional probability of a new *event* during $[t, t+dt]$, given the history of the process over $[0, t]$.

Example 2.235. (Poisson process). A stochastic intensity does exist for a Poisson process with intensity $(\lambda_t)_{t \in \mathbb{R}_+}$ and, in fact, is identically equal to the latter (hence deterministic).

A direct consequence of Theorem 2.231 and of the previous definitions is the following theorem.

Theorem 2.236 (Karr 1986, p. 64). *Let $(N_t)_{t \in \mathbb{R}_+}$ be a point process satisfying assumption (A^*) of Definition 2.226 and admitting stochastic intensity $(\lambda_t)_{t \in \mathbb{R}_+}$. Assume further that the innovation martingale*

$$M_t = N_t - \int_0^t \lambda_s ds, \qquad t \in \mathbb{R}_+$$

is an L^2-martingale. Then for any $t \in \mathbb{R}_+$:

$$\langle M \rangle_t = \int_0^t \lambda_s ds.$$

An important theorem that further explains the role of the stochastic intensity for counting processes is as follows (Karr 1986, p. 71).

Theorem 2.237. *Let* $(\Omega, \mathcal{F}, P)$ *be a probability space over which a simple point process with an $\mathcal{F}_t$-stochastic intensity $(\lambda_t)_{t \in \mathbb{R}_+}$ is defined. Suppose that P_0 is another probability measure on $(\Omega, \mathcal{F})$ with respect to which $(N_t)_{t \in \mathbb{R}_+}$ is a stationary Poisson process with rate 1. Then $P << P_0$, and for any $t \in \mathbb{R}_+$ we have*

$$\frac{dP}{dP_0}|_{\mathcal{F}_t} = \exp\left\{ \int_0^t (1 - \lambda_s) ds + \int_0^t \ln \lambda_s dN_s \right\}. \qquad (2.52)$$

Conversely, if P_0 is as above and P a probability measure on $(\Omega, \mathcal{F})$, absolutely continuous with respect to P_0, then there exists a predictable process λ such that N has stochastic intensity λ with respect to P [and (2.52) holds].

2.11.2 Marked Counting Processes

We will now consider a locally compact Polish space endowed with its Borel σ-algebra $(E, \mathcal{B}_E)$ and introduce a sequence of $(E, \mathcal{B}_E)$-valued random variables $(Z_n)_{n \in \mathbb{N}^*}$ in addition to the sequence of random times $(\tau_n)_{n \in \mathbb{N}^*}$, which are $\bar{\mathbb{R}}_+$-valued random variables.

Definition 2.238.   The random measure on $\bar{\mathbb{R}}_+ \times E$,

$$N = \sum_{n \in \mathbb{N}^*} \epsilon_{(\tau_n, Z_n)},$$

is called a *marked counting process* with *mark space* $(E, \mathcal{B}_E)$. The random variable Z_n is called the *mark* of the event occurring at time τ_n. The process

$$N_t = N([0,t] \times E), \qquad t \in \mathbb{R}_+,$$

is called the *underlying counting process* of the process N. As usual, we assume that the process $(N_t)_{t \in \mathbb{R}_+}$ is simple.

For $B \in \mathcal{B}_E$ the process

$$N_t(B) := N([0,t] \times B) = \sum_{n \in \mathbb{N}^*} I_{[\tau_n \leq t, Z_n \in B]}(t), \qquad t \in \mathbb{R}_+,$$

represents the counting process of events occurring up to time t with marks in $B \in \mathcal{B}_E$. The *history* of the process up to time t is denoted as

$$\mathcal{F}_t := \sigma(N_s(B)|0 \le s \le t, B \in \mathcal{B}_E).$$

We will assume throughout that the filtered space $(\Omega, \mathcal{F}, (\mathcal{F}_t)_{t \in \mathbb{R}_+}, P)$ satisfies the usual hypotheses (Definition 2.35).

Remark 2.239. Note that, for any $n \in \mathbb{N}^*$, while τ_n is $\mathcal{F}_{\tau_n-}$-measurable, Z_n is $\mathcal{F}_{\tau_n}$-measurable but not $\mathcal{F}_{\tau_n-}$-measurable; indeed

$$\mathcal{F}_{\tau_n} = \sigma\left((\tau_1, Z_1), \ldots, (\tau_n, Z_n)\right),$$

whereas

$$\mathcal{F}_{\tau_n-} = \sigma\left((\tau_1, Z_1), \ldots, (\tau_{n-1}, Z_{n-1}), \tau_n\right).$$

Hence τ_n is an $(\mathcal{F}_t)_{t \in \mathbb{R}_+}$ stopping time.

By a reasoning similar to that employed for regular conditional probabilities in Chap. 1, the following theorem can be proved, which provides an extension of Theorem 2.229 to marked counting processes.

Theorem 2.240 *Bremaud (1981); Karr (1986); Last and Brandt (1995). Let N be a marked counting process such that the underlying counting process $(N_t)_{t \in \mathbb{R}_+}$ satisfies the assumptions of Proposition 2.227. Then there exists a unique random measure Λ on $\mathbb{R}_+ \times E$ such that*

1. *For any $B \in \mathcal{B}_E$, the process $\Lambda([0, t] \times B)$ is $\mathcal{F}_t$-predictable.*
2. *For any nonnegative $\mathcal{F}_t$-predictable process C on $\mathbb{R}_+ \times E$:*

$$E\left[\int C(t, z) N(dt \times dz)\right] = E\left[\int C(t, z) \Lambda(dt \times dz)\right].$$

The random measure Λ introduced in the preceding theorem, i.e. the $\mathcal{F}_t$-compensator of the process N, is called *intensity measure* of the process. The preceding theorem again suggests that formally the following holds:

$$\Lambda(dt \times dz) = E\left[N(dt \times dz)|\mathcal{F}_{t-}\right].$$

Definition 2.241. Let N be a marked counting process as in Definition 2.238. If the sequence of marks $(Z_n)_{n \in \mathbb{N}^*}$ is made of i.i.d. random variables, independent of the underlying counting process $(N_t)_{t \in \mathbb{R}_+}$, with common probability law Q_Z, then N is called a marked counting process with *independent marking*.

Proposition 2.242 *(Independent marking) Consider a simple marked counting process N on $\mathbb{R}_+$, with independent marking given by a sequence $(Z_n)_{n \in \mathbb{N}^*}$ of i.i.d. random variables taking values in $(E, \mathcal{B}_E)$. Let Q_Z be the common distribution of the marks, and let ν be the (random) intensity measure of the underlying counting process $(N_t)_{t \in \mathbb{R}_+}$. Then the intensity measure of the marked counting process N is given by*

$$\Lambda(dt \times dz) = \nu(dt)Q_Z(dz).$$

Proof.

See, e.g., Karr (1986, p. 17), Mikosch (2009, p. 246).   □

The following proposition extends the previous one to the case of dependent marking.

Proposition 2.243 *Consider a simple marked counting process N on $\mathbb{R}_+$, defined as in Definition 2.238, with marks given by a sequence $(Z_n)_{n \in \mathbb{N}^*}$ random variables taking values in $(E, \mathcal{B}_E)$. Assume that*

a) *$(Z_n)_{n \in \mathbb{N}^*}$ is an independent sequence conditional upon $\mathcal{F}^\tau := \sigma\{\tau_n, \, n \in \mathbb{N}^*\}$;*

b) *for any $n \in \mathbb{N}^*$, the random variable Z_n is independent of τ_m, $m \in \mathbb{N}^*$, $m \neq n$;*

c) *for any $n \in \mathbb{N}^*$, and any $B \in \mathcal{B}_E$, and any $t \in \mathbb{R}_+$,*

$$P(Z_n \in B \mid \tau_n = t) = \Phi(t, B),$$

where $\Phi : (\mathbb{R}_+ \times \mathcal{B}_E) \to [0,1]$, such that, for any $B \in \mathcal{B}_E$, the map $t \in \mathbb{R}_+ \mapsto \Phi(t, B)$ is measurable and, for any $t \in \mathbb{R}_+$, $\Phi(t, \cdot)$ is a probability measure on $\mathcal{B}_E$.

If ν is the (random) intensity measure of the underlying counting process $(N_t)_{t \in \mathbb{R}_+}$, then the intensity measure of the marked counting process N is given by

$$\Lambda(dt \times dz) = \nu(dt)\Phi(t, dz). \tag{2.53}$$

Proof.

See, e.g., Last and Brandt (1995, p. 177), or Karr (1986, p. 66).   □

We may remark that, thanks to the fact that, for any $t \in \mathbb{R}_+$, $\Phi(t, \cdot)$ is a probability measure on $\mathcal{B}_E$,

$$\Lambda(dt \times E) = \nu(dt),$$

as expected.

It may happen that ν is absolutely continuous with respect to the usual Lebesgue measure on $\mathbb{R}_+$, so that it admits a density $\lambda \in \mathcal{L}^1_+(\mathbb{R}_+)$, such that, for any $a, b \in \mathbb{R}_+$,

$$\nu(dt) = \int_a^b \lambda(t)dt.$$

Then, from (2.53) we get

$$\Lambda(dt \times dz) = \lambda(t)\Phi(t, dz)dt,$$

and, for any $B \in \mathcal{B}_E$,

$$\Lambda(dt \times B) = \lambda(t)\Phi(t, B)dt.$$

If we set

$$\lambda_t(B) := \lambda(t)\Phi(t, B),$$

we have that, for any $B \in \mathcal{B}_E$,

$$A_t(B) := \Lambda([0, t] \times B) = \int_0^t \lambda_s(B)ds,$$

i.e. for any $B \in \mathcal{B}_E$, $A_t(B)$ is itself absolutely continuous with respect to the usual Lebesgue measure on $\mathbb{R}_+$. Under these circumstances we say that the marked counting process N admits the $\mathcal{F}_t-$ stochastic intensity $(\lambda_t(B))_{t \in \mathbb{R}_+, B \in \mathcal{B}_E}$.

In general we may give the following definition.

Definition 2.244. Let N be a marked counting process on $\mathbb{R}_+ \times E$. We say that $(\lambda_t(B))_{t \in \mathbb{R}_+, B \in \mathcal{B}_E}$ is the $\mathcal{F}_t$-*stochastic intensity* of N provided that

1. For any $t \in \mathbb{R}_+$ the map

$$B \in \mathcal{B}_E \rightarrow \lambda_t(B) \in \mathbb{R}_+$$

is a random measure on $\mathcal{B}_E$;

2. for any $t \in \mathbb{R}_+, B \in \mathcal{B}_E$

$$A_t(B) = \int_0^t \lambda_s(B)ds;$$

i.e., for any $B \in \mathcal{B}_E$, the process $(\lambda_t(B))_{t \in \mathbb{R}_+}$ is the stochastic intensity of the counting process

$$N_t(B) = \sum_{n \in \mathbb{N}^*} I_{[\tau_n \leq t, Z_n \in B]}(t).$$

In this case the process $(A_t(B))_{t \in \mathbb{R}_+, B \in \mathcal{B}_E}$ is known as the *cumulative stochastic intensity* of N.

The following propositions mimic the corresponding results for the unmarked counting processes.

Proposition 2.245. *For any $B \in \mathcal{B}_E$, the process*

$$M_t(B) := N_t(B) - \Lambda([0, t] \times B), \qquad t \in \mathbb{R}_+,$$

is a zero-mean martingale.

We will call the process $M = (M_t(B))_{t \in \mathbb{R}_+, B \in \mathcal{B}_E}$ the innovation process of N.

Proposition 2.246 (Karr 1986, p. 65). *Let N be a marked counting process on $\mathbb{R}_+ \times E$, with compensator Λ, and let B_1 and B_2 be two disjoint sets in $\mathcal{B}_E$ for which $M_t(B_1)$ and $M_t(B_2)$ are L^2-martingales. Then*

$$\langle M_t(B_1), M_t(B_2) \rangle_t = -\int_0^t \Delta A_s(B_1) \Delta A_s(B_2) ds.$$

Hence, if $(A_t(B))_{t \in \mathbb{R}_+}$ is continuous in t for any $B \in \mathcal{B}_E$, then the two martingales $M_t(B_1)$ and $M_t(B_2)$ are orthogonal, i.e.

$$\langle M_t(B_1), M_t(B_2) \rangle_t = 0, \quad \text{for any} \quad t \in \mathbb{R}_+.$$

Representation of Counting Process Martingales

Let N be a point process on $\mathbb{R}_+$ with $\mathcal{F}$-compensator A. From the section on martingales we know that if $M = N - A$ is the innovation martingale of N and H is a bounded predictable process, then

$$\tilde{M}_t = \int_0^t H(s) dM_s, \quad t \in \mathbb{R}_+,$$

is also a martingale. In fact, the converse also holds, as stated by the following theorem, which extends an analogous result for Wiener processes to marked counting processes.

Theorem 2.247 (Martingale representation). *Let N be a marked counting process on $\mathbb{R}_+$ with mark space $(E, \mathcal{B}_E)$, and let M be its innovation process with respect to the internal history $(\mathcal{F}_t)_{t \in \mathbb{R}_+}$. Suppose the assumptions of Proposition 2.227 are satisfied, and let $(\tilde{M}_t)_{t \in \mathbb{R}_+}$ be a right-continuous and uniformly integrable $\mathcal{F}_t$-martingale. Then there exists a process $(H(t, x))_{t \in \mathbb{R}_+, x \in E}$ such that*

$$\tilde{M}_t = \tilde{M}_0 + \int_{[0,t] \times E} H(s, x) M_s(dx).$$

Proof. See Last and Brandt (1995). $\qquad\square$

2.11.3 The Marked Poisson Process

Definition 2.248. Given a σ-finite deterministic measure Λ on $\mathbb{R}_+ \times E$, N is a *marked Poisson process* if, for any $s, t \in \mathbb{R}_+, s < t$ and any $B \in \mathcal{E}$,

$$P(N(]s, t] \times B) = k | \mathcal{F}_s) = \frac{(\Lambda(]s, t] \times B))^k}{k!} \exp\{-\Lambda(]s, t] \times B)\},$$

for $k \in \mathbb{N}$, almost surely with respect to P.

In the preceding case the intensity measure Λ is such that

$$\Lambda(]s,t] \times B) = E[N(]s,t] \times B)]$$

for any $s,t \in \mathbb{R}_+, s < t$ and any $B \in \mathcal{B}_E$. It is the (deterministic) compensator of the marked Poisson process, formally:

$$\Lambda(dt \times dx) = E[N(dt \times dx)|\mathcal{F}_{t-}] = E[N(dt \times dx)],$$

thereby confirming the independence of increments for the marked Poisson process.

Given a Borel set $B \in \mathcal{E}$, we will denote by $N(\cdot \times B)$ the counting process $(N_t(B))_{t \in \mathbb{R}_+}$, and by $\Lambda(\cdot \times B)$ the measure on $\mathcal{B}_{\mathbb{R}_+}$ defined by $\Lambda(\cdot \times B)((a,b]) = \Lambda((a,b] \times B)$, for any $a,b \in \mathbb{R}_+, a < b$.

The following theorem is a consequence of the definitions.

Theorem 2.249. *Let N be a marked Poisson process and $B_1, \ldots, B_m \in \mathcal{B}_E$ for $m \in \mathbb{N}^*$ mutually disjoint sets. Then $N(\cdot \times B_1), \ldots, N(\cdot \times B_m)$ are independent Poisson processes with intensity measures $\Lambda(\cdot \times B_1), \ldots, \Lambda(\cdot \times B_m)$, respectively.*

Proof. See Last and Brandt (1995).   □

The underlying counting process of a marked Poisson process $N(]0,t] \times E)$ is itself a univariate Poisson process with intensity measure $\bar{\Lambda}(]s,t]) = \Lambda(]s,t] \times E)$ for any $s,t \in \mathbb{R}_+, s < t$. The intensity measure may be chosen to be continuous, in which case $\bar{\Lambda}(\{t\}) = 0$, or even absolutely continuous with respect to the Lebesgue measure on $\mathbb{R}_+$, so that

$$\bar{\Lambda}([0,t]) = \int_0^t \lambda(s)ds,$$

where $\lambda \in \mathcal{L}^1(\mathbb{R}_+)$.

Time-Homogenous Marked Poisson Process

A particular case of interest for our subsequent analysis is the following one.

Definition 2.250. A marked Poisson process N on $\mathbb{R}_+ \times E$ is time-homogenous if there exists a σ-finite deterministic measure ν on $\mathcal{B}_E$ such that the intensity measure Λ of N is given by

$$\Lambda(]s,t] \times B) = E[N(]s,t] \times B)] = (t-s)\,\nu(B)$$

for any $s,t \in \mathbb{R}_+, s < t$ and any $B \in \mathcal{B}_E$.

Formally:

$$E[N(dt \times dx)|\mathcal{F}_{t-}] = E[N(dt \times dx)] = dt\,\nu(dx).$$

We will have in particular

$$E[N([0,t] \times B)] = t \nu(B)$$

for any $t \in \mathbb{R}_+$ and any $B \in \mathcal{B}_E$ and

$$\nu(B) = E[N_1(B)] = E[N([0,1] \times B)]$$

for any $B \in \mathcal{B}_E$.

Proposition 2.251. *For any $B \in \mathcal{B}_E$, the process*

$$M_t(B) := N_t(B) - t\nu(B), \qquad t \in \mathbb{R}_+$$

is a zero-mean martingale.

The random measure $N(dt \times dx) - dt\,\nu(dx)$ is usually called the *compensated Poisson measure*. The measure ν is called the *characteristic* of the time-homogeneous marked Poisson measure.

As a consequence of the preceding theorems we may state the following.

Proposition 2.252. *If N is a Poisson random measure with intensity measure $\Lambda(]s,t] \times B) = E[N(]s,t] \times B)] = (t-s)\,\nu(B)$, then for any $B \in \mathcal{B}_E$,*

- $N_t(B) = \int_0^t \int_B N(dt \times dx)$, $t \in \mathbb{R}_+$ *is a Poisson process, with intensity $t\nu(B)$.*
- *For any finite family of disjoint Borel sets $B_1, \ldots, B_k, k \in \mathbb{N} \setminus \{0,1\}$, the random variables $N_t(B_1), \ldots, N_t(B_k)$ are independent, for any $t \in \mathbb{R}_+$.*

Theorem 2.253. *Given a deterministic σ-finite measure ν on a Polish space $(E, \mathcal{B}_E)$, there exists a time-homogeneous marked Poisson process N on $\mathbb{R}_+ \times E$ having characteristic measure ν.*

Proof. See, e.g., Ikeda and Watanabe (1989).   □

2.11.4 Time-space Poisson Random Measures

In the sequel, it will be of interest the particular case of a time-homogeneous marked Poisson process N with marks on $\mathbb{R}$, endowed with its usual Borel sigma algebra $\mathcal{B}_{\mathbb{R}}$, having intensity measure Λ on $\mathcal{B}_{\mathbb{R}_+} \otimes \mathcal{B}_{\mathbb{R}}$, such that, for $(a,b] \subset [0, +\infty)$, and $B \in \mathcal{B}_{\mathbb{R}}$,

$$\Lambda((a,b] \times B) = (b-a)\nu(B), \tag{2.54}$$

where ν is a σ-finite measure on $\mathcal{B}_{\mathbb{R}}$ concentrated on $\mathbb{R} \setminus \{0\}$.

They will be of particular interest integrals of the form

$$X_t := \int_{[0,t]} \int_B x N(ds \times dx), \quad t \in \mathbb{R}_+ \tag{2.55}$$

for $t \in \mathbb{R}_+$, and $B \in \mathcal{B}_{\mathbb{R}}$.

By taking into account Theorem 2.224, we may state that the integral in (2.55) is absolutely convergent if and only if

$$\int_B (1 \wedge |x|) \nu(dx) < +\infty, \tag{2.56}$$

which guarantees that also the characteristic function of X_t is well defined.

Further, in order that $E[X_t]$ be finite, we have to require that

$$\int_B |x| \nu(dx) < +\infty; \tag{2.57}$$

while for a finite second order moment of X_t we need to require that

$$\int_B x^2 \nu(dx) < +\infty. \tag{2.58}$$

This justifies the assumptions in the following Lemmas.

Lemma 2.254. *Within the above framework, if we take a $B \in \mathcal{B}_{\mathbb{R}}$ such that $0 < \nu(B) < +\infty$, then the process*

$$X_t := \int_{[0,t]} \int_B x N(ds \times dx), \quad t \in \mathbb{R}_+ \tag{2.59}$$

is a compound Poisson process with arrival rate $\nu(B)$, and common jump distribution $\dfrac{\nu|_B(dx)}{\nu(B)}$.

Proof. Since $0 < \nu(B) < +\infty$, the integral in 2.60 is reduced to a finite sum thanks to Proposition 2.222, so that the process $(X_t)_{t \in \mathbb{R}_+}$ is RCLL. By construction it has independent and stationary increments. Further, according to (2.43) in Theorem 2.224, we have that the characteristic function of X_t, for $t \in \mathbb{R}_+$, is given by

$$\phi_{X_t}(u) = E[e^{iuX_t}] = \exp\left\{-t \int_B (1 - e^{iux}) \nu(dx)\right\}, \quad u \in \mathbb{R},$$

which is the characteristic function of a compound Poisson process with arrival rate $\nu(B)$, and common jump distribution $\dfrac{\nu|_B(dx)}{\nu(B)}$. $\qquad\square$

Lemma 2.255. *Within the above framework, if we now take $B \in \mathcal{B}_{\mathbb{R}}$ such that $0 < \nu(B) < +\infty$, and $\int_B |x| \nu(dx) < +\infty$, then the compensated compound Poisson process*

$$M_t := \int_{[0,t]} \int_B x N(ds \times dx) - t \int_B x\nu(dx), \quad t \in \mathbb{R}_+ \qquad (2.60)$$

is a martingale.

Further, if $\int_B x^2 \nu(dx) < +\infty$, then $(M_t)_{t \in \mathbb{R}_+}$ *is a square integrable martingale.*

Proof. See, e.g., Kyprianou (2014) p. 47.   □

2.12 White Noise

Very often in applications we read about *white noise*, typically associated with a property of the Wiener process. Actually we may quote two major cases of white noise in stochastic processes, Gaussian and Poisson.

2.12.1 Gaussian white noise

The Gaussian white noise is associated with a Wiener process $(W_t)_{t \in \mathbb{R}_+}$ which is a Gaussian process. If we take two time intervals of the same duration Δt, but shifted by a nontrivial quantity $h \neq 0$, the corresponding time increments of the standard Wiener process are $W_{t+\Delta t} - W_t$, and $W_{t+h+\Delta t} - W_{t+h}$.

The covariance of the corresponding incremental ratios is given by

$$C_{\Delta t}(h) := Cov\left[\frac{W_{t+\Delta t} - W_t}{\Delta t}, \frac{W_{t+h+\Delta t} - W_{t+h}}{\Delta t}\right] =$$

$$= \frac{1}{(\Delta t)^2} E[(W_{t+\Delta t} - W_t)(W_{t+h+\Delta t} - W_{t+h})]. \qquad (2.61)$$

By the property of independent increments of the Wiener process, we may claim that

$$C_{\Delta t}(h) = \frac{1}{(\Delta t)^2} \begin{cases} 0, & \text{if } h \leq -\Delta t, \\ \Delta t + h, & \text{if } -\Delta t \leq h \leq 0, \\ \Delta t - h, & \text{if } 0 \leq h \leq \Delta t, \\ 0, & \text{if } h \geq \Delta t; \end{cases} \qquad (2.62)$$

i.e.

$$C_{\Delta t}(h) = \frac{1}{(\Delta t)^2}(\Delta t - |h|) I_{[-\Delta t, \Delta t]}(h). \qquad (2.63)$$

Anticipating the result, let us compute the following integral, where g is a sufficiently smooth test function,

$$\int_{-\infty}^{+\infty} C_{\Delta t}(h) g(h) dh = \int_{-\Delta t}^{+\Delta t} \frac{1}{(\Delta t)^2}(\Delta t - |h|) g(h) dh =$$

$$= \int_{-1}^{+1} (1 - |u|) g(u\Delta t) du, \qquad (2.64)$$

which implies

$$\lim_{\Delta t \downarrow 0^+} \int_{-\infty}^{+\infty} C_{\Delta t}(h) g(h) dh = g(0) \int_{-1}^{+1} (1 - |u|) du = g(0). \qquad (2.65)$$

Hence we may claim that, in a generalized sense,

$$C_{dt}(h) = \delta_0(h), \qquad (2.66)$$

having denoted by δ_0 the Dirac delta function centered at 0. This is the reason why the Wiener process is known as a delta-correlated Gaussian noise.

2.12.2 Poissonian white noise

Similarly to the Gaussian case, for a standard Poisson process $(P_t)_{t \in \mathbb{R}_+}$, with parameter 1, proceeding as above, we get

$$C_{\Delta t}(h) = \frac{1}{(\Delta t)^2} (\Delta t - |h|) I_{[-\Delta t, \Delta t]}(h), \qquad (2.67)$$

so that, as above

$$C_{dt}(h) = \lim_{\Delta t \downarrow 0^+} C_{\Delta t}(h) = \delta_0(h). \qquad (2.68)$$

In either case, the name *white noise* derives from the fact that the Fourier transform of the Dirac delta is a constant

$$\widehat{\delta_0}(\omega) = \frac{1}{\pi} \int_{-\infty}^{+\infty} e^{-i\omega h} \delta_0(h) dh = \frac{1}{\pi}, \quad \text{for any} \quad \omega \in \mathbb{R}. \qquad (2.69)$$

2.13 Lévy Processes

Definition 2.256. Let $(X_t)_{t \in \mathbb{R}_+}$ be an adapted process with $X_0 = 0$ almost surely. If X_t

1. has independent increments,
2. has stationary increments,
3. is continuous in probability so that $X_s \xrightarrow[s \to t]{P} X_t$,

then it is a *Lévy process*.

Proposition 2.257. *Both the Wiener and the Poisson processes are Lévy processes.*

Proposition 2.258. *The compound Poisson process is a Lévy process.*

Proof. See Exercise 2.18. $\qquad \qquad \square$

Theorem 2.259. *Let $(X_t)_{t \in \mathbb{R}_+}$ be a Lévy process. Then it has an RCLL version $(Y_t)_{t \in \mathbb{R}_+}$, which is also a Lévy process.*

Proof. See, e.g., Kallenberg (1997, p. 235). □

For Lévy processes we can invoke examples of filtrations that satisfy the usual hypotheses.

Theorem 2.260. *Let $(X_t)_{t \in \mathbb{R}_+}$ be a Lévy process and $\mathcal{G}_t = \sigma(\mathcal{F}_t, \mathcal{N})$, where $(\mathcal{F}_t)_{t \in \mathbb{R}_+}$ is the natural filtration of X_t and $\mathcal{N}$ the family of P-null sets of $\mathcal{F}_t$. Then $(\mathcal{G}_t)_{t \in \mathbb{R}_+}$ is right-continuous.*

Proof. See, e.g., Protter (2004, p. 22). □

Remark 2.261. Because, by Theorem 2.259, every Lévy process has an RCLL version, by Proposition 2.199, the only type of discontinuity it may admit is jumps.

Theorem 2.262. *Let $(X_t)_{t \in \mathbb{R}_+}$ be a Lévy process. Then it has an RCLL version without fixed jumps (Proposition 2.199).*

Proof. See, e.g., Kallenberg (1997). □

Definition 2.263. Taking the left limit $X_{t-} = \lim_{s \to t} X_s$, $s < t$, we define

$$\Delta X_t = X_t - X_{t-}$$

as the jump at t. If $\sup_t |\Delta X_t| \leq c$ almost surely, $c \in \mathbb{R}_+$, constant and nonrandom, then X_t is said to have *bounded jumps*.

Theorem 2.264. *Let $(X_t)_{t \in \mathbb{R}_+}$ be a Lévy process with bounded jumps. Then*

$$E[|X_t|^p] < \infty, \qquad i.e., \ X_t \in \mathcal{L}^p \qquad for \ any \ p \in \mathbb{N}^*.$$

Proof. See, e.g., Protter (2004, p. 25). □

Proposition 2.265.

(i) *Let $X = (X_t)_{t \in \mathbb{R}_+}$ be a Lévy process. For any $t \in \mathbb{R}_+$, the distribution of X_t is infinitely divisible.*
(ii) *For any infinitely divisible law P one can construct a Lévy process $X = (X_t)_{t \in \mathbb{R}_+}$ such that X_1 has law P.*

Proof. Proof of (i).

It is a trivial consequence of the following remark; by definition, we have $X_0 = 0$. Now, for any $n \in \mathbb{N}^*$, we may decompose

$$X_t = X_{t/n} + (X_{t\,2/n} - X_{t/n}) + \cdots + (X_t - X_{t\,(n-1)/n}),$$

where, again by definition, all random variables $X_{t/n}, X_{t\,2/n} - X_{t/n}, \ldots, X_t - X_{t\,(n-1)/n}$ all have the same distribution and are independent.

Proof of (ii).

We postpone to Theorem 2.268. $\square$

Proposition 2.266. *The characteristic function of a Lévy process* $X = (X_t)_{t \in \mathbb{R}_+}$ *at time* $t \in \mathbb{R}_+$ *admits the following representation, for any* $t \in \mathbb{R}_+$:

$$\phi_{X_t}(u) = (\phi_{X_1}(u))^t, \quad u \in \mathbb{R}. \tag{2.70}$$

Proof. Consider the complex function

$$\psi_t(u) := -\ln E[e^{iuX_t}], \quad u \in \mathbb{R}; \tag{2.71}$$

it is such that

$$\exp\{-\psi_t(u)\} := E[e^{iuX_t}] =: \phi_{X_t}(u), \quad u \in \mathbb{R} \tag{2.72}$$

which is the characteristic function of X_t.

Thanks to the above decomposition, we may state that, for any two integers $m, n \in \mathbb{N}^*$

$$m\,\psi_1(u) = \psi_m(u) = n\psi_{m/n}(u), \quad u \in \mathbb{R}; \tag{2.73}$$

i.e., for any rational $t \in \mathbb{Q}_+^*$, we may state

$$\psi_t(u) = t\,\psi_1(u), \quad u \in \mathbb{R}. \tag{2.74}$$

If now $t \in \mathbb{R}_+^*$, we may always take a sequence of rational numbers $(t_n)_{n \in \mathbb{N}}$, such that $t_n \downarrow t$ as n tends to ∞. We know that we can choose a version of the Lévy process which is right-continuous; by the Dominated Convergence Theorem, we may then claim that Equation (2.74) holds for any $t \in \mathbb{R}_+^*$. As a consequence Equation (2.70) holds true, with

$$\phi_{X_1}(u) = \exp\{-\psi_1(u)\}, \quad u \in \mathbb{R}. \tag{2.75}$$

$\square$

To summarize, the previous result states that the characteristic function of a Lévy process, for any $t \in \mathbb{R}_+^*$, admits the representation

$$\phi_{X_t}(u) = \exp\{-t\psi(u)\}, \quad u \in \mathbb{R}, \tag{2.76}$$

where $\psi(u)$, $u \in \mathbb{R}$, is the characteristic exponent of X_1.

From now on $\psi(u)$, $u \in \mathbb{R}$, will be called the *characteristic exponent of the Lévy process* $(X_t)_{t \in \mathbb{R}_+^*}$.

A trivial consequence of the above is the following theorem.

Theorem 2.267. *Let $(X_t)_{t \in \mathbb{R}_+}$ be a Lévy process. Then*

(i) If $X_t \in \mathcal{L}^1$ for some $t \in \mathbb{R}_+$, then $X_t \in \mathcal{L}^1$ for any $t \in \mathbb{R}_+$, and

$$E[X_t] = tE[X_1].$$

(ii) If $X_t \in \mathcal{L}^2$ for some $t \in \mathbb{R}_+$, then $X_t \in \mathcal{L}^2$ for any $t \in \mathbb{R}_+$, and

$$Var[X_t] = tVar[X_1].$$

Proof. See, e.g., Mikosch (2009, p. 338). □

We now complete the proof of (ii) of Proposition 2.265, by proving the following theorem.

Theorem 2.268 (Lévy–Khintchine formula for Lévy processes).
Given an infinitely divisible law P with characteristic triplet (μ, σ^2, ν), where $\mu \in \mathbb{R}$, $\sigma \in \mathbb{R}$, and ν is a measure on $\mathbb{R}$, concentrated on $\mathbb{R}^*$ satisfying

$$\int_{\mathbb{R}^*} \min \left\{ x^2, 1 \right\} \nu(dx) < +\infty,$$

there exists a probability space $(\Omega, \mathcal{F}, P)$ on which a Lévy process $X = (X_t)_{t \in \mathbb{R}_+}$ can be defined such that its characteristic function at time $t \in \mathbb{R}_+$ is given by

$$\phi_{X_t}(u) = E\left[e^{iuX_t} \right] = \exp\left\{ -t\psi(u) \right\}, \quad u \in \mathbb{R},$$

where ψ is the characteristic exponent of X_1, whose probability law is P,

$$\psi(u) = \frac{1}{2}\sigma^2 u^2 - i\mu u + \int_{\{|x|<1\}} (1 - \exp\{iux\} + iux)\nu(dx) \qquad (2.77)$$

$$+ \int_{\{|x|\geq 1\}} (1 - \exp\{iux\})\nu(dx), \qquad (2.78)$$

Further, the triplet (μ, σ^2, ν) characterizes the probability law of the Lévy process $(X_t)_{t \in \mathbb{R}_+}$ in a unique way.

Proof.
    Given an i.d. law, from Theorem 1.204 we know that it is characterized by its generating triplet (μ, σ^2, ν), satisfying the given conditions. We may then consider first the process

$$X_t^{(1)} := \sigma W_t - \mu t, \quad t \in \mathbb{R}_+, \qquad (2.79)$$

where $(W_t)_{t \in \mathbb{R}_+}$ is a standard Wiener process, on a suitable probability space $(\Omega^{(1)}, \mathcal{F}^{(1)}, P^{(1)})$.

Now, given the Lévy measure ν, we know by Theorem 2.221 that a Poisson random measure $N(dt \times dx)$ exists, on a suitable probability space $(\Omega^{(2)}, \mathcal{F}^{(2)}, P^{(2)})$, having intensity measure $dt\nu(dx)$ on $(\mathbb{R}_+ \times \mathbb{R}, \mathcal{B}_{\mathbb{R}_+} \otimes \mathcal{B}_{\mathbb{R}})$. Then introduce the process

$$X_t^{(2)} := \int_0^t \int_{|x| \geq 1} x N(ds \times dx), \quad t \in \mathbb{R}_+. \tag{2.80}$$

The Lévy measure ν either is trivial, in which case we take $X_t^{(2)} \equiv 0$, or

$$0 < \nu(\mathbb{R} \setminus (-1, 1)) < +\infty,$$

in which case we know by Lemma 2.254 that $(X_t^{(2)})_{t \in \mathbb{R}_+}$ is a compound Poisson process with arrival rate $\nu(\mathbb{R} \setminus (-1, 1))$ and jump distribution $\dfrac{\nu \mid_{\mathbb{R} \setminus (-1,1)}}{\nu(\mathbb{R} \setminus (-1, 1))}$.

We wish now to introduce an additional Lévy process having only small jumps.

Take $0 < \varepsilon < 1$, and consider the compensated compound Poisson process defined as

$$X_t^{(3,\varepsilon)} := \int_0^t \int_{\varepsilon \leq |x| < 1} x \widetilde{N}(ds \times dx), \quad t \in \mathbb{R}_+, \tag{2.81}$$

with

$$\widetilde{N}(ds \times dx) = N(ds \times dx) - ds\nu(dx). \tag{2.82}$$

Its characteristic exponent will be

$$\psi^{(3,\varepsilon)}(u) := \int_{\varepsilon \leq |x| < 1} (1 - e^{iux} + iux)\nu(dx), \quad u \in \mathbb{R}. \tag{2.83}$$

Under the above assumptions, we may claim that $\int_{(-1,1)} x^2 \nu(dx) < +\infty$, hence (see, e.g., Kyprianou (2014), p. 48) we may state that, on a suitable probability space $(\Omega^{(3)}, \mathcal{F}^{(3)}, P^{(3)})$, there exists a Lévy process $(X_t^{(3)})_{t \in \mathbb{R}_+}$ (which is a square integrable martingale) to which $(X_t^{(3,\varepsilon)})_{t \in \mathbb{R}_+}$ converges a.s. uniformly on any $[0, T]$, for $\varepsilon \downarrow 0$. This process will have characteristic exponent

$$\psi^{(3)}(u) := \int_{0 \leq |x| < 1} (1 - e^{iux} + iux)\nu(dx), \quad u \in \mathbb{R}. \tag{2.84}$$

Due to the independence of N over $[0, t] \times \{\mathbb{R} \setminus (-1, 1)\}$, and $[0, t] \times (-1, 1)$, we may claim that the two processes $(X_t^{(2)})_{t \in \mathbb{R}_+}$ and $(X_t^{(3)})_{t \in \mathbb{R}_+}$ are independent; we may further take $(X_t)_{t \in \mathbb{R}_+}$ independent of N.

For our scopes, we may finally take the process $(X_t)_{t \in \mathbb{R}_+} = (X_t^{(1)})_{t \in \mathbb{R}_+} + (X_t^{(2)})_{t \in \mathbb{R}_+} + (X_t^{(3)})_{t \in \mathbb{R}_+}$.

It is indeed a Lévy process on the probability space $(\Omega, \mathcal{F}, P) = (\Omega^{(1)}, \mathcal{F}^{(1)}, P^{(1)}) \otimes (\Omega^{(2)}, \mathcal{F}^{(2)}, P^{(2)}) \otimes (\Omega^{(3)}, \mathcal{F}^{(3)}, P^{(3)})$, with characteristic exponent

$$\psi(u) = \psi^{(1)}(u) + \psi^{(2)}(u) + \psi^{(3)}(u)$$

$$= i\mu u + \frac{1}{2}\sigma^2 u^2 + \int_{\mathbb{R}} (1 - e^{iux} + iux\,\mathbb{I}_{|x|<1})\nu(dx), \quad u \in \mathbb{R}. \quad (2.85)$$

$\square$

This is why the triplet (μ, σ^2, ν) is called the *characteristic triplet* of the Lévy process $X = (X_t)_{t \in \mathbb{R}_+}$.

Remark 2.269. It is worth noticing that, as a subproduct of the proof of the convergence of the sequence of $\psi^{(3,\varepsilon)}$ to $\psi^{(3)}$, one may realize for the corresponding sequence of random processes $(X_t^{(3,\varepsilon)})_{t \in \mathbb{R}_+}$ that the following holds (rigorously in L^2) along a specific subsequence

$$\lim_{k\uparrow\infty} X_t^{(3,2^{-k})} = \lim_{k\uparrow\infty} \int_0^t \int_{2^{-k}\leq|x|<1} xN(ds \times dx) - t\int_{2^{-k}\leq|x|<1} x\nu(dx)$$

$$= \lim_{k\uparrow\infty} \sum_{i=0}^{k-1} \left\{ \int_0^t \int_{2^{-(i+1)}\leq|x|<2^{-i}} xN(ds \times dx) \right.$$

$$\left. - t\int_{2^{-(i+1)}\leq|x|<2^{-i}} x\nu(dx) \right\}, \quad (2.86)$$

according to which we may state that $(X_t^{(3)})_{t \in \mathbb{R}_+}$ represents the superposition of an infinite sequence of compound Poisson processes (see, e.g., Kyprianou (2014), p. 48).

Remark 2.270. We may remark that the integral in Equation (2.85) is always integrable with respect to ν, since

i) outside a neighborhood of 0 it is bounded;
ii) for $|x| \to 0$

$$1 - e^{iux} + iux\mathbb{I}_{|x|<1} = O(x^2), \quad (2.87)$$

and, for the Lévy measure ν, we have assumed that

$$\int_{|x|<1} x^2\nu(dx) < +\infty.$$

Remark 2.271. A very important additional remark regarding the process $(X_t^{(3)})_{t \in \mathbb{R}_+}$, is the following. In order to ensure the convergence of the integral

$$\int_0^t \int_{|x|<1} xN(ds \times dx), \quad t \in \mathbb{R}_+, \quad (2.88)$$

a necessary and sufficient condition is

$$\int_{|x|<1} |x|\nu(dx) < +\infty, \tag{2.89}$$

which, in our framework, is equivalent to

$$\int_{\mathbb{R}} (1 \wedge |x|)\nu(dx) < +\infty. \tag{2.90}$$

It is clear that condition (2.90) is more stringent than the usual condition on ν as a Lévy measure, i.e. $\int_{\mathbb{R}} (1 \wedge x^2)\nu(dx) < +\infty$. Under these more restrictive circumstances, we may then write explicitly

$$X_t^{(3)} = \int_0^t \int_{|x|<1} xN(ds \times dx) - t\int_{|x|<1} x\nu(dx), \quad t \in \mathbb{R}_+. \tag{2.91}$$

Consequently we may state the following theorem.

Theorem 2.272 (Lévy-Itô decomposition). *Let $(X_t)_{t\in\mathbb{R}_+}$ be a Lévy process with characteristic triplet (μ, σ^2, ν). Suppose that the Lévy measure ν satisfies the additional assumption*

$$\int_{\mathbb{R}} \min\{1, |x|\}\,\nu(dx) < +\infty. \tag{2.92}$$

Then, for any $t \in \mathbb{R}_+$,

$$X_t = \sigma W_t + \mu t + \int_{\{|x|<1\}} x[N_t(dx) - t\nu(dx)] + \int_{\{|x|\geq 1\}} xN_t(dx),$$

where

(i)   *W_t is a standard Brownian motion;*
(ii) *$N_t(dx) = \int_0^t N(dt \times dx)$,*
       $N(dt \times dx)$ being a random Poisson measure with intensity $dt\nu(dx)$ on $(\mathbb{R}_+ \times \mathcal{B}_{\mathbb{R}_+} \otimes \mathcal{B}_{\mathbb{R}})$.

Proof. See, e.g., Sato (1999, p. 119), Kyprianou (2014, p. 56),Mikosch (2009, p.353) .   □

Remark 2.273. In the preceding formula we may recall that the process

$$\int_{\{|x|\geq 1\}} xN_t(dx), \quad t \in \mathbb{R}_+,$$

describing the "large jumps" of a Levy process is a compound Poisson process.

We may further notice that while the process

$$X_t - \int_{\{|x| \geq 1\}} x N_t(dx), \quad t \in \mathbb{R}_+,$$

has finite moments of any order, the process $\int_{\{|x| \geq 1\}} x N_t(dx)$, $t \in \mathbb{R}_+$, may have no finite moments (e.g., Applebaum (2004), p. 110).

The following result is a natural consequence of Theorems 2.272 and 2.268.

Theorem 2.274. *Any Lévy process $X = (X_t)_{t \in \mathbb{R}_+}$ can be decomposed as follows:*

$$X_t = \mu t + \sigma W(t) + S(t), \quad t \in \mathbb{R}_+, \tag{2.93}$$

where $\mu \in \mathbb{R}$ and $\sigma \in \mathbb{R}_+$ are constants and $W(t)$ is a standard Brownian motion independent of the process S.

Notice that the component $(\mu t + \sigma W(t))_{t \in \mathbb{R}_+}$ has continuous sample paths almost surely, while $(S(t))_{t \in \mathbb{R}_+}$ has almost surely discontinuous sample paths (for any $t \in \mathbb{R}_+$, $S(t)$ can be obtained as the weak limit of a sequence of compound Poisson random variables); this is why S is called the *pure jump process* of X, and the Lévy measure ν is known as the *jump measure* of X.

Corollary 2.275. *Let $X = (X_t)_{t \in \mathbb{R}_+}$ be a Lévy process. If $X_t \in \mathcal{L}^2$ for $t \in \mathbb{R}_+$, then*

$$\int_{|x| \geq 1} |x|^2 \nu(dx) < +\infty,$$

and the representation (2.93) can be written more explicitly as

$$X(t) = \mu_1 t + \sigma W(t) + \int_{\mathbb{R} - \{0\}} x[N_t(dx) - t\nu(dx)], \quad t \in \mathbb{R}_+, \tag{2.94}$$

with $\mu_1 = \mu + \int_{|x| \geq 1} x \nu(dx)$.

Proof. See, e.g., Di Nunno et al. (2009, p. 162). ☐

We may introduce the compensated measure

$$\widetilde{N}(dt \times dx) = N(dt \times dx) - dt\nu(dx).$$

Accordingly, (2.94) may be rewritten as

$$X(t) = \mu_1 t + \sigma W(t) + \int_0^t \int_{\mathbb{R} - \{0\}} x \widetilde{N}(ds \times dx), \quad t \in \mathbb{R}_+. \tag{2.95}$$

Subordinators

We may remark that $(X_t^{(2)})_{t \in \mathbb{R}_+}$, being a compound Poisson process, has always paths of bounded variation, but the process $(X_t^{(3)})_{t \in \mathbb{R}_+}$ will be of bounded variation if and only if condition (2.90) applies. Then we may state the following (Kyprianou (2014) p. 56)

Proposition 2.276 *A Lévy process with characteristic triplet (μ, σ^2, ν) has paths of bounded variation if and only if $\sigma = 0$, and (2.90) holds.*

We may notice that, under the assumptions of Proposition 2.276, we may rewrite $(X_t)_{t \in \mathbb{R}_+}$ as follows

$$X(t) = \delta t + \int_0^t \int_{\mathbb{R}} x N(ds \times dx), \quad t \in \mathbb{R}_+. \tag{2.96}$$

having characteristic exponent

$$\psi(u) = -i\delta u + \int_{\mathbb{R}} (1 - e^{iux}) \nu(dx), \quad u \in \mathbb{R}, \tag{2.97}$$

where

$$\delta = -\left(\mu + \int_{|x|<1} x\nu(dx) \right). \tag{2.98}$$

In this case we may recognize that, apart from δ, known as the *drift* of the Lévy process, $(X_t)_{t \in \mathbb{R}_+}$ reduces to a compound Poisson process.

If now $\nu((-\infty, 0)) = 0$, the Lévy process $(X_t)_{t \in \mathbb{R}_+}$ does not have negative jumps, hence, under the assumptions of Proposition 2.276, the process given by (2.96) has nondecreasing paths. Viceversa if it has nondecreasing paths it is of bounded variation, so that by Proposition 2.276 it must be $\sigma = 0$, and satisfy condition (2.90), which leads to $\delta \geq 0$.

Definition 2.277. A *subordinator* is a Lévy process whose paths are (a.s.) nondecreasing.

Proposition 2.278 *A Lévy process with characteristic triplet (μ, σ^2, ν) has paths of bounded variation if and only if $\sigma = 0$, $\nu((-\infty, 0)) = 0$, (2.90) holds, and $\delta := -\left(\mu + \int_{|x|<1} x\nu(dx) \right) \geq 0$.*

Proposition 2.279 *A subordinator has bounded variation. Any Lévy process of bounded variation can be written as the difference of two independent subordinators.*

Proof. See Exercise 2.35 $\qquad\qquad\qquad\qquad\qquad\qquad\qquad\qquad\qquad\qquad$ □

In terms of its characteristic triplet (μ, σ^2, ν), a subordinator has $\mu \geq 0$, $\sigma = 0$, and a Lévy measure ν concentrated on $(0, +\infty)$.

The following theorem holds

Theorem 2.280 *If $Z = (Z_t)_{t \in \mathbb{R}_+}$ is a subordinator, then its characteristic triplet is of the form $(\mu, 0, \nu)$, with $\mu \geq 0$, and the Lévy measure ν satisfying the additional requirements*

A.1 $\nu(-\infty, 0) = 0$;
A.2 $\int_0^{+\infty} (x \wedge 1)\nu(dx) < +\infty$.

Its characteristic exponent takes the form

$$\psi(u) = i\mu u + \int_0^{+\infty} (e^{iux} - 1)\nu(dx), \quad u \in \mathbb{R}, \tag{2.99}$$

Proof. See, e.g., Bertoin (1996, Theorem 1.2). $\qquad\square$

Example 2.281. A Poisson process is clearly a subordinator.

A Compound Poisson process is a subordinator if and only if all random variables Y_k in Definition 2.218 are valued in $\mathbb{R}_+$.

Example 2.282. Consider the Gaussian process $C_t = W_t + \gamma t$, $t \in \mathbb{R}_+$, where $(W_t)_{t \in \mathbb{R}_+}$ is a standard Wiener process, and $\gamma \in \mathbb{R}$. The *Inverse Gaussian subordinator* is the real valued process $(S_t)_{t \in \mathbb{R}_+}$ defined by

$$S_t := \inf\{s > 0 \,|\, C_s > t\}, \quad t \in \mathbb{R}_+, \tag{2.100}$$

i.e. the first time the process $(C_t)_{t \in \mathbb{R}_+}$ crosses the level $t \in \mathbb{R}_+$.

Thanks to the continuity of the Wiener process we may take the equivalent definition

$$S_t := \inf\{s > 0 \,|\, C_s = t\}, \quad t \in \mathbb{R}_+. \tag{2.101}$$

The process $(S_t)_{t \in \mathbb{R}_+}$ is a family of Markov times for the process $(C_t)_{t \in \mathbb{R}_+}$ which, as far as we know, is a strong Markov process; as a consequence, the process $(C_{t+S_s} - s)_{t \in \mathbb{R}_+}$ has the same distribution of the process $(C_t)_{t \in \mathbb{R}_+}$, for any $s \in \mathbb{R}_+$, $s < t$. Hence, for all $s, t \in \mathbb{R}_+$, $s < t$,

$$S_t = S_s + \widetilde{S}_{t-s}, \tag{2.102}$$

where $\widetilde{S}_{t-s}$ is independent of S_s, and admits the same distribution of S_{t-s}.

It is not difficult to show the other characteristic properties of a subordinator. It is left to the reader to show (see Exercise 2.39) that

(i) the characteristic exponent of $(S_t)_{t \in \mathbb{R}_+}$ is

$$\psi(u) = -\gamma + \sqrt{\gamma - 2iu}, \quad u \in \mathbb{R}; \tag{2.103}$$

(ii) the Lévy measure of $(S_t)_{t \in \mathbb{R}_+}$ on $\mathbb{R}_+$ is

$$\nu(dx) = \frac{1}{\sqrt{2\pi x^3}} e^{-\frac{\gamma^2}{2} x} dx, \tag{2.104}$$

(iii) the drift of $(S_t)_{t \in \mathbb{R}_+}$ is

$$\mu = -\frac{2}{\gamma} \int_0^\gamma \frac{1}{\sqrt{2\pi}} e^{-y^2/2} \, dy; \tag{2.105}$$

(iv) the pdf of S_t, for any $t \in \mathbb{R}_+$ is

$$f_{S_t}(x) = \frac{t}{\sqrt{2\pi x^3}} e^{t\gamma} e^{-\frac{1}{2}(\frac{t^2}{x} + \gamma^2 x)}, \quad x \in \mathbb{R}_+. \tag{2.106}$$

The following proposition can be useful in applications.

Proposition 2.283 *Let* $X = (X_t)_{t \in \mathbb{R}_+}$ *be a Lévy process with characteristic triplet* (μ, σ^2, ν), *and let* $Z = (Z_t)_{t \in \mathbb{R}_+}$ *be a subordinator with drift* β *and Lévy measure* ν_Z, *independent of* X. *Then the subordinated process* $(X_{Z_t})_{t \in \mathbb{R}_+}$ *is a Lévy process with characteristic exponent*

$$\psi(u) = -i\widetilde{\mu}u + \frac{1}{2}\widetilde{\sigma}^2 u^2 + \int_{\mathbb{R}} \left(1 - \exp(iux) + iux I_{\{|x| \leq 1\}}\right) \widetilde{\nu}(dx), \tag{2.107}$$

where

$$\widetilde{\mu} = \beta\mu + \int_0^{+\infty} \int_{\mathbb{R}} \nu_Z(ds) I_{\{|x| \leq 1\}} x P_{X_s}(dx); \tag{2.108}$$

$$\widetilde{\sigma}^2 = \beta\sigma^2; \tag{2.109}$$

$$\widetilde{\nu}(dx) = \beta\nu(dx) + \int_0^{+\infty} \nu_Z(ds) P_{X_s}(dx). \tag{2.110}$$

Proof.
See, e.g., Sato (1999, Theorem 30.1).   □

Translation-Invariant Semigroup

Let, for $a \in \mathbb{R}$, τ_a denote the translation operator

$$f \in BC(\mathbb{R}) \mapsto \tau_a f \in BC(\mathbb{R})$$

such that

$$(\tau_a f)(x) = f(x - a), \quad \text{for any} \quad x \in \mathbb{R}.$$

Definition 2.284. Given a one-parameter semigroup $(T_t)_{t \in \mathbb{R}_+}$ on $BC(\mathbb{R})$, we say that it is translation invariant if, for any $a \in \mathbb{R}$ and any $t \in \mathbb{R}_+$,

$$T_t \tau_a = \tau_a T_t.$$

According to Definition 2.65 and the following remark, since a Lévy process $X = (X_t)_{t \in \mathbb{R}_+}$ is a time-homogenous process with independent increments (with $X_0 = 0$), it is completely characterized by the convolution semigroup of the probability laws $\mu_t = P_{X_t}$.

Thanks to Theorem 2.73 we may state the following proposition.

Proposition 2.285. *Any Lévy process is a Markov process; further, it is a Feller process.*

Proof. See, e.g., Applebaum (2004, p. 126).   □

Theorem 2.286. *Let $(X_t)_{t \in \mathbb{R}_+}$ be a Lévy process and T a stopping time. Then the process $(Y_t)_{t \in \mathbb{R}_+}$, given by*

$$Y_t = X_{T+t} - X_T,$$

is a Lévy process on the set $]T, \infty[$, adapted to $\mathcal{F}_{T+t}$. Furthermore, Y_t is independent of $\mathcal{F}_T$ and has the same distribution as X_t.

Proof. See, e.g., Protter (2004, p. 23).   □

The following result is a consequence of the stationarity and independence of increments of a Lévy process.

Theorem 2.287. *The one-parameter semigroup $(T_t)_{t \in \mathbb{R}_+}$ associated with a Feller process X such that $X(0) = 0$ is translation invariant if and only if X is a Lévy process.*

Proof. See, e.g., Bauer (1981, p. 410) and Applebaum (2004, p. 137).   □

In particular, if $(\mu_t)_{t \in \mathbb{R}_+}$ denotes the convolution semigroup of probability measures associated with the Lévy process $(X_t)_{t \in \mathbb{R}_+}$,

$$\mu_t(B) = P(X_t \in B), \quad B \in \mathcal{B}_{\mathbb{R}}, \quad t \in \mathbb{R}_+,$$

then the following relation holds:

$$\mu_t(B - x) = P(X_{t+s} \in B | X_s = x), \quad B \in \mathcal{B}_{\mathbb{R}}, \quad x \in \mathbb{R}, \quad s, t \in \mathbb{R}_+.$$

As a consequence, the transition semigroup of a Lévy process $X = (X_t)_{t \in \mathbb{R}_+}$ is a one-parameter contraction semigroup $(T_t)_{t \in \mathbb{R}_+}$ given by

$$(T(t)f)(x) = \int_{\mathbb{R}} f(x + y)\mu_t(dx), \qquad x \in \mathbb{R},$$

for any $BC(\mathbb{R})$ (e.g., Bauer (1981, p. 405)).

If the characteristic triplet of the Lévy process is (μ, σ^2, ν), then one can show (e.g., Sato (1999, p. 208) and the references therein) that the infinitesimal generator $\mathcal{A}$ of the semigroup $(T_t)_{t \in \mathbb{R}_+}$ is well defined on $BC(\mathbb{R}) \cap C^2(\mathbb{R})$, and it is given by

$$(\mathcal{A}f)(x) = -\mu f'(x) + \frac{1}{2}\sigma^2 f''(x)$$

$$+ \int_{\mathbb{R}} (f(x+y) - f(x) - I_{[|y|<1]}\, y\, f'(x))\, \nu(dy), \quad x \in \mathbb{R}. \ (2.111)$$

The following examples are trivial consequences of (2.111).

Example 2.288. For the standard Brownian motion the triplet is $(0, 1, 0)$, so that the infinitesimal generator is

$$(\mathcal{A}f)(x) = \frac{1}{2}f''(x), \qquad x \in \mathbb{R},$$

for $f \in BC(\mathbb{R}) \cap C^2(\mathbb{R})$.

Example 2.289. For a Brownian motion with drift the triplet is of the form $(\mu, \sigma^2, 0)$, so that the infinitesimal generator is

$$(\mathcal{A}f)(x) = -\mu f'(x) + \frac{1}{2}\sigma^2 f''(x), \qquad x \in \mathbb{R},$$

for $f \in BC(\mathbb{R}) \cap C^2(\mathbb{R})$.

Example 2.290. For a Poisson process with triplet $(0, 0, \lambda\epsilon_1)$ the infinitesimal generator is

$$(\mathcal{A}f)(x) = \lambda(f(x+y) - f(x)), \qquad x \in \mathbb{R},$$

for $f \in C_0(\mathbb{R})$.

Example 2.291. For a compound Poisson process with triplet $(0, 0, \nu)$ the infinitesimal generator is

$$(\mathcal{A}f)(x) = \int_{\mathbb{R}} (f(x+y) - f(x))\nu(dy), \qquad x \in \mathbb{R},$$

for $f \in C_0(\mathbb{R})$.

Stable Lévy Processes

As a particular important subclass of Lévy processes we will briefly mention the case of stable Lévy processes.

Definition 2.292. A Lévy process $X = (X_t)_{t \in \mathbb{R}_+}$ is stable if, for any $t \in \mathbb{R}_+$, X_t is a stable random variable.

According to Definition 1.214 and Proposition 1.217, symmetric stable distributions have a characteristic function of the form

$$\phi(u) = \exp\left\{-\sigma^\alpha |u|^\alpha\right\}, \quad u \in \mathbb{R},$$

for $\sigma \in \mathbb{R}^*$ and $\alpha \in (0,2]$. The case $\alpha = 2$ corresponds to the Normal distribution $N(0, 2\sigma^2)$.

Corollary 2.293. *A Lévy symmetric stable process $(X_t)_{t \in \mathbb{R}_+}$ has the scaling property, i.e., for some $\alpha \in (0,2]$, the rescaled process $(k^{-\frac{1}{\alpha}} X_{kt})_{t \in \mathbb{R}_+}$ has the same probability law as $(X_t)_{t \in \mathbb{R}_+}$, for any $k > 0$.*

For a Lévy process this is equivalent to state that, for each $t > 0$, the random variables X_t, and $t^{\frac{1}{\alpha}} X_1$ have the same probability law.

This is a generalization of the specific case for the Wiener process (for which $\alpha = 2$) of Proposition 2.186.

Proof. See, e.g., Bertoin (1996, pp. 13, 216), and Rogers and Williams (1994, p. 77). □

This is one reason why Lévy symmetric stable processes are so important in applications. Another reason why general Lévy stable processes are important in applications is that they exhibit *heavy tails*, i.e., for any $t \in \mathbb{R}_+$, $P(|X_t| > y) \propto y^{-\alpha}$ for $y \to +\infty$, as opposed to the exponential decay of the Gaussian case (e.g., Applebaum (2004) and the references therein).

2.14 Exercises and Additions

2.1. Let $(\mathcal{F}_t)_{t \in \mathbb{R}_+}$ be a filtration on the measurable space $(\Omega, \mathcal{F})$. Show that $\mathcal{F}_{t^+} = \bigcap_{u > t} \mathcal{F}_u$ is a σ-algebra (Theorem 2.84 and Remark 2.85).

2.2. Prove that two processes that are modifications of each other are equivalent.

2.3. A real-valued stochastic process, indexed in $\mathbb{R}$, is *strictly stationary* if and only if all its joint finite-dimensional distributions are invariant under a parallel time shift, i.e.,

$$F_{X_{t_1}, \ldots, X_{t_n}}(x_1, \ldots, x_n) = F_{X_{t_1 + h}, \ldots, X_{t_n + h}}(x_1, \ldots, x_n)$$

for any $n \in \mathbb{N}$, any choice of $t_1, \ldots, t_n \in \mathbb{R}$ and $h \in \mathbb{R}$, and any $x_1, \ldots, x_n \in \mathbb{R}$.

1. Prove that a process of i.i.d. random variables is strictly stationary.
2. Prove that a time-homogeneous process with independent increments is strictly stationary.
3. Prove that a Gaussian process $(X_t)_{t \in \mathbb{R}}$ is strictly stationary if and only if the following two conditions hold:
   (a) $E[X_t] = constant$ for any $t \in \mathbb{R}$.
   (b) $Cov[s, t] = K(t - s)$ for any $s, t \in \mathbb{R}$, $s < t$.

2.4. An L^2 real-valued stochastic process indexed in $\mathbb{R}$ is *weakly stationary* if and only if the following two conditions hold:

(a) $E[X_t] = constant$ for any $t \in \mathbb{R}$.

(b) $Cov[s,t] = K(t-s)$ for any $s,t \in \mathbb{R}$, $s < t$.

   1. Prove that an L^2 strictly stationary process is also weakly stationary.

   2. Prove that a weakly stationary Gaussian process is also strictly stationary.

2.5. Show that Brownian motion is not stationary.

2.6. Compute the mean function and the covariance function of the Brownian bridge.

2.7 (Prediction). Let $(X_{r-j}, \ldots, X_r)$ be a family of random variables representing a sample of a (weakly) stationary stochastic process in L^2. We know that the best approximation in L^2 of an additional random variable X_{r+s}, for any $s \in \mathbb{N}^*$, in terms of $(X_{r-j}, \ldots, X_r)$ is given by $E[Y|X_{r-j}, \ldots, X_r]$. The evaluation of this quantity is generally a hard task. On the other hand, the problem of the best linear approximation can be handled in terms of the covariances of the random variables $X_{r-j}, \ldots, X_r, X_{r+s}$, as follows.

   Prove that the best approximation of X_{r+s} in terms of a linear function of $(X_{r-j}, \ldots, X_r)$ is given by

$$\widehat{X}_{r+s} = \sum_{k=0}^{j} a_k X_{r-k},$$

where the a_k satisfy the linear system

$$\sum_{k=0}^{j} a_k c(|k-i|) = c(s+i) \text{ for } 0 \leq i \leq j.$$

Here we have denoted $c(m) = Cov[X_i, X_{i+m}]$.

2.8. Refer to Proposition 2.47. Prove that $\mathcal{F}_T$ is a σ-algebra of the subsets of Ω.

2.9. Prove all the statements of Theorem 2.48.

2.10. Prove Lemma 2.97 by considering the sequence

$$T_n = \sum_{k=1}^{\infty} k 2^{-n} I_{(k-1)2^{-n} \leq T \leq k2^{-n}}.$$

2.11. Let $(\mathcal{F}_t)_{t \in \mathbb{R}_+}$ be a filtration and prove that T is a stopping time if and only if the process $X_t = I_{\{T \leq t\}}$ is adapted to $(\mathcal{F}_t)_{t \in \mathbb{R}_+}$. Show that if T and S are stopping times, then so is $T + S$.

2.12. Show that any (sub- or super-) martingale remains a (sub- or super-) martingale with respect to the induced filtration.

2.13. Let $(X_t)_{t \in \mathbb{R}_+}$ be a martingale in L^2. Show that its increments on nonoverlapping intervals are orthogonal.

2.14. Prove Proposition 2.70. (*Hint:* To prove that 1⇒2, it suffices to use the indicator function on B; to prove that 2⇒1, it should first be shown for simple measurable functions, and then the theorem of approximation of measurable functions through elementary functions is invoked.)

2.15. Prove Remark 2.88.

2.16. Verify Example 2.112.

2.17. Determine the infinitesimal generator of a time-homogeneous Poisson process.

2.18. Show that the compound Poisson process $(Z_t)_{t \in \mathbb{R}_+}$ is a stochastic process with time-homogeneous (stationary) independent increments.

2.19. Show that

1. The Brownian motion and the compound Poisson process are both almost surely continuous at any $t \geq 0$.
2. The Brownian motion is sample continuous, but the compound Poisson process is not sample continuous.

Hence almost sure continuity does not imply sample continuity.

2.20. In the compound Poisson process, assume that the random variables Y_n are i.i.d. with common distribution

$$P(Y_n = a) = P(Y_n = -a) = \frac{1}{2},$$

where $a \in \mathbb{R}_+^*$.

1. Find the characteristic function ϕ of the process $(Z_t)_{t \in \mathbb{R}_+}$.
2. Discuss the limiting behavior of the characteristic function ϕ when $\lambda \to +\infty$ and $a \to +\infty$ in such a way that the product λa^2 is constant.

2.21. An integer-valued stochastic process $(N_t)_{t \in \mathbb{R}_+}$ with stationary (time-homogeneous) independent increments is called a *generalized Poisson process*.

1. Show that the characteristic function of a generalized Poisson process necessarily has the form

$$\phi_{N_t}(u) = e^{\lambda t [\phi(u) - 1]}$$

for some $\lambda \in \mathbb{R}_+^*$ and some characteristic function ϕ of a nonnegative integer-valued random variable. The Poisson process corresponds to the degenerate case $\phi(u) = e^{iu}$.

2. Let $(N_t^{(k)})_{t\in\mathbb{R}_+}$ be a sequence of independent Poisson processes with respective parameters λ_k. Assume that $\lambda = \sum_{k=1}^{+\infty} \lambda_k < +\infty$. Show that the process

$$N_t^{(k)} = \sum_{k=1}^{+\infty} k N_t^{(k)}, \ t \in \mathbb{R}_+,$$

is a generalized Poisson process, with characteristic function

$$\phi(u) = \sum_{k=1}^{+\infty} \frac{\lambda_k}{\lambda} e^{iku}.$$

3. Show that any generalized Poisson process can be represented as a compound Poisson process. Conversely, if the random variables Y_k in the compound Poisson process are integer-valued, then the process is a generalized Poisson process.

2.22. Let $(X_n)_{n\in\mathbb{N}} \subset E$ be a *Markov chain*, i.e., a discrete-time Markov jump process, where E is a countable set. Let $i, j \in E$ be *states* of the process; j is said to be *accessible* from state i if for some integer $n \geq 0$, $p_{ij}(n) > 0$, i.e., state j is accessible from state i if there is positive probability that in a finite number of transition states j can be reached starting from state i. Two states i and j, each accessible to the other, are said to *communicate*, and we write $i \leftrightarrow j$. If two states i and j do not communicate, then

$$p_{ij}(n) = 0 \quad \forall n \geq 0,$$

$$p_{ji}(n) = 0 \quad \forall n \geq 0,$$

or both relations are true.

We define the period of state i, written $d(i)$, as the greatest common divisor of all integers $n \geq 1$ for which $p_{ii}(n) > 0$ (if $p_{ii}(n) = 0$ for all $n \geq 1$, define $d(i) = 0$).

1. Show that the concept of communication is an equivalence relationship.
2. Show that if $i \leftrightarrow j$, then $d(i) = d(j)$.
3. Show that if state i has period $d(i)$, then there exists an integer N depending on i such that for all integers $n \geq N$

$$p_{ii}(nd(i)) > 0.$$

2.23.

1. Consider two urns A and B containing a total of N balls. A ball is selected at random (all selections are equally likely) at time $t = 1, 2, \ldots$ from among the N balls. The drawn ball is placed with probability p in urn A and with probability $q = 1 - p$ in urn B. The state of the system at each trial is represented by the number of balls in A. Determine the transition matrix for this Markov chain.

2. Assume that at each time t there are exactly k balls in A. At time $t+1$ an urn is selected at random proportionally to its content (i.e., A is chosen with probability k/N and B with probability $(N-k)/N$). Then a ball is selected either from A with probability p or from B with probability $1-p$ and placed in the previously chosen urn. Determine the transition matrix for this Markov chain.

3. Now assume that at time $t+1$ a ball and an urn are chosen with probability depending on the contents of the urn (i.e., a ball is chosen from A with probability $p = k/N$ or from B with probability q; urn A is chosen with probability p and B with probability q). Determine the transition matrix of the Markov chain.

4. Determine the equivalence classes in parts 1, 2, and 3.

2.24. Let $(X_n)_{n \in \mathbb{N}}$ be a Markov chain whose transition probabilities are $p_{ij} = 1/[e(j-i)!]$ for $i = 0, 1, \ldots$ and $j = i, i+1, \ldots$, Verify the martingale property for

- $Y_n = X_n - n$
- $U_n = Y_n^2 - n$
- $V_n = \exp\{X_n - n(e-1)\}$

2.25. Let $(X_t)_{t \in \mathbb{R}_+}$ be a process with the following properties:

- $X_0 = 0$.
- For any $0 \le t_0 < t_1 < \cdots < t_n$ the random variables $X_{t_k} - X_{t_{k-1}}$ ($1 \le k \le n$) are independent.
- If $0 \le s < t$, $X_t - X_s$ is normally distributed with

$$E(X_t - X_s) = (t-s)\mu, \qquad E\left[(X_t - X_s)^2\right] = (t-s)\sigma^2$$

where μ, σ are real constants ($\sigma \ne 0$).

The process $(X_t)_{t \in \mathbb{R}_+}$ is called a Brownian motion with *drift* μ and *variance* σ^2. (Note that if $\mu = 0$ and $\sigma = 1$, then X_t is the so-called standard Brownian motion.) Show that $Cov(X_t, X_s) = \sigma^2 \min\{s, t\}$ and $(X_t - \mu t)/\sigma$ is a standard Brownian motion.

2.26. Show that if $(X_t)_{t \in \mathbb{R}_+}$ is a Brownian motion, then the processes

$$Y_t = c X_{t/c^2} \quad \text{for fixed} \quad c > 0,$$

$$U_t = \begin{cases} t X_{1/t} & \text{for } t > 0, \\ 0 & \text{for } t = 0, \end{cases}$$

and

$$V_t = X_{t+h} - X_h \quad \text{for fixed} \quad h > 0$$

are all Brownian motions.

2.27. Let $(X_t)_{t \in \mathbb{R}_+}$ be a Brownian motion; given $a \in \mathbb{R}$, let τ_a denote the first passage time of the process started at $X(0) = 0$.

1. Show that for any $t \in \mathbb{R}_+$

$$P(\tau_a \leq t) = \frac{2}{\sqrt{2\pi t}} \int_a^{+\infty} \exp\left\{-\frac{y^2}{2t}\right\} dy.$$

Conclude that the first passage time through any given point is a.s. finite, but its mean value is infinite.

2. Use the preceding result to show that one-dimensional Brownian motion is *recurrent*, in the sense that, for any $x \in \mathbb{R}$ and any $T \in \mathbb{R}_+$,

$$P(W_t = x \quad \text{for some} \quad t > T) = 1.$$

This means that a Brownian motion returns to every point infinitely many times, for arbitrary large times.

2.28. Let $(X_t)_{t \in \mathbb{R}_+}$ be a Brownian motion, and let $M_t = \max_{0 \leq s \leq t} X_s$. Prove that $Y_t = M_t - X_t$ is a continuous-time Markov process. (*Hint:* Note that for $t' < t$

$$Y(t) = \max\left\{\max_{t' \leq s \leq t} \{(X_s - X_{t'})\}, Y_{t'}\right\} - (X_t - X_{t'}).)$$

2.29. Let T be a stopping time for a Brownian motion $(X_t)_{t \in \mathbb{R}_+}$. Then the process

$$Y_t = X_{t+T} - X_T, \qquad t \geq 0,$$

is a Brownian motion, and $\sigma(Y_t, t \geq 0)$ is independent of $\sigma(X_t, 0 \leq t \leq T)$.

(*Hint:* At first consider T constant. Then suppose that the range of T is a countable set and finally approximate T by a sequence of stopping times such as in Lemma 2.97.)

2.30. Let $(X_t)_{t \in \mathbb{R}_+}$ be an n-dimensional Brownian motion starting at 0, and let $U \in \mathbb{R}^{n \times n}$ be a (constant) orthogonal matrix, i.e., $UU^T = I$. Prove that

$$\widetilde{X}_t \equiv U X_t$$

is also a Brownian motion.

2.31. Let $(X_t)_{t \in \mathbb{R}_+}$ be a Lévy process:

1. Show that the characteristic function of X_t is infinitely divisible.
2. Suppose that the law of X_1 is $P_{X_1} = \mu$. Then, for any $t > 0$ the law of X_t is $P_{X_1} = \mu^t$.
3. Given two Lévy processes $(X_t)_{t \in \mathbb{R}_+}$ and $(X'_t)_{t \in \mathbb{R}_+}$, if $P_{X_1} = P_{X'_1}$, then the two processes are identical in law.

We call $\mu = P_{X_1}$ the infinitely divisible distribution of the Lévy process $(X_t)_{t \in \mathbb{R}_+}$.

2.32. Consider a Lévy process $X = (X_t)_{t \in \mathbb{R}_+}$. Show that the finite-dimensional distributions of X are determined by the one-dimensional marginal distributions.

2.33. Show that a Lévy process $(X_t)_{t \in \mathbb{R}_+}$ is a subordinator if and only if $X_1 \geq 0$ almost surely.

2.34. Show Proposition 2.279.

2.35. Let $W = (W_t)_{t \in \mathbb{R}_+}$ be a standard Wiener process, $(X_t)_{t \in \mathbb{R}_+}$ given by $X_t = \mu t + \sigma W_t$, $t \in \mathbb{R}_+$, and $(N_t)_{t \in \mathbb{R}_+}$ a Poisson process independent of W. Prove that the process $\{Z_t = X_{N_t}, \ t \in \mathbb{R}_+\}$ is a compound Poisson process. Give the law of Z_1.

2.36. Show that Brownian motions with drift, i.e.,

$$X_t = \sigma W_t + \alpha t \text{ for } \alpha, \sigma \in \mathbb{R},$$

are the only Lévy processes with continuous paths.

2.37. Consider two sequences of real numbers $(\alpha_k)_{k \in \mathbb{N}}$ and $(\beta_k)_{k \in \mathbb{N}}$ such that $\sum_{k \in \mathbb{N}} \beta_k^2 \alpha_k < +\infty$. Let N_t^k be a sequence of Poisson processes with intensities α_k and $k \in \mathbb{N}$, respectively.

Then the process

$$X_t = \sum_{k \in \mathbb{N}} \beta_k (N_t^k - \alpha_k t), \ t \in \mathbb{R}_+,$$

is a Lévy process having ν as its Lévy measure.

2.38. Show that

1. Any Lévy process is a Markov process.
2. Conversely, any stochastically continuous and temporarily homogeneous Markov process on $\mathbb{R}$ is a Lévy process.

2.39. Show that properties (2.103)–(2.106) hold for an Inverse Gaussian subordinator (see, e.g., Applebaum (2004, p. 51), and Kyprianou (2014, p. 29)).

2.40. According to, e.g., Jacod and Shiryaev (2003) (see also Grigoriu (2002)), we define as a classical *semimartingale* any adapted RCLL process X_t that admits the following decomposition:

$$X_t = X_0 + M_t + A_t,$$

where M_t is a local martingale and A_t is a finite variation (on compacts) RCLL process such that $M_0 = A_0 = 0$.

1. Show that any Lévy process is a semimartingale.
2. Show that the Poisson process is a semimartingale.
3. Show that the square of a Wiener process is a semimartingale.

2.41 (Poisson process and order statistics). Let $X_1, \ldots, X_n$ denote a *sample*, i.e., a family of nondegenerate i.i.d. random variables with common cumulative distribution function F. We define an *ordered sample* as the family

$$X_{n,n} \leq \cdots \leq X_{1,n},$$

so that $X_{n,n} = \min\{X_1, \ldots, X_n\}$ and $X_{1,n} = \max\{X_1, \ldots, X_n\}$. The random variable $X_{k,n}$ is called the *k-order statistic*.

Let $N = (N_t)_{t \in \mathbb{R}_+}$ be a homogeneous Poisson process with intensity $\lambda > 0$. Prove that the arrival times T_i of N in $]0, t]$, conditionally upon $\{N_t = n\}$, have the same distribution as the order statistics of a uniform sample on $]0, t[$ of size n, i.e., for all Borel sets A in $\mathbb{R}_+$ and any $n \in \mathbb{N}$ we have

$$P((T_1, T_2, \ldots, T_{N_t}) \in A | N_t = n) = P((U_{n,n}, \ldots, U_{1,n}) \in A).$$

2.42. (Self-similarity). A real-valued stochastic process $(X_t)_{t \in \mathbb{R}_+}$ is said to be *self-similar* with *index* $H > 0$ (H-ss) if its finite-dimensional distributions satisfy the relation

$$(X_{at_1}, \ldots, X_{at_n}) \overset{d}{=} a^H (X_{t_1}, \ldots, X_{t_n})$$

for any choice of $a > 0$ and $t_1, \ldots, t_n \in \mathbb{R}_+$. Show that a Gaussian process with mean function $m_t = E[X_t]$ and covariance function $K(s, t) = Cov(X_s, X_t)$ is H-ss for some $H > 0$ if and only if

$$m_t = ct^H, \quad \text{and} \quad K(s, t) = s^{2H} C(t/s, 1)$$

for some constant $c \in \mathbb{R}$ and some nonnegative definite function C. As a consequence, show that the standard Brownian motion is $1/2$-ss. Also, show that any α-stable process is $1/\alpha$-ss.

2.43 (Affine processes). Let $\Phi = (\Phi_t)_{t \in \mathbb{R}_+}$ be a process on a given probability space $(\Omega, \mathcal{F}, P)$ such that $E[\|\Phi_t\|] < +\infty$ for each $t \in \mathbb{R}_+$. The past-future filtration associated with Φ is defined as the family

$$\mathcal{F}_{s,T} = \sigma\{\Phi_u | u \in [0, s] \cup [T, +\infty[\}.$$

We shall call Φ an *affine process* if it satisfies

$$E[\Phi_t | \mathcal{F}_{s,T}] = \frac{T - t}{T - s} \Phi_s + \frac{t - s}{T - s} \Phi_T, \quad s < t < T.$$

Show that the preceding condition is equivalent to the property that for $s \leq t < t' \leq u$ the quantity

$$E\left[\frac{\Phi_t - \Phi_{t'}}{t - t'} \middle| \mathcal{F}_{s,u} \right] = \frac{\Phi_u - \Phi_s}{u - s}$$

and, hence, does not depend on the pair (t, t').

2.44. Prove that Brownian motion is an affine process.

2.45. Let $X = (X_t)_{t \in \mathbb{R}_+}$ be a Lévy process such that $E[\|X_t\|] < +\infty$ for each $t \in \mathbb{R}_+$. Show that X is an affine process.

2.46. Consider a process $M = (M_t)_{t \in \mathbb{R}_+}$ that is adapted to the filtration $(\mathcal{F}_t)_{t \in \mathbb{R}_+}$ on a probability space $(\Omega, \mathcal{F}, P)$ and satisfies

$$E[\|M_t\|] < +\infty \text{ and } E\left[\int_0^t du |M_u|\right] < +\infty \text{ for any } t > 0.$$

Prove that the following two conditions are equivalent:

1. M is an $\mathcal{F}_t$-martingale.
2. For every $t > s$,

$$E\left[\frac{1}{t-s}\int_s^t du M_u \,\middle|\, \mathcal{F}_s\right] = M_s.$$

2.47 (Empirical process and Brownian bridge). Let $U_1, \ldots, U_n, \ldots$, be a sequence of i.i.d. random variables uniformly distributed on $[0, 1]$. Define the stochastic process $b^{(n)}$ on the interval $[0, 1]$ as follows:

$$b^{(n)}(t) = \sqrt{n}\left(\frac{1}{n}\sum_{k=1}^n I_{[0,t]}(U_k) - t\right), \quad t \in [0, 1].$$

1. For any s and t in $[0, 1]$, compute $E[b^{(n)}(t)]$ and $Cov[b^{(n)}(s), b^{(n)}(t)]$.
2. Prove that, as $n \to \infty$, the finite-dimensional distributions of the process $(b^{(n)}(t))_{t \in [0,1]}$ converge weakly toward those of a Gaussian process on $[0, 1]$ whose mean and covariance functions are the same as those of $b^{(n)}$.

3

The Itô Integral

3.1 Definition and Properties

The remaining chapters on the theory of stochastic processes will focus primarily on Brownian motion, as it is by far the most useful and applicable model and allows for many explicit calculations and, as was demonstrated in the pollen grain example, arises naturally. Continuing the formal analysis of this example, suppose that a small amount of liquid flows with the macroscopic velocity $a(t, u(t))$ [where $u(t)$ is its position at time t]. Then a microscopic particle that is suspended in this liquid will, as mentioned, display evidence of Brownian motion. The change in the particle's position $u(t + dt) - u(t)$ over the time interval $[t, t + dt[$ is due to, first, the macroscopic flow of the liquid, with the latter's contribution given by $a(t, u(t))dt$. But, second, there is the additional molecular bombardment of the particle, which contributes to its dynamics with the term $b(t, u(t))[W_{t+dt} - W_t]$, where $(W_t)_{t \geq 0}$ is Brownian motion. Summing the terms results in the equation

$$du(t) = a(t, u(t))dt + b(t, u(t))dW_t,$$

which, however, in the current form does not make sense because the trajectories of $(W_t)_{t \geq 0}$ are not differentiable. Instead, we will try to interpret it in the form

$$\forall \omega \in \Omega: \qquad u(t) - u(0) = \int_0^t a(s, u(s))ds + \int_0^t b(s, u(s))dW_s,$$

which requires us to give meaning to an integral $\int_a^b f(t)dW_t$ that, as will be demonstrated, is not of the Lebesgue–Stieltjes[6] or, hence, of the Riemann–Stieltjes type.

---

[6] For a revision, see the appendix A or, in addition, e.g., Kolmogorov and Fomin (1961).

© Springer Science+Business Media New York 2015
V. Capasso, D. Bakstein, *An Introduction to Continuous-Time Stochastic Processes*, Modeling and Simulation in Science, Engineering and Technology, DOI 10.1007/978-1-4939-2757-9_3

Definition 3.1. Let $F : [a, b] \to \mathbb{R}$ be a function and Π the set of partitions $\pi : a = x_0 < x_1 < \cdots < x_n = b$ of the interval $[a, b]$. Putting

$$\forall \pi \in \Pi : \qquad V_F(\pi) = \sum_{i=1}^{n} |F(x_i) - F(x_{i-1})|,$$

then F is of *bounded variation* if

$$\sup_{\pi \in \Pi} V_F(\pi) < \infty.$$

Also, $V_F(a, b) = \sup_{\pi \in \Pi} V_F(\pi)$ is called the *total variation* of F in the interval $[a, b]$.

Remark 3.2. If $F : [a, b] \to \mathbb{R}$ is monotonic, then F is of bounded variation and

$$V_F(a, b) = |F(b) - F(a)|.$$

Lemma 3.3. *Let $F : [a, b] \to \mathbb{R}$. Then the following two statements are equivalent:*

1. *F is of bounded variation.*
2. *There exists an $F_1 : [a, b] \to \mathbb{R}$, and there exists an $F_2 : [a, b] \to \mathbb{R}$ monotonically increasing, such that $F = F_1 - F_2$.*

Lemma 3.4. *If $F : [a, b] \to \mathbb{R}$ is monotonically increasing, then F is λ almost everywhere differentiable in $[a, b]$ (where λ is the Lebesgue measure).*

Corollary 3.5. *If $F : [a, b] \to \mathbb{R}$ is of bounded variation, then F is differentiable almost everywhere.*

Definition 3.6. Let $f : [a, b] \to \mathbb{R}$ be continuous and $F : [a, b] \to \mathbb{R}$ of bounded variation, for all $\pi \in \Pi, \pi : a = x_0 < x_1 < \cdots < x_n = b$. We will fix points ξ_i arbitrarily in $[x_{i-1}, x_i[$, $i = 1, \ldots, n$ and construct the sum

$$S_n = \sum_{i=1}^{n} f(\xi_i)[F(x_i) - F(x_{i-1})].$$

If for $\max_{i \in \{1, \ldots, n\}} (x_i - x_{i-1}) \to 0$ the sum S_n tends to a limit (which depends neither on the choice of the partition nor on the selection of the points ξ_i within the partial intervals of the partition), then this limit is the *Riemann–Stieltjes integral of f with respect to the function F* over $[a, b]$ and is denoted by the symbol $\int_a^b f(x) dF(x)$.

Remark 3.7. By Theorem 2.183 and by Corollary 3.5, it can be shown that a Wiener process is not of bounded variation, and hence $\int_a^b f(t) dW_t$ cannot be interpreted in the sense of Riemann–Stieltjes.

Definition 3.8. Let $(W_t)_{t\geq 0}$ be a Wiener process defined on the probability space $(\Omega, \mathcal{F}, P)$ and $\mathcal{C}$ the set of functions $f(t,\omega) : [a,b] \times \Omega \to \mathbb{R}$ satisfying the following conditions:

1. f is $\mathcal{B}_{[a,b]} \otimes \mathcal{F}$-measurable
2. For all $t \in [a,b]$, $f(t,\cdot) : \Omega \to \mathbb{R}$ is $\mathcal{F}_t$-measurable, where $\mathcal{F}_t = \sigma(W_s, 0 \leq s \leq t)$
3. For all $t \in [a,b]$, $f(t,\cdot) \in L^2(\Omega, \mathcal{F}, P)$ and $\int_a^b E[|f(t)|^2]dt < \infty$

Remark 3.9. Condition 2 of Definition 3.8 stresses the nonanticipatory nature of f through the fact that it only depends on the present and the past history of the Brownian motion, but not on the future.

   Some authors instead of the above condition 2. require the stronger condition that the process $(f_t)_{t\in[a,b]}$ is progressively measurable, in order to guarantee desirable properties for the Itô integral defined below. Actually (see Proposition 2.43) if the process is right-continuous with left limits (cádlág or RCLL) or left-continuous with right limits, it is progressively measurable. Since we usually refer to either of these two cases, the restriction becomes irrelevant (see, e.g., Klenke (2014, p.566)).

Definition 3.10. Let $f \in \mathcal{C}$. If there exist both a partition π of $[a,b]$, $\pi : a = t_0 < t_1 < \cdots < t_n = b$, and some real-valued random variables $f_0, \ldots, f_{n-1}$ defined on $(\Omega, \mathcal{F}, P)$, such that

$$f(t,\omega) = \sum_{i=0}^{n-1} f_i(\omega) I_{[t_i, t_{i+1}[}(t)$$

(with the convention that $[t_{n-1}, t_n[= [t_{n-1}, b])$, then f is a *piecewise function*.

Remark 3.11. By condition 2 of Definition 3.8 it follows that, for all $i \in \{0, \ldots, n\}$, f_i is $\mathcal{F}_{t_i}$-measurable.

Definition 3.12. If $f \in \mathcal{C}$, with $f(t,\omega) = \sum_{i=0}^{n-1} f_i(\omega) I_{[t_i, t_{i+1}[}(t)$, is a piecewise function, then the real random variable $\Phi(f)$ is a *(stochastic) Itô integral of process f*, where

$$\forall \omega \in \Omega : \qquad \Phi(f)(\omega) = \sum_{i=0}^{n-1} f_i(\omega)(W_{t_{i+1}}(\omega) - W_{t_i}(\omega)).$$

$\Phi(f)$ is denoted by the symbol $\int_a^b f(t)dW_t$, henceforth suppressing the explicit dependence on the trajectory ω wherever obvious.

Lemma 3.13. *Let $f, g \in C$ be piecewise functions. Then they have the properties that*

1. $E[\int_a^b f(t)dW_t] = 0,$
2. $E[\int_a^b f(t)dW_t \int_a^b g(t)dW_t] = \int_a^b E[f(t)g(t)]dt.$

Proof.

1. Let $f(t,\omega) = \sum_{i=0}^{n-1} f_i(\omega)I_{[t_i,t_{i+1}[}(t)$. Then

$$
E\left[\int_a^b f(t)dW_t\right] = E\left[\sum_{i=0}^{n-1} f_i(W_{t_{i+1}} - W_{t_i})\right]
$$

$$
= E\left[\sum_{i=0}^{n-1} E[f_i(W_{t_{i+1}} - W_{t_i})|\mathcal{F}_{t_i}]\right]
$$

$$
= E\left[\sum_{i=0}^{n-1} f_i E[W_{t_{i+1}} - W_{t_i}|\mathcal{F}_{t_i}]\right],
$$

where the last step follows from Remark 3.11. Now, because $(W_t)_{t \geq 0}$ has independent increments, $(W_{t_{i+1}} - W_{t_i})$ is independent of $\mathcal{F}_{t_i}$. Hence $E[W_{t_{i+1}} - W_{t_i}|\mathcal{F}_{t_i}] = E[W_{t_{i+1}} - W_{t_i}]$, and the completion of the proof follows from the fact that the Wiener process has mean zero.

2. The piecewise functions f and g can be represented by means of the same partition $a = t_0 < t_1 < \cdots < t_n = b$ of the interval $[a,b]$. For this purpose it suffices to choose the union of the partitions associated with f and g, respectively. Thus let $f(t,\omega) = \sum_{i=0}^{n-1} f_i(\omega)I_{[t_i,t_{i+1}[}(t)$ and $g(t,\omega) = \sum_{i=0}^{n-1} g_i(\omega)I_{[t_i,t_{i+1}[}(t)$. Then

$$
E\left[\int_a^b f(t)dW_t \int_a^b g(t)dW_t\right]
$$

$$
= E\left[\sum_{i=0}^{n-1} f_i(W_{t_{i+1}} - W_{t_i}) \sum_{j=0}^{n-1} g_j(W_{t_{j+1}} - W_{t_j})\right]
$$

$$
= E\left[\sum_{i=0}^{n-1}\sum_{j=0}^{n-1} f_i g_j(W_{t_{i+1}} - W_{t_i})(W_{t_{j+1}} - W_{t_j})\right]
$$

$$
= E\left[\sum_{i=0}^{n-1}\sum_{j=0}^{n-1} E[f_i g_j(W_{t_{i+1}} - W_{t_i})(W_{t_{j+1}} - W_{t_j})|\mathcal{F}_{t_i \vee t_j}]\right],
$$

where $t_i \vee t_j = \max\{t_i, t_j\}$. If $i < j$, then $t_i < t_j$, and therefore $\mathcal{F}_{t_i} \subset \mathcal{F}_{t_j}$, resulting in f_i being $\mathcal{F}_{t_j}$-measurable (already being $\mathcal{F}_{t_i}$-measurable) and $(W_{t_{i+1}} - W_{t_i})$ being $\mathcal{F}_{t_j}$-measurable (already being $\mathcal{F}_{t_{i+1}}$-measurable with $t_{i+1} \leq t_j$). Finally, by Remark 3.11, g_j is $\mathcal{F}_{t_j}$-measurable. Thus

$$E[f_i g_j (W_{t_{i+1}} - W_{t_i})(W_{t_{j+1}} - W_{t_j})|\mathcal{F}_{t_j}]$$
$$= f_i g_j (W_{t_{i+1}} - W_{t_i}) E[W_{t_{j+1}} - W_{t_j}|\mathcal{F}_{t_j}]$$
$$= 0,$$

given that $(W_{t_{j+1}} - W_{t_j})$ is independent of $\mathcal{F}_{t_j}$ $((W_t)_{t \geq 0}$ having independent increments) and $E[W_{t_{j+1}} - W_{t_j}] = 0$.
Instead, if $i = j$, then

$$E[f_i g_i (W_{t_{i+1}} - W_{t_i})(W_{t_{i+1}} - W_{t_i})|\mathcal{F}_{t_i}] = f_i g_i E[(W_{t_{i+1}} - W_{t_i})^2|\mathcal{F}_{t_i}]$$
$$= f_i g_i E[(W_{t_{i+1}} - W_{t_i})^2].$$

But since $(W_{t_{i+1}} - W_{t_i})$ is normally distributed as $N(0, t_{i+1} - t_i)$,

$$E[(W_{t_{i+1}} - W_{t_i})^2] = t_{i+1} - t_i$$

and therefore

$$E[f_i g_i (W_{t_{i+1}} - W_{t_i})^2|\mathcal{F}_{t_i}] = f_i g_i (t_{i+1} - t_i).$$

Putting parts together, we obtain

$$E\left[\int_a^b f(t) dW_t \int_a^b g(t) dW_t\right] = E\left[\sum_{i=0}^{n-1} f_i g_i (t_{i+1} - t_i)\right]$$
$$= \sum_{i=0}^{n-1} E[f_i g_i](t_{i+1} - t_i) = \int_a^b E[f(t)g(t)]dt.$$

$\square$

Corollary 3.14. *If $f \in \mathcal{C}$ is a piecewise function, then*

$$E\left[\left(\int_a^b f(t) dW_t\right)^2\right] = \int_a^b E\left[(f(t))^2\right] dt < \infty.$$

Lemma 3.15. *If $\mathcal{S}$ denotes the space of piecewise functions belonging to the class $\mathcal{C}$, then $\mathcal{S} \subset L^2([a,b] \times \Omega)$ and $\Phi : \mathcal{S} \to L^2(\Omega)$ is linearly continuous.*

Proof. By point 3 of the characterization of the class $\mathcal{C}$, it follows that $\mathcal{S} \subset L^2([a,b] \times \Omega)$, whereas by Corollary 3.14, it follows that Φ takes values in $L^2(\Omega)$. The linearity and continuity of Φ can be inferred from Definition 3.12 and, again, from Corollary 3.14, respectively, the latter by observing that if $f \in \mathcal{S}$, then

$$\|f\|^2_{L^2([a,b] \times \Omega)} = \int_a^b E\left[(f(t))^2\right] dt,$$

$$\|\Phi(f)\|_{L^2(\Omega)}^2 = E\left[(\Phi(f))^2\right] = E\left[\left(\int_a^b f(t)dW_t\right)^2\right].$$

Thus $\|\Phi(f)\|_{L^2(\Omega)}^2 = \|f\|_{L^2([a,b]\times\Omega)}^2$, which guarantees the continuity of the linear mapping Φ.[7] $\square$

Lemma 3.16. *$\mathcal{C}$ is a closed subspace of the Hilbert space $L^2([a,b]\times\Omega)$ and is therefore a Hilbert space as well. The scalar product is defined as*

$$\langle f,g\rangle = \int_a^b \int_\Omega f(t,\omega)g(t,\omega)dP(\omega)dt = \int_a^b E[f(t)g(t)]dt.$$

Hence Φ has a unique linear continuous extension in the closure of $\mathcal{S}$ in $\mathcal{C}$ (which we will continue to denote by Φ), i.e., $\Phi : \bar{\mathcal{S}} \to L^2(\Omega)$.

Lemma 3.17. *$\mathcal{S}$ is dense in $\mathcal{C}$.*

Proof. See, e.g., Dieudonné (1960). $\square$

Theorem 3.18. *The (stochastic) Itô integral $\Phi : \mathcal{S} \to L^2(\Omega)$ has a unique linear continuous extension in $\mathcal{C}$. If $f \in \mathcal{C}$, then we denote $\Phi(f)$ by $\int_a^b f(t)dW_t$.*

Remark 3.19. Due to Theorem 3.18, if $f \in \mathcal{C}$ and $(f_n)_{n\in\mathbb{N}} \in \mathcal{S}^\mathbb{N}$ is such that $f_n \overset{n}{\to} f$ in $L^2([a,b]\times\Omega)$, then

1. $\Phi(f_n) \overset{n}{\to} \Phi(f)$ in $L^2(\Omega)$ (by the continuity of Φ)
2. $\Phi(f_n) \overset{n}{\to} \Phi(f)$ in $L^1(\Omega)$
3. $\Phi(f_n) \overset{n}{\to} \Phi(f)$ in probability

In fact, as was already mentioned, with P being a finite measure, convergence in $L^2(\Omega)$ implies convergence in $L^1(\Omega)$ and, furthermore, convergence in $L^1(\Omega)$ implies convergence in probability, by Theorem 1.180.

Proposition 3.20. *If $f,g \in \mathcal{C}$, then*

1. *$E[\int_a^b f(t)dW_t] = 0$*
2. *$E[\int_a^b f(t)dW_t \int_a^b g(t)dW_t] = \int_a^b E[f(t)g(t)]dt$*
3. *$E[(\int_a^b f(t)dW_t)^2] = \int_a^b E[(f(t))^2]dt$ (Itô isometry)*

Proof.

1. Let $f \in \mathcal{C}$; then there exists $(f_n)_{n\in\mathbb{N}} \in \mathcal{S}^\mathbb{N}$ such that $\lim_{n\to\infty} f_n = f$ in $L^2([a,b]\times\Omega)$. Because of Remark 3.19 we also have that

---

[7] For this classical result of analysis, see, e.g., Kolmogorov and Fomin (1961).

$$\lim_{n\to\infty} E\left[\left|\int_a^b (f(t) - f_n(t))dW_t\right|\right] = 0,$$

from which it follows that

$$\lim_{n\to\infty} E\left[\int_a^b (f(t) - f_n(t))dW_t\right] = 0,$$

and from the linearity of both the stochastic integral and its expectation we obtain

$$\lim_{n\to\infty} E\left[\int_a^b f_n(t)dW_t\right] = E\left[\int_a^b f(t)dW_t\right].$$

Now item 1 follows from point 1 of Lemma 3.13.

2. Let $f, g \in \mathcal{C}$. Then

$$\exists (f_n)_{n\in\mathbb{N}} \in \mathcal{S}^\mathbb{N} \text{ such that } f_n \xrightarrow{n} f \text{ in } L^2([a,b] \times \Omega);$$
$$\exists (g_n)_{n\in\mathbb{N}} \in \mathcal{S}^\mathbb{N} \text{ such that } g_n \xrightarrow{n} g \text{ in } L^2([a,b] \times \Omega).$$

By the continuity of the scalar product (in $L^2([a,b] \times \Omega)$),

$$\langle f_n, g_n \rangle \xrightarrow{n} \langle f, g \rangle,$$

and thus

$$\lim_{n\to\infty} \int_a^b E[f_n(t)g_n(t)]dt = \int_a^b E[f(t)g(t)]dt. \tag{3.1}$$

Moreover, by point 2 of Lemma 3.13,

$$\int_a^b E[f_n(t)g_n(t)]dt = E\left[\int_a^b f_n(t)dW_t \int_a^b g_n(t)dW_t\right]. \tag{3.2}$$

From the fact that $f_n \xrightarrow{n} f$ in $L^2([a,b] \times \Omega)$ it also follows that $\Phi(f_n) \xrightarrow{n} \Phi(f)$ in $L^2(\Omega)$ (by the continuity of Φ) and, analogously, since $g_n \xrightarrow{n} g$ in $L^2([a,b] \times \Omega)$, it follows that $\Phi(g_n) \xrightarrow{n} \Phi(g)$ in $L^2(\Omega)$. Then, by the continuity of the scalar product in $L^2(\Omega)$, we get

$$\langle \Phi(f_n), \Phi(g_n) \rangle \xrightarrow{n} \langle \Phi(f), \Phi(g) \rangle,$$

and hence

$$\lim_{n\to\infty} E\left[\int_a^b f_n(t)dW_t \int_a^b g_n(t)dW_t\right] = E\left[\int_a^b f(t)dW_t \int_a^b g(t)dW_t\right]. \tag{3.3}$$

The assertion finally follows from (3.1)–(3.3).

Point 3 is a direct consequence of point 2.   □

An extension of the concept of a stochastic integral is as follows.

Definition 3.21. Let $\mathcal{C}_1$ be the set of functions $f : [a, b] \times \Omega \to \mathbb{R}$ such that conditions 1 and 2 of the characterization of the class $\mathcal{C}$ are satisfied, but instead of condition 3, we have

$$P\left(\int_a^b |f(t)|^2 dt < \infty\right) = 1. \tag{3.4}$$

Remark 3.22. It is obvious that $\mathcal{C} \subset \mathcal{C}_1$, and thus $\mathcal{S} \subset \mathcal{C}_1$. We will show that it is also possible to define a stochastic integral in $\mathcal{C}_1$ that, in $\mathcal{C}$, is identical to the (stochastic) Itô integral as defined above.

Lemma 3.23. *If $f \in \mathcal{S} \subset \mathcal{C}_1$, then for all $c > 0$ and for all $N > 0$:*

$$P\left(\left|\int_a^b f(t)dW_t\right| > c\right) \leq P\left(\int_a^b |f(t)|^2 dt > N\right) + \frac{N}{c^2}. \tag{3.5}$$

Proof. See, e.g., Friedman (1975).   □

Lemma 3.24. *If $f \in \mathcal{C}_1$, then there exists $(f_n)_{n \in \mathbb{N}} \in \mathcal{S}^{\mathbb{N}}$ such that*

$$\lim_{n \to \infty} \int_a^b |f(t) - f_n(t)|^2 dt = 0 \qquad \text{a.s.}$$

Proof. See, e.g., Friedman (1975).   □

Remark 3.25. Resorting to the same notation as in the preceding lemma, we also have that $P - \lim_{n \to \infty} \int_a^b |f_n(t) - f(t)|^2 dt = 0$ because almost sure convergence implies convergence in probability. Let $f \in \mathcal{C}_1$. Then, by the preceding lemma, there exists $(f_n)_{n \in \mathbb{N}} \in \mathcal{S}^{\mathbb{N}}$ such that $\lim_{n \to \infty} \int_a^b |f(t) - f_n(t)|^2 dt = 0$ a.s. Let $(n, m) \in \mathbb{N} \times \mathbb{N}$. Then, because $(a + b)^2 \leq 2(a^2 + b^2)$, we obtain

$$\int_a^b |f_n(t) - f_m(t)|^2 dt \leq 2\left(\int_a^b |f_n(t) - f(t)|^2 dt + \int_a^b |f_m(t) - f(t)|^2 dt\right),$$

and hence $\lim_{m,n \to \infty} \int_a^b |f_n(t) - f_m(t)|^2 dt = 0$ a.s. Consequently

$$P - \lim_{n \to \infty} \int_a^b |f_n(t) - f_m(t)|^2 dt = 0.$$

But $(f_n - f_m) \in \mathcal{S} \cap \mathcal{C}_1$ (for all $n, m \in \mathbb{N}$), and by Lemma 3.23, for all $\rho > 0$ and all $\epsilon > 0$

$$P\left(\left|\int_a^b (f_n(t) - f_m(t))dW_t\right| > \epsilon\right) \leq P\left(\int_a^b |f_n - f_m|^2 dt > \rho\epsilon^2\right) + \rho.$$

Finally, by the arbitrary nature of ρ, we have that

$$\lim_{m,n\to\infty} P\left(\left|\int_a^b (f_n(t) - f_m(t))dW_t\right| > \epsilon\right) = 0.$$

Hence the sequence of random variables $(\int_a^b f_n(t)dW_t)_{n\in\mathbb{N}}$ is Cauchy in probability and therefore admits a limit in probability [see, e.g., Baldi (1984) for details]. This limit will be denoted by $\int_a^b f(t)dW_t$.

Definition 3.26. If $f \in \mathcal{C}_1$ and $(f_n)_{n\in\mathbb{N}} \in \mathcal{S}^{\mathbb{N}}$ such that $\lim_{n\to\infty} \int_a^b |f(t) - f_n(t)|^2 dt = 0$ a.s., then the limit in probability to which the sequence of random variables $(\int_a^b f_n(t)dW_t)_{n\in\mathbb{N}}$ converges is the *(stochastic) Itô integral of f*.

Remark 3.27. The preceding definition is well posed because it can be shown that $\int_a^b f(t)dW_t$ is independent of the particular approximating sequence $(f_n)_{n\in\mathbb{N}}$. [See, e.g., Baldi (1984) for details.]

Theorem 3.28. *If $f \in \mathcal{C}_1$, then (3.5) applies again.*

Proof. See, e.g., Friedman (1975). □

Theorem 3.29. *Let $f \in \mathcal{C}_1$ and $(f_n)_{n\in\mathbb{N}} \in \mathcal{C}_1^{\mathbb{N}}$. If*

$$P - \lim_{n\to\infty} \int_a^b |f_n(t) - f(t)|^2 dt = 0,$$

then

$$P - \lim_{n\to\infty} \int_a^b f_n(t)dW_t = \int_a^b f(t)dW_t.$$

Proof. Fixing $c > 0, \rho > 0$, by Theorem 3.28, we obtain

$$P\left(\left|\int_a^b (f_n(t) - f(t))dW_t\right| > c\right) \leq P\left(\int_a^b |f_n(t) - f(t)|^2 dt > c^2\rho\right) + \rho.$$

Now, the proof follows for $n \to \infty$. □

Now we are able to show that the stochastic integral in $\mathcal{C}_1$ of Definition 3.26 is identical to that of Theorem 3.18 in $\mathcal{C}$. In fact, for $f \in \mathcal{C}$, because $\mathcal{S}$ is dense in $\mathcal{C}$, there exists $(f_n)_{n\in\mathbb{N}} \in \mathcal{S}^{\mathbb{N}}$ such that

$$\lim_{n\to\infty} E\left[\int_a^b |f_n(t) - f(t)|^2 dt\right] = 0. \tag{3.6}$$

Putting $X_n = \int_a^b |f_n(t) - f(t)|^2 dt$ for all $n \in \mathbb{N}$, by the Markov inequality we obtain

$$\forall \lambda > 0: \qquad P(X_n \geq \lambda E[X_n]) \leq \frac{1}{\lambda} \qquad (n \in \mathbb{N}),$$

and thus $P(X_n \geq \epsilon) \leq \frac{E[X_n]}{\epsilon}$ for $\epsilon = \lambda E[X_n]$. But by (3.6), $\lim_{n \to \infty} E[X_n] = 0$, and therefore also $\lim_{n \to \infty} P(X_n \geq \epsilon) = 0$ and

$$P - \lim_{n \to \infty} \int_a^b |f_n(t) - f(t)|^2 dt = 0. \tag{3.7}$$

From (3.7) and by Theorem 3.29, it follows that

$$P - \lim_{n \to \infty} \int_a^b f_n(t) dW_t = \int_a^b f(t) dW_t, \tag{3.8}$$

where the limit $\int_a^b f(t) dW_t$ is the stochastic integral of f in $\mathcal{C}_1$. But, on the other hand, (3.6) implies, by point 2 of Remark 3.19, that $\Phi(f_n) \overset{n}{\to} \Phi(f)$ in probability (Φ is the linear continuous extension in $\mathcal{C}$), and thus again

$$P - \lim_{n \to \infty} \int_a^b f_n(t) dW_t = \int_a^b f(t) dW_t. \tag{3.9}$$

Now by (3.8) and (3.9), as well as the uniqueness of the limit, the proof is complete.

Remark 3.30. If $f \in \mathcal{C}_1$ and $P(\int_a^b |f(t)|^2 dt = 0) = 1$, then

$$\forall N > 0: \qquad P\left(\int_a^b |f(t)|^2 dt > N \right) = 0$$

and, by Theorem 3.28,

$$P\left(\left| \int_a^b f(t) dW_t \right| > c \right) = 0 \qquad \forall c > 0,$$

so that

$$P\left(\left| \int_a^b f(t) dW_t \right| = 0 \right) = 1.$$

Theorem 3.31. *If $f \in \mathcal{C}_1$ and continuous for almost every ω, then, for every sequence $(\pi_n)_{n \in \mathbb{N}}$ of the partitions $\pi_n : a = t_0^{(n)} < t_1^{(n)} < \cdots < t_n^{(n)} = b$ of the interval $[a,b]$ such that*

$$|\pi_n| = \sup_{k \in \{0, \ldots, n\}} \left| t_{k+1}^{(n)} - t_k^{(n)} \right| \overset{n}{\to} 0,$$

we have

$$P - \lim_{n\to\infty} \sum_{k=0}^{n-1} f\left(t_k^{(n)}\right)\left(W_{t_{k+1}^{(n)}} - W_{t_k^{(n)}}\right) = \int_a^b f(t)dW_t.$$

Proof. Consider the piecewise function

$$f_n(t,\omega) = \sum_{k=0}^{n-1} f\left(t_k^{(n)},\omega\right) I_{[t_k^{(n)},t_{k+1}^{(n)}[}(t).$$

By definition of Itô integral, we have that

$$\sum_{k=0}^{n-1} f\left(t_k^{(n)}\right)\left(W_{t_{k+1}^{(n)}} - W_{t_k^{(n)}}\right) = \int_a^b f_n(t)dW_t.$$

Now by Theorem 3.29 all that needs to be shown is that

$$P - \lim_{n\to\infty} \int_a^b |f_n(t) - f(t)|^2 dt = 0,$$

which follows by the continuity of f for almost every ω. $\qquad\square$

Proposition 3.32. *Let $(\pi_n)_{n\in\mathbb{N}}$ be a sequence of the partitions $\pi_n : a = t_0^{(n)} < t_1^{(n)} < \cdots < t_n^{(n)} = b$ of the interval $[a,b]$ such that $|\pi_n| \xrightarrow{n} 0$, and for all $n \in \mathbb{N}$, let $S_n = \sum_{j=0}^{n-1}(W_{t_{j+1}^{(n)}} - W_{t_j^{(n)}})^2$, i.e., the quadratic variation of $(W_t)_{t\in[a,b]}$ with respect to the partition π_n. Then we have that*

1. *$E[S_n] = b - a$ for all $n \in \mathbb{N}$;*
2. *$Var[S_n] = E[(S_n - (b-a))^2] \xrightarrow{n} 0.$*

Proof.

1.

$$E[S_n] = \sum_{j=0}^{n-1} E\left[\left(W_{t_{j+1}^{(n)}} - W_{t_j^{(n)}}\right)^2\right] = \sum_{j=0}^{n-1} Var\left[W_{t_{j+1}^{(n)}} - W_{t_j^{(n)}}\right]$$

$$= \sum_{j=0}^{n-1}\left(t_{j+1}^{(n)} - t_j^{(n)}\right) = b - a.$$

2. Because Brownian motion, by definition, has independent increments, we have that

$$Var[S_n] = \sum_{j=0}^{n-1} Var\left[\left(W_{t_{j+1}^{(n)}} - W_{t_j^{(n)}}\right)^2\right].$$

Writing $\delta_j = W_{t_{j+1}^{(n)}} - W_{t_j^{(n)}}$, then, by (1.6),

$$\sum_{j=0}^{n-1} Var\left[(\delta_j)^2\right] = \sum_{j=0}^{n-1} \left(E\left[(\delta_j)^4\right] - \left(E\left[(\delta_j)^2\right]\right)^2\right) \leq \sum_{j=0}^{n-1} E\left[(\delta_j)^4\right].$$

Now, by the definition of Brownian motion, the increments δ_j are Gaussian, i.e., $N(0, t_{j+1} - t_j)$, and direct calculation results in

$$E\left[(\delta_j)^4\right] = \int_{-\infty}^{+\infty} (\delta_j)^4 \frac{\exp\left\{-\frac{(\delta_j)^2}{2(t_{j+1}-t_j)}\right\}}{\sqrt{2\pi(t_{j+1}-t_j)}} d\delta_j = 3(t_{j+1}-t_j)^2 \xrightarrow{n} 0.$$

$\square$

Remark 3.33.   Given the hypotheses of the preceding proposition, by the Chebyshev inequality,

$$P(|S_n - (b-a)| > \epsilon) \leq \frac{Var[S_n]}{\epsilon^2} \xrightarrow{n} 0 \qquad (\epsilon > 0).$$

It follows that $P - \lim_{n\to\infty} S_n = b - a$. On the other hand, if we compare it to the classical Lebesgue integral, we obtain

$$\lim_{n\to\infty} \sum_{j=0}^{n-1} \left(t_{j+1}^{(n)} - t_j^{(n)}\right)^2 \leq \lim_{n\to\infty} |\pi_n| \sum_{j=0}^{n-1} \left(t_{j+1}^{(n)} - t_j^{(n)}\right) = \lim_{n\to\infty} |\pi_n|(b-a) = 0.$$

Remark 3.34.   Since the Brownian motion $(W_t)_{t\geq 0}$ is a continuous square-integrable martingale, due to Proposition 3.32, we may state that its quadratic variation process is

$$[W](t) = \langle W\rangle(t) = t, \quad t \geq 0.$$

Remark 3.35.   Because the Brownian motion $(W_t)_{t\geq 0}$ is continuous for almost every ω, we can apply Theorem 3.31 with $f(t) = W_t$, obtaining the result of Proposition 3.36.

Proposition 3.36. $\int_a^b W_t dW_t = \frac{1}{2}(W_b^2 - W_a^2) - \frac{b-a}{2}$.

Proof. Let $(\pi_n)_{n\in\mathbb{N}}$ be a sequence of the partitions $\pi_n : a = t_0^{(n)} < t_1^{(n)} < \cdots < t_n^{(n)} = b$ of the interval $[a, b]$ such that $|\pi_n| \xrightarrow{n} 0$. Then, by Theorem 3.31, we have

$$\int_a^b W_t dW_t = P - \lim_{n\to\infty} \sum_{k=0}^{n-1} W_{t_k^{(n)}} \left(W_{t_{k+1}^{(n)}} - W_{t_k^{(n)}}\right). \qquad (3.10)$$

Because, in general, $a(b-a) = \frac{1}{2}(b^2 - a^2 - (b-a)^2)$, therefore

$$W_{t_k^{(n)}}\left(W_{t_{k+1}^{(n)}} - W_{t_k^{(n)}}\right) = \frac{1}{2}\left(W_{t_{k+1}^{(n)}}^2 - W_{t_k^{(n)}}^2 - \left(W_{t_{k+1}^{(n)}} - W_{t_k^{(n)}}\right)^2\right).$$

Substitution into (3.10) results in

$$\int_a^b W_t dW_t = P - \lim_{n \to \infty} \frac{1}{2} \sum_{k=0}^{n-1} \left(W_{t_{k+1}^{(n)}}^2 - W_{t_k^{(n)}}^2 - \left(W_{t_{k+1}^{(n)}} - W_{t_k^{(n)}} \right)^2 \right)$$

$$= \frac{1}{2}(W_b^2 - W_a^2) - P - \lim_{n \to \infty} \frac{1}{2} \sum_{k=0}^{n-1} \left(W_{t_{k+1}^{(n)}} - W_{t_k^{(n)}} \right)^2$$

$$= \frac{1}{2}(W_b^2 - W_a^2) - P - \lim_{n \to \infty} \frac{1}{2} S_n = \frac{1}{2}(W_b^2 - W_a^2) - \frac{b-a}{2},$$

by Remark 3.33. $\qquad\square$

Remark 3.37. The classical Lebesgue integral results in $\int_a^b t \, dt = \frac{b^2 - a^2}{2}$. However, in the (stochastic) Itô integral we obtain an additional term $(-\frac{b-a}{2})$. Generally, in certain practical applications involving stochastic models, the Stratonovich integral is employed. In the latter, $t_k^{(n)}$ is replaced by $r_k^{(n)} = \frac{t_k^{(n)} + t_{k+1}^{(n)}}{2}$, thereby eliminating the additional term $(-\frac{b-a}{2})$. Therefore, in general, one obtains a new family of integrals by varying the chosen point of the partition. In particular, the Stratonovich integral has the advantage that its rules of calculus are identical with those of the classical integral. But, nonetheless, the Itô integral is often a more appropriate model for many applications.

3.2 Stochastic Integrals as Martingales

Theorem 3.38. *If $f \in C$ and, for all $t \in [a,b]$, $X(t) = \int_a^t f(s) dW_s$, then $(X_t)_{t \in [a,b]}$ is a martingale with respect to $\mathcal{F}_t = \sigma(W_s, 0 \le s \le t)$.*

Proof. Initially, let $f \in C \cap S$. Then there exists a π, a partition of $[a,b]$, $\pi : a = t_0 < t_1 < \cdots < t_n = b$, such that

$$f(t, \omega) = \sum_{i=0}^{n-1} f(t_i, \omega) I_{[t_i, t_{i+1}[}(t), \qquad t \in [a,b], \omega \in \Omega,$$

and for all $t \in [a,b]$

$$X(t) = \int_a^t f(s) dW_s = \sum_{i=0}^{k-1} f(t_i)(W_{t_{i+1}} - W_{t_i}) + f(t_k)(W_t - W_{t_k})$$

for k such that $t_k \le t < t_{k+1}$. Because for all $i \in \{0, \ldots, k\}$, $f(t_i)$ is $\mathcal{F}_{t_i}$-measurable (by Remark 3.11), $X(t)$ is obviously $\mathcal{F}$-measurable for all $t \in [a,b]$. Now, let $(s,t) \in [a,b] \times [a,b]$ and $s < t$. Then it needs to be shown that

$$E[X(t)|\mathcal{F}_s] = X(s) \text{ a.s.}$$

and thus

$$E[X(t) - X(s)|\mathcal{F}_s] = 0 \text{ a.s.} \tag{3.11}$$

We observe that

$$X(t) - X(s)$$

$$= \int_s^t f(u)dW_u = \sum_{i=0}^{k-1} f(t_i)(W_{t_{i+1}} - W_{t_i})$$

$$+ f(t_k)(W_t - W_{t_k}) - \sum_{j=0}^{h-1} f(t_j)(W_{t_{j+1}} - W_{t_j}) - f(t_h)(W_s - W_{t_h})$$

if $t_h \leq s < t_{h+1}$ and $t_k \leq t < t_{k+1}$, where $h \leq k$. Therefore,

$$X(t) - X(s)$$

$$= \sum_{i=h}^{k-1} f(t_i)(W_{t_{i+1}} - W_{t_i}) + f(t_k)(W_t - W_{t_k}) - f(t_h)(W_s - W_{t_h})$$

$$= \sum_{i=h+1}^{k-1} f(t_i)(W_{t_{i+1}} - W_{t_i}) + f(t_k)(W_t - W_{t_k}) + f(t_h)(W_{t_{h+1}} - W_s).$$

Because $s < t_i$, for $i = h+1, \ldots, k$, thus $\mathcal{F}_s \subset \mathcal{F}_{t_i}$, and by the properties of conditional expectations we obtain

$$E[X(t) - X(s)|\mathcal{F}_s]$$

$$= \sum_{i=h+1}^{k-1} E[f(t_i)(W_{t_{i+1}} - W_{t_i})|\mathcal{F}_s]$$

$$+ E[f(t_k)(W_t - W_{t_k})|\mathcal{F}_s] + E[f(t_h)(W_{t_{h+1}} - W_{t_s})|\mathcal{F}_s]$$

$$= \sum_{i=h+1}^{k-1} E[E[f(t_i)(W_{t_{i+1}} - W_{t_i})|\mathcal{F}_{t_i}]|\mathcal{F}_s]$$

$$+ E[E[f(t_k)(W_t - W_{t_k})|\mathcal{F}_{t_k}]|\mathcal{F}_s] + E[f(t_h)(W_{t_{h+1}} - W_{t_s})|\mathcal{F}_s]$$

$$= \sum_{i=h+1}^{k-1} E[f(t_i)E[W_{t_{i+1}} - W_{t_i}|\mathcal{F}_{t_i}]|\mathcal{F}_s]$$

$$+ E[f(t_k)E[W_t - W_{t_k}|\mathcal{F}_{t_k}]|\mathcal{F}_s] + f(t_h)E[W_{t_{h+1}} - W_{t_s}|\mathcal{F}_s]$$

$$= \sum_{i=h+1}^{k-1} E[f(t_i)E[W_{t_{i+1}} - W_{t_i}]|\mathcal{F}_s]$$

$$+ E[f(t_k)E[W_t - W_{t_k}]|\mathcal{F}_s] + f(t_h)E[W_{t_{h+1}} - W_{t_s}]$$

$$= 0$$

since $E[W_t] = 0$ for all t and $(W_t)_{t \geq 0}$ has independent increments. This completes the proof for the case $f \in \mathcal{C} \cap \mathcal{S}$.

Now, let $f \in \mathcal{C}$; then $\exists (f_n)_{n \in \mathbb{N}} \in (\mathcal{C} \cap \mathcal{S})^{\mathbb{N}}$ such that $\lim_{n \to \infty} \int_a^b |f(t) - f_n(t)|^2 dt = 0$ a.s., by Lemma 3.24. We put

$$X_n(t) = \int_a^t f_n(s) dW_s \qquad \forall n \in \mathbb{N}, \forall t \in [a, b],$$

for which we have just shown that $((X_n(t))_{t \in [a,b]})_{n \in \mathbb{N}}$ is a sequence of martingales. Now, let $(s, t) \in [a, b] \times [a, b]$ and $s < t$; it will be shown that

$$E[X(t) - X(s)|\mathcal{F}_s] = 0 \text{ a.s.} \tag{3.12}$$

We obtain for all $n \in \mathbb{N}$

$$E[X(t) - X(s)|\mathcal{F}_s]$$
$$= E[X(t) - X_n(t)|\mathcal{F}_s] + E[X_n(t) - X_n(s)|\mathcal{F}_s] + E[X_n(s) - X(s)|\mathcal{F}_s].$$

Because $(X_n(t))_{t \in [a,b]}$ is a martingale, $E[X_n(t) - X_n(s)|\mathcal{F}_s] = 0$. We also observe that

$$E[(E[X(t) - X_n(t)|\mathcal{F}_s])^2]$$
$$\leq E\left[E\left[|X(t) - X_n(t)|^2 \Big| \mathcal{F}_s\right]\right] = E\left[|X(t) - X_n(t)|^2\right]$$
$$= E\left[\left|\int_a^t (f(u) - f_n(u)) dW_u\right|^2\right] = \int_a^t E\left[|f(u) - f_n(u)|^2\right] du \xrightarrow{n} 0,$$

following the properties of conditional expectations and by point 3 of Proposition 3.20. Hence $E[X(t) - X_n(t)|\mathcal{F}_s]$ converges to zero in $L^2(\Omega)$, as it does, analogously, $E[X(s) - X_n(s)|\mathcal{F}_s]$, thus proving (3.12). Finally, we need to show that $X(t)$ is $\mathcal{F}_t$-measurable for $t \in [a, b]$. This follows from the fact that $X_n(t)$ is $\mathcal{F}_t$-measurable for $n \in \mathbb{N}$ and, moreover,

$$E\left[|X(t) - X_n(t)|^2\right] = \int_a^t E\left[|f(u) - f_n(u)|^2\right] du \xrightarrow{n} 0,$$

following the preceding derivation. Hence $X_n(t) \to X(t)$ in $L^2(\Omega)$. $\qquad \square$

Remark 3.39. We may remark here that Equation (3.12) can be rewritten more explicitly as

$$E\left[\int_s^t f(u) dW_u | \mathcal{F}_s\right] = 0, \text{ a.s.,} \tag{3.13}$$

for any $(s, t) \in [a, b] \times [a, b]$, $s < t$.

Similarly it can be shown (see Exercise 3.3)

$$E\left[\left(\int_s^t f(u) dW_u\right)^2 | \mathcal{F}_s\right] = \int_s^t E[(f(u))^2 | \mathcal{F}_s] du, \text{ a.s.,} \tag{3.14}$$

for any $(s, t) \in [a, b] \times [a, b]$, $s < t$.

Note that the integral in $\mathcal{C}_1$ is not in general a martingale. It is, however, a local martingale (see, e.g., Karatzas and Shreve (1991, p. 146)) .

Proposition 3.40. *Resorting to the notation of the preceding theorem, the martingale $(X_t)_{t \in [a,b]}$ is continuous (in $L^2(\Omega)$, and in probability).*

Proof. If $t, s \in [a, b]$, then

$$\lim_{t \to s} E\left[|X(t) - X(s)|^2\right] = \lim_{t \to s} E\left[\left|\int_s^t f(u)dW_u\right|^2\right]$$

$$= \lim_{t \to s} \int_s^t E\left[(f(u))^2\right] du = 0,$$

by point 3 of Proposition 3.20 and following the continuity of the Lebesgue integral.   □

Theorem 3.41. *If $f \in \mathcal{C}_1$, then $(X_t)_{t \in [a,b]}$ admits a continuous version and thus admits a modified form, with almost every trajectory being continuous.*

Proof. See, e.g., Baldi (1984) or Friedman (1975).   □

Following Theorems 2.28 and 3.41, henceforth we can always consider continuous and separable versions of $(X_t)_{t \in [a,b]}$. If $f \in \mathcal{C}$ and $X(t) = \int_a^t f(u)dW_u$, $t \in [a, b]$, then because (by Theorem 3.38) $(X_t)_{t \in [a,b]}$ is a martingale, it satisfies Doob's inequality (Proposition 2.125), and the following proposition holds.

Proposition 3.42. *If $f \in \mathcal{C}$, then*

1. $E[\max_{a \le s \le b} |\int_a^s f(u)dW_u|^2] \le 4E[|\int_a^b f(u)dW_u|^2] = 4E[\int_a^b |f(u)|^2 du]$;
2. $P(\max_{a \le s \le b} |\int_a^s f(u)dW_u| > \lambda) \le \frac{1}{\lambda^2} E[\int_a^b |f(u)|^2 du]$, $\lambda > 0$.

Proof. Point 1 follows directly from point 2 of Proposition 2.121 with $p = 2$.
    Point 2 follows by continuity

$$\left(\max_{a \le s \le b} \left|\int_a^s f(u)dW_u\right|\right)^2 = \max_{a \le s \le b} \left|\int_a^s f(u)dW_u\right|^2 ;$$

therefore

$$P\left(\max_{a \le s \le b} \left|\int_a^s f(u)dW_u\right| > \lambda\right) = P\left(\left(\max_{a \le s \le b} \left|\int_a^s f(u)dW_u\right|\right)^2 > \lambda^2\right)$$

$$= P\left(\max_{a \le s \le b} \left|\int_a^s f(u)dW_u\right|^2 > \lambda^2\right)$$

and the proof follows from point 1 of Proposition 2.125.   □

Remark 3.43. Generally, $\max_{a \le s \le b} X_s$, almost everywhere with respect to P, is defined due to the continuity of Brownian motion.

Stochastic Integrals with Stopping Times

Let $f \in C_1([0,T])$, $(W_t)_{t \in \mathbb{R}_+}$ a Wiener process, and τ_1 and τ_2 two random variables representing stopping times such that $0 \leq \tau_1 \leq \tau_2 \leq T$. Then

$$\int_{\tau_1}^{\tau_2} f(t)dW_t = \int_0^{\tau_2} f(t)dW_t - \int_0^{\tau_1} f(t)dW_t.$$

Lemma 3.44. *Defining the characteristic function as*

$$\chi_i(t) = \begin{cases} 1 & \text{if } t < \tau_i, \\ 0 & \text{if } t \geq \tau_i, \end{cases} \quad i = 1,2,$$

we have that

1. $\chi_i(t)$ *is* $\mathcal{F}_t = \sigma(W_s, 0 \leq s \leq t)$-*measurable* $(i = 1,2)$;
2. $\int_{\tau_1}^{\tau_2} f(t)dW_t = \int_0^T \chi_2(t)f(t)dW_t - \int_0^T \chi_1(t)f(t)dW_t$.

Proof. See, e.g., Friedman (1975).   □

Theorem 3.45. *Let* $f \in C_1([0,T])$ *and let* τ_1 *and* τ_2 *be two stopping times such that* $0 \leq \tau_1 \leq \tau_2 \leq T$. *Then*

1. $E[\int_{\tau_1}^{\tau_2} f(t)dW_t] = 0$
2. $E[(\int_{\tau_1}^{\tau_2} f(t)dW_t)^2] = E[\int_{\tau_1}^{\tau_2} |f(t)|^2 dt]$

Proof. By Lemma 3.44, we get

$$\int_{\tau_1}^{\tau_2} f(t)dW_t = \int_0^T (\chi_2(t) - \chi_1(t))f(t)dW_t,$$

and after applying Proposition 3.20 the proof is completed. This theorem is, in fact, just a generalization of Proposition 3.20.   □

The following lemma is very useful in applications.

Lemma 3.46. *For any* $T > 0$, *let* $f \in C_1([0,T])$, *with respect to a Wiener process* $(W_t)_{t \in \mathbb{R}_+}$, *and assume further that*

$$P\left(\int_0^\infty f^2(t)dW_t = \infty \right) = 1.$$

Then the random process

$$z_s = \int_0^{\tau_s} f_t dW_t, \qquad s \in \mathbb{R}_+,$$

with $\tau_s = \inf \left\{ t \in \mathbb{R}_+ | \int_0^t f^2(u)du > s \right\}$ *is a Wiener process, and*

$$P\left(\lim_{t \to +\infty} \frac{\int_0^t f_u dW_u}{\int_0^t f^2(u)du} = 0 \right) = 1.$$

Proof. See, e.g., Lipster and Shiryaev (2010), p. 235.   □

3.3 Itô Integrals of Multidimensional Wiener Processes

We denote by $\mathbb{R}^{mn}$ all real-valued $m \times n$ matrices and by

$$\mathbf{W}(t) = (W_1(t), \ldots, W_n(t))', \qquad t \geq 0,$$

an n-dimensional Wiener process. Let $[a, b] \subset [0, +\infty[$, and we put

$$\mathcal{C}_{\mathbf{W}}([a, b])$$
$$= \left\{ f : [a, b] \times \Omega \to \mathbb{R}^{mn} | \forall 1 \leq i \leq m, \forall 1 \leq j \leq n : f_{ij} \in \mathcal{C}_{W_j}([a, b]) \right\},$$
$$\mathcal{C}_{1\mathbf{W}}([a, b])$$
$$= \left\{ f : [a, b] \times \Omega \to \mathbb{R}^{mn} | \forall 1 \leq i \leq m, \forall 1 \leq j \leq n : f_{ij} \in \mathcal{C}_{1W_j}([a, b]) \right\},$$

where $\mathcal{C}_{W_j}([a, b])$ and $\mathcal{C}_{1W_j}([a, b])$ correspond to the classes $\mathcal{C}([a, b])$ and $\mathcal{C}_1([a, b])$, respectively, as defined in Sect. 3.1.

Definition 3.47. If $f : [a, b] \times \Omega \to \mathbb{R}^{mn}$ belongs to $\mathcal{C}_{1\mathbf{W}}([a, b])$, then the *stochastic integral with respect to* $\mathbf{W}$ is the m-dimensional vector defined by

$$\int_a^b f(t) d\mathbf{W}(t) = \left(\sum_{j=1}^n \int_a^b f_{ij}(t) dW_j(t) \right)'_{1 \leq i \leq m},$$

where each of the integrals on the right-hand side is defined in the sense of Itô.

Proposition 3.48. *If* $(i, j) \in \{1, \ldots, n\}^2$ *and*

$$f_i : [a, b] \times \Omega \to \mathbb{R} \text{ belongs to } \mathcal{C}_{W_i}([a, b]) \text{and}$$
$$f_j : [a, b] \times \Omega \to \mathbb{R} \text{ belongs to } \mathcal{C}_{W_j}([a, b]),$$

then

$$E\left[\int_a^b f_i(t) dW_i(t) \int_a^b f_j(t) dW_j(t) \right] = \delta_{ij} E\left[\int_a^b f_i(t) f_j(t) dt \right],$$

where $\delta_{ij} = 1$, *if* $i = j$ *or* $\delta_{ij} = 0$, *if* $i \neq j$, *is the Kronecker delta.*

Proof. Suppose $i \neq j$. Then the processes $(W_i(t))_{t \geq 0}$ and $(W_j(t))_{t \geq 0}$ are independent, as are hence, the σ-algebras $\mathcal{F}^{(i)} = \sigma(W_i(s), s \geq 0)$ and $\mathcal{F}^{(j)} = \sigma(W_j(s), s \geq 0)$. Moreover, for all $t \in [a, b]$: $f_i(t)$ is $\mathcal{F}^{(i)}$-measurable, $f_j(t)$ is $\mathcal{F}^{(j)}$-measurable, and $\mathcal{F}_t^{(i)} = \sigma(W_i(s), 0 \leq s \leq t) \subset \mathcal{F}^{(i)}$ as well as $\mathcal{F}_t^{(j)} = \sigma(W_j(s), 0 \leq s \leq t) \subset \mathcal{F}^{(j)}$. Therefore, $f_i = (f_i(t))_{t \in [a, b]}$ and $f_j = (f_j(t))_{t \in [a, b]}$ are independent, as are $\int_a^b f_i(t) dW_i(t)$ and $\int_a^b f_j(t) dW_j(t)$, and therefore

$$E\left[\int_a^b f_i(t)dW_i(t) \int_a^b f_j(t)dW_j(t)\right]$$

$$= E\left[\int_a^b f_i(t)dW_i(t)\right] E\left[\int_a^b f_j(t)dW_j(t)\right] = 0,$$

by Proposition 3.20. If instead $i = j$, then the proof immediately follows by Proposition 3.20. $\qquad\square$

Proposition 3.49. *Let* $f : [a,b] \times \Omega \to \mathbb{R}^{mn}$ *and* $g : [a,b] \times \Omega \to \mathbb{R}^{mn}$. *Then*

1. If $f \in \mathcal{C}_{\mathbf{W}}([a,b])$, *then*

$$E\left[\int_a^b f(t)d\mathbf{W}(t)\right] = \mathbf{0} \in \mathbb{R}^m;$$

2. If $f,g \in \mathcal{C}_{\mathbf{W}}([a,b])$, *then*

$$E\left[\left(\int_a^b f(t)d\mathbf{W}(t)\right)\left(\int_a^b g(t)d\mathbf{W}(t)\right)'\right] = E\left[\int_a^b (f(t))(g(t))'dt\right];$$

3. If $f \in \mathcal{C}_{\mathbf{W}}([a,b])$, *then*

$$E\left[\left|\int_a^b f(t)d\mathbf{W}(t)\right|^2\right] = E\left[\int_a^b |f(t)|^2 dt\right],$$

where

$$|f|^2 = \sum_{i=1}^m \sum_{j=1}^n (f_{ij})^2$$

and

$$\left|\int_a^b f(t)d\mathbf{W}(t)\right|^2 = \sum_{i=1}^m \left(\sum_{j=1}^n \int_a^b f_{ij}(t)dW_j(t)\right)^2.$$

Proof.

1. Let $f \in \mathcal{C}_{\mathbf{W}}([a,b])(\subset \mathcal{C}_{1\mathbf{W}}([a,b]))$. Then

$$E\left[\int_a^b f(t)d\mathbf{W}(t)\right] = \left(E\left[\sum_{j=1}^n \int_a^b f_{ij}(t)dW_j(t)\right]\right)'_{1\le i\le m}$$

$$= \left(\sum_{j=1}^n E\left[\int_a^b f_{ij}(t)dW_j(t)\right]\right)'_{1\le i\le m} = \mathbf{0} \in \mathbb{R}^m,$$

by Proposition 3.20.

2. Let $f, g \in \mathcal{C}_{\mathbf{W}}([a,b])$ and $(1,k) \in \{1, \ldots, m\}^2$. Then

$$E\left[\left(\int_a^b f(t)d\mathbf{W}(t)\right)\left(\int_a^b g(t)d\mathbf{W}(t)\right)'\right]_{lk}$$

$$= E\left[\left(\sum_{j=1}^n \int_a^b f_{lj}(t)dW_j(t)\right)\left(\sum_{j'=1}^n \int_a^b g_{kj'}(t)dW_{j'}(t)\right)\right]$$

$$= \sum_{j=1}^n \sum_{j'=1}^n E\left[\int_a^b f_{lj}(t)dW_j(t)\int_a^b g_{kj'}dW_{j'}(t)\right] = \sum_{j=1}^n E\left[\int_a^b f_{lj}(t)g_{kj}(t)dt\right]$$

$$= E\left[\sum_{j=1}^n \int_a^b f_{lj}(t)g_{kj}(t)dt\right] = E\left[\int_a^b \sum_{j=1}^n (f_{lj}(t)g_{kj}(t))dt\right]$$

$$= E\left[\int_a^b ((f(t))(g(t))')_{lk}dt\right],$$

by Proposition 3.48. With each of the components verified, the proof of point 2 is complete.

3. Let $f \in \mathcal{C}_{\mathbf{W}}([a,b])$. Then by point 2 we have

$$E\left[\left(\int_a^b f(t)d\mathbf{W}(t)\right)\left(\int_a^b f(t)d\mathbf{W}(t)\right)'\right] = E\left[\int_a^b (f(t))(f(t))'dt\right].$$

$$\tag{3.15}$$

Furthermore, it is easily verified that if a generic $b \in \mathbb{R}^{mn}$, then

$$|b|^2 = \sum_{i=1}^m \sum_{j=1}^n (b_{ij})^2 = trace(bb'),$$

and if a generic $\mathbf{a} \in \mathbb{R}^m$, then

$$|\mathbf{a}|^2 = \sum_{i=1}^m (a_i)^2 = trace(\mathbf{aa}').$$

Therefore, if in (3.15) we consider the trace of both the former and the latter term, we obtain point 3.

$\square$

3.4 The Stochastic Differential

Definition 3.50. Let $(u(t))_{0 \le t \le T}$ be a process such that for every $(t_1, t_2) \in [0,T] \times [0,T], t_1 < t_2$:

$$u(t_2) - u(t_1) = \int_{t_1}^{t_2} a(t)dt + \int_{t_1}^{t_2} b(t)dW_t, \tag{3.16}$$

where $(a)^{1/2} \in \mathcal{C}_1([0,T])$ and $b \in \mathcal{C}_1([0,T])$. Then $u(t)$ is said to have the *stochastic differential*

$$du(t) = a(t)dt + b(t)dW_t \tag{3.17}$$

on $[0,T]$.

Remark 3.51. If $u(t)$ has the stochastic differential in the form of (3.17), then for all $t > 0$, we have

$$u(t) = u(0) + \int_0^t a(s)ds + \int_0^t b(s)dW_s.$$

Hence

1. The trajectories of $(u(t))_{0 \leq t \leq T}$ are continuous almost everywhere (see Theorem 3.41).
2. For $t \in [0,T]$, $u(t)$ is $\mathcal{F}_t = \sigma(W_s, 0 \leq s \leq t)$-measurable, thus $u(t) \in \mathcal{C}_1([0,T])$.

Example 3.52. The stochastic differential of $(W_t^2)_{t \geq 0}$ is given by

$$dW_t^2 = dt + 2W_t dW_t. \tag{3.18}$$

In fact, if $0 \leq t_1 < t_2$, then, by Proposition 3.36, it follows that

$$\int_{t_1}^{t_2} W_t dW_t = \frac{1}{2}(W_{t_2}^2 - W_{t_1}^2) - \frac{t_2 - t_1}{2}.$$

Therefore, $W_{t_2}^2 - W_{t_1}^2 = t_2 - t_1 + 2 \int_{t_1}^{t_2} W_t dW_t$, which is of the form (3.16) with $a(t) = 1$ and $b(t) = 2W_t$, $t \geq 0$.

Example 3.53. The stochastic differential of the process $(tW_t)_{t \geq 0}$ is given by

$$d(tW_t) = W_t dt + t dW_t. \tag{3.19}$$

Let $0 \leq t_1 < t_2$ and $(\pi_n)_{n \in \mathbb{N}}$ be a sequence of partitions of $[t_1, t_2]$, where $\pi_n : t_1 = r_1^{(n)} < \cdots < r_n^{(n)} = t_2$, such that $|\pi_n| \xrightarrow{n} 0$. Then, by Theorem 3.31,

$$\int_{t_1}^{t_2} t dW_t = P - \lim_{n \to \infty} \sum_{k=1}^{n-1} r_k^{(n)} \left(W_{r_{k+1}^{(n)}} - W_{r_k^{(n)}} \right). \tag{3.20}$$

Moreover, because $(W_t)_{t \geq 0}$ is continuous almost surely, we can consider $\int_{t_1}^{t_2} W_t dt$, obtaining

$$\int_{t_1}^{t_2} W_t dt = \lim_{n \to \infty} \sum_{k=1}^{n-1} W_{r_{k+1}^{(n)}} \left(r_{k+1}^{(n)} - r_k^{(n)} \right) \text{ a.s.}$$

But since almost sure convergence implies convergence in probability, we have

$$\int_{t_1}^{t_2} W_t dt = P - \lim_{n\to\infty} \sum_{k=1}^{n-1} W_{r_{k+1}^{(n)}} \left(r_{k+1}^{(n)} - r_k^{(n)} \right). \tag{3.21}$$

Combining the relevant terms of (3.20) and (3.21) we obtain

$$\int_{t_1}^{t_2} t dW_t + \int_{t_1}^{t_2} W_t dt = P - \lim_{n\to\infty} \sum_{k=1}^{n-1} \left(r_{k+1}^{(n)} W_{r_{k+1}^{(n)}} - r_k^{(n)} W_{r_k^{(n)}} \right)$$

$$= t_2 W_{t_2} - t_1 W_{t_1},$$

which is of the form (3.16) with $a(t) = W_t$ and $b(t) = t$, proving (3.19).

Proposition 3.54. *If the stochastic differential of $(u_i(t))_{t\in[0,T]}$ is given by*

$$du_i(t) = a_i(t)dt + b_i(t)dW_t, \qquad i = 1, 2,$$

then $(u_1(t)u_2(t))_{t\in[0,T]}$ has the stochastic differential

$$d(u_1(t)u_2(t)) = u_1(t)du_2(t) + u_2(t)du_1(t) + b_1(t)b_2(t)dt, \tag{3.22}$$

and thus, for all $0 \le t_1 < t_2 < T$

$$u_1(t_2)u_2(t_2) - u_1(t_1)u_2(t_1)$$

$$= \int_{t_1}^{t_2} u_1(t)a_2(t)dt + \int_{t_1}^{t_2} u_1(t)b_2(t)dW_t$$

$$+ \int_{t_1}^{t_2} u_2(t)a_1(t)dt + \int_{t_1}^{t_2} u_2(t)b_1(t)dW_t + \int_{t_1}^{t_2} b_1(t)b_2(t)dt. \tag{3.23}$$

Proof. (See, e.g., Baldi (1984)):

Case 1: a_i, b_i constant on $[t_1, t_2]$, i.e., $a_i(t) = a_i$, $b_i(t) = b_i$, for all $t \in [t_1, t_2], i = 1, 2, a_i, b_i$ in $\mathcal{C}_1([0, T])$. Then

$$u_1(t_2) = u_1(t_1) + a_1(t_2 - t_1) + b_1(W_{t_2} - W_{t_1}), \tag{3.24}$$

$$u_2(t_2) = u_2(t_1) + a_2(t_2 - t_1) + b_2(W_{t_2} - W_{t_1}). \tag{3.25}$$

The proof of formula (3.23) is complete by employing Equations (3.18), (3.19), (3.24), and (3.25) and the definitions of both the Lebesgue and stochastic integrals.

Case 2: It can be shown that (3.23) holds for a_i, b_i, $i = 1, 2$, which are piecewise functions.

Case 3: Ultimately it can be shown that (3.23) holds for any a_i, b_i ($a_i, b_i \in \mathcal{C}$, $i = 1, 2$).

$\square$

Remark 3.55. Generally, if $(u(t))_{t \in [0,T]}$ admits a stochastic differential and $b(t) \in \mathcal{C}_1([0,T])$, then, by the Hölder inequality (see Proposition 1.170), $u(t)b(t) \in \mathcal{C}_1([0,T])$ and so $\int_0^T u(t)b(t)dW_t$ is well defined (see Friedman (2004, p.79)).

Remark 3.56. If $f : \mathbb{R} \to \mathbb{R}$ is a continuous function, then $f(W_t) \in \mathcal{C}_1([0,T])$; in fact, the trajectories of $(f(W_t))_{t \in [0,T]}$ are continuous almost everywhere, and thus condition (3.4) is certainly verified. In particular, we have

$$(W_t^n)_{t \in [0,T]} \in \mathcal{C}_1([0,T]) \qquad \forall n \in \mathbb{N}^*.$$

Corollary 3.57. *For every integer $n \geq 2$ we get*

$$d(W_t^n) = nW_t^{n-1}dW_t + \frac{1}{2}(n-1)nW_t^{n-2}dt.$$

Proof. The proof follows from Proposition 3.54 by induction. $\qquad\square$

Corollary 3.58. *For every polynomial $P(x)$:*

$$dP(W_t) = P'(W_t)dW_t + \frac{1}{2}P''(W_t)dt. \tag{3.26}$$

Remark 3.59. The second derivative of $P(W_t)$ is required for its differential.

Proposition 3.60. *If $f \in C^2(\mathbb{R})$, then*

$$df(W_t) = f'(W_t)dW_t + \frac{1}{2}f''(W_t)dt.$$

Proof. Since $f \in C^2(\mathbb{R})$ it can be shown (see Friedman (2004, Problem 11, p.95)) that there exists a sequence of polynomials $(Q_n(x))_{n \in \mathbb{N}}$ such that

$$Q_n(x) \overset{n}{\to} f, \qquad Q_n'(x) \overset{n}{\to} f', \qquad Q_n''(x) \overset{n}{\to} f'' \tag{3.27}$$

uniformly on compact subsets of $\mathbb{R}$. Take $0 \leq t_1 < t_2 \leq T$; for any $Q_n(x)$ we have by (3.26)

$$Q_n(W_{t_2}) - Q_n(W_{t_1}) = \int_{t_1}^{t_2} Q_n'(W_t)dW_t + \frac{1}{2}\int_{t_1}^{t_2} Q_n''(W_t)dt. \tag{3.28}$$

By the above quoted result it is clear that

$$\frac{1}{2}\int_{t_1}^{t_2} Q_n''(W_t)dt \overset{n}{\to} \frac{1}{2}\int_{t_1}^{t_2} f''(W_t)dt \text{ a.s. (and thus in probability)},$$

we also have

$$\int_{t_1}^{t_2} [Q'_n(W_t) - f'(t)]^2 \, dt \xrightarrow{n} 0 \text{ a.s. (and thus in probability)},$$

so that by Theorem 3.29 we have

$$P - \lim_{n \to \infty} \int_{t_1}^{t_2} Q'_n(W_t) dW_t = \int_{t_1}^{t_2} f'(W_t) dW_t.$$

Finally, taking $n \to \infty$ in (3.28), we have

$$f(W_{t_2}) - f(W_{t_1}) = \int_{t_1}^{t_2} f'(W_t) dW_t + \frac{1}{2} \int_{t_1}^{t_2} f''(W_t) dt.$$

$\square$

3.5 Itô's Formula

As one of the most important topics on Brownian motion, Itô's formula represents the stochastic equivalent of Taylor's theorem about the expansion of functions. It is the key concept that connects classical and stochastic theory.

Proposition 3.61. *If $u(t, x) : [0, T] \times \mathbb{R} \to \mathbb{R}$ is continuous with the derivatives $u_x, u_{xx},$ and u_t, then*

$$du(t, W_t) = \left(u_t(t, W_t) + \frac{1}{2} u_{xx}(t, W_t) \right) dt + u_x(t, W_t) dW_t. \qquad (3.29)$$

Proof.

Case 1: We suppose $u(t, x) = g(t)\psi(x)$, with $g \in C^1([0, T])$ and $\psi \in C^2(\mathbb{R})$. Then, by Proposition 3.60,

$$d\psi(W_t) = \psi'(W_t) dW_t + \frac{1}{2} \psi''(W_t) dt,$$

and, by formula (3.22), we obtain an expression for (3.29), namely,

$$d(g(t)\psi(W_t)) = g(t)\psi'(W_t) dW_t + \frac{1}{2} g(t)\psi''(W_t) dt + \psi(W_t) g'(t) dt.$$

Case 2: If

$$u(t, x) = \sum_{i=1}^{n} g_i(t)\psi_i(x), \qquad g \in C^1([0, T]), \psi \in C^2(\mathbb{R}), i = 1, \ldots, n,$$

$$(3.30)$$

then (3.29) is an immediate consequence of the first case.

Case 3: If u is a generic function, satisfying the hypotheses of the proposition, then it can be shown that there exists $(u_n)_{n\in\mathbb{N}}$, a sequence of functions of type (3.30), such that for all $K > 0$

$$\lim_{n\to\infty} \sup_{|x|\leq K} \sup_{t\in[0,T]} \{|u_n - u| + |(u_n)_t - u_t| + |(u_n)_x$$
$$-u_x| + |(u_n)_{xx} - u_{xx}|\} = 0.$$

Therefore, we can approximate u uniformly through the sequence u_n, and the proof follows from the second case.

$\square$

Remark 3.62. Contrary to what is obtained for an ordinary differential, Equation (3.29) contains the additional term $\frac{1}{2}u_{xx}(t, W_t)dt$. This is due to the presence of Brownian motion.

Remark 3.63. If $u(t, z, \omega) : [0, T] \times \mathbb{R} \times \Omega \to \mathbb{R}$ is continuous with the derivatives u_z, u_{zz}, and u_t such that, for all (t, z), $u(t, z, \cdot)$ is $\mathcal{F}_t = \sigma(W_s, 0 \leq s \leq t)$-measurable, then formula (3.29) holds for every $\omega \in \Omega$.

Theorem 3.64 (Itô's formula). *If $du(t) = a(t)dt + b(t)dW_t$, and if $f(t, x) : [0, T] \times \mathbb{R} \to \mathbb{R}$ is continuous with the derivatives f_x, f_{xx}, and f_t, then the stochastic differential of the process $f(t, u(t))$ is given by*

$$df(t, u(t)) = \left(f_t(t, u(t)) + \frac{1}{2}f_{xx}(t, u(t))b^2(t) + f_x(t, u(t))a(t) \right) dt$$
$$+ f_x(t, u(t))b(t)dW_t. \tag{3.31}$$

Proof. See, e.g., Karatzas and Shreve (1991). $\square$

3.6 Martingale Representation Theorem

Theorem 3.38 stated that, given a process $(f_t)_{t\in[0,T]} \in \mathcal{C}([0, T])$, the Itô integral $\int_0^t f_s dW_s$ is a zero-mean $\mathcal{L}^2$-martingale. The martingale representation theorem establishes the relationship between a martingale and the existence of a process converse.

Theorem 3.65 (Martingale representation theorem I). *Let $(M_t)_{t\in[0,T]}$ be an $\mathcal{L}^2$-martingale with respect to the Wiener process $(W_t)_{t\in[0,T]}$ and to $(\mathcal{F}_t)_{t\in[0,T]}$, its natural filtration. Then there exists a unique process $(f_t)_{t\in[0,T]} \in \mathcal{C}([0, T])$, so that*

$$\forall t \in [0, T]: \qquad M(t) = M(0) + \int_0^t f(s)dW_s \qquad a.s. \tag{3.32}$$

Theorem 3.66   (Martingale representation theorem II).   *Let*
$(M_t)_{t\in[0,T]}$ *be a martingale with respect to the Wiener process* $(W_t)_{t\in[0,T]}$
and to $(\mathcal{F}_t)_{t\in[0,T]}$, *its natural filtration. Then there exists a unique process*
$(f_t)_{t\in[0,T]} \in \mathcal{C}_1([0,T])$, *so that (3.32) holds.*

The martingale representation theorems are a direct consequence of the
following theorem (Øksendal 1998).

Theorem 3.67 (Itô representation theorem).   *Let* $X \in L^2(\Omega, \mathcal{F}_T, P)$ *be a
random variable. Then there exists a unique process* $(f_t)_{t\in[0,T]} \in \mathcal{C}([0,T])$ *such
that*

$$X = E[X] + \int_0^T f(s)dW_s.$$

For the proof of the Itô representation theorem we require the following
lemma.

Lemma 3.68.   *The linear span of random variables of the Doléans exponential
type*

$$\exp\left\{ \int_0^T h(t)dW_t - \frac{1}{2} \int_0^T (h(t))^2 dt \right\}$$

for a deterministic process $(h_t)_{t\in[0,T]} \in L^2([0,T])$ *is dense in* $L^2(\Omega, \mathcal{F}_T, P)$.

Proof of Theorem 3.67.   Initially suppose that X has the Doléans exponential
form

$$X = \exp\left\{ \int_0^T h(s)dW_s - \frac{1}{2} \int_0^T (h(s))^2 ds \right\} \qquad \forall t \in [0,T]$$

for a deterministic process $(h_t)_{t\in[0,T]} \in L^2([0,T])$. Also define

$$Y(t) = \exp\left\{ \int_0^t h(s)dW_s - \frac{1}{2} \int_0^t (h(s))^2 ds \right\} \qquad \forall t \in [0,T].$$

Then, invoking Itô's formula, we obtain

$$dY(t)$$
$$= Y(t)\left(h(t)dW_t - \frac{1}{2}(h(t))^2 dt \right) + \frac{1}{2}Y(t)(h(t))^2 dt = Y(t)h(t)dW_t.$$

$$(3.33)$$

Therefore,

$$Y(t) = 1 + \int_0^t Y(s)h(s)dW_s, \qquad t \in [0,T],$$

and in particular

$$X = Y(T) = 1 + \int_0^T Y(s)h(s)dW_s,$$

so that, after taking expectations, we obtain $E[X] = 1$. By Lemma 3.68, the proof can be extended to any arbitrary $X \in L^2(\Omega, \mathcal{F}_T, P)$. To prove that the process $(f_t)_{t \in [0,T]}$ is unique, suppose that two processes $f_t^1, f_t^2 \in \mathcal{C}([0,T])$ exist with

$$X(T) = E[X] + \int_0^T f^1(t)dW_t = E[X] + \int_0^T f^2(t)dW_t.$$

Subtracting the two integrals and taking expectation of the squared difference, we obtain

$$E\left[\left(\int_0^T \left(f^1(t) - f^2(t)\right) dW_t\right)^2\right] = 0,$$

and using the Itô isometry we obtain

$$\int_0^T E\left[f^1(t) - f^2(t)\right]^2 dt = 0,$$

implying that $f_t^1 = f_t^2$ a.s. for all $t \in [0,T]$. $\qquad\square$

Proof of Theorem 3.65. Take $T \equiv t$ and $X \equiv M_t$. By Theorem 3.67, there exists a unique process $(f^{(t)}) \in \mathcal{C}([0,T])$ such that

$$M_t = E[M_t] + \int_0^t f^{(t)}(s)dW_s.$$

Since, by assumption, M is a martingale, we may then claim

$$M_t = E[M_0] + \int_0^t f^{(t)}(s)dW_s.$$

In particular, this will apply to any time t_2, for some $f^{(t_2)}$,

$$M_{t_2} = E[M_0] + \int_0^{t_2} f^{(t_2)}(s)dW_s.$$

Take now an additional time $t_1 < t_2$; by the martingale property, we get

$$M_{t_1} = E[M_{t_2} \mid \mathcal{F}_{t_1}] = E[M_0] + E\left[\int_0^{t_2} f^{(t_2)}(s)dW_s \mid \mathcal{F}_{t_1}\right]$$

$$= E[M_0] + \int_0^{t_1} f^{(t_2)}(s)dW_s,$$

but we will also have

$$M_{t_1} = E[M_0] + \int_0^{t_1} f^{(t_1)}(s)dW_s$$

for some $f^{(t_1)}$. Thus, because of uniqueness, up to t_1, $f^{(t_1)} = f^{(t_2)}$. We may then take $f = f^{(T)}$ to obtain, finally,

$$M_t = E[M_0] + \int_0^t f(s)dW_s. \tag{3.34}$$

To conclude, we may notice that from (3.34) we easily derive

$$M_0 = E[M_0] + 0,$$

which completes the proof.   □

3.7 Multidimensional Stochastic Differentials

Definition 3.69. Let $(\mathbf{u}_t)_{0 \le t \le T}$ be an m-dimensional process and

$$\mathbf{a} : [0,T] \times \Omega \to \mathbb{R}^m, \mathbf{a} \in \mathcal{C}_{1\mathbf{W}}([0,T]),$$
$$b : [0,T] \times \Omega \to \mathbb{R}^{mn}, b \in \mathcal{C}_{1\mathbf{W}}([0,T]).$$

The *stochastic differential* $d\mathbf{u}(t)$ of $\mathbf{u}(t)$ is given by

$$d\mathbf{u}(t) = \mathbf{a}(t)dt + b(t)d\mathbf{W}(t)$$

if, for all $0 \le t_1 < t_2 \le T$,

$$\mathbf{u}(t_2) - \mathbf{u}(t_1) = \int_{t_1}^{t_2} \mathbf{a}(t)dt + \int_{t_1}^{t_2} b(t)d\mathbf{W}(t).$$

Remark 3.70. Under the assumptions of the preceding definition, we obtain for $1 \le i \le m$

$$du_i(t) = a_i(t)dt + \sum_{j=1}^n (b_{ij}(t)dW_j(t)).$$

Example 3.71. Suppose that the coefficients a_{11} and a_{12} of the system

$$\begin{cases} du_1(t) = a_{11}(t)u_1(t)dt + a_{12}(t)u_2(t)dt, \\ du_2(t) = a_{21}(t)u_1(t)dt + a_{22}(t)u_2(t)dt \end{cases} \tag{3.35}$$

are subject to the noise

$$a_{11}(t)dt = a_{11}^0(t)dt + \tilde{a}_{11}(t)dW_1(t),$$
$$a_{12}(t)dt = a_{12}^0(t)dt + \tilde{a}_{12}(t)dW_2(t).$$

The first equation of (3.35) becomes

$$du_1(t)$$
$$= (a_{11}^0(t)u_1(t) + a_{12}^0(t)u_2(t))dt + \tilde{a}_{11}(t)u_1(t)dW_1(t) + \tilde{a}_{12}(t)u_2(t)dW_2(t)$$
$$= \bar{a}_1(t)dt + b_{11}(t)dW_1(t) + b_{12}(t)dW_2(t),$$

where the meaning of the new parameters $\bar{a}_1, b_{11},$ and b_{12} is obvious. Now, if both a_{21} and a_{22} are affected by the noise

$$a_{21}(t)dt = a_{21}^0(t)dt + \tilde{a}_{21}(t)dW_3(t),$$
$$a_{22}(t)dt = a_{22}^0(t)dt + \tilde{a}_{22}(t)dW_4(t),$$

then the second equation of (3.35) becomes

$$du_2(t) = \bar{a}_2(t)dt + b_{23}(t)dW_3(t) + b_{24}(t)dW_4(t).$$

In this case the matrix

$$b = \begin{pmatrix} b_{11} & b_{12} & 0 & 0 \\ 0 & 0 & b_{23} & b_{24} \end{pmatrix}$$

is of order 2×4, but, in general, it is possible that $m > n$.

Theorem 3.72 (Multidimensional Itô formula). *Let* $f(t, \mathbf{x}) : \mathbb{R}_+ \times \mathbb{R}^m \to \mathbb{R}$ *be continuous with the derivatives* $f_{x_i}, f_{x_i x_j},$ *and* f_t. *Let* $\mathbf{u}(t)$ *be an m-dimensional process, endowed with the stochastic differential*

$$d\mathbf{u}(t) = \mathbf{a}(t)dt + b(t)d\mathbf{W}(t),$$

where $\mathbf{a} = (a_1, \ldots, a_m)' \in \mathcal{C}_\mathbf{W}([0, T])$ *and* $b = (b_{ij})_{1 \le i \le m, 1 \le j \le n} \in \mathcal{C}_\mathbf{W}([0, T])$. *Then* $f(t, \mathbf{u}(t))$ *has the stochastic differential*

$$df(t, \mathbf{u}(t)) = \left(f_t(t, \mathbf{u}(t)) + \sum_{i=1}^m f_{x_i}(t, \mathbf{u}(t))a_i(t) \right.$$

$$\left. + \frac{1}{2} \sum_{l=1}^n \sum_{i,j=1}^m f_{x_i x_j}(t, \mathbf{u}(t))b_{il}(t)b_{jl}(t) \right) dt$$

$$+ \sum_{l=1}^n \sum_{i=1}^m f_{x_i}(t, \mathbf{u}(t))b_{il}(t)dW_l(t). \tag{3.36}$$

If we put $a_{ij} = (bb')_{ij}$, $i, j = 1, \ldots, m$, introduce the differential operator

$$\mathcal{L} = \frac{1}{2} \sum_{i,j=1}^{m} a_{ij} \frac{\partial^2}{\partial x_i \partial x_j} + \sum_{i=1}^{m} a_i \frac{\partial}{\partial x_i} + \frac{\partial}{\partial t},$$

and introduce the gradient operator

$$\nabla_{\mathbf{x}} = \left(\frac{\partial}{\partial x_1}, \ldots, \frac{\partial}{\partial x_m} \right)',$$

then, in vector notation, (3.36) can be written as

$$df(t, \mathbf{u}(t)) = \mathcal{L}f(t, \mathbf{u}(t))dt + \nabla_{\mathbf{x}}f(t, \mathbf{u}(t)) \cdot b(t)d\mathbf{W}(t), \qquad (3.37)$$

where $\nabla_{\mathbf{x}}f(t, \mathbf{u}(t)) \cdot b(t)d\mathbf{W}(t)$ is the scalar product of two m-dimensional vectors.

Proof. Employing the following two lemmas the proof is similar to the one-dimensional case. [See, e.g., Baldi (1984).] $\square$

Lemma 3.73. *If $(W_1(t))_{t \geq 0}$ and $(W_2(t))_{t \geq 0}$ are two independent Wiener processes, then*

$$d(W_1(t)W_2(t)) = W_1(t)dW_2(t) + W_2(t)dW_1(t). \qquad (3.38)$$

Proof. Since $W_1(t)$ and $W_2(t)$ are independent, it is easily shown that $W_t = \frac{1}{\sqrt{2}}(W_1(t) + W_2(t))$ is also a Wiener process. Moreover, for a Wiener process $W(t)$ we have

$$dW^2(t) = dt + 2W(t)dW(t). \qquad (3.39)$$

Hence from

$$W_1(t)W_2(t) = W^2(t) - \frac{1}{2}W_1^2(t) - \frac{1}{2}W_2^2(t)$$

it follows that $W_1(t)W_2(t)$ is endowed with the differential

$$d(W_1(t)W_2(t))$$
$$= dW^2(t) - \frac{1}{2}dW_1^2(t) - \frac{1}{2}dW_2^2(t)$$
$$= dt + 2W(t)dW(t) - \frac{1}{2}dt - W_1(t)dW_1(t) - \frac{1}{2}dt - W_2(t)dW_2(t)$$
$$= 2\left(\frac{1}{2}W_1(t)dW_1(t) + \frac{1}{2}W_1(t)dW_2(t) + \frac{1}{2}W_2(t)dW_1(t) + \frac{1}{2}W_2(t)dW_2(t) \right)$$
$$- W_1(t)dW_1(t) - W_2(t)dW_2(t),$$

completing the proof. $\square$

Lemma 3.74. *If* $W_1, \ldots, W_n$ *are independent Wiener processes and*

$$du_i(t) = a_i(t)dt + \sum_{j=1}^{n}(b_{ij}(t)dW_j(t)), \qquad i = 1, 2,$$

then

$$d(u_1 u_2)(t) = u_1(t)du_2(t) + u_2(t)du_1(t) + \sum_{j=1}^{n} b_{1j}b_{2j}dt. \qquad (3.40)$$

Proof. It is analogous to the proof of Proposition 3.54 (e.g., Baldi 1984). Use (3.38), (3.39), and (3.19) and approximate the resulting polynomials. $\square$

Remark 3.75. Equation (3.39) is not a particular case of (3.38) (in the latter, independence is not given), whereas (3.40) generalizes both.

Remark 3.76. The multidimensional Itô formula (3.37) asserts that the processes

$$f(t, \mathbf{u}(t)) - f(0, \mathbf{u}(0))$$

and

$$\int_0^t \mathcal{L}f(s, \mathbf{u}(s))ds + \int_0^t \nabla_{\mathbf{x}} f(s, \mathbf{u}(s)) \cdot b(s)d\mathbf{W}(s)$$

are stochastically equivalent. They are both continuous, and so their trajectories coincide a.s. Taking expectations on both sides, we therefore get

$$E[f(t, \mathbf{u}(t))] - E[f(0, \mathbf{u}(0))] = E\left[\int_0^t \mathcal{L}f(s, \mathbf{u}(s))ds\right].$$

3.8 The Itô Integral with Respect to Lévy Processes

Motivated by the representation (2.41) of a Lévy process, it seems natural to consider more general processes $(X(t))_{t \in \mathbb{R}_+}$, admitting an integral representation of the form

$$X(t) = x + \int_0^t \alpha(s)ds + \int_0^t \beta(s)dW_s + \int_0^t \int_{\mathbb{R}-\{0\}} \gamma(s, z)\tilde{N}(ds, dz), \quad t \in \mathbb{R}_+,$$
$$(3.41)$$

where $x \in \mathbb{R}$ for a suitable choice of the stochastic processes α, β, and γ.

Here $(W_t)_{t \in \mathbb{R}_+}$ is a usual Wiener process, while $\tilde{N}(ds, dz)$ is a compensated random Poisson measure.

Since we already know about the stochastic integral with respect to the Wiener process, we shall concentrate on the stochastic integrals

$$\int_0^t \int_{\mathbb{R}-\{0\}} \gamma(s,z)\tilde{N}(ds,dz)$$

by adopting the approach by Gihman and Skorohod (1972, p. 253ff) [see also Skorohod (1982, p. 34)].

Since we are dealing with $\mathbb{R} - \{0\}$, for the associated Borel σ-algebra we take $\mathcal{B}_0 = \sigma\left(\bigcup_{\varepsilon>0} \mathcal{B}_\varepsilon\right)$, where, for any $\varepsilon > 0$, we take

$$\mathcal{B}_\varepsilon = \sigma\left\{B \in \mathcal{B}_\mathbb{R} | B \subset \left\{x \in \mathbb{R} | \varepsilon \leq | x | \leq \frac{1}{\varepsilon}\right\}\right\}.$$

Assume we are given a probability space $(\Omega, \mathcal{F}, P)$ and a filtration $(\mathcal{F}_t)_{t \in \mathbb{R}_+}$ subject to the usual assumptions.

As for the compensated random Poisson measure, we take first a random Poisson measure $N(ds, dz)$ such that

(a) For any $B \in \mathcal{B}_0$ and for any $t \in \mathbb{R}_+$ the random variable $N([0,t], B) = \int_0^t \int_B N(ds, dz)$ is $\mathcal{F}_t$-measurable.

(b) The compensating measure $\nu(B) = E[N([0,1], B)]$ is finite on any $B \in \mathcal{B}_\mathbb{R}$ such that $0 \notin \bar{B}$.

(c) For any $t \in \mathbb{R}_+$ the family of random variables

$$\{N((t, t+h), B); B \in \mathcal{B}_0, h > 0\}$$

is independent of $\mathcal{F}_t$.

Take now a given time $T > 0$. As far as the integrand process

$$\{\gamma(t, z); t \in \mathbb{R}_+, z \in \mathbb{R}\}$$

is concerned, we assume that the function $\{\gamma(\omega, t, z); \omega \in \Omega, t \in \mathbb{R}_+, z \in \mathbb{R}\}$ is measurable with respect to all variables ω, t, z.

(d) For any $z \in \mathbb{R}$, $\gamma(t, z)$ is $\mathcal{F}_t$-measurable;

further, either

(e)

$$\int_0^T \int_{\mathbb{R}-\{0\}} E[\gamma^2(t, z)]\nu(dz)) < +\infty$$

or

(e')

$$P(\int_0^T \int_{\mathbb{R}-\{0\}} \gamma^2(t, z)\nu(dz)) < +\infty) = 1.$$

In cases (d) and (e), we say that $\gamma \in H(\nu)$; in cases (d) and (e'), we say that $\gamma \in H_2(\nu)$. In either case, γ can be approximated by a sequence $(\gamma_n)_{n\in\mathbb{N}}$ of stepwise processes with respect to time, all belonging to $H(\nu)$, in the mean square topology, i.e.,

$$\int_0^T \int_{\mathbb{R}-\{0\}} E[|\gamma(t,z) - \gamma_n(t,z)|^2]\nu(dz)) \xrightarrow{n} 0.$$

For any $n \in \mathbb{N}$ the integral

$$\int_0^T \int_{\mathbb{R}-\{0\}} \gamma_n(t,z)\tilde{N}(dt,dz) = \sum_{k=1}^n \int_{\mathbb{R}-\{0\}} \gamma_n(t_{k-1},z)\tilde{N}(t_k - t_{k-1}, dz)$$

is well defined, so that we may extend this to any $\gamma \in H(\nu)$ as follows:

$$\int_0^T \int_{\mathbb{R}-\{0\}} \gamma(t,z)\tilde{N}(dt,dz) = \overset{L^2}{\underset{n\to\infty}{\lim}} \int_0^T \int_{\mathbb{R}-\{0\}} \gamma_n(t,z)\tilde{N}(dt,dz).$$

For the more general case $\gamma \in H_2(\nu)$, provided $(\gamma_n)_{n\in\mathbb{N}}$ in $H(\nu)$ approximates γ in the following sense

$$\int_0^T \int_{\mathbb{R}-\{0\}} [|\gamma(t,z) - \gamma_n(t,z)|^2\nu(dz)) \xrightarrow[n\to\infty]{P} 0, \qquad (3.42)$$

we may define

$$\int_0^T \int_{\mathbb{R}-\{0\}} \gamma(t,z)\tilde{N}(dt,dz) = \overset{P}{\underset{n\to\infty}{\lim}} \int_0^T \int_{\mathbb{R}-\{0\}} \gamma_n(t,z)\tilde{N}(dt,dz). \qquad (3.43)$$

Clearly, if $(\gamma_n)_{n\in\mathbb{N}}$ approximates γ in $H(\nu)$ in the mean square sense, then (3.42) will also hold, and consequently (3.43) applies, too. The following result holds.

Theorem 3.77. *If $\gamma \in H(\nu)$, then the process*

$$t \in [0,T] \mapsto \Phi(t) := \int_0^t \int_{\mathbb{R}-\{0\}} \gamma(s,z)\tilde{N}(ds,dz)$$

is an $\mathcal{F}_t$-martingale, and for any $s,t \in \mathbb{R}_+$, $s < t$,

$$E[|\Phi(t) - \Phi(s)|^2|\mathcal{F}_s] = \int_s^t d\tau \int_{\mathbb{R}-\{0\}} \nu(dz)E[\gamma^2(\tau,z)|\mathcal{F}_s].$$

Remark 3.78. The definition of Φ allows its extension to a separable process, a.s. bounded, without discontinuities of the second kind.

Further, the following inequalities can be proved.

Theorem 3.79. *If $\gamma \in H(\nu)$, then, for any $a > 0$,*

$$P(\sup_{0 \leq t \leq T} |\Phi(t)| > a) \leq \frac{1}{a^2} \int_0^T dt \int_{\mathbb{R}-\{0\}} \nu(dz) E[\gamma^2(t,z)],$$

$$E(\sup_{0 \leq t \leq T} |\Phi(t)|^2) \leq 4 \int_0^T dt \int_{\mathbb{R}-\{0\}} \nu(dz) E[\gamma^2(t,z)].$$

Theorem 3.80. *If $\gamma \in H_2(\nu)$, then, for any $a > 0, K > 0$,*

$$P(\sup_{0 \leq t \leq T} |\Phi(t)| > a) \leq \frac{K}{a^2} + P(\int_0^T dt \int_{\mathbb{R}-\{0\}} \nu(dz)\gamma^2(t,z) > K).$$

3.9 The Itô–Lévy Stochastic Differential and the Generalized Itô Formula

Within the framework established in the previous section, we are now ready to generalize the concept of stochastic differential so as to give a rigorous meaning to (3.41).

We are given an underlying probability space $(\Omega, \mathcal{F}, P)$ and a filtration $(\mathcal{F}_t)_{t \in \mathbb{R}_+}$ subject to the usual assumptions. Consider a real-valued Wiener process $(W_t)_{t \in \mathbb{R}_+}$ such that for any $t \in \mathbb{R}_+$, W_t is $\mathcal{F}_t$-measurable; consider a random Poisson measure $N(ds, dz)$ satisfying Conditions (a)–(c) of the previous section, with respect to the filtration $(\mathcal{F}_t)_{t \in \mathbb{R}_+}$.

Consider now the processes $(\alpha(t))_{t \in \mathbb{R}_+}$, $(\beta(t))_{t \in \mathbb{R}_+}$ such that $(\alpha)^{1/2}$ and β belong to $\mathcal{C}_1$, with respect to the filtration $(\mathcal{F}_t)_{t \in \mathbb{R}_+}$; further, take a process $\gamma \in H_2(\nu)$, where ν is the compensating measure of the random Poisson measure $N(ds, dz)$. Finally, take $X(0)$, a given random variable on the probability space $(\Omega, \mathcal{F}, P)$, independent of $(\mathcal{F}_t)_{t \in \mathbb{R}_+}$.

Thanks to the definition of the Itô–Lévy integral introduced in the previous section, it makes sense to consider the process $(X(t))_{t \in [0,T]}$ defined, for any $t \in [0, T]$, by

$$X(t) = X(0) + \int_0^t \alpha(s)ds + \int_0^t \beta(s)dW_s + \int_0^t \int_{\mathbb{R}-\{0\}} \gamma(s,z)\tilde{N}(ds, dz). \quad (3.44)$$

Any process defined as in (3.44) is called an *Itô–Lévy process*.

Under the preceding assumptions, if the process $(X(t))_{t \in [0,T]}$ satisfies (3.44), then we say that it admits, on $[0, T]$, the (generalized) stochastic differential

$$dX(t) = \alpha(t)dt + \beta(t)dW_t + \int_{\mathbb{R}-\{0\}} \gamma(t,z)\tilde{N}(dt, dz).$$

We are now ready to state a main theorem of the Lévy–Itô calculus.

Theorem 3.81. *Under the preceding assumptions, let the process* $(X(t))_{t\in[0,T]}$ *satisfy (3.44), let* $f \in C^{1,2}(\mathbb{R}^* \times \mathbb{R}$, *and define*

$$Y(t) := f(t, X(t)), \quad t \in \mathbb{R}_+ .$$

Then the process $(Y(t))_{t\in\mathbb{R}_+}$ *is itself an Itô–Lévy process, and its stochastic differential is given by*

$$dY(t) = \frac{\partial f}{\partial t}(t, X(t))dt + \frac{\partial f}{\partial x}(t, X(t))\alpha(t)dt + \frac{\partial f}{\partial x}(t, X(t))\beta(t)dW_t$$

$$+ \frac{1}{2}\frac{\partial^2 f}{\partial x^2}(t, X(t))\beta^2(t)dt$$

$$+ \int_{\mathbb{R}-\{0\}}\left[f(t, X(t) + \gamma(t, z)) - f(t, X(t)) - \frac{\partial f}{\partial x}(t, X(t))\gamma(t, z)\right]\nu(dz)dt$$

$$+ \int_{\mathbb{R}-\{0\}}\left[f(t, X(t^-) + \gamma(t, z)) - f(t, X(t^-))\right]\tilde{N}(dt, dz).$$

Proof. See, e.g., Gihman and Skorohod (1972, p. 272), Medvegyev (2007, p. 394), Di Nunno et al. (2009, p. 163). □

For the extension of the Itô–Lévy formula to the multidimensional case, see, e.g., Medvegyev (2007, p. 394), or Di Nunno et al. (2009, p. 164).

3.10 Fractional Brownian Motion

In recent years the well-studied stochastic calculus based on the usual Wiener process turns out not being adequate for describing many phenomena, in telecommunication, financial mathematics, bioengineering, etc. where processes with *long memory* are better suited. A Wiener process has no memory (see Section 2.12), so that various authors started paying attention to the so-called *fractional Brownian motion*, thanks to the impulse due to the pioneering work by Mandelbrot and van Ness (1968) (see, e.g., Franke et al. (2011, p. 347)).

Definition 3.82. (Nualart (2006), Mishura (2008)) The *fractional Brownian motion (fBm)* with Hurst index $H \in (0, 1)$ is a Gaussian process $B^H = (B_t^H)_{t\in\mathbb{R}_+}$ on a probability space $(\Omega, \mathcal{F}, P)$ which satisfies the following properties

(i) $B_0^H = 0$,   a.s.;

(ii) $E[B_t^H] = 0$, $t \in \mathbb{R}_+$;

(iii) $E[B_t^H B_s^H] = \frac{1}{2}\left(t^{2H} + s^{2H} - (t-s)^{2H}\right) =: R_H(t, s)$, $s, t \in \mathbb{R}_+$, $s \leq t$.

It is clear that the usual Brownian motion (the standard Wiener process) $W = (W_t)_{t\in\mathbb{R}_+}$ is then obtained for $H = \frac{1}{2}$.

For $H = 1$, it is usually taken $B_t^1 = tZ$, $t \in \mathbb{R}_+$, where Z is a standard normal random variable.

As a consequence of the axioms, for $0 \le s \le t$,

$$E[(B_t^H - B_s^H)^2] = (t - s)^{2H};\tag{3.45}$$

then the process has stationary increments, though the process itself is not stationary.

Further, for any integer $n \ge 1$, and for any $0 \le s \le t$,

$$E[(B_t^H - B_s^H)^n] = \frac{2^{\frac{n}{2}}}{\pi^{\frac{1}{2}}}\Gamma\left(\frac{n+1}{2}\right)(t-s)^{nH},\tag{3.46}$$

so that we may apply the Kolmogorov Continuity Theorem 2.173 in Chapter 2, and claim that a fBm admits a continuous modification.

An fBm keeps the property of non-differentiability, as from the following proposition (see, e.g., Biagini et al (2007, p. 12)).

Proposition 3.83 *For $H \in (0,1)$, a fBm $(B_t^H)_{t\in\mathbb{R}_+}$ is a.s. nowhere differentiable.*

The Hurst index plays the role of parameter of self-similarity according to the following proposition Mandelbrot and van Ness (1968).

Proposition 3.84 *For $H \in (0,1)$, a fractional Brownian motion is self-similar with Hurst parameter H, i.e., for any $a > 0$ the processes $(B_{at}^H)_{t\in\mathbb{R}_+}$ and $(a^H B_t^H)_{t\in\mathbb{R}_+}$ have the same system of finite-dimensional distributions.*

We now compute the covariance of the increments of a fBm, for $0 \le s \le t < u \le v$,

$$E[(B_t^H - B_s^H)(B_v^H - B_u^H)] =$$
$$= \frac{1}{2}\left((u-s)^{2H} + (v-t)^{2H} - (u-t)^{2H} - (v-s)^{2H}\right),\tag{3.47}$$

from which we may recognize that, while for $H = \frac{1}{2}$ (standard Wiener process) the increments are non-correlated, hence independent (by Gaussianity), they are correlated for $H \in (0,\frac{1}{2}) \cup (\frac{1}{2},1)$.

Moreover, for $H \in (0,\frac{1}{2}) \cup (\frac{1}{2},1)$, and $t_1 < t_2 < t_3 < t_4$, from (3.47) we have

$$E[(B_{t_4}^H - B_{t_3}^H)(B_{t_2}^H - B_{t_1}^H)] = 2\alpha H \int_{t_1}^{t_2}\int_{t_3}^{t_4}(u-v)^{2\alpha-1}dudv,\tag{3.48}$$

with $\alpha = H - \frac{1}{2}$.

Hence the increments are positively correlated for $H \in (\frac{1}{2},1)$, and negatively correlated for $H \in (0,\frac{1}{2})$.

Take now $n \in \mathbb{Z} \setminus \{0\}$, and consider the autocovariance function

$$r(n) := E[B_1^H (B_{n+1}^H - B_n^H)] =$$

$$= 2\alpha H \int_0^1 \int_n^{n+1} (u-v)^{2\alpha-1} du dv$$

$$\sim 2\alpha H \mid n \mid^{2\alpha-1}, \text{ as } \mid n \mid \to \infty. \tag{3.49}$$

Hence

for $H \in (0, \frac{1}{2})$: $\sum_{n\in\mathbb{Z}} |r(n)| \sim \sum_{n\in\mathbb{Z}\setminus\{0\}} |n|^{2\alpha-1} < +\infty, \tag{3.50}$

while

for $H \in (\frac{1}{2}, 1)$: $\sum_{n\in\mathbb{Z}} |r(n)| \sim \sum_{n\in\mathbb{Z}\setminus\{0\}} |n|^{2\alpha-1} = +\infty; \tag{3.51}$

this shows that for $H \in (\frac{1}{2}, 1)$ a fBm B^H has the property of long range autocorrelation.

3.10.1 Integral with respect to a fBm

Let $[0, T] \subset \mathbb{R}_+$ be a bounded time interval, and consider a fBm $(B_t^H)_{t \in [0,T]}$ with Hurst index $H \in (0, 1)$.

Let $\mathcal{S}$ denote the set of step functions on $[0, T]$, and $\mathcal{H}$ the Hilbert space obtained by the closure of $\mathcal{S}$ with respect to the scalar product

$$R_H(t, s) := \langle I_{[0,t]}, I_{[0,s]} \rangle = \int_0^T I_{[0,t]}(u) I_{[0,s]}(u) du, \tag{3.52}$$

for any $s, t \in [0, T]$.

Let c_H denote the following constant

$$c_H = \left[\frac{(2H + \frac{1}{2}) \Gamma (\frac{1}{2} - H)}{\Gamma (\frac{1}{2} + H) \Gamma (2 - 2H)} \right]^{\frac{1}{2}}, \tag{3.53}$$

and consider the kernel $K_H(t, s)$ given by

$$K_H(t, s) = c_H \left[\left(\frac{t}{s} \right)^{H-\frac{1}{2}} (t-s)^{H-\frac{1}{2}} \right.$$

$$\left. - (H - \frac{1}{2}) s^{\frac{1}{2}-H} \int_s^t u^{H-\frac{3}{2}} (u-s)^{H-\frac{1}{2}} du \right] \tag{3.54}$$

for $s < t$, and

$$K_H(t, s) = 0, \tag{3.55}$$

for $s \geq t$.

It is known (Mandelbrot and van Ness (1968)) that an fBm $(B_t^H)_{t \in \mathbb{R}_+}$ can be represented as

$$B_t^H = \int_0^t K_H(t, s)dW_s,$$ (3.56)

where $(W_t)_{t \in \mathbb{R}_+}$ is a standard Wiener process.

The operator defined by

$$K_H^* : \mathcal{S} \to L^2([0, T])$$
$$(K_H^* I_{[0,t]})(s) = K_H(t, s), \ s \in [0, t],$$

is a linear isometry that can be extended to the whole Hilbert space $\mathcal{H}$. In fact, for any $s, t \in [0, T]$, we have

$$\langle K_H^* I_{[0,t]}, K_H^* I_{[0,s]} \rangle_{L^2([0,T])} =$$
$$= \langle K_H(t, \cdot), K_H(s, \cdot) \rangle_{L^2([0,T])}$$
$$= \int_0^{t \wedge s} K_H(t, u)K_H(s, u)du$$
$$= \langle I_{[0,t]}, I_{[0,s]} \rangle_{\mathcal{H}} = R_H(t, s).$$ (3.57)

The above allows the construction of a stochastic integral of a deterministic function in $\mathcal{H}$ with respect to an fBm B^H by means of the following equality, which takes into account the representation (3.56):

$$\int_0^T \varphi(t)dB_t^H := \int_0^T (K_H^* \varphi)(t)dW_s.$$ (3.58)

The representation (3.58) requires additional care with respect to the usual Itô integral, since B^H is not a semimartingale for $H \neq \frac{1}{2}$.

We may just report that for $H > \frac{1}{2}$ the operator K_H^* can be expressed in terms of *fractional integrals*, while for $H < \frac{1}{2}$ it can be expressed in terms of *fractional derivatives* (Mishura (2008)).

Major problems arise when we wish to extend directly formula (3.58) to the integration of stochastic processes, as required in stochastic differential equations driven by fractional Brownian motions. This is due to the fact that even if a stochastic process $u = (u(t))_{t \in \mathbb{R}_+}$ is adapted to the filtration generated by a fBm B^H (which coincides with the filtration of the Wiener process, as from the representation (3.56)), the related process $K_H^* u$ is no longer adapted, because the operator K_H^* does not preserve adaptability. Therefore the definition of stochastic integrals of random processes with respect to a fBm requires further analysis which is beyond the background required in this introductory volume; we refer to Biagini et al (2007) and Mishura (2008), and the references therein, for a detailed presentation.

In financial applications long range dependence has a long history (see, e.g., Williger et al. (1999)), since it may provide an explanation of the empirical law commonly known as the *Hurst effect*. We refer again to Biagini et al (2007) and

Mishura (2008), and the references therein, for an extensive presentation of the mathematical problems and issues which are relevant in financial applications.

Further mathematical problems arise in connection with stochastic differential equations driven by fBm's, and in particular the extension of Itô's formula. For a discussion about these concepts and the application to the Black-Scholes model for a fractional Brownian noise, the reader may refer, e.g., to Dai-Heyde (1996), Zähle (2002), and the references therein. In Shevchenko (2014) stochastic differential equations including Wiener noise, fractional Brownian motion, and jumps are discussed together with their relevance in economic and financial models.

3.11 Exercises and Additions

3.1. Show that, if W is a scalar Wiener process, and $f \in \mathcal{C}([a,b])$ is a given process on $[a,b] \subset \mathbb{R}$, then for all $c > 0$

$$P\left(\left|\int_a^b f(t)dW_t\right| > c\right) \leq \frac{\int_a^b E[|f(t)|^2]dt}{c^2}.$$

3.2. Show that, if W is a scalar Wiener process, and $f \in \mathcal{L}^2([a,b])$ is a deterministic real-valued function on $[a,b] \subset \mathbb{R}$, then

$$\int_a^b f(t)dW_t \sim N\left(0, \int_a^b |f(t)|^2 dt\right).$$

Hint: The proof is based on the following result [see also Arnold (1974)]. Let $(X_n)_{n\in\mathbb{N}}$ be a sequence of Gaussian random variables, i.e., $X_n \sim N(\mu_n, \sigma_n^2)$, for any $n \in \mathbb{N}$. If

$$(\mu_n, \sigma_n^2) \xrightarrow{n} (\mu, \Sigma),$$

with $\mu \in \mathbb{R}$, and $\Sigma > 0$, then

$$X_n \underset{n\to\infty}{\Rightarrow} N(\mu, \Sigma).$$

3.3. Let $W = (W_t)_{t\in\mathbb{R}_+}$ be a scalar Wiener process, and let $(\mathcal{F}_t)_{t\in\mathbb{R}_+}$ be the filtration generated by W. Let $f \in \mathcal{C}_W([a,b])$ be a given process on $[a,b] \subset \mathbb{R}$. Show that the following holds

$$E[(\int_s^t f(u)dW_u)^2|\mathcal{F}_s] = \int_s^t E[(f(u))^2|\mathcal{F}_s]du, \text{ a.s.,} \qquad (3.59)$$

for any $(s,t) \in [a,b] \times [a,b]$, $s < t$.

3.4. Let $(X_t)_{t\in\mathbb{R}_+}$ be the Itô integral

$$X_t = \int_0^t f(s)dW_s,$$

where $(W_t)_{t\in\mathbb{R}_+}$ is a Brownian motion, and $f \in \mathcal{C}_W(0,T)$.

1. Give an example to show that X_t^2, in general, is not a martingale.
2. Prove that

$$M_t = X_t^2 - \int_0^t |f(s)|^2 ds$$

is a martingale. Hence the process $\langle X \rangle_t = \int_0^t |f(s)|^2 ds$ is the *quadratic variation process* of the martingale X_t.

3.5. Prove Lemma 3.46.

3.6. Let $W = (W_t)_{t \in \mathbb{R}_+}$ be a scalar Wiener process, and let $(\mathcal{F}_t)_{t \in \mathbb{R}_+}$ be the filtration generated by W. Let $f \in \mathcal{C}_W([a,b])$ be a given process on $[a,b] \subset \mathbb{R}$; prove that, for any bounded real-valued $\mathcal{F}_a$−measurable random variable Y, the following holds

(i) $Yf \in \mathcal{C}_W([a,b])$;
(ii) $\int_a^b Y f(t) dW_t = Y \int_a^b f(t) dW_t$.

3.7. Let $(X_t)_{t \in \mathbb{R}_+}$ be a Brownian motion in $\mathbb{R}$, $X_0 = 0$. Prove directly from the definition of Itô integrals that

$$\int_0^t X_s^2 dX_s = \frac{1}{3} X_t^3 - \int_0^t X_s ds.$$

3.8. Prove Corollary 3.57.

3.9. Prove Lemma 3.74.

3.10. Prove the multidimensional Itô formula (3.37).

3.11. Let $(W_t)_{t \in \mathbb{R}_+}$ denote an n-dimensional Brownian motion, and let $f : \mathbb{R}^n \to \mathbb{R}$ be C^2. Use Itô's formula to prove that

$$df(W_t) = \nabla f(W_t) dW_t + \frac{1}{2} \triangle f(W_t) dt,$$

where ∇ denotes the gradient, and $\triangle = \sum_{i=1}^n \frac{\partial^2}{\partial x_i^2}$ is the Laplace operator.

3.12. Let $(W_t)_{t \in \mathbb{R}_+}$ be a one-dimensional Brownian motion with $W_0 = 0$. Using Itô's formula, show that

$$E[W_t^k] = \frac{1}{2} k(k-1) \int_0^t E[W_s^{k-2}] ds, \qquad k \geq 2, \ t \geq 0.$$

Further, for any $k \in \mathbb{N}$,

$$E[W_t^{2k}] = O(t^k),$$

whereas

$$E[W_t^{2k+1}] = 0.$$

3.13. Use Itô's formula to write the following stochastic process u_t in the standard form

$$du_t = \mathbf{a}(t)dt + b(t)dW_t$$

for a suitable choice of $\mathbf{a} \in \mathbb{R}^n$, $b \in \mathbb{R}^{nm}$, and dimensions n, m:

1. $u_1(t, W_1(t)) = 3 + 2t + e^{2W_1(t)}$ [$W_1(t)$ is one-dimensional];
2. $u_2(t, \mathbf{W}_t) = W_2^2(t) + W_3^2(t)$ [$\mathbf{W}_t = (W_2(t), W_3(t))$ is two-dimensional];
3. $u_3(t, \mathbf{W}_t) = \ln(u_1(t)u_2(t))$;
4. $u_4(t, \mathbf{W}_t) = \exp\left\{\frac{u_1(t)}{u_2(t)}\right\}$;
5. $u_5(t, W_t) = (5 + t, t + 4W_t)$ (W_t is one-dimensional);
6. $u_6(t, \mathbf{W}_t) = (W_1(t) + W_2(t) - W_3(t), W_2^2(t) - W_1(t)W_2(t) + W_3(t))$ [$\mathbf{W}_t = (W_1(t), W_2(t), W_3(t))$ is three-dimensional].

3.14. Let $(\mathbf{W}_t)_{t \in \mathbb{R}_+}$ be an n-dimensional Brownian motion starting at $x \neq 0$. Are the processes

$$u_t = \ln\left(|\mathbf{W}_t|^2\right)$$

and

$$v_t = \frac{1}{|\mathbf{W}_t|}$$

martingales? If not, find two processes $(\overline{u}_t)_{t \in \mathbb{R}_+}$, $(\overline{v}_t)_{t \in \mathbb{R}_+}$ such that

$$u_t - \overline{u}_t$$

and

$$v_t - \overline{v}_t$$

are martingales.

3.15 (Exponential martingales). Let $dZ_t = \alpha dt + \beta dW_t$, $Z_0 = 0$ where α, β are constants and $(W_t)_{t \in \mathbb{R}_+}$ is a one-dimensional Brownian motion. Define

$$M_t = \exp\left\{Z_t - \left(\alpha + \frac{1}{2}\beta^2\right)t\right\} = \exp\left\{-\frac{1}{2}\beta^2 t + \beta W_t\right\}.$$

Use Itô's formula to prove that

$$dM_t = \beta M_t dW_t.$$

In particular, $M = (M_t)_{t \in \mathbb{R}_+}$ is a martingale.

3.16. Let $(W_t)_{t \in \mathbb{R}_+}$ be a one-dimensional Brownian motion, and let $\phi \in L^2_{loc}[0, T]$ for any $T \in \mathbb{R}_+$. Show that for any $\theta \in \mathbb{R}$

$$X_t := \exp\left\{i\theta \int_0^t \phi(s)dW_s + \frac{1}{2}\theta^2 \int_0^t \phi^2(s)ds\right\}$$

is a local martingale.

3.17. With reference to the preceding Problem 3.16, assume now that

$$P\left(\int_0^{+\infty} \phi^2(s)ds = +\infty\right) = 1,$$

and let

$$\tau_t := \min\left\{u\mathbb{R}_+| \int_0^u \phi^2(s)ds \geq t\right\}, \quad t \in \mathbb{R}_+.$$

Show that $(X_{\tau_t})_{t\in\mathbb{R}_+}$ is an $\mathcal{F}_{\tau_t}$-martingale.

3.18. With reference to Problem 3.17, let

$$Z_t := \int_0^t \phi(s)dW_s, \quad t \in \mathbb{R}_+.$$

Show that $(Z_{\tau_t})_{t\in\mathbb{R}_+}$ has independent increments and $Z_{\tau_t} - Z_{\tau_s} \sim N(0, t-s)$ for any $0 < s < t < +\infty$. (*Hint:* Show that if $\mathcal{F}' \subset \mathcal{F}'' \subset \mathcal{F}$ are σ-fields on the probability space $(\Omega, \mathcal{F}, P)$ and Z is an $\mathcal{F}''$-measurable random variable such that

$$E\left[e^{i\theta Z}\,\middle|\,\mathcal{F}'\right] = e^{-\theta^2\sigma^2/2},$$

then Z is independent of $\mathcal{F}'$ and $Z \sim N(0, \sigma^2)$.)

3.19. With reference to Problem 3.18, show that the process $(Z_{\tau_t})_{t\in\mathbb{R}_+}$ is a standard Brownian motion.

3.20. Let $(W_t)_{t\in\mathbb{R}_+}$ be a one-dimensional Brownian motion. Formulate suitable conditions on u, v such that the following holds:

Let $dZ_t = u_t dt + v_t dW_t$, $Z_0 = 0$ be a stochastic integral with values in $\mathbb{R}$. Define

$$M_t = \exp\left\{Z_t - \int_0^t \left[u_s + \frac{1}{2}v_s v_s'\right] ds\right\}.$$

Then $M = (M_t)_{t\in\mathbb{R}_+}$ is a martingale.

3.21. Let $(W_t)_{t\in\mathbb{R}_+}$ be a one-dimensional Brownian motion. Show that for any real function that is continuous up to its second derivative the process

$$\left(f(W_t) - \frac{1}{2}\int_0^t f''(W_s)ds\right)_{t\in\mathbb{R}_+}$$

is a local martingale.

3.22. Let $\mathbf{X}$ be a time-homogeneous Markov process with transition probability measure $P_t(\mathbf{x}, A)$, $\mathbf{x} \in \mathbb{R}^d$, $A \in \mathcal{B}_{\mathbb{R}^d}$, with $d \geq 1$. Given a test function φ, let

$$u(t, \mathbf{x}) := E_{\mathbf{x}}[\varphi(\mathbf{X}(t))] = \int_{\mathbb{R}^d} \varphi(\mathbf{y})P_t(\mathbf{x}, d\mathbf{y}), \quad t \in \mathbb{R}_+, \; \mathbf{x} \in \mathbb{R}^d.$$

Show that, under rather general assumptions, the function u satisfies the so-called Kolmogorov equation

$$\frac{\partial}{\partial t} u(t, \mathbf{x})$$

$$= \frac{1}{2} \sum_{i,j=1}^{d} q_{ij}(\mathbf{x}) \frac{\partial^2}{\partial x_i \partial x_j} u(t, \mathbf{x})$$

$$+ \sum_{j=1}^{d} f_j(\mathbf{x}) \frac{\partial}{\partial x_j} u(t, \mathbf{x})$$

$$+ \int_{\mathbb{R}^d} \left(u(t, \mathbf{x} + \mathbf{y}) - u(t, \mathbf{x}) - \frac{1}{1 + |\mathbf{y}|^2} \sum_{j=1}^{d} y_j \frac{\partial}{\partial x_j} u(t, \mathbf{x}) \right) \nu(\mathbf{x}, d\mathbf{y})$$

for $t > 0$, $\mathbf{x} \in \mathbb{R}^d$, subject to the initial condition

$$u(0, \mathbf{x}) = \phi(\mathbf{x}), \ \mathbf{x} \in \mathbb{R}^d.$$

Here f and Q are functions with values being, respectively, vectors in $\mathbb{R}^d$ and symmetric, nonnegative $d \times d$ matrices, $f : \mathbb{R}^d \to \mathbb{R}^d$ and $Q := (q_{ij}) : \mathbb{R}^d \to L_+(\mathbb{R}^d, \mathbb{R}^d)$, and ν is a Lévy measure, i.e., $\nu : \mathbb{R}^d \to \mathcal{M}(\mathbb{R}^d \setminus \{0\})$, with $\mathcal{M}(\mathbb{R}^d \setminus \{0\})$ being the set of nonnegative measures on $\mathbb{R}^d \setminus \{0\}$ such that

$$\int_{\mathbb{R}^d} (|\mathbf{y}^2| \wedge 1) \nu(\mathbf{x}, d\mathbf{y}) < +\infty.$$

The functions f, Q, and ν are known as the *drift vector, diffusion matrix*, and *jump measure*, respectively. Show that the process $\mathbf{X}$ has continuous trajectories whenever $\nu \equiv 0$.

4

Stochastic Differential Equations

4.1 Existence and Uniqueness of Solutions

Let $(W_t)_{t \in \mathbb{R}_+}$ be a Wiener process on the probability space $(\Omega, \mathcal{F}, P)$, equipped with its natural filtration $(\mathcal{F}_t)_{t \in \mathbb{R}_+}$, $\mathcal{F}_t = \sigma(W_s, 0 \leq s \leq t)$. Furthermore, let $a(t,x)$, $b(t,x)$ be deterministic measurable functions in $[t_0, T] \times \mathbb{R}$ for some $t_0 \in \mathbb{R}_+$. Finally, consider a real-valued random variable u^0; we will denote by $\mathcal{F}_{u^0}$ the σ-algebra generated by u^0, and we assume that $\mathcal{F}_{u^0}$ is independent of $(\mathcal{F}_t)$ for $t \in (t_0, +\infty)$. We will denote by $\mathcal{F}_{u^0,t}$ the σ-algebra generated by the union of $\mathcal{F}_{u^0}$ and $\mathcal{F}_t$ for $t \in (t_0, +\infty)$.

Definition 4.1. The stochastic process $(u(t))_{t \in [t_0, T]}$ $(T \in (t_0, +\infty))$ is said to be a *solution of the stochastic differential equation* (SDE)

$$du(t) = a(t, u(t))dt + b(t, u(t))dW_t, \qquad t_0 \leq t \leq T, \qquad (4.1)$$

subject to the initial condition

$$u(t_0) = u^0 \text{ a.s. }, \qquad (4.2)$$

if

1. $u(t)$ is measurable with respect to the σ-algebra $\mathcal{F}_{u^0,t}$, $\qquad t_0 \leq t \leq T$.
2. $|a(\cdot, u(\cdot))|^{\frac{1}{2}}, b(\cdot, u(\cdot)) \in \mathcal{C}_1([t_0, T])$.
3. The stochastic differential of $u(t)$ in $[t_0, T]$ is
   $$du(t) = a(t, u(t))dt + b(t, u(t))dW_t,$$
   thus
   $$u(t) = u(t_0) + \int_{t_0}^t a(s, u(s))ds + \int_{t_0}^t b(s, u(s))dW_s, \; t \in [t_0, T].$$

Remark 4.2. If $u(t)$ is the solution of (4.1) and (4.2), then it is nonanticipating (by point 3 of the preceding definition and as already observed in Remark 3.51).

© Springer Science+Business Media New York 2015   231
V. Capasso, D. Bakstein, *An Introduction to Continuous-Time Stochastic Processes*, Modeling and Simulation in Science, Engineering and Technology, DOI 10.1007/978-1-4939-2757-9_4

The following lemma will be of interest for the proof of uniqueness of the solution of (4.1).

Lemma 4.3. (Gronwall).   *If $\phi(t)$ is an integrable, nonnegative function, defined on $t \in [0, T]$, with*

$$\phi(t) \leq \alpha(t) + L \int_0^t \phi(s)ds, \tag{4.3}$$

where L is a positive constant and $\alpha(t)$ is an integrable function, then

$$\phi(t) \leq \alpha(t) + L \int_0^t e^{L(t-s)}\alpha(s)ds.$$

Proof. Putting $\psi(t) = L \int_0^t \phi(s)ds$, as well as $z(t) = \psi(t)e^{-Lt}$, then $z(0) = \psi(0) = 0$, and, moreover,

$$\begin{aligned} z'(t) &= \psi'(t)e^{-Lt} - L\psi(t)e^{-Lt} = L\phi(t)e^{-Lt} - L\psi(t)e^{-Lt} \\ &\leq L\alpha(t)e^{-Lt} + L\psi(t)e^{-Lt} - L\psi(t)e^{-Lt}. \end{aligned}$$

Therefore, $z'(t) \leq L\alpha(t)e^{-Lt}$ and, after integration, $z(t) \leq L \int_0^t \alpha(s)e^{-Ls}ds$. Hence

$$\psi(t)e^{-Lt} \leq L \int_0^t \alpha(s)e^{-Ls}ds \Rightarrow \psi(t) \leq L \int_0^t e^{L(t-s)}\alpha(s)ds,$$

but, by (4.3), $\psi(t) = L \int_0^t \phi(s)ds \geq \phi(t) - \alpha(t)$, completing the proof.   □

In the sequel, unless explicitly specified, we take $t_0 = 0$ to reduce the complexity of notations.

Theorem 4.4 (Existence and uniqueness).   *Suppose constants K^*, K exist such that the following conditions are satisfied:*

1. *For all $t \in [0, T]$ and all $(x, y) \in \mathbb{R} \times \mathbb{R}$: $|a(t, x) - a(t, y)| + |b(t, x) - b(t, y)| \leq K^*|x - y|$.*
2. *For all $t \in [0, T]$ and all $x \in \mathbb{R}$: $|a(t, x)| \leq K(1 + |x|)$, $|b(t, x)| \leq K(1 + |x|)$.*
3. *$E[|u^0|^2] < \infty$.*

Then there exists a unique $(u(t))_{t \in [0,T]}$ solution of (4.1) and (4.2) such that

- *$(u(t))_{t \in [0,T]}$ is continuous almost surely (thus almost every trajectory is continuous).*
- *$(u(t))_{t \in [0,T]} \in \mathcal{C}([0, T])$.*

Remark 4.5.   If $(u_1(t))_{t \in [0,T]}$ and $(u_2(t))_{t \in [0,T]}$ are two solutions of (4.1) and (4.2) that belong to $\mathcal{C}([0, T])$, then the uniqueness of a solution is understood in the sense that

$$P\left(\sup_{0\leq t\leq T}|u_1(t)-u_2(t)|=0\right)=1.$$

Proof of Theorem 4.4.

STEP 1: *Uniqueness.* Let $u_1(t)$ and $u_2(t)$ be solutions of (4.1) and (4.2) that belong to $\mathcal{C}([0,T])$. Then, by point 3 of Definition 4.1,

$$u_1(t)-u_2(t)$$

$$=\int_0^t[a(s,u_1(s))-a(s,u_2(s))]ds+\int_0^t[b(s,u_1(s))-b(s,u_2(s))]dW_s$$

$$=\int_0^t\tilde{a}(s)ds+\int_0^t\tilde{b}(s)dW_s,\qquad t\in]0,T],$$

where $\tilde{a}(s)=a(s,u_1(s))-a(s,u_2(s))$ and $\tilde{b}(s)=b(s,u_1(s))-b(s,u_2(s))$. Because, in general, $(a+b)^2\leq 2(a^2+b^2)$, we obtain

$$|u_1(t)-u_2(t)|^2\leq 2\left(\int_0^t\tilde{a}(s)ds\right)^2+2\left(\int_0^t\tilde{b}(s)dW_s\right)^2,$$

and by the Cauchy–Schwarz inequality,

$$\left(\int_0^t\tilde{a}(s)ds\right)^2\leq t\left(\int_0^t|\tilde{a}(s)|^2ds\right),$$

therefore

$$E\left[\left(\int_0^t\tilde{a}(s)ds\right)^2\right]\leq tE\left[\int_0^t|\tilde{a}(s)|^2ds\right].$$

Moreover, by assumption 2,

$$E\left[\int_0^T(b(s,u_i(s)))^2ds\right]\leq E\left[\int_0^T(K(1+|u_i(s)|))^2ds\right]$$

$$\leq 2K^2E\left[\int_0^T(1+|u_i(s)|^2)ds\right]<+\infty$$

for $i=1,2$ and because $u_i(s)\in\mathcal{C}$. Now, this shows $b(s,u_i(s))\in\mathcal{C}$ for $i=1,2$, and thus $\tilde{b}(s)\in\mathcal{C}$. Then, by Proposition 3.20,

$$E\left[\left(\int_0^t\tilde{b}(s)dW_s\right)^2\right]=E\left[\int_0^t(\tilde{b}(s))^2ds\right],$$

from which it follows that

$$E[(u_1(t) - u_2(t))^2] \leq 2tE\left[\int_0^t (\tilde{a}(s))^2 ds\right] + 2E\left[\int_0^t (\tilde{b}(s))^2 ds\right].$$

By assumption 1, we have that

$$|\tilde{a}(s)|^2 \leq (K^*)^2 |u_1(s) - u_2(s)|^2,$$
$$|\tilde{b}(s)|^2 \leq (K^*)^2 |u_1(s) - u_2(s)|^2,$$

and therefore, by Fubini's theorem,

$$E[|u_1(t) - u_2(t)|^2]$$

$$\leq 2t(K^*)^2 \int_0^t E[|u_1(t) - u_2(t)|^2] ds + 2(K^*)^2 \int_0^t E[|u_1(t) - u_2(t)|^2] ds$$

$$\leq 2T(K^*)^2 \int_0^t E[|u_1(t) - u_2(t)|^2] ds + 2(K^*)^2 \int_0^t E[|u_1(t) - u_2(t)|^2] ds$$

$$= 2(K^*)^2 (T + 1) \int_0^t E[|u_1(t) - u_2(t)|^2] ds.$$

Since, by Gronwall's Lemma 4.3,

$$E[|u_1(t) - u_2(t)|^2] = 0 \qquad \forall t \in [0, T],$$

we get

$$u_1(t) - u_2(t) = 0, \ P\text{-a.s.} \ \forall t \in [0, T]$$

or, equivalently, for all $t \in [0, T]$:

$$\exists N_t \subset \Omega, P(N_t) = 0 \text{ such that } \forall \omega \notin N_t \colon u_1(t)(\omega) - u_2(t)(\omega) = 0.$$

Because we are considering separable processes, there exists an $M \subset [0, T]$, a separating set of $(u_1(t) - u_2(t))_{t \in [0,T]}$, countable and dense in $[0, T]$, such that for all $t \in [0, T]$

$$\exists (t_n)_{n \in \mathbb{N}} \in M^{\mathbb{N}} \text{ such that } \lim_n t_n = t,$$

and

$$\lim_n (u_1(t_n) - u_2(t_n)) = u_1(t) - u_2(t), \ P\text{-a.s.}$$

(and the empty set A, where this does not hold, does not depend on t). Putting $N = \bigcup_{t \in M} N_t$, we obtain $P(N) = 0$ and

$$\forall \omega \notin N \colon u_1(t) - u_2(t) = 0, t \in M,$$

hence

$$\forall t \in [0, T], \forall \omega \notin N \cup A \colon u_1(t) - u_2(t) = 0,$$

and thus

$$P\left(\sup_{0 \le t \le T} |u_1(t) - u_2(t)| = 0\right) = 1.$$

STEP 2: *Existence.* We will prove the existence of a solution $u(t)$ by the method of sequential approximations. We define

$$\begin{cases} u_0(t) = u^0, \\ u_n(t) = u^0 + \int_0^t a(s, u_{n-1}(s))ds + \int_0^t b(s, u_{n-1}(s))dW_s, \; \forall t \in [0, T], n \in \mathbb{N}^*. \end{cases}$$

Take u^0 $\mathcal{F}_0$-measurable; by assumption 3, it is obvious that $u^0 \in \mathcal{C}([0, T])$. By induction, we will now show both that

$$\forall n \in \mathbb{N} : E[|u_{n+1}(t) - u_n(t)|^2] \le \frac{(ct)^{n+1}}{(n+1)!}, \tag{4.4}$$

where $c = \max\{4K^2(T+1)(1 + E[|u^0|^2]), 2(K^*)^2(T+1)\}$, and

$$\forall n \in \mathbb{N} : u_{n+1} \in \mathcal{C}([0, T]). \tag{4.5}$$

By assumptions 1 and 2, we obtain

$$E[|b(s, u^0)|^2] \le E[K^2(1 + |u^0|)^2] \le 2K^2(1 + E[|u^0|^2]) < +\infty,$$

where we make use of the generic inequality

$$(|x| + |y|)^2 \le 2|x|^2 + 2|y|^2, \tag{4.6}$$

and thus

$$b(s, u^0) \in \mathcal{C}([0, T]).$$

Analogously, $a(s, u^0) \in \mathcal{C}([0, T])$, resulting in u_1 being nonanticipatory and well posed. As a further result of (4.6) we have

$$|u_1(t) - u^0|^2 = \left|\int_0^t a(s, u^0)ds + \int_0^t b(s, u^0)dW_s\right|^2$$
$$\le 2\left|\int_0^t a(s, u^0)ds\right|^2 + 2\left|\int_0^t b(s, u^0)dW_s\right|^2,$$

and by the Schwarz inequality,

$$\left|\int_0^t a(s, u^0)ds\right|^2 \le t\int_0^t |a(s, u^0)|^2 ds \le T\int_0^t |a(s, u^0)|^2 ds.$$

Moreover, by Itô's isometry (Proposition 3.20), we have

$$E\left[\left|\int_0^t b(s, u^0)dW_s\right|^2\right] = E\left[\int_0^t |b(s, u^0)|^2 ds\right].$$

Therefore, as a conclusion and by assumption 2,

$$
\begin{aligned}
E[|u_1(t) - u^0|^2] &\leq 2TE\left[\int_0^t |a(s, u^0)|^2 ds\right] + 2E\left[\int_0^t |b(s, u^0)|^2 ds\right] \\
&\leq 2TE\left[\int_0^t K^2(1 + |u^0|)^2 ds\right] + 2E\left[\int_0^t K^2(1 + |u^0|)^2 ds\right] \\
&= (2TK^2 + 2K^2)E\left[\int_0^t (1 + |u^0|)^2 ds\right] \\
&= 2K^2(T + 1)tE[(1 + |u^0|)^2] \\
&\leq 4K^2(T + 1)t(1 + E[|u^0|^2]) = ct,
\end{aligned}
$$

where the last inequality is a direct result of (4.6). Hence (4.4) holds for $n = 1$, from which it follows that $u_1 \in \mathcal{C}([0, T])$.

STEP 3: Suppose now that (4.4) and (4.5) hold for $n \in \mathbb{N}$; we will show that this implies that they also hold for $n + 1$. By the induction hypotheses, $u_n \in \mathcal{C}([0, T])$. Then, by assumption 2, and proceeding as before, we obtain that

$$
a(s, u_n(s)) \in \mathcal{C}([0, T]) \text{ and } b(s, u_n(s)) \in \mathcal{C}([0, T]).
$$

Therefore, u_{n+1} is well posed and nonanticipatory. We thus get

$$
\begin{aligned}
|u_{n+1}(t) - u_n(t)|^2 &\leq \left(\int_0^t |a(s, u_n(s)) - a(s, u_{n-1}(s))| ds \right. \\
&\quad \left. + \left|\int_0^t (b(s, u_n(s)) - b(s, u_{n-1}(s))) dW_s\right|\right)^2 \\
&\leq 2\left(\int_0^t |a(s, u_n(s)) - a(s, u_{n-1}(s))| ds\right)^2 \\
&\quad + 2\left(\int_0^t (b(s, u_n(s)) - b(s, u_{n-1}(s))) dW_s\right)^2,
\end{aligned}
$$

and by the Schwarz inequality,

$$
\begin{aligned}
\left(\int_0^t |a(s, u_n(s)) - a(s, u_{n-1}(s))| ds\right)^2 &\leq t\int_0^t |a(s, u_n(s)) - a(s, u_{n-1}(s))|^2 ds \\
&\leq T(K^*)^2 \int_0^t |u_n(s) - u_{n-1}(s)|^2 ds,
\end{aligned}
$$

where the last inequality is due to assumption 2. Moreover, by Itô's isometry (Proposition 3.20),

$$E\left[\left(\int_0^t (b(s,u_n(s)) - b(s,u_{n-1}(s)))dW_s\right)^2\right]$$

$$= E\left[\int_0^t |b(s,u_n(s)) - b(s,u_{n-1}(s))|^2 ds\right]$$

$$\leq (K^*)^2 E\left[\int_0^t |u_n(s) - u_{n-1}(s)|^2 ds\right],$$

again by point 1. Now we obtain

$$E[|u_{n+1}(t) - u_n(t)|^2] \leq 2T(K^*)^2 E\left[\int_0^t |u_n(s) - u_{n-1}(s)|^2 ds\right]$$

$$+2(K^*)^2 E\left[\int_0^t |u_n(s) - u_{n-1}(s)|^2 ds\right]$$

$$\leq cE\left[\int_0^t |u_n(s) - u_{n-1}(s)|^2 ds\right]$$

$$\leq c\int_0^t \frac{(cs)^n}{n!} ds = \frac{(ct)^{n+1}}{(n+1)!},$$

where the last inequality is due to the induction hypotheses. Hence the proof of (4.4) is complete, and so $u_{n+1} \in \mathcal{C}([0,T])$.

STEP 4: From (4.6) it follows that

$$\sup_{0\leq t\leq T} |u_{n+1}(t) - u_n(t)|^2 \leq 2 \sup_{0\leq t\leq T} \left(\int_0^t |a(s,u_n(s)) - a(s,u_{n-1}(s))|ds\right)^2$$

$$+2 \sup_{0\leq t\leq T} \left|\int_0^t (b(s,u_n(s)) - b(s,u_{n-1}(s)))dW_s\right|^2,$$

where, after taking expectations and recalling point 1 of Proposition 3.42,

$$E\left[\sup_{0\leq t\leq T} |u_{n+1}(t) - u_n(t)|^2\right] \leq 2E\left[\left(\int_0^T |a(s,u_n(s)) - a(s,u_{n-1}(s))|ds\right)^2\right]$$

$$+8E\left[\int_0^T |b(s,u_n(s)) - b(s,u_{n-1}(s))|^2 ds\right]$$

$$\leq 2T(K^*)^2 E\left[\int_0^T |u_n(s) - u_{n-1}(s)|^2 ds\right]$$

$$+8(K^*)^2 E\left[\int_0^T |u_n(s) - u_{n-1}(s)|^2 ds\right]$$

$$= 2T(K^*)^2 \int_0^T E[|u_n(s) - u_{n-1}(s)|^2] ds$$

$$+8(K^*)^2 \int_0^T E[|u_n(s) - u_{n-1}(s)|^2]ds$$

$$\leq \frac{(cT)^n}{n!}(2(K^*)T^2 + 8(K^*)^2T),$$

where the last equality is due to assumption 1, as well as the Schwarz inequality, and the last inequality is due to (4.4). Hence

$$E\left[\sup_{0 \leq t \leq T}|u_{n+1}(t) - u_n(t)|^2\right] \leq c^* \frac{(cT)^n}{n!}, \tag{4.7}$$

with $c^* = 2(K^*)^2T^2 + 8(K^*)^2T$. Because the terms are positive,

$$\sup_{0 \leq t \leq T}|u_{n+1}(t) - u_n(t)|^2 = \left(\sup_{0 \leq t \leq T}|u_{n+1}(t) - u_n(t)|\right)^2,$$

and therefore

$$P\left(\sup_{0 \leq t \leq T}|u_{n+1}(t) - u_n(t)| > \frac{1}{2^n}\right) = P\left(\sup_{0 \leq t \leq T}|u_{n+1}(t) - u_n(t)|^2 > \frac{1}{2^{2n}}\right)$$

$$\leq E\left[\sup_{0 \leq t \leq T}|u_{n+1}(t) - u_n(t)|^2\right]2^{2n}$$

$$\leq c^* \frac{(cT)^n}{n!}2^{2n},$$

where the last two inequalities are due to the Markov inequality and (4.7), respectively. Because the series $\sum_{n=1}^{\infty} \frac{(cT)^n}{n!}2^{2n}$ converges, so does

$$\sum_{n=1}^{\infty} P\left(\sup_{0 \leq t \leq T}|u_{n+1}(t) - u_n(t)| > \frac{1}{2^n}\right),$$

and, by the Borel–Cantelli Lemma 1.161, we have that

$$P\left(\limsup_n \left\{\sup_{0 \leq t \leq T}|u_{n+1}(t) - u_n(t)| > \frac{1}{2^n}\right\}\right) = 0.$$

Therefore, putting $A = \limsup_n \{\sup_{0 \leq t \leq T}|u_{n+1}(t) - u_n(t)| > \frac{1}{2^n}\}$, for all $\omega \in (\Omega - A)$:

$$\exists N = N(\omega) \text{ such that } \forall n \in N, n \geq N(\omega) \Rightarrow \sup_{0 \leq t \leq T}|u_{n+1}(t) - u_n(t)| \leq \frac{1}{2^n},$$

and $u^0 + \sum_{n=0}^{\infty}(u_{n+1}(t) - u_n(t))$ converges uniformly on $t \in [0, T]$ with probability 1. Thus, given the sum $u(t)$ and observing that $u^0 + \sum_{k=0}^{n-1}(u_{k+1}(t) - u_k(t)) = u_n(t)$, it follows that the sequence $(u_n(t))_n$ of the nth partial sum of $u^0 + \sum_{n=0}^{\infty}(u_{n+1}(t) - u_n(t))$ has the limit

$$\lim_{n \to \infty} u_n(t) = u(t), \ P\text{-a.s., uniformly on } t \in [0, T]. \tag{4.8}$$

Analogous to the property of the processes u_n, it follows that the trajectories of $u(t)$ are continuous a.s. and nonanticipatory.

STEP 5: We will now demonstrate that $u(t)$ is the solution of (4.1) and (4.2).
By point 1 of the same theorem, we have

$$\left| \int_0^t a(s, u_{n-1}(s))ds - \int_0^t a(s, u(s))ds \right| \leq K^* \int_0^t |u_{n-1}(s) - u(s)|ds,$$

and since we can take the limit of (4.8) inside the integral sign,

$$\int_0^t a(s, u_{n-1}(s))ds \xrightarrow{n} \int_0^t a(s, u(s))ds, \ P\text{-a.s., uniformly on } t \in [0, T],$$

and therefore also in probability. Moreover,

$$|b(s, u_{n-1}(s)) - b(s, u(s))|^2 \leq (K^*)^2 |u_{n-1}(s) - u(s)|^2,$$

and thus

$$\int_0^t |b(s, u_{n-1}(s)) - b(s, u(s))|^2 ds \xrightarrow{n} 0, \ P\text{-a.s., uniformly on } t \in [0, T],$$

and therefore also in probability. Hence, by Theorem 3.29, we also have

$$P - \lim_{n \to \infty} \int_0^t b(s, u_{n-1}(s))dW_s = \int_0^t b(s, u(s))dW_s.$$

Then if we take the limit $n \to \infty$ of

$$u_n(t) = u^0 + \int_0^t a(s, u_{n-1}(s))ds + \int_0^t b(s, u_{n-1}(s))dW_s, \qquad (4.9)$$

by the uniqueness of the limit in probability, we obtain

$$u(t) = u^0 + \int_0^t a(s, u(s))ds + \int_0^t b(s, u(s))dW_s,$$

with $u(t)$ as the solution of (4.1) and (4.2).

STEP 6: It remains to show that

$$E[u^2(t)] < \infty, \text{ for all } t \in [0, T].$$

Because, in general, $(a + b + c)^2 \leq 3(a^2 + b^2 + c^2)$, by (4.9), it follows that

$$E[u_n^2(t)] \leq 3 \left(E[(u^0)^2] + E\left[\left| \int_0^t a(s, u_{n-1}(s))ds \right|^2 \right] \right.$$

$$\left. + E\left[\left| \int_0^t b(s, u_{n-1}(s))dW_s \right|^2 \right] \right)$$

$$\leq 3 \left(E[(u^0)^2] + TE\left[\int_0^t |a(s, u_{n-1}(s))|^2 ds \right] \right.$$

$$\left. + E\left[\int_0^t |b(s, u_{n-1}(s))|^2 ds \right] \right),$$

where the last relation holds due to the Schwarz inequality as well as point 3 of Proposition 3.20. From assumption 2 and inequality (4.6) it further follows that

$$|a(s, u_{n-1}(s))|^2 \leq K^2(1 + |u_{n-1}(s)|)^2 \leq 2K^2(1 + |u_{n-1}(s)|^2),$$
$$|b(s, u_{n-1}(s))|^2 \leq K^2(1 + |u_{n-1}(s)|)^2 \leq 2K^2(1 + |u_{n-1}(s)|^2).$$

Therefore,

$$E[u_n^2(t)] \leq 3\left(E[(u^0)^2] + 2K^2(T+1)\int_0^t (1 + E[|u_{n-1}(s)|^2])ds\right)$$
$$\leq 3\left(E[(u^0)^2] + 2K^2T(T+1) + 2K^2(T+1)\int_0^t E[|u_{n-1}(s)|^2]ds\right)$$
$$\leq c(1 + E[(u^0)^2]) + c\int_0^t E[|u_{n-1}(s)|^2]ds,$$

where c is a constant that only depends on K and T. Continuing with the induction, we have

$$E[u_n^2(t)] \leq \left(c + c^2t + c^3\frac{t^2}{2} + \cdots + c^{n+1}\frac{t^n}{n!}\right)(1 + E[(u^0)^2]),$$

and taking the limit $n \to \infty$,

$$\lim_{n\to\infty} E\left[u_n^2(t)\right] \leq ce^{ct}\left(1 + E\left[(u^0)^2\right]\right) \leq ce^{cT}\left(1 + E\left[(u^0)^2\right]\right).$$

Therefore, by Fatou's Lemma A.27 and by assumption 3, we obtain

$$E\left[u^2(t)\right] \leq ce^{cT}\left(1 + E\left[(u^0)^2\right]\right) < +\infty, \tag{4.10}$$

and hence $(u(t))_{t\in[0,T]} \in \mathcal{C}([0,T])$, completing the proof.

$\square$

Remark 4.6. By (4.10), it also follows that

$$\sup_{0\leq t\leq T} E\left[u^2(t)\right] \leq ce^{cT}\left(1 + E\left[(u^0)^2\right]\right) < +\infty.$$

Remark 4.7. Theorem 4.4 continues to hold if its Hypothesis 1 is substituted by the following local condition.

1'. For all $n > 0$ there exists a $K_n > 0$ such that, for all $(x_1, x_2) \in \mathbb{R}^2, |x_i| \leq n$ $i = 1, 2$:

$$|a(t, x_1) - a(t, x_2)| \leq K_n|x_1 - x_2|,$$
$$|b(t, x_1) - b(t, x_2)| \leq K_n|x_1 - x_2|.$$

See, e.g., Friedman 2004 or Gihman and Skorohod 1972, pp. 45–47.

Remark 4.8. If the functions a and b in Theorem 4.4 are defined on the whole $[t_0, +\infty)$, and if the assumptions of the theorem hold on every bounded subinterval $[t_0, T]$, then the SDE (4.1) admits a unique solution defined on the entire half-line $[t_0, +\infty)$. In this case we say that the SDE (4.1) admits a *global solution* (in time) (see, e.g., Arnold 1974, p. 113).

The assumptions in Remark 4.8 hold in particular in the case of autonomous SDEs, i.e., in the case where the coefficients a and b do not depend explicitly on time.

Proposition 4.9 *Consider the following autonomous SDE:*

$$du(t) = a(u(t))dt + b(u(t))dW_t, \qquad t_0 \leq t \leq +\infty, \qquad (4.11)$$

subject to the initial condition

$$u(t_0) = u^0, \qquad (4.12)$$

where u^0 is a random variable satisfying the same assumptions as in Theorem 4.4 and a and b satisfy the following global Lipschitz condition; there exists a positive real constant K such that for any $x, y \in \mathbb{R}$

$$|a(x) - a(y)| + |b(x) - b(y)| \leq K|x - y|.$$

Then (4.11), subject to the initial condition (4.12), admits a unique global solution on the entire $[t_0, +\infty)$.

Proof. See, e.g., Arnold (1974, p. 113). $\qquad\qquad\qquad\qquad\qquad\qquad$ □

Example 4.10. We suppose that in (4.1) $a(t, u(t)) = 0$ and $b(t, u(t)) = g(t)u(t)$. Then the SDE

$$\begin{cases} u(t_0) = u^0, \\ du(t) = g(t)u(t)dW_t \end{cases}$$

has the solution

$$u(t) = u^0 \exp\left\{ \int_{t_0}^t g(s)dW_s - \frac{1}{2}\int_{t_0}^t g^2(s)ds \right\}.$$

In fact, by introducing

$$X(t) = \int_0^t g(s)dW_s - \frac{1}{2}\int_0^t g^2(s)ds$$

and $Y(t) = \exp\{X(t)\} = f(X(t))$, then $u(t) = u^0 Y(t)$ and, thus, $du(t) = u^0 dY(t)$. We will further show that $u^0 dY(t) = g(t)u(t)dW_t$. Because

$$dX(t) = -\frac{1}{2}g^2(t)dt + g(t)dW_t,$$

with the help of Itô's formula, we obtain

$$dY(t) = \left(-\frac{1}{2}g^2(t)f_x(X(t)) + \frac{1}{2}g^2(t)f_{xx}(X(t))\right)dt + g(t)f_x(X(t))dW_t$$

$$= \left(-\frac{1}{2}g^2(t)\exp\{X(t)\} + \frac{1}{2}g^2(t)\exp\{X(t)\}\right)dt + g(t)\exp\{X(t)\}dW_t$$

$$= Y(t)g(t)dW_t,$$

resulting in $du(t) = u^0 Y(t)g(t)dW_t = u(t)g(t)dW_t$.

Example 4.11. (*Linear time-homogeneous stochastic differential equations*). Three important SDEs that have wide applicability, for instance in financial modeling, are

1. Arithmetic Brownian motion:

$$du(t) = a\,dt + b\,dW_t.$$

2. Geometric Brownian motion:

$$du(t) = au(t)dt + bu(t)dW_t.$$

3. The (mean-reverting) Ornstein–Uhlenbeck process:

$$du(t) = (a - bu(t))dt + c\,dW_t.$$

Since all three cases are time-homogeneous, we may assume 0 as the initial time, impose an initial condition $u(0) = u^0$, and look for solutions in $\mathbb{R}_+$. The derivations of the solutions of 1–3 resort to a number of standard solution techniques for SDEs.

1. Direct integration gives

$$u(t) = u^0 + at + bW_t,$$

so that we can take the expectation and variance directly to obtain

$$E[u(t)] = u^0 + at, \qquad Var[u(t)] = b^2 t.$$

2. We calculate the stochastic differential $d\ln u(t)$ with the help of Itô's formula (3.31) and obtain

$$d\ln u(t) = \left(a - \frac{1}{2}b^2\right)dt + b\,dW_t.$$

We can integrate both sides directly, which results in

$$\ln u(t) = \ln u^0 + \left(a - \frac{1}{2}b^2\right)t + bW_t$$

or

$$u(t) = u^0 \exp\left\{\left(a - \frac{1}{2}b^2\right)t + bW_t\right\}. \tag{4.13}$$

To calculate its expectation, we will require the expected value of $\tilde{u}(t) = \exp\{bW_t\}$. We apply Itô's formula to calculate the latter's differential as

$$d\exp\{bW_t\} = d\tilde{u}(t) = b\tilde{u}(t)dW_t + \frac{1}{2}b^2\tilde{u}(t)dt,$$

which after direct integration, rearrangement, and the taking of expectations results in

$$E[\tilde{u}(t)] = 1 + \int_0^t \frac{1}{2}b^2 E[\tilde{u}(s)]ds.$$

Differentiating both sides with respect to t gives

$$\frac{dE[\tilde{u}(t)]}{dt} = \frac{1}{2}b^2 E[\tilde{u}(t)],$$

which, after rearrangement and integration, results in

$$E[\tilde{u}(t)] = e^{\frac{1}{2}b^2 t}.$$

Therefore, for a deterministic initial condition, the expectation of (4.13) is

$$E[u(t)] = u^0 e^{\left(a - \frac{1}{2}b^2\right)t} E\left[e^{bW_t}\right] = u^0 e^{at}.$$

For the variance we employ the standard general result (1.6), so that we only need to calculate $E[(u(t))^2]$:

$$(u(t))^2 = (u^0)^2 \exp\left\{(2a - b^2)t + 2bW_t\right\},$$

as previously $E[\exp\{2bW_t\}] = \exp\{2b^2 t\}$, and by easy computations the variance of (4.13) is

$$Var[u(t)] = E[u(t)^2] - E[u(t)]^2 = (u^0)^2 \exp\{2at\}[\exp\{b^2 t\} - 1].$$

3. To find the solution of the Ornstein–Uhlenbeck process, we require an integrating factor $\phi = \exp\{bt\}$, so that

$$d(\phi u(t)) = \phi(bu(t) + du(t)) = \phi(adt + cdW_t).$$

Because the drift term, which depended on $u(t)$, has dropped out, we can integrate directly and, after rearrangement, obtain

$$u(t) = \frac{a}{b} + u^0 \exp\{-bt\} + c \int_0^t \exp\{-b(t-s)\}\, dW_s. \qquad (4.14)$$

Therefore, for a deterministic initial condition the expectation of (4.14) is

$$E[u(t)] = \frac{a}{b} + u^0 \exp\{-bt\},$$

and for the variance we again resort to (1.6), so that we require $E[(u(t))^2]$. Squaring (4.14) and taking expectations yields

$$E[(u(t))^2] = \left(\frac{a}{b} + u^0 e^{-bt}\right)^2 + E\left[\left(c \int_0^t e^{-b(t-s)} dW_s\right)^2\right]$$

$$= (E[u(t)])^2 + c^2 \int_0^t e^{-2b(t-s)} ds,$$

where the last step is due to the Itô isometry (point 3 of Proposition 3.20). Hence the variance of (1.6) is

$$Var[u(t)] = (E[u(t)])^2 + c^2 \int_0^t \exp\{-2b(t-s)\}\, ds - (E[u(t)])^2$$

$$= \frac{c^2}{2b}(1 - \exp\{-2bt\}).$$

Proposition 4.12. *Let $(X_t)_t$ be a process that is continuous in probability, stationary, Gaussian, and Markovian. Then it is of the form $Y_t + c$, where Y_t is an Ornstein–Uhlenbeck process and c is a constant.*

Proof. See Breiman (1968).   □

Example 4.13. (A generalized Ornstein–Uhlenbeck time-inhomogeneous SDE. see, e.g., Kloeden and Platen 1999, p. 110ff). Consider the SDE

$$du(t) = (a_1(t)u(t) + a_2(t))dt + c(t)dW_t. \qquad (4.15)$$

Consider the linear ODE

$$dz(t) = a_1(t)z(t)dt;$$

its fundamental solution is

$$\Phi(t, t_0) = \exp\left\{\int_{t_0}^t a_1(s)ds\right\}.$$

Apply the Itô formula to $\Phi(t, t_0)u(t)$, where $u(t)$ is the solution of (4.15), so to obtain

$$d(\Phi^{-1}(t,t_0)u(t)) = \left(\frac{d\Phi^{-1}(t,t_0)}{dt}u(t) + (a_1(t)u(t) + a_2(t))\Phi^{-1}(t,t_0) \right) dt$$

$$+ c(t)\Phi^{-1}(t,t_0)dW_t$$

$$= a_2(t)\Phi^{-1}(t,t_0)dt + c(t)\Phi^{-1}(t,t_0)dW_t \qquad (4.16)$$

since

$$\frac{d\Phi^{-1}(t,t_0)}{dt} = -\Phi^{-1}(t,t_0)a_1(t).$$

By integration of (4.16) we obtain

$$u(t) = \Phi(t,t_0) \left(u(t_0) + \int_{t_0}^{t} a_2(s)\Phi^{-1}(t,t_0)ds + \int_{t_0}^{t} c(s)\Phi^{-1}(s,t_0)dW_s \right)$$

$$(4.17)$$

as the solution of (4.15), subject to an initial condition $u(t_0)$.

Remark 4.14. The solution (4.17) is a Gaussian process whenever the initial condition $u(t_0)$ is either deterministic or a Gaussian random variable (Problem 4.16).

We saw in the proof of Theorem 4.4 that if $E[(u^0)^2] < +\infty$, then $E[(u(t))^2] < +\infty$. This result can be generalized as follows.

Theorem 4.15. *Given the hypotheses of Theorem 4.4, if $E[(u^0)^{2n}] < +\infty$ for $n \in \mathbb{N}$, then*

1. $E[(u(t))^{2n}] \leq (1 + E[(u^0)^{2n}])e^{ct}$
2. $E[\sup_{0 \leq s \leq t} |u(s) - u^0|^{2n}] \leq \bar{c}(1 + E[(u^0)^{2n}])t^n e^{ct}$

where c and $\bar{c}$ are constants that only depend on K, T, and n.

Proof. For all $N \in \mathbb{N}$ we put

$$u_N^0(\omega) = \begin{cases} u_0(\omega) & \text{for } |u^0(\omega)| \leq N, \\ N\text{sgn}\{u^0(\omega)\} & \text{for } |u^0(\omega)| > N; \end{cases}$$

$$a_N(t,x) = \begin{cases} a(t,x) & \text{for } |x| \leq N, \\ a(t,N\text{sgn}\{x\}) & \text{for } |x| > N; \end{cases}$$

$$b_N(t,x) = \begin{cases} b(t,x) & \text{for } |x| \leq N, \\ b(t,N\text{sgn}\{x\}) & \text{for } |x| > N, \end{cases}$$

and we denote by $u_N(t)$ the solution of

$$\begin{cases} u_N(0) = u_N^0, \\ du_N(t) = a_N(t,u_N(t))dt + b_N(t,u_N(t))dW_t \end{cases}$$

(the solution will exist due to Theorem 4.4). Then, applying Itô's formula to $f(u_N(t)) = (u_N(t))^{2n}$, we obtain

$$d(u_N(t))^{2n}$$
$$= (n(2n-1)(u_N(t))^{2n-2}b_N^2(t, u_N(t))$$
$$+2n(u_N(t))^{2n-1}a_N(t, u_N(t)))dt + 2n(u_N(t))^{2n-1}b_N(t, u_N(t))dW_t.$$

Hence

$$(u_N(t))^{2n}$$
$$= (u_N^0)^{2n} + n(2n-1)\int_0^t (u_N(s))^{2n-2}b_N^2(s, u_N(s))ds$$
$$+2n\int_0^t (u_N(s))^{2n-1}a_N(s, u_N(s))ds + 2n\int_0^t (u_N(s))^{2n-1}b_N(s, u_N(s))dW_s.$$

Since $u_N(t) = u_N^0 + \int_0^t a_N(s, u_N(s))ds + \int_0^t b_N(s, u_N(s))dW_s$, $E[(u_N^0)^{2n}] < +\infty$ and both $a_N(t, x)$ and $b_N(t, x)$ are bounded, we have

$$E[(u_N(t))^{2n}] < +\infty,$$

meaning[8] $(u_N(t))^n \in \mathcal{C}([0, T])$. By 2 of Theorem 4.4 and by $(a+b)^2 \le 2(a^2+b^2)$ it follows that

$$|a_N(s, u_n(s))| \le K(1 + |u_N(s)|),$$
$$|b_N(s, u_n(s))|^2 \le 2K^2(1 + |u_N(s)|^2).$$

Moreover, because $(u_N(t))^n \in \mathcal{C}([0, T])$, we have

$$E\left[2n\int_0^t |u_N(s)|^{2n-1}|b_N(s, u_N(s))|dW_s\right] = 0,$$

and therefore

$$E[u_N(t)^{2n}] = E[(u_N^0)^{2n}] + \int_0^t E[(2nu_N(s)a_N(s, u_N(s))$$
$$+n(2n-1)b_N^2(s, u_N(s)))u_N(s)^{2n-2}]ds$$

---

[8] It suffices to make use of the following theorem for $E\left[\left(\int_0^t b_N(s, u_N(s))dW_s\right)^{2n}\right]$:

Theorem. *If $f^n \in \mathcal{C}([0, T])$ for $n \in \mathbb{N}^*$, then*

$$E\left[\left(\int_0^T f(t)dW_t\right)^{2n}\right] \le [n(2n-1)]^n T^{n-1} E\left[\int_0^T f^{2n}(t)dt\right].$$

Proof. See, e.g., Friedman (1975). $\qquad\qquad\square$

$$\leq E[(u_N^0)^{2n}] + n(2n+1) \int_0^t E[(u_N(s)a_N(s, u_N(s))$$

$$+ b_N^2(s, u_N(s)))u_N(s)^{2n-2}]ds$$

$$\leq E[(u_N^0)^{2n}] + n(2n+1)K^2 \int_0^t E[(1 + u_N^2(s))u_N(s)^{2n-2}]ds,$$

where the first inequality follows when condition 2 of Theorem 4.4 is substituted by $xa(t, x) + b^2(t, x) \leq K^2(1 + x^2)$ for all $t \in [0, T]$, and all $x \in \mathbb{R}$. Now since, in general, $x^{2n-2} \leq 1 + x^{2n}$, we have

$$u_N(s)^{2n-2}(1 + u_N^2(s)) \leq 1 + 2u_N(s)^{2n}.$$

Therefore,

$$E[u_N(t)^{2n}] \leq E[(u_N^0)^{2n}] + n(2n+1)K^2 \int_0^t E[1 + 2u_N(s)^{2n}]ds,$$

and, by putting $\phi(t) = E[u_N(t)^{2n}]$, we can write

$$\phi(t) \leq \phi(0) + n(2n+1)K^2 \int_0^t (1 + 2\phi(s))ds$$

$$= \phi(0) + n(2n+1)K^2 t + 2n(2n+1)K^2 \int_0^t \phi(s) = \alpha(t) + L \int_0^t \phi(s)ds,$$

where $\alpha(t) = \phi(0) + n(2n+1)K^2 t$ and $L = 2n(2n+1)K^2$. By Gronwall's Lemma 4.3, we have that

$$\phi(t) \leq \alpha(t) + L \int_0^t e^{L(t-s)}\alpha(s)ds,$$

and thus

$$E[u_N(t)^{2n}]$$

$$\leq E[(u_N^0)^{2n}] + \frac{L}{2}t + L \int_0^t e^{L(t-s)} \left(E[(u_N^0)^{2n}] + \frac{L}{2}s \right) ds$$

$$= E[(u_N^0)^{2n}] + \frac{L}{2}t - E[(u_N^0)^{2n}] + E[(u_N^0)^{2n}]e^{Lt} + Le^{Lt} \int_0^t e^{-Ls}\frac{L}{2}sds$$

$$= \frac{L}{2}t + E[(u_N^0)^{2n}]e^{Lt} - \frac{L}{2}t - \frac{1}{2}e^{Lt}(e^{-Lt} - 1) \leq e^{Lt}(1 + E[(u_N^0)^{2n}]).$$

Therefore, point 1 holds for $u_N(t)$ ($N \in \mathbb{N}^*$) and, taking the limit $N \to \infty$, it also holds for $u(t)$. For the proof of 2, see, e.g., Gihman and Skorohod (1972).

□

Remark 4.16. The preceding theory provides only sufficient conditions for the global (in time) existence of solutions to SDEs. It is interesting to realize that, for example, the linear growth condition on the coefficients is not necessary for global existence. We will report here two interesting examples from the literature, one regarding an application to finance and the other to biology.

Application: A Highly Sensitive Mean-Reverting Process in Finance

According to Chan et al. (1992) and Nowman (1997), a large class of models regarding the short-term riskless interest rate $R(t)$ are included in the following model:

$$dR(t) = \lambda(\mu - R(t))dt + \sigma R^\gamma(t)dW_t, \tag{4.18}$$

for suitable choices of the parameters λ, μ, σ, and γ; W denotes a standard Wiener noise.

Statistical analysis carried out by the aforementioned authors evidence that the most suited models of this kind are those that allow a high sensitivity of the volatility to the interest rate, i.e., $\gamma > 1$. The problem is that these values for γ imply that the diffusion coefficient does not satisfy the linear growth condition, so that the preceding theory cannot be applied to show global existence at the time of the solutions to (4.18) and the boundedness of its moments. However, Wu et al. (2008) have been able to show the following important result.

Theorem 4.17. *For any set of parameters $\lambda > 0$, $\mu > 0$, $\sigma > 0$, $\gamma > 1$, and any nontrivial initial condition $R(0) > 0$, there exists a unique solution $R(t)$, global in time, of (4.18) that stays, almost surely, in $\mathbb{R}_+^*$, for all times $t \in \mathbb{R}_+$.*

Further, the solution $R(t)$ of (4.18) satisfies the bounds

$$E[R(t)] \leq R(0) + \mu, \quad t \in \mathbb{R}_+$$

and

$$\limsup_{t \to +\infty} E[R(t)] \leq \mu.$$

Application: A General Lotka–Volterra Model

A mathematical problem of interest in population dynamics in presence of Wiener noise has been discussed in Mao et al. (2002).

A general Lotka–Volterra model for a system of $n \in \mathbb{N} - \{0\}$ interacting components appears in the form

$$\dot{x}(t) = diag(x_1(t), \ldots, x_n(t))[b + Ax(t)], \tag{4.19}$$

where $x(t) = (x_1(t), \ldots, x_n(t))'$ is the vector of the component populations, $b = (b_1, \ldots, b_n)'$ is a vector of real coefficients, and $A = (a_{i,j})_{1 \leq i,j \leq n}$ is the real-valued interaction matrix; $diag(x_1(t), \ldots, x_n(t))$ denotes a diagonal $n \times n$ matrix having on the diagonal the elements $x_1(t), \ldots, x_n(t)$.

If we assume that the matrix parameters are perturbed by a Wiener noise W, i.e.,

$$a_{ij}dt \to a_{ij}dt + \sigma_{ij}dW_t,$$

System (4.19) becomes a system of SDEs

$$dx(t) = diag(x_1(t), \ldots, x_n(t))[b + Ax(t)]dt + \sigma x(t)dW_t, \qquad (4.20)$$

where $\sigma = (\sigma_{i,j})_{1 \leq i,j \leq n}$ is subject to the following condition:

(H1) $\sigma_{i,i} > 0$ for $1 \leq i \leq n$; and $\sigma_{i,j} \geq 0$ for $1 \leq i, j \leq n$, $i \neq j$.

Now, system (4.20) does not satisfy the linear growth condition, though it satisfies the local Lipschitz condition; according to the theory, the solution of (4.20) might explode in a finite time, and indeed the authors, in Mao et al. (2002), have shown that this is the case as far as the solution of the corresponding deterministic system (4.19) is concerned, but they have also shown the following interesting result, according to which they might claim that environmental noise suppresses explosion.

Theorem 4.18. *Under Hypothesis (H1), for any system parameters $b \in \mathbb{R}^n$, $A \in \mathbb{R}^{n \times n}$ and any nontrivial initial condition, there exists a unique solution, global in time, of system (4.20) that stays, almost surely, in $\mathbb{R}_+^n$ for all times $t \in \mathbb{R}_+$.*

Remark 4.19. (Weak Solutions of Stochastic Differential Equations).
In the analysis carried out in Sect. 4.1 with respect to the SDE

$$du(t) = a(t, u(t))dt + b(t, u(t))dW_t,$$

the probability space $(\Omega, \mathcal{F}, P)$, the Wiener process W, and the coefficients $a(t, x)$ and $b(t, x)$ were all given in advance, and then conditions were given for the existence and uniqueness of the solution. Such a solution is known as a *strong solution*. Existence and uniqueness theorems may also hold by assigning only the coefficients of the SDE, leaving open the problem of finding a suitable Wiener process and the corresponding filtration. In this case we speak of a *weak solution*. Clearly a strong solution is also a weak solution, but the converse does not hold (Rogers and Williams 1994).

Two solutions (either strong or weak) are called *weakly unique* if they possess the same probability law, i.e., if their finite-dimensional distributions are equal.

4.2 Markov Property of Solutions

In the preceding section we showed that if $a(t, x)$ and $b(t, x)$ are measurable functions on $(t, x) \in [0, T] \times \mathbb{R}$ that satisfy conditions 1 and 2 of Theorem 4.4, then there exists a unique solution in $\mathcal{C}([0, T])$ of

$$\begin{cases} u(0) = u^0 \text{ a.s.,} \\ du(t) = a(t, u(t))dt + b(t, u(t))dW_t, \end{cases} \tag{4.21}$$

provided that the random variable u^0 is independent of $\mathcal{F}_T = \sigma(W_s, 0 \le s \le T)$ and $E[(u^0)^2] < +\infty$. Analogously, for all $s \in\,]0, T]$, there exists a unique solution in $\mathcal{C}([s, T])$ of

$$\begin{cases} u(s) = u_s \text{ a.s.,} \\ du(t) = a(t, u(t))dt + b(t, u(t))dW_t, \end{cases} \tag{4.22}$$

provided that the random variable u_s is independent of $\mathcal{F}_{s,T} = \sigma(W_t - W_s, t \in [s, T])$ and $E[(u_s)^2] < +\infty$. (The proof is left to the reader as a useful exercise.)

Now, let $t_0 \ge 0$ and c be a random variable with $u(t_0) = c$ almost surely and, moreover, c be independent of $\mathcal{F}_{t_0,T} = \sigma(W_t - W_{t_0}, t_0 \le t \le T)$ as well as $E[c^2] < +\infty$. Under conditions 1 and 2 of Theorem 4.4 there exists a unique solution $\{u(t), t \in [t_0, T]\}$ of (4.21) with the initial condition $u(t_0) = c$ almost surely, and the following holds.

Lemma 4.20. *If $h(x, \omega)$ is a real-valued function defined for all $(x, \omega) \in \mathbb{R} \times \Omega$ such that*

1. *h is $\mathcal{B}_{\mathbb{R}} \otimes \mathcal{F}$-measurable.*
2. *h is bounded.*
3. *For all $x \in \mathbb{R} : h(x, \cdot)$ is independent of $\mathcal{F}_s$ for all $s \in [t_0, T]$,*

then

$$\forall s \in [t_0, T]: \qquad E[h(u(s), \cdot)|\mathcal{F}_s] = E[h(u(s), \cdot)|u(s)] \text{ a.s.} \tag{4.23}$$

Proof. We limit ourselves to the case of h being decomposable of the form

$$h(x, \omega) = \sum_{i=1}^{n} Y_i(x) Z_i(\omega), \tag{4.24}$$

with the Z_i independent of $\mathcal{F}_s$. In that case

$$E[h(u(s), \cdot)|\mathcal{F}_s] = \sum_{i=1}^{n} E[Y_i(u(s)) Z_i(\cdot)|\mathcal{F}_s] = \sum_{i=1}^{n} Y_i(u(s)) E[Z_i(\cdot)|\mathcal{F}_s],$$

because $Y_i(u(s))$ is $\mathcal{F}_s$-measurable. Therefore

$$E[h(u(s), \cdot)|\mathcal{F}_s] = \sum_{i=1}^{n} Y_i(u(s)) E[Z_i(\cdot)],$$

and recapitulating, because $\sigma(u(s)) \subset \mathcal{F}_s$, we have

$$E[h(u(s), \cdot)|\mathcal{F}_s] = \sum_{i=1}^{n} Y_i(u(s))E[Z_i(\cdot)|u(s)]$$

$$= \sum_{i=1}^{n} E[Y_i(u(s))Z_i(\cdot)|u(s)] = E[h(u(s), \cdot)|u(s)].$$

It can be shown that every h that satisfies conditions 1, 2, and 3 can be approximated by functions that are decomposable as in (4.24). $\qquad\square$

Theorem 4.21. *If $(u(t))_{t\in[t_0,T]}$ is the solution of the SDE problem (4.22) on $[t_0, T]$, then it is a Markov process with respect to the filtration $\mathcal{U}_t = \sigma(u(s), t_0 \leq s \leq t)$, i.e., it satisfies the condition*

$$\forall B \in \mathcal{B}_{\mathbb{R}}, \forall s \in [t_0, t[:\ P(u(t) \in B|\mathcal{U}_s) = P(u(t) \in B|u(s))\ a.s. \qquad (4.25)$$

Proof. Putting $\mathcal{F}_t = \sigma(c, W_s, t_0 \leq s \leq t)$, then $u(t)$ is $\mathcal{F}_t$-measurable, as can be deduced from Theorem 4.4. Therefore, $\sigma(u(t)) \subset \mathcal{F}_t$ and thus $\mathcal{U}_t \subset \mathcal{F}_t$. In order to prove (4.25), it is now sufficient to show that

$$\forall B \in \mathcal{B}_{\mathbb{R}}, \forall s \in [t_0, t[:\ P(u(t) \in B|\mathcal{F}_s) = P(u(t) \in B|u(s))\ \text{a.s.} \qquad (4.26)$$

Fixing $B \in \mathcal{B}_{\mathbb{R}}$ and $s < t$, we denote by $u(t, s, x)$ the solution of (4.22) with the initial condition $u(s) = x$ a.s. ($x \in \mathbb{R}$), and we define the mapping $h : \mathbb{R} \times \Omega \to \mathbb{R}$ as

$$h(x, \omega) = I_B(u(t, s, x; \omega))\ \text{for}\ (x, \omega) \in \mathbb{R} \times \Omega.$$

h is bounded and, moreover, for all $x \in \mathbb{R}$, $h(x, \cdot)$ is independent of $\mathcal{F}_s$, because so is $u(t, s, x; \omega)$ (given that $u(s) = x \in \mathbb{R}$ is a certain event). Furthermore, observe that if $t_0 < s$, $s \in [0, T]$, we obtain by uniqueness

$$u(t, t_0, c) = u(t, s, u(s, t_0, c))\ \text{for}\ t \geq s, \qquad (4.27)$$

where c is the chosen random value. Equation (4.27) states the fact that the solution of (4.21) with the initial condition $u(t_0) = c$ is identical to the solution of the same equations with the initial condition $u(s) = u(s, t_0, c)$ for $t \geq s$ (e.g., Baldi 1984). Equation (4.27) is called the *semigroup property* or *dynamic system*. (The proof of the property is left to the reader as an exercise.) Now, because $h(x, \omega) = I_B(u(t, s, x; \omega))$ satisfies conditions 1, 2, and 3 of Lemma 4.20 and by (4.27), we have $h(u(s), \omega) = I_B(u(t; \omega))$. Then, by (4.23), we obtain

$$P(u(t) \in B|\mathcal{F}_s) = P(u(t) \in B|u(s))\ \text{a.s.},$$

completing the proof. $\qquad\square$

Remark 4.22. By (4.27) and (4.26), we also have

$$P(u(t) \in B|u(s)) = P(u(t,s,u(s)) \in B|u(s))$$

and, in particular,

$$P(u(t) \in B|u(s) = x) = P(u(t,s,u(s)) \in B|u(s) = x), \qquad x \in \mathbb{R}.$$

Hence

$$P(u(t) \in B|u(s) = x) = P(u(t,s,x) \in B), \qquad x \in \mathbb{R}. \qquad (4.28)$$

Theorem 4.23. *If $(u(t))_{t \in [t_0,T]}$ is the solution of*

$$\begin{cases} u(t_0) = c \ a.s., \\ du(t) = a(t,u(t))dt + b(t,u(t))dW_t, \end{cases}$$

defining, for all $B \in \mathcal{B}_{\mathbb{R}}$, all $t_0 \leq s < t \leq T$ and all $x \in \mathbb{R}$:

$$p(s,x,t,B) = P(u(t) \in B|u(s) = x) = P(u(t,s,x) \in B),$$

then p is a transition probability (of the Markov process $u(t)$).

Proof. We have to show that conditions 1, 2, and 3 of Definition 2.75 are satisfied.

1. Fixing s and t such that $t_0 \leq s < t \leq T$ and $B \in \mathcal{B}_{\mathbb{R}}$,

$$p(s,x,t,B) = P(u(t) \in B|u(s) = x) = E[I_B(u(t))|u(s) = x], \qquad x \in \mathbb{R}.$$

Then, as a property of conditional probabilities, $p(s,\cdot,t,B)$ is $\mathcal{B}_{\mathbb{R}}$-measurable.
2. is true by the definition of $p(s,x,t,B)$.
3. Fixing s and t such that $t_0 \leq s < t \leq T$ and $x \in \mathbb{R}$, $p(s,x,t,B) = P(u(t,s,x) \in B)$, for all $B \in \mathcal{B}_{\mathbb{R}}$. This is the induced probability P of $u(t,s,x)$. Therefore, if $\psi : \mathbb{R} \to \mathbb{R}$ is a bounded $\mathcal{B}_{\mathbb{R}}$-measurable function, then

$$\int_{\mathbb{R}} \psi(y)p(s,x,t,dy) = \int_{\Omega} \psi(u(t,s,x,\omega))dP(\omega).$$

Now, let $\psi(y) = p(r,y,t,B)$ with $B \in \mathcal{B}_{\mathbb{R}}$, $y \in \mathbb{R}$, $t_0 \leq r < t \leq T$. Then, for $s < r$, we have

$$\int_{\mathbb{R}} p(r,y,t,B)p(s,x,r,dy)$$

$$= \int_{\Omega} p(r,u(r,s,x,\omega),t,B)dP(\omega)$$

$$= E[p(r,u(r,s,x),t,B)] = E[P(u(t) \in B|u(r) = u(r,s,x))]$$

$$= E[E[I_B(u(t))|u(r) = u(r,s,x)]] = E[I_B(u(t))|u(s) = x]$$

$$= P(u(t,s,x) \in B) = p(s,x,t,B).$$

In fact, $u(t)$ satisfies (4.21) with the initial condition $u(s) = x$.   □

Remark 4.24. By Theorem 2.78, the knowledge of the solution $u(t)$ of

$$\begin{cases} u(t_0) = c \text{ a.s.}, \\ du(t) = a(t, u(t))dt + b(t, u(t))dW_t \end{cases}$$

is equivalent to assigning the transition probability p to the process $u(t)$ and the distribution P_0 of c.

Remark 4.25. Every SDE generates Markov processes in the sense that every solution is a Markov process.

Implicit in the underlying hypotheses of Theorem 4.4 is the following theorem.

Theorem 4.26. *If the SDE is autonomous of form (4.11), then the Markov process $\{u(t, t_0, c), t \in [t_0, T]\}$ is homogeneous.*

Remark 4.27. The transition measure of the homogeneous process $(u_t)_{t \in [t_0, T]}$ is time-homogeneous, i.e.,

$$P(u(t + s) \in A | u(t) = x) = P(u(s) \in A | u(0) = x) \text{almost surely}$$

for any $s, t \in \mathbb{R}_+$, $x \in \mathbb{R}$, and $A \in \mathcal{B}_{\mathbb{R}}$.

Theorem 4.28. *If for*

$$\begin{cases} u(t_0) = c \text{ a.s.}, \\ du(t) = a(t, u(t))dt + b(t, u(t))dW_t, \end{cases}$$

the hypotheses of Theorem 4.4 are satisfied, with $a(t, x)$ and $b(t, x)$ being continuous in $(t, x) \in [0, \infty] \times \mathbb{R}$, then the solution $u(t)$ is a diffusion process with drift coefficient $a(t, x)$ and diffusion coefficient $b^2(t, x)$.

Proof. We prove point 1 of Lemma 2.106. Let $u(t, s, x)$ be a solution of the problem with initial value

$$u(s) = x \text{ a.s.}, x \in \mathbb{R} \text{ (fixed)}, t \geq s \text{ (s fixed)}.$$

By (4.28):

$$p(t, x, t + h, A) = P(u(t + h, t, u(t)) \in A | u(t) = x) = P(u(t + h, t, x) \in A).$$

Hence $p(t, x, t + h, A)$ is the probability distribution of the random variable $u(t + h, t, x)$, and thus

$$E[f(u(t + h, t, x) - x)] = \int_{\mathbb{R}} f(y - x)p(t, x, t + h, dy)$$

for every function $f(z)$ such that[9] $|f(z)| \leq \dot{K}(1 + |z|^{2n})$, with $\alpha \geq 1$, $K > 0$, and $f(z)$ continuous. It is now sufficient to prove that

$$\lim_{h \downarrow 0} \frac{1}{h} E[|u(t + h, t, x) - x|^4] = 0.$$

Given that z^4 is of the preceding form $f(z)$, the preceding limit follows from

$$\frac{1}{h} E[|u(t + h, t, x) - x|^4] \leq \frac{1}{h} K h^2 (1 + |x|^4)$$

by point 2 of Theorem 4.15.

Now we prove point 2 of Lemma 2.106. This is equivalent to showing that

$$\lim_{h \downarrow 0} \frac{1}{h} E[u(t + h, t, x) - x] = a(t, x).$$

Because $u(t, t, x) = x$ almost surely, due to the definition of the stochastic differential we obtain

$$E[u(t + h, t, x) - x] = E\left[\int_t^{t+h} a(s, u(s, t, x))ds + \int_t^{t+h} b(s, u(s, t, x))dW_s\right].$$

But since $E[\int_t^{t+h} b(s, u(s, t, x))dW_s] = 0$, we get

$$E[u(t + h, t, x) - x] = E\left[\int_t^{t+h} a(s, u(s, t, x))ds\right]$$

$$= E\left[\int_t^{t+h} (a(s, u(s, t, x)) - a(s, x))ds\right] + \int_t^{t+h} a(s, x)ds$$

$$= \int_t^{t+h} E[a(s, u(s, t, x)) - a(s, x)]ds + \int_t^{t+h} a(s, x)ds,$$

after adding and subtracting the term $a(s, x)$. Moreover, since $|\cdot|$ is a convex function, by the Schwarz inequality,

$$\left|\int_t^{t+h} E[a(s, u(s, t, x)) - a(s, x)]ds\right|$$

$$\leq \int_t^{t+h} E[|a(s, u(s, t, x)) - a(s, x)|]ds$$

---

[9] The assumption $|f(z)| \leq K(1 + |z|^{2n})$ implies that $E[|f(z)|] \leq K(1 + E[|z|^{2n}])$ and, by Theorem 4.15, $E[|u(t + h, t, x)|^{2n}] < +\infty$. Therefore, $f(u(t + h, t, x) - x)$ is integrable.

$$\leq h^{\frac{1}{2}} \left(\int_t^{t+h} (E[|a(s,u(s,t,x)) - a(s,x)|])^2 ds \right)^{\frac{1}{2}}$$

$$\leq h^{\frac{1}{2}} \left(\int_t^{t+h} E[|a(s,u(s,t,x)) - a(s,x)|^2] ds \right)^{\frac{1}{2}}.$$

Then, by Hypothesis 1 of Theorem 4.4,

$$|a(s,u(s,t,x)) - a(s,x)|^2 \leq (K^*)^2 |u(s,t,x) - x|^2,$$

and, by point 2 of Theorem 4.15,

$$E[|u(s,t,x) - x|^2] \leq Kh(1 + |x|^2), \ K \ \text{constant, positive,}$$

and thus for $h \downarrow 0$

$$\frac{1}{h} \left| \int_t^{t+h} E[a(s,u(s,t,x)) - a(s,x)] ds \right| \leq \frac{1}{h} h^{\frac{1}{2}} K^* (hKh(1 + |x|^2))^{\frac{1}{2}} \to 0.$$

Hence, as a conclusion, by the mean value theorem for $t \leq r \leq t+h$,

$$\lim_{h \downarrow 0} \frac{1}{h} E[u(t+h,t,x) - x] = \lim_{h \downarrow 0} \frac{1}{h} \int_t^{t+h} a(s,x) ds = \lim_{h \downarrow 0} \frac{1}{h} a(t,x)h = a(t,x).$$

Lastly, we have to show that the assumptions of Lemma 2.106 are satisfied (e.g., Friedman 1975). □

Strong Markov Property of Solutions of Stochastic Differential Equations

Lemma 4.29. *By Hypotheses 1 and 2 of Theorem 4.4, we have that*

$$\forall R > 0, \forall T > 0: \quad E \left[\sup_{r \leq t \leq T} |u(t,s,x) - u(t,r,y)|^2 \right] \leq C(|x - y|^2 + |s - r|)$$

for $|x| \leq R$, $|y| \leq R$, $0 \leq s \leq r \leq T$, where C is a constant that depends on R and T.

Proof. See, e.g., Friedman (1975). □

Theorem 4.30. *By Hypotheses 1 and 2 of Theorem 4.4, $(u(t,s,x))_{t \in [s,T]}$, the solution of*

$$du(t) = a(t,u(t)) dt + b(t,u(t)) dW_t$$

satisfies the Feller property and, hence, the strong Markov property.

Proof. Let $f \in BC(\mathbb{R})$. By the Lebesgue theorem, we have

$$E[f(u(t+r,s,x))] \to E[f(u(t+s,s,x))] \text{ for } r \to s.$$

Moreover, by Lemma 4.29, and again by the Lebesgue theorem,

$$E[f(u(t+r,r,y))] - E[f(u(t+r,s,x))] \to 0 \text{ for } y \to x, r \to s;$$

therefore,

$$E[f(u(t+r,r,y))] - E[f(u(t+s,s,x))] \to 0 \text{ for } y \to x, r \to s.$$

Hence $(s,x) \to \int_{\mathbb{R}} p(s,x,s+t,dy)f(y)$ is continuous, and so $(u(t,s,x))_{t \in [s,T]}$ satisfies the Feller property and, by Theorem 2.98 (because it is continuous) has the strong Markov property. $\qquad\square$

4.3 Girsanov Theorem

The Girsanov theorem is an interesting result in that it states that the addition of a drift to a standard Brownian motion with respect to a law P leads to a Brownian motion with respect to a new probability law Q that is absolutely continuous with respect to P.

Lemma 4.31. *Let Z be a strictly positive random variable on $(\Omega, \mathcal{F}, P)$ with $E[Z] \equiv E_P[Z] = 1$. Furthermore, define the random measure $dQ = ZdP$. If $\mathcal{G}$ is a σ-algebra with $\mathcal{G} \subseteq \mathcal{F}$, then for any random variable X on $(\Omega, \mathcal{F})$ such that $X \in \mathcal{L}^1(Q)$ we have that*

$$E_Q[X|\mathcal{G}] = \frac{E[XZ|\mathcal{G}]}{E[Z|\mathcal{G}]}.$$

Proof. This easily follows from the definition of conditional expectation of a random variable with respect to a σ-algebra (e.g., Øksendal 1998, p. 152). $\square$

Lemma 4.32. *Let $(\mathcal{F}_t)_{t \in [0,T]}$, for $T > 0$, be a filtration on the probability space $(\Omega, \mathcal{F}, P)$, and let $(Z_t)_{t \in [0,T]}$ be a strictly positive $\mathcal{F}_t$-martingale with respect to the probability measure P such that $E_P[Z_t] = 1$ for any $t \in [0,T]$. A sufficient condition for an adapted stochastic process $(Y_t)_{t \in [0,T]}$ to be an $\mathcal{F}_t$-martingale with respect to the measure $dQ = Z_T dP$ is that the process $(Z_t Y_t)_{t \in [0,T]}$ is an $\mathcal{F}_t$-martingale with respect to P.*

Proof. Because $(Z_t Y_t)_{t \in [0,T]}$ is an $\mathcal{F}_t$-martingale with respect to P, for $s \le t \le T$, by the tower law of probability, we have that

$$E[Z_T Y_t|\mathcal{F}_s] = E[E[Z_T Y_t|\mathcal{F}_t]|\mathcal{F}_s] = E[Y_t E[Z_T|\mathcal{F}_t]|\mathcal{F}_s] = E[Y_t Z_t|\mathcal{F}_s]$$
$$= Y_s Z_s.$$

As a consequence we have that

$$E_Q[Y_t|\mathcal{F}_s] = \frac{E[Z_T Y_t|\mathcal{F}_s]}{E[Z_T|\mathcal{F}_s]} = \frac{Z_s Y_s}{Z_s} = Y_s.$$

$\square$

Proposition 4.33

1. Let $h_t \in L^2([0,T])$ be a deterministic function and $W_t(\omega)$ a Brownian motion, and define

$$Y_t(\omega) = \exp\left\{\int_0^t h_s dW_s(\omega) - \frac{1}{2}\int_0^t h_s^2 ds\right\}, \qquad t \in [0,T].$$

Then, by Itô's formula [see (3.33)],

$$dY_t = Y_t h_t dW_t.$$

2. Let $\vartheta \in \mathcal{C}([0,T])$, with $T \leq \infty$, and define

$$Z_t(\omega) = \exp\left\{\int_0^t \vartheta_s(\omega)dW_s(\omega) - \frac{1}{2}\int_0^t \vartheta_s^2(\omega)ds\right\}, \qquad t \in [0,T].$$

Then, by Itô's formula,

$$dZ_t = Z_t \vartheta_t dW_t.$$

Lemma 4.34. (Novikov condition). *Under the assumptions of point 2 of Proposition 4.33, if*

$$E\left[\exp\left\{\frac{1}{2}\int_0^T |\vartheta(s)|^2 ds\right\}\right] < +\infty,$$

then $(Z_t)_{t\in[0,T]}$ *is a martingale and* $E[Z_t] = E[Z_0] = 1$.

Theorem 4.35 (Girsanov). *Let* $(Z_t)_{t\in[0,T]}$ *be a P-martingale (i.e., let* ϑ_s *satisfy the Novikov condition). Then the process*

$$Y_t = W_t - \int_0^t \vartheta_s ds$$

is a Brownian motion with respect to the measure $dQ = Z_T dP$.

Proof. We resort to the Lévy characterization of Brownian motion proving point 2 in Theorem 2.177. Let $M_t = Z_t Y_t$. Then, by Lemma 4.32, to prove that $(Y_t)_{t\in[0,T]}$ is a Q-martingale, it is sufficient to show that $(M_t)_{t\in[0,T]}$ is a P-martingale. Assuming that $(\vartheta_t)_{t\in[0,T]}$ satisfies the Novikov condition and that $(Z_t)_{t\in[0,T]}$ is a martingale with $E[Z_t] = 1$, by Itô's formula, we obtain

$$dM_t = Z_t dY_t + Y_t dZ_t + Z_t \vartheta_t dt = Z_t(dW_t - \vartheta_t dt) + Y_t Z_t \vartheta_t dW_t + Z_t \vartheta_t dt$$
$$= Z_t(dW_t + Y_t \vartheta_t dW_t) = Z_t(1 + \vartheta_t Y_t)dW_t.$$

Hence $(M_t)_{t \in [0,T]}$ is a martingale. Further showing that $Y_t^2 - t$ is a martingale is left as an exercise. $\square$

Remark 4.36. The Girsanov theorem implies that for all $F_1, \ldots, F_n \in \mathcal{B}$, where $\mathcal{B}$ is the Borel σ-algebra on the state space of the processes, and for all $t_1, \ldots, t_n \in [0, T]$:

$$Q(Y_{t_1} \in F_1, \ldots, Y_{t_k} \in F_k) = P(W_{t_1} \in F_1, \ldots, W_{t_k} \in F_k)$$

and $Q \ll P$ as well as with the Radon–Nikodym derivative

$$\frac{dQ}{dP} = Z_T, \qquad \text{on } \mathcal{F}_T.$$

Furthermore, because by the Radon–Nikodym Theorem A.54

$$Q(F) = \int_F Z_T(\omega)P(d\omega)$$

and $Z_T > 0$, we have that

$$Q(F) > 0 \Rightarrow P(F) > 0$$

and vice versa, so that

$$Q(F) = 0 \Rightarrow P(F) = 0,$$

and thus $P \ll Q$. Therefore, the two measures are equivalent.

Application to the Statistics of Stochastic Differential Equations

Consider an SDE of the form

$$dXt = \vartheta a(X(t))dt + dW_t, \tag{4.29}$$

where the parameter is unknown. We wish to estimate such a parameter based on the observation of the process $X(t)$, solution of (4.29), during a time interval $[0, T]$.

If we assume that the real-valued function a belongs to $\mathcal{C}_1$ and satisfies a global Lipschitz condition, then Proposition 4.9 assures that for any given square-integrable random variable X_0 independent of the Wiener process, taken as initial condition, SDE (4.29) admits a unique global solution $X \in \mathcal{C}([0, +\infty))$.

If we assume that, for any given $T > 0$, the function a satisfies the Novikov condition

$$E\left[\exp\left\{\frac{1}{2}\vartheta^2 \int_0^T a^2(X((s))ds\right\}\right] < +\infty,$$

we may apply Girsanov's Theorem 4.35 and state that the process $X(t)$ solution of (4.29) in the interval $[0, T]$, subject to the initial condition X_0, is a standard Wiener process with respect to the probability measure

$$Q = Z_T P_X^\vartheta$$

if P_X^ϑ is the law of the process $X(t)$ and

$$Z_t = \exp\left\{-\vartheta \int_0^t a(X(s))dW_s - \frac{1}{2}\vartheta^2 \int_0^t a^2(X(s))ds\right\}, \quad t \in [0, T]. \quad (4.30)$$

If we take into account (4.29), then we may rewrite (4.30) in the following form:

$$Z_t = \exp\left\{-\vartheta \int_0^t a(X(s))dX(s) + \frac{1}{2}\vartheta^2 \int_0^t a^2(X(s))ds\right\}, \quad t \in [0, T].$$

We may now notice that, since $Z_T > 0$, we may state (Lipster and Shiryaev 1977, p. 237) that the two measures P_X^ϑ and Q are equivalent on the time interval $[0, T]$, and Z_T^{-1} is the density of P_X^ϑ with respect to Q, i.e.,

$$\frac{dP_X^\vartheta}{dQ} = Z_T^{-1}.$$

Since the measure Q is now independent of the unknown parameter ϑ, while P_X^ϑ is the measure associated with the process and depends upon the parameter, we can claim that Z_T^{-1} may play the role of the likelihood function with respect to ϑ, i.e., given a trajectory of the process $\{X(t), t \in [0, T]\}$, during a finite interval of time, the corresponding (random) likelihood function is

$$L_T(\vartheta, X) = \exp\left\{\vartheta \int_0^T a(X(s))dX(s) - \frac{1}{2}\vartheta^2 \int_0^T a^2(X(s))ds\right\}.$$

By maximization, we obtain the maximum likelihood estimator of ϑ

$$\hat{\Theta}(X, T) = \frac{\int_0^T a(X(s))dX(s)}{\int_0^T a^2(X(s))ds}$$

as a random variable depending on the random process up to time T.

The following theorem holds.

Theorem 4.37. *Under the preceding assumptions, the maximum likelihood estimator $\hat{\Theta}(X, T)$ is consistent for ϑ, i.e., for any $\vartheta \in \mathbb{R}$,*

$$P_\vartheta(\lim_{T \to \infty} \hat{\Theta}(X, T) = \vartheta) = 1.$$

Proof. See Lipster and Shiryaev (2010, p. 234). □

For a more detailed account of this topic, the reader may refer to p. 225 of Lipster and Shiryaev (2010).

4.4 Kolmogorov Equations

This section is devoted to establishing evolution equations for the transition probabilities of Markov processes that are solutions of SDEs; as a natural fallout we obtain the infinitesimal generators of the evolution operators of such processes.

We will consider the SDE

$$du(t) = a(t, u(t))dt + b(t, u(t))dW_t \tag{4.31}$$

and suppose that the coefficients a and b satisfy the assumptions of the existence and uniqueness Theorem 4.4. We will denote by $u(t, x)$, for $s \leq t \leq T$, the solution of (4.31) subject to the initial condition

$$u(s, s, x) = x \text{ a.s. } (x \in \mathbb{R}).$$

Remark 4.38. Under assumptions 1 and 2 of Theorem 4.4 on the coefficients a and b, if $f(t, x)$ is continuous in both variables as well as $|f(t, x)| \leq K(1 + |x|^m)$ with k, m positive constants, then it can be shown that

$$\lim_{h \downarrow 0} \frac{1}{h} \int_h^{t+h} E[f(s, u(s, t, x))]ds = f(t, x), \tag{4.32}$$

$$\lim_{h \downarrow 0} \frac{1}{h} \int_{t-h}^{t} E[f(s, u(s, t, x))]ds = f(t, x).$$

The proof employs arguments similar to the proofs of Theorems 4.28 and 4.15.

Lemma 4.39. *If $f : \mathbb{R} \to \mathbb{R}$ is a twice continuously differentiable function, and if there exist $C > 0$ and $m > 0$ such that*

$$|f(x)| + |f'(x)| + |f''(x)| \leq C(1 + |x|^m), \qquad x \in \mathbb{R},$$

and if the coefficients $a(t, x)$ and $b(t, x)$ satisfy assumptions 1 and 2 of Theorem 4.4, then

$$\lim_{h \downarrow 0} \frac{1}{h} \left(E[f(u(t, t - h, x))] - f(x)\right) = a(t, x)f'(x) + \frac{1}{2}b^2(t, x)f''(x). \tag{4.33}$$

Proof. By Itô's formula, we get

$$f(u(t, t - h, x)) - f(x) = \int_{t-h}^{t} a(s, u(s, t - h, x)) f'(u(s, t - h, x)) ds$$

$$+ \int_{t-h}^{t} \frac{1}{2} b^2(s, u(s, t - h, x)) f''(u(s, t - h, x)) ds$$

$$+ \int_{t-h}^{t} b(s, u(s, t - h, x)) f'(u(s, t - h, x)) dW_s,$$

and after taking expectations

$$E[f(u(t, t - h, x))] - f(x)$$

$$= E\left[\int_{t-h}^{t} a(s, u(s, t - h, x)) f'(u(s, t - h, x)) ds \right.$$

$$\left. + \int_{t-h}^{t} \frac{1}{2} b^2(s, u(s, t - h, x)) f''(u(s, t - h, x)) ds \right],$$

hence

$$\frac{1}{h} \left(E[f(u(t, t - h, x))] - f(x) \right)$$

$$= \frac{1}{h} \int_{t-h}^{t} E[a(s, u(s, t - h, x)) f'(u(s, t - h, x))] ds$$

$$+ \int_{t-h}^{t} E\left[\frac{1}{2} b^2(s, u(s, t - h, x)) f''(u(s, t - h, x)) \right] ds.$$

Then (4.33) follows from (4.32) because $u(t, t, x) = x$. $\qquad\square$

Remark 4.40. Resorting to the notation of Definitions 2.81 and 2.87, (4.33) can also be written as

$$\mathcal{A}_t f = \lim_{h \downarrow 0} \frac{T_{t-h, t} f - f}{h} = a(t, \cdot) f' + \frac{1}{2} b^2(t, \cdot) f''.$$

Moreover, by Theorem 4.28 and Proposition 2.107, we also have

$$\mathcal{A}_s f = \lim_{h \downarrow 0} \frac{T_{s, s+h} f - f}{h} = a(s, \cdot) f' + \frac{1}{2} b^2(s, \cdot) f'' \qquad (4.34)$$

if $f \in BC(\mathbb{R}) \cap C^2(\mathbb{R})$. On the other hand, in the time-homogeneous case we have

$$\mathcal{A} f = \lim_{h \downarrow 0} \frac{T_h f - f}{h} = a(\cdot) f' + \frac{1}{2} b^2(\cdot) f''.$$

Theorem 4.41. *If $u(t)$ is the Markovian solution of the homogeneous SDE (4.11) and $\mathcal{A}$ the associated infinitesimal generator, then for $f \in BC(\mathbb{R}) \cap C^2(\mathbb{R})$ the process*

$$M_t = f(u(t)) - \int_0^t [\mathcal{A}f](u(s))ds \qquad (4.35)$$

is a martingale.

Proof. By Itô's formula, we have that

$$f(u(t)) = f(u^0) + \int_0^t [\mathcal{A}f](u(s))ds + \int_0^t b(u(s))f'(u(s))dW_s,$$

which, substituted into (4.35), results in

$$M_t = f(u^0) + \int_0^t b(u(s))f'(u(s))dW_s.$$

Since an Itô integral is a martingale with respect to filtration $(\mathcal{F}_t)_{t\in\mathbb{R}_+}$ generated by the Wiener process $(W_t)_{t\in\mathbb{R}_+}$, therefore

$$E[M_t|\mathcal{F}_s] = M_s.$$

If we now consider the filtration $(\mathcal{M}_t)_{t\in[0,T]}$, generated by $(M_t)_{t\in[0,T]}$, then

$$E[M_t^-|\mathcal{M}_s] = E[E[M_t|\mathcal{F}_s]|\mathcal{M}_s] = E[M_t|\mathcal{F}_s] = M_s,$$

because $\mathcal{F}_s \subset \mathcal{M}_s$. □

Furthermore, we note that it is valid to reverse the argumentation of Theorem 4.28, as the following theorem states.

Theorem 4.42. *If $(u(t))_{t\in[0,T]}$ is a diffusion process with drift $a(t,x)$ and diffusion coefficient $c(t,x)$, where*

1. *$a(t,x)$ is continuous in both variables as well as $|a(t,x)| \leq K(1+|x|)$, K a positive constant.*
2. *$c(t,x)$ is continuous in both variables and has continuous as well as bounded derivatives $\frac{\partial}{\partial t}c(t,x)$ and $\frac{\partial}{\partial x}c(t,x)$, and, moreover, $\frac{1}{c(t,x)}$ is bounded.*
3. *There exists a function $\psi(x)$ that is independent of t and where*

$$\psi(x) > 1 + |x|, \qquad \sup_{0\leq t\leq T} E[\psi(u(t))] < +\infty,$$

as well as

$$\left|\int_\Omega (y-x)p(t,x,t+h,dy)\right| + \left|\int_\Omega (y-x)^2 p(t,x,t+h,dy)\right| \leq \psi(x)h,$$

$$\int_\Omega (|y|+y^2)p(t,x,t+h,dy) \leq \psi(x),$$

then there exists a Wiener process W_t, so that $u(t)$ satisfies the SDE

$$du(t) = a(t, u(t))dt + \sqrt{c(t, u(t))}dW_t.$$

Proof. See, e.g., Gihman and Skorohod (1972).   □

Remark 4.43. Equation (4.34) can also be shown by Itô's formula, as in the proof of Lemma 4.39.

Proposition 4.44. *Let $f(x)$ be r times differentiable and suppose that there exists an $m > 0$ such that $|f^{(k)}(x)| \leq L(1 + |x|^m)$. If $a(t, x)$ and $b(t, x)$ both satisfy the assumptions of Theorem 4.4 and there exist $\frac{\partial^k}{\partial x^k}a(t, x), \frac{\partial^k}{\partial x^k}b(t, x)$, $k = 1, \ldots, r$, that are continuous, as well as*

$$\left|\frac{\partial^k}{\partial x^k}a(t, x)\right| + \left|\frac{\partial^k}{\partial x^k}b(t, x)\right| \leq C_k(1 + |x|^{m_k}), \qquad k = 1, \ldots, r$$

(with C_k and m_k being positive constants), then the function

$$\phi_s(z) = E[f(u(t, s, z))]$$

is r times differentiable with respect to z (i.e., with respect to the initial condition).

Proof. See, e.g., Gihman and Skorohod (1972).   □

Theorem 4.45. *If the coefficients $a(t, x)$ and $b(t, x)$ are continuous and have continuous partial derivatives $a_x(t, x)$, $a_x(t, x)$, $b_x(t, x)$, and $b_{xx}(t, x)$; and, moreover, if there exist a $k > 0$ and an $m > 0$ such that*

$$|a(t, x)| + |b(t, x)| \leq k(1 + |x|),$$
$$|a_x(t, x)| + |a_{xx}(t, x)| + |b_x(t, x)| + |b_{xx}(t, x)| \leq k(1 + |x|^m),$$

and furthermore if the function $f(x)$ is twice continuously differentiable with

$$|f(x)| + |f'(x)| + |f''(x)| \leq k(1 + |x|^m),$$

then the function

$$q(t, x) \equiv E[f(u(s, t, x))], \qquad 0 < t < s, \qquad x \in \mathbb{R}, s \in]0, T[,$$

satisfies the equation

$$\frac{\partial}{\partial t}q(t, x) + a(t, x)\frac{\partial}{\partial x}q(t, x) + \frac{1}{2}b^2(t, x)\frac{\partial^2}{\partial x^2}q(t, x) = 0, \qquad (4.36)$$

subject to the condition

$$\lim_{t \uparrow s} q(t, x) = f(x). \qquad (4.37)$$

Equation (4.36) is called Kolmogorov's backward differential equation.

Proof. Since, by the semigroup property, $u(s, t - h, x) = u(s, t, u(t, t - h, x))$, and in general $E[f(Y(\cdot, X))|X = x] = E[f(Y(\cdot, x))]$, we have

$$q(t - h, x) = E[f(u(s, t - h, x))] \tag{4.38}$$
$$= E[E[f(u(s, t - h, x))|u(t, t - h, x)]]$$
$$= E[E[f(u(s, t, u(t, t - h, x)))|u(t, t - h, x)]]$$
$$= E[E[f(u(s, t, u(t, t - h, x)))]] = E[q(t, u(t, t - h, x))].$$

By Proposition 4.44, $q(t, x)$ is twice differentiable with respect to x, and, by Lemma 4.39, we get

$$\lim_{h \downarrow 0} \frac{E[q(t, u(t, t - h, x))] - q(t, x)}{h} = a(t, x) \frac{\partial}{\partial x} q(t, x) + \frac{1}{2} b^2(t, x) \frac{\partial^2}{\partial x^2} q(t, x).$$

Therefore, by (4.38), the limit

$$\lim_{h \downarrow 0} \frac{q(t, x) - q(t - h, x)}{h} = \lim_{h \downarrow 0} \frac{q(t, x) - E[q(t, u(t, t - h, x))]}{h},$$

and thus

$$\frac{\partial}{\partial t} q(t, x) = \lim_{h \downarrow 0} \frac{q(t, x) - q(t - h, x)}{h} = -a(t, x) \frac{\partial}{\partial x} q(t, x) - \frac{1}{2} b^2(t, x) \frac{\partial^2}{\partial x^2} q(t, x).$$

It can further be shown that $\frac{\partial}{\partial t} q(t, x)$ is continuous in t, as are $\frac{\partial q}{\partial x}$ as well as $\frac{\partial^2 q}{\partial x^2}$. We observe that

$$|E[f(u(s, t, x)) - f(x)]| \leq E[|f(u(s, t, x)) - f(x)|],$$

and, by Lagrange's theorem (also known as the *mean value theorem*),

$$|f(u(s, t, x)) - f(x)| = |u(s, t, x) - x||f'(\xi)|,$$

with ξ related to $u(s, t, x)$ and x through the assumptions $|f'(\xi)| \leq k(1 + |\xi|^m)$ and

$$(1 + |\xi|^m) \leq \begin{cases} 1 + |x|^m & \text{if } u(s, t, x) \leq \xi \leq x, \\ 1 + |u(s, t, x)|^m & \text{if } x \leq \xi \leq u(s, t, x). \end{cases}$$

Therefore, by both the Schwarz inequality and the fact that

$$E[(u(s, t, x) - x)^2] \leq \tilde{L}(1 + |x|^2)(s - t)^2,$$

we obtain

$$|E[f(u(s,t,x)) - f(x)]|$$
$$\leq LE[|u(s,t,x) - x|(1 + |x|^m + |u(s,t,x)|^m)]$$
$$\leq L(E[(u(s,t,x) - x)^2])^{\frac{1}{2}} (E[(1 + |x|^m + |u(s,t,x)|^m)^2])^{\frac{1}{2}},$$

where L is a positive constant. Since $\tilde{L}(1+|x|^2)(s-t)^2 \to 0$ for $t \uparrow s$, it follows that

$$\lim_{t \uparrow s} E[f(u(s,t,x))] = f(x).$$

$\square$

Remark 4.46. If we put $\tilde{t} = s - t$ for $0 < t < s$, then $\frac{\partial}{\partial t} = -\frac{\partial}{\partial \tilde{t}}$ and the limit $\lim_{t \uparrow s}$ is equivalent to $\lim_{\tilde{t} \downarrow 0}$. Hence (4.36) takes us back to a classic parabolic differential equation with initial condition (4.37) given by $\lim_{\tilde{t} \downarrow 0} q(\tilde{t}, x) = f(x)$.

Theorem 4.47 (Feynman–Kac formula). *Under the assumptions of Theorem 4.45, let c be a real-valued, nonnegative continuous function in $]0, T[\times \mathbb{R}$. Then the function, for $x \in \mathbb{R}$,*

$$q(t,x) = E\left[f(u(s,t,x)) e^{-\int_t^s c(u(\tau,t,x),\tau)d\tau} \right], \qquad 0 < t < s < T, \qquad (4.39)$$

satisfies the equation

$$\frac{\partial}{\partial t} q(t,x) + a(t,x)\frac{\partial}{\partial x} q(t,x) + \frac{1}{2}b^2(t,x)\frac{\partial^2}{\partial x^2} q(t,x) - c(t,x)q(t,x) = 0,$$

subject to the boundary condition $\lim_{t \uparrow s} q(t,x) = f(x)$. Equation (4.39) is called the Feynman–Kac formula.

Proof. The proof is a direct consequence of Theorem 4.45 and Itô's formula, considering that the process

$$Z(t) = e^{-\int_t^s c(\tau, u(\tau,t,x))d\tau}, \qquad 0 < t < s < T, x \in \mathbb{R},$$

satisfies the SDE

$$dZ(t) = -c(t, u(t,t_0,x))Z(t)dt$$

with initial condition $Z(t_0) = 1$ (see e.g. Pascucci (2008)). $\square$

Remark 4.48. We can interpret the exponential term in the Feynman–Kac formula as due to a *killing* process (e.g., Schuss 2010). Suppose that at any time $\tau > t$ the trajectory $u(\tau,t,x)$ of a particle subject to the SDE (4.31), with initial condition $u(t,t,x) = x$, may terminate at a rate $c(\tau, u(\tau,t;x))$ (probability per unit time independent of past history $\mathcal{F}_\tau$); hence, the *killing* probability over an interval $]\tau, \tau + dt]$ will be equal to $c(\tau, u(\tau,t,x))dt + o(dt)$. Then the survival probability until s is given by

$$(1-c(t_1,u(t_1,t,x))dt)(1-c(t_2,u(t_2,t,x))dt)\cdots(1-c(t_n,u(t_n,t,x))dt)+o(dt),$$
$$(4.40)$$

where $t = t_0 < t_1 < \cdots < t_n = s$, $dt = t_{i+1} - t_i$, $i = 0,1,\ldots,n-1$. As $dt \to 0$, (4.40) tends to

$$e^{-\int_t^s c(\tau,u(\tau,t,x))d\tau}.$$

Hence for any function $f \in BC(\mathbb{R})$

$$\begin{aligned}
q(t,x) &= E[f(u(s,t,x)), \text{ killing time} > s] \\
&= E[f(u(s,t,x))]P(\text{killing time} > s) \\
&= E\left[f(u(s,t,x))e^{-\int_t^s c(\tau,u(\tau,t,x))d\tau}\right].
\end{aligned}$$

Introduce the following operator as from (4.36):

$$L_0[\cdot] = \frac{1}{2}b^2(t,x)\frac{\partial^2}{\partial x^2} + a(t,x)\frac{\partial}{\partial x},$$

and suppose that (Appendix C)

(A_1) There exists a $\mu > 0$ such that $b(x,t) \geq \mu$ for all $(x,t) \in \mathbb{R} \times [0,T]$.

(B_1) a and b are bounded in $[0,T] \times \mathbb{R}$ and uniformly Lipschitz in (t,x) on compact subsets of $[0,T] \times \mathbb{R}$.

(B_2) b is Hölder continuous in x and uniform with respect to (t,x) on $[0,T] \times \mathbb{R}$.

Proposition 4.49. *Consider the Cauchy problem:*

$$\begin{cases} L_0[q] + \frac{\partial q}{\partial t} = 0 & \text{in } [0,T[\times\mathbb{R}, \\ \lim_{t\uparrow T} q(t,x) = \phi(x) & \text{in } \mathbb{R}, \end{cases} \tag{4.41}$$

where $\phi(x)$ is a continuous function on $\mathbb{R}$, and there exist $A > 0, a > 0$ such that

$$|\phi(x)| \leq A(1 + |x|^a). \tag{4.42}$$

Under conditions (A_1), (B_1), and (B_2), the Cauchy problem (4.41) admits a unique solution $q(t,x)$ in $[0,T] \times \mathbb{R}$ such that

$$|q(t,x)| \leq C(1 + |x|^a),$$

where C is a constant.

If we denote by $\Gamma_0^*(x,s;y,t)$ the fundamental solution of $L_0 + \frac{\partial}{\partial s}$ ($s < t$), the solution of the Cauchy problem (4.41) can be expressed as follows:

$$q(t,x) = \int_{\mathbb{R}} \Gamma_0^*(x,t;y,T)\phi(y)dy. \tag{4.43}$$

Proof. The uniqueness is shown through Corollary D.7, and existence follows from Theorem D.10. Then (4.41) follows, by Theorem D.9, with $m = 0$.

The representation (4.43) follows from Theorem D.10, by replacing t by $T - t$ (Friedman 2004, Chap. 6). □

By a direct comparison of the Cauchy problem (4.41) and problem (4.36), (4.37), because of the uniqueness of the solution of (4.41), we may finally state the following.

Theorem 4.50. *Under the assumptions of Proposition 4.49, the solution of the Cauchy problem (4.41) is given by*

$$q(t, x) = E[\phi(u(T, t, x))] \equiv E_{t,x}[\phi(u(T))]. \tag{4.44}$$

From (4.44) and (4.43) it then follows that

$$E[\phi(u(t, s, x))] = \int_{\mathbb{R}} \Gamma_0^*(x, s; y, t)\phi(y)dy$$

or, equivalently,

$$\int_{\mathbb{R}} \phi(y)p(s, x, t, dy) = \int_{\mathbb{R}} \Gamma_0^*(x, s; y, T)\phi(y)dy, \tag{4.45}$$

and because (4.45) holds for an arbitrary ϕ that satisfies (4.42), we may state the following theorem.

Theorem 4.51. *Under conditions (A_1) of Appendix D and (B_1), the transition probability $p(s, x, t, A) = P(u(t, s, x) \in A)$ of the Markov process $u(t, s, x)$ [the solution of differential equation (4.31)] admits a density. The latter is given by $\Gamma_0^*(x, s; y, t)$, and thus*

$$p(s, x, t, A) = \int_A \Gamma_0^*(x, s; y, t)dy \qquad (s < t), \text{ for all } A \in \mathcal{B}_{\mathbb{R}}. \tag{4.46}$$

Definition 4.52. The density $\Gamma_0^*(x, s; y, t)$ of $p(s, x, t, A)$ is the *transition density* of the solution $u(t)$ of (4.31).

Remark 4.53. By the definition of fundamental solution, we may realize that the transition density $\Gamma_0^*(x, s; y, t)$ of the Markov process associated with SDE (4.31) obeys itself to the following Kolmogorov backward equation:

$$\frac{1}{2}b^2(t, x)\frac{\partial^2}{\partial x^2}\Gamma_0^*(x, t; y, T) + a(t, x)\frac{\partial}{\partial x}\Gamma_0^*(x, t; y, T) + \frac{\partial}{\partial t}\Gamma_0^*(x, t; y, T) = 0, \tag{4.47}$$

for $x \in \mathbb{R}$, $t \in [0, T]$, subject to

$$\lim_{t \to T} \Gamma_0^*(x, t; y, T) = \delta(x - y).$$

where we recall that δ denotes the Dirac δ function centered at 0.

As a direct consequence of (4.46), the transition density $\Gamma_0^*(x, s; y, t)$ satisfies the Chapman–Kolmogorov equation.

Corollary 4.54. *For any* $s, r, t \in [0, T]$ *such that* $s < r < t$, *the following holds:*

$$\Gamma_0^*(x, s; y, t) = \int_{\mathbb{R}} dz \; \Gamma_0^*(x, s; z, r) \Gamma_0^*(z, r; y, t).$$

Example 4.55. The Brownian motion $(W_t)_{t \geq 0}$ is the solution of

$$\begin{cases} du(t) = dW_t, \\ u(0) = 0 \text{ a.s.} \end{cases}$$

We define the operator L_0 by $\frac{1}{2}\Delta$, where Δ is the Laplacian $\frac{\partial^2}{\partial x^2}$. The fundamental solution $\Gamma_0^*(x, s; y, t)$ of the operator $\frac{1}{2}\Delta + \frac{\partial}{\partial t}$, $s < t$, corresponds to the fundamental solution $\Gamma_0(y, t; x, s)$ of the operator $\frac{1}{2}\Delta - \frac{\partial}{\partial t}$, $s < t$, which, apart from the coefficient $\frac{1}{2}$, is the diffusion or heat operator. We therefore find that

$$\Gamma_0^*(x, s; y, t) = \Gamma(y, t; x, s) = \frac{1}{\sqrt{2\pi(t - s)}} e^{-\frac{(x-y)^2}{2(t-s)}},$$

the probability density function of $W_t - W_s$.

Under the assumptions of Theorem 4.51, the transition probability

$$p(s, x, t, A) = P(u(t, s, x) \in A)$$

of the Markov diffusion process $u(t, s, x)$, the latter being the solution of the SDE (4.31), subject to the initial condition $u(s, s, x) = x$ a.s. ($x \in \mathbb{R}$), admits a density $\Gamma_0^*(x, s; y, t)$, which is the solution of system (4.47). Under these conditions the following theorem also holds (Gihman and Skorohod 1974, p. 374$f\!f$):

Theorem 4.56. *In addition to the assumptions of Theorem 4.51, if the transition density* $\Gamma_0^*(x, s; y, t)$ *is sufficiently regular so that there exist continuous derivatives*

$$\frac{\partial \Gamma_0^*}{\partial t}(x, s; y, t), \qquad \frac{\partial}{\partial y}(a(t, y) \Gamma_0^*(x, s; y, t)), \qquad \frac{\partial^2}{\partial y^2}(b^2(t, y) \Gamma_0^*(x, s; y, t)),$$

then $\Gamma_0^*(x, s; y, t)$, *as a function of* t *and* y, *satisfies the equation*

$$\frac{\partial \Gamma_0^*}{\partial t}(x, s; y, t) + \frac{\partial}{\partial y}(a(t, y) \Gamma_0^*(x, s; y, t)) - \frac{1}{2}\frac{\partial^2}{\partial y^2}(b^2(t, y) \Gamma_0^*(x, s; y, t)) = 0 \tag{4.48}$$

in the region $t \in]s, T]$, $y \in \mathbb{R}$, *subject to*

$$\lim_{t \to s} \Gamma_0^*(x, s; y, t) = \delta(x - y).$$

Proof. Let $g \in C_0^2(\mathbb{R})$ denote a sufficiently smooth function with compact support. By proceeding as in Lemma 4.39 [see also (4.34)],

$$\lim_{h \to 0} \frac{1}{h} \left(\int g(y) \Gamma_0^*(x, t; y, t + h) dy - g(x) \right) = a(t, x) g'(x) + \frac{1}{2} b^2(t, y) g''(x)$$

uniformly with respect to x. The Chapman–Kolmogorov equation for the transition densities is

$$\Gamma_0^*(x, t_1; y, t_3) = \int \Gamma_0^*(x, t_1; z, t_2) \, \Gamma_0^*(z, t_2; y, t_3) dz \qquad \text{for } t_1 < t_2 < t_3.$$

If we take $t_1 = s$, $t_2 = t$, $t_3 = t + h$, then we obtain

$$\frac{\partial}{\partial t} \int \Gamma_0^*(x, s; y, t) g(y) dy$$

$$= \int \frac{\partial}{\partial t} \Gamma_0^*(x, s; y, t) g(y) dy$$

$$= \lim_{h \to 0} \frac{1}{h} \left(\int g(y) \Gamma_0^*(x, s; y, t + h) dy - \int g(z) \Gamma_0^*(x, s; z, t) dz \right)$$

$$= \lim_{h \to 0} \frac{1}{h} \left(\int \Gamma_0^*(x, s; z, t) \left(\int g(y) \Gamma_0^*(z, s; y, t + h) dy - g(z) \right) \right) dz$$

$$= \int \Gamma_0^*(x, s; z, t) \left(a(t, z) g'(z) + \frac{1}{2} b(t, z) g''(z) \right) dz.$$

An integration by parts leads to

$$\int \frac{\partial}{\partial t} \Gamma_0^*(x, s; y, t) g(y) dy$$

$$= \int \left(-\frac{\partial}{\partial y} (a(t, y) \Gamma_0^*(x, s; y, t)) + \frac{1}{2} \frac{\partial^2}{\partial y^2} (b^2(t, y) \Gamma_0^*(x, s; y, t)) \right) g(y) dy,$$

which represents a weak formulation of (4.48). $\qquad\square$

Equation (4.48) is known as the *forward Kolmogorov equation* or *Fokker–Planck equation*. While the forward equation has a more intuitive interpretation than the backward equation, the regularity conditions on the functions a and b are more stringent than those needed in the backward case. The problem of existence and uniqueness of the solution of the Fokker–Planck equation is not of an elementary nature, especially in the presence of boundary conditions. This suggests that the backward approach is more convenient than the forward approach from the viewpoint of analysis. For a discussion on the subject, we refer the reader to Feller (1971, p. 326*ff*), Sobczyk (1991, p. 34), and Taira (1988, p. 9). An extended treatment of the Fokker–Planck equation with a view on applications can be found in Risken (1989). A discussion on the Fokker–Planck equation associated with a Langevin system, and its Smoluchowski approximation can be found in Appendix C.

4.5 Multidimensional Stochastic Differential Equations

Let $\mathbf{a}(t, \mathbf{x}) = (a_1(t, \mathbf{x}), \dots, a_m(t, \mathbf{x}))'$ and $b(t, \mathbf{x}) = (b_{ij}(t, \mathbf{x}))_{i=1,\dots,m,j=1,\dots,n}$ be measurable functions with respect to $(t, \mathbf{x}) \in [0, T] \times \mathbb{R}^n$. An m-dimensional SDE is of the form

$$d\mathbf{u}(t) = \mathbf{a}(t, \mathbf{u}(t))dt + b(t, \mathbf{u}(t))d\mathbf{W}(t), \qquad (4.49)$$

with the initial condition

$$\mathbf{u}(0) = \mathbf{u}^0 \text{ a.s.},$$

where $\mathbf{u}^0$ is a fixed m-dimensional random vector. The entire theory of the one-dimensional case translates to the multidimensional case, with the norms redefined as

$$|\mathbf{a}|^2 = \sum_{i=1}^m |a_i|^2 \text{ if } \mathbf{a} \in \mathbb{R}^m,$$

$$|b|^2 = \sum_{i=1}^m \sum_{j=1}^n |b_{ij}|^2 \text{ if } b \in \mathbb{R}^{mn}.$$

Further, for $\boldsymbol{\alpha} = (\alpha_1, \dots, \alpha_m)$, we introduce the notation

$$D_{\mathbf{x}}^{|\alpha|} = \frac{\partial^{\alpha}}{\partial \mathbf{x}^{\alpha}} = \frac{\partial^{\alpha_1 + \dots + \alpha_m}}{\partial x_1^{\alpha_1} \cdots \partial x_m^{\alpha_m}}, \qquad |\alpha| = \alpha_1 + \dots + \alpha_m,$$

which, as an application of Itô's formula, gives the following result.

Theorem 4.57. *If for a system of SDEs the conditions of the existence and uniqueness theorem (analogous to Theorem 4.4) are satisfied and if*

1. *There exist $D_{\mathbf{x}}^{\alpha}\mathbf{a}(t, \mathbf{x})$ and $D_{\mathbf{x}}^{\alpha}b(t, \mathbf{x})$ continuous for $|\alpha| \leq 2$, with*

$$|D_{\mathbf{x}}^{\alpha}\mathbf{a}(t, \mathbf{x})| + |D_{\mathbf{x}}^{\alpha}b(t, \mathbf{x})| \leq k_0(1 + |\mathbf{x}|^{\beta}), \qquad |\alpha| \leq 2,$$

   where k_0, β are strictly positive constants.
2. *$f : \mathbb{R}^m \to \mathbb{R}$ is a function endowed with continuous derivatives to second order, with*

$$|D_{\mathbf{x}}^{\alpha}f(\mathbf{x})| \leq c(1 + |\mathbf{x}|^{\beta'}), \qquad |\alpha| \leq 2,$$

   where c, β' are strictly positive constants;

then, putting $q(t, \mathbf{x}) = E[f(\mathbf{u}(s, t, \mathbf{x}))]$ for $\mathbf{x} \in \mathbb{R}^m$ and $t \in]0, s[$, we have that $q_t, q_{x_i}, q_{x_i x_j}$ are continuous in $(t, \mathbf{x}) \in]0, s[\times \mathbb{R}^m$ and q satisfies the backward Kolmogorov equation

$$\mathcal{L}q(\mathbf{x}, t) = 0 \quad in \;]0, s[\times \mathbb{R}^m,$$
$$\lim_{t \uparrow s} q(t, \mathbf{x}) = f(\mathbf{x}) \quad in \; \mathbb{R}^m,$$

where

$$\mathcal{L} := \frac{\partial}{\partial t} + \sum_{i=1}^{m} a_i \frac{\partial}{\partial x_i} + \frac{1}{2} \sum_{i,j=1}^{m} (bb')_{ij} \frac{\partial^2}{\partial x_i \partial x_j}. \tag{4.50}$$

Applications of Itô's Formula: Dynkin's Formula

Following Theorem 3.72, if $\phi : (\mathbf{x}, t) \in \mathbb{R}^m \times \mathbb{R}_+ \to \phi(\mathbf{x}, t) \in \mathbb{R}$ is sufficiently regular, then we may apply Itô's formula to obtain

$$d\phi(\mathbf{u}(t), t) = \mathcal{L}\phi(\mathbf{u}(t), t)dt + \nabla_{\mathbf{x}}\phi(\mathbf{u}(t), t) \cdot b(t, \mathbf{u}(t))d\mathbf{W}(t).$$

By integration on the interval $[s, t] \subset \mathbb{R}$, we obtain

$$\phi(\mathbf{u}(t), t) - \phi(\mathbf{u}(s), s) = \int_s^t \mathcal{L}\phi(\mathbf{u}(\tau), \tau)d\tau + \int_s^t \nabla_{\mathbf{x}}\phi(\mathbf{u}(\tau), \tau) \cdot b(\tau, \mathbf{u}(\tau))d\mathbf{W}(\tau). \tag{4.51}$$

Since the Itô integral is a zero-mean martingale by Theorem 3.45, by applying expected values to both sides of the preceding formula, we get

$$E[\phi(\mathbf{u}(t), t)] - E[\phi(\mathbf{u}(s), s)] = E\left[\int_s^t \mathcal{L}\phi(\mathbf{u}(\tau), \tau)d\tau\right].$$

In particular, if $\mathbf{u}(t)$ is the solution of (4.49) subject to the initial condition $\mathbf{u}(s) = \mathbf{x}$, almost surely, for $s \in \mathbb{R}, \mathbf{x} \in \Omega$, then

$$\phi(\mathbf{x}, s) = E[\phi(\mathbf{u}(t), t)] - E\left[\int_s^t \mathcal{L}\phi(\mathbf{u}(\tau), \tau)d\tau\right].$$

Applications of Itô's Formula: First Hitting Times

Let $\Omega \subset \mathbb{R}^m$ and $\mathbf{u}(t)$ be the solution of (4.49) with the initial condition $\mathbf{u}(s) = \mathbf{x}$, almost surely, for $s \in \mathbb{R}, \mathbf{x} \in \Omega$. Putting

$$\tau_{\mathbf{x},s} = \inf\{t \geq s | \mathbf{u}(t) \in \partial\Omega\},$$

then $\tau_{\mathbf{x},s}$ is the *first hitting time of the boundary of Ω* or the *first exit time from Ω*. Because $\partial\Omega$ is a closed set, by Theorem 2.78, $\tau_{\mathbf{x},s}$ is a stopping time.

Following Theorem 3.72, if $\phi : (\mathbf{x}, t) \in \mathbb{R}^m \times \mathbb{R} \to \phi(\mathbf{x}, t) \in \mathbb{R}$ is sufficiently regular, then by applying (4.51) on the interval $[s, \tau_{\mathbf{x},s}]$,

$$\phi(\mathbf{u}(\tau_{\mathbf{x},s}), \tau_{\mathbf{x},s}) = \phi(\mathbf{x}, s) + \int_s^{\tau_{\mathbf{x},s}} \mathcal{L}\phi(\mathbf{u}(t'), t')dt'$$

$$+ \int_s^{\tau_{\mathbf{x},s}} \nabla_{\mathbf{x}}\phi(\mathbf{u}(t'), t') \cdot b(t', \mathbf{u}(t'))d\mathbf{W}(t'),$$

and after taking expectations

$$E[\phi(\mathbf{u}(\tau_{\mathbf{x},s}), \tau_{\mathbf{x},s})] = \phi(\mathbf{x}, s) + E\left[\int_s^{\tau_{\mathbf{x},s}} \mathcal{L}\phi(\mathbf{u}(t'), t')dt'\right]. \tag{4.52}$$

If we now suppose that ϕ satisfies the conditions

$$\begin{cases} \mathcal{L}\phi(\mathbf{x},t) = -1 \ \forall t \geq s, \forall \mathbf{x} \in \Omega, \\ \phi(\mathbf{x},t) = 0 \quad \forall \mathbf{x} \in \partial\Omega, \end{cases} \tag{4.53}$$

then, by (4.52), we get

$$E[\phi(\mathbf{u}(\tau_{\mathbf{x},s}),\tau_{\mathbf{x},s})] = \phi(\mathbf{x},s) - E[\tau_{\mathbf{x},s}] + s,$$

and by (4.53),

$$E[\phi(\mathbf{u}(\tau_{\mathbf{x},s}),\tau_{\mathbf{x},s})] = 0.$$

Thus

$$E[\tau_{\mathbf{x},s}] = s + \phi(\mathbf{x},s). \tag{4.54}$$

Equation (4.54) states in particular that, if $\phi(\mathbf{x},s)$ is a finite solution of problem (4.53) at point $(\mathbf{x},s) \in \Omega \times \mathbb{R}_+$, then the mean value $E[\tau_{\mathbf{x},s}]$ of the first exit time from Ω, for a trajectory of (4.49) started at point $\mathbf{x} \in \Omega$ at time $s \in \mathbb{R}_+$, is finite.

Based on this information, it makes sense to consider the problem of finding a stochastic representation of the solution $\psi(\mathbf{x},s)$ of the following problem:

$$\begin{cases} \mathcal{L}[\psi](\mathbf{x},t) = 0 \ \forall t \geq s, \forall \mathbf{x} \in \Omega, \\ \psi(\mathbf{x},t) = f(\mathbf{x}) \ \forall \mathbf{x} \in \partial\Omega. \end{cases}$$

By (4.52), we obtain

$$\phi(\mathbf{x},s) = E[f(\mathbf{u}(\tau_{\mathbf{x},s}),\tau_{\mathbf{x},s})], \tag{4.54bis}$$

which is *Kolmogorov's formula*.

Time-Homogeneous Case

If (4.49) is time-homogeneous [i.e., $\mathbf{a} = \mathbf{a}(\mathbf{x})$ and $b = b(\mathbf{x})$ do not explicitly depend on time], then the process $\mathbf{u}(t)$, namely, the solution of (4.49), is time-homogeneous. Without loss of generality we can assume that $s = 0$. Then (4.54) becomes

$$E[\tau_{\mathbf{x}}] = \phi(\mathbf{x}), \qquad \mathbf{x} \in \Omega, \tag{4.55}$$

which is *Dynkin's formula*. Notably, in this case, $\phi(\mathbf{x})$ is the solution of the elliptic problem

$$\begin{cases} L_0[\phi] = -1 \text{ in } \Omega, \\ \phi = 0 \quad\quad \text{ on } \partial\Omega, \end{cases} \tag{4.56}$$

where $L_0 = \sum_{i=1}^m a_i \frac{\partial}{\partial x_i} + \frac{1}{2}\sum_{i,j=1}^m (bb')_{ij} \frac{\partial^2}{\partial x_i \partial x_j}$.

As before, (4.55) states in particular that, if $\phi(\mathbf{x})$ is a finite solution of problem (4.56) at point $\mathbf{x} \in \Omega$, then the mean value $E[\tau_\mathbf{x}]$ of the first exit time from Ω, for a trajectory of (4.49) started at point $\mathbf{x} \in \Omega$, at time 0, is finite.

Based on this information, it makes sense to consider the problem of finding a stochastic representation of the solution $\psi(\mathbf{x})$ of the following elliptic problem:

$$\begin{cases} L_0[\psi] = 0 \text{ in } \Omega, \\ \psi = f \quad \text{on } \partial\Omega. \end{cases}$$

For the time-homogenous case (4.52) leads to

$$\psi(\mathbf{x}) = E[f(\mathbf{u}(\tau_\mathbf{x}))], \qquad \mathbf{x} \in \Omega. \tag{4.56bis}$$

Equations (4.39), (4.44), (4.54bis), and (4.56bis) may suggest the so-called Montecarlo methods for the numerical solution of PDE's by means of the approximations of expected values via suitable laws of large numbers (see, e.g., Lapeyre et al. 2003).

For a general reference on Dynkin's formula, and diffusion processes the reader may refer to Ventcel' (1996).

4.6 Itô–Lévy Stochastic Differential Equations

Within the framework established in Sects. 3.8 and 3.9, we are now ready to generalize the concept of SDE with a general Lévy noise (e.g., Gihman and Skorohod 1972, p. 273).

We may consider SDEs of the following form:

$$du(t) = a(t, u(t))dt + b(t, u(t))dW_t + \int_{\mathbb{R}-\{0\}} f(t, u(t), z)\tilde{N}(dt, dz), \tag{4.57}$$

subject to an initial condition

$$u(t_0) = u^0 \text{ a.s.},$$

where u^0 is a real-valued random variable.

The well-posedness of the preceding problem can be established under frame conditions inherited from the definition of the Itô–Lévy stochastic differential in Sect. 3.9, i.e., $(W_t)_{t \in \mathbb{R}_+}$ is a standard Wiener process:

$$\tilde{N}(dt, dz) = N(dt, dz) - dt\nu(dz),$$

where $N(dt, dz)$ is a Poisson random measure, independent of the Wiener process, and $dt\nu(dz)$ is its compensator.

Further we assume that $a(t, x)$, $b(t, x)$, and $f(t, x, z)$ are deterministic real-valued functions such that

1. An $L > 0$ exists for which

$$|a(t,x)|^2 + |b(t,x)|^2 + \int_{\mathbb{R}_0} |f(t,x,z)|^2 \nu(dz) \leq L(1 + |x|^2)$$

for $t \in [0,T]$, $x \in \mathbb{R}$.

2. They satisfy a local Lipschitz condition, i.e., for any arbitrary $R > 0$ a constant C_R exists for which

$$|a(t,x) - a(t,y)|^2 + |b(t,x) - b(t,y)|^2 + \int_{\mathbb{R}_0} |f(t,x,z) - f(t,y,z)|^2 \nu(dz)$$
$$\leq C_R(|x - y|^2),$$

for $t \in [0,T]$, $x, y \in \mathbb{R}, |x|, |y| < R$.

3. There exist $K > 0$ and a function $g(h)$ such that $g(h) \downarrow 0$ as $h \to 0$, for which

$$|a(t+h,x) - a(t,x)|^2 + |b(t+h,x) - b(t,x)|^2$$
$$+ \int_{\mathbb{R}_0} |f(t+h,x,z) - f(t,x,z)|^2 \nu(dz) \leq K(1+|x|^2)g(h)$$

for $x \in \mathbb{R}$ and $t \in [0,T]$, $h \in \mathbb{R}_+$, such that $t + h \in [0,T]$.

Theorem 4.58. *Under conditions 1, 2, and 3, (4.57), subject to an initial condition u^0 independent of both the Wiener process and the random Poisson measure, admits a unique solution that is right-continuous with probability 1. If f is identically zero, then the solution is continuous with probability 1.*

Proof. See, e.g., Gihman and Skorohod (1972, p. 274).   □

As far as the moments of the solution are concerned, the following theorem holds.

Theorem 4.59. *Under the assumptions of Theorem 4.58, if, in addition, for $m \in \mathbb{N} - \{0\}$,*

$$\int_{\mathbb{R}_0} |f(t,x,z)|^p \nu(dz) \leq L(1 + |x|^p),$$

for $p = 2, 3, \ldots, 2m$, $t \in [0,T]$, $x \in \mathbb{R}$, then

$$E[|u(t)|^{2p}] \leq L_p(1 + |x|^{2p})$$

for $p = 2, 3, \ldots, 2m$, $t \in [0,T]$, where L_p depends only on L, T, and p.

Proof. See, e.g., Gihman and Skorohod (1972, p. 275).   □

SDEs of the general type (4.57) are very important in applications; an example from neurosciences is discussed in Sect. 7.4, and an additional case can be found in Champagnat et al. (2006).

4.6.1 Markov Property of Solutions of Itô–Lévy Stochastic Differential Equations

By methods already taken into account for SDEs with only the Wiener noise in Sect. 4.2, the following theorem holds.

Theorem 4.60. *Under the assumptions of Theorem 4.58, the solution of the Itô–Lévy SDE (4.57) is a Markov process. Its infinitesimal generator is*

$$(\mathcal{A}_s\phi)(x) = \frac{\partial\phi}{\partial x}(x)a(s,x) + \frac{1}{2}\frac{\partial^2\phi}{\partial x^2}(x)b^2(s,x)$$

$$+ \int_{\mathbb{R}-\{0\}}\left[\phi(x + f(s,x,z)) - \phi(x) - \frac{\partial\phi}{\partial x}(x)f(s,x,z)\right]\nu(dz)$$

for $\phi \in C^{1,2}([0,T],\mathbb{R})$.

4.7 Exercises and Additions

4.1. Prove Remark 4.7.

4.2. Prove Remark 4.12.

4.3. Prove that if $a(t,x)$ and $b(t,x)$ are measurable functions in $[0,T]\times\mathbb{R}$ that satisfy conditions 1 and 2 of Theorem 4.4, then, for all $s \in]0,T]$, there exists a unique solution in $\mathcal{C}([s,T])$ of

$$\begin{cases} u(s) = u_s \text{ a.s.,} \\ du(t) = a(t,u(t))dt + b(t,u(t))dW_t, \end{cases}$$

provided that the random variable u_s is independent of $\mathcal{F}_{s,T} = \sigma(W_t - W_s, t \in [s,T])$ and $E[(u_s)^2] < \infty$.

4.4. Complete the proof of Theorem 4.21 by proving the *semigroup property*: If $t_0 < s$, $s \in [0,T]$, denote by $u(t,s,x)$ the solution of

$$\begin{cases} u(s) = x \text{ a.s.,} \\ du(t) = a(t,u(t))dt + b(t,u(t))dW_t. \end{cases}$$

Then

$$u(t,t_0,c) = u(t,s,u(s,t_0,c)) \text{ for } t \geq s,$$

where x is a fixed real number and c is a random variable.

4.5. Complete the proof of Theorem 4.35 (Girsanov) showing that $(Y_t^2 - t)_{t\in[0,T]}$ is a martingale where

$$Y_t = W_t - \int_0^t \vartheta_s ds,$$

$(W_t)_{t\in[0,T]}$ is a Brownian motion, and $(\vartheta_t)_{t\in[0,T]}$ satisfies the Novikov condition.

4.6. Show that

$$\Gamma_0^*(x, s; y, t) = \int_{\mathbb{R}} \Gamma_0^*(x, s; z, r) \Gamma_0^*(z, r; y, t) dz \qquad (s < r < t). \qquad (4.58)$$

Expression (4.58) is in general true for the fundamental solution $\Gamma(x, t; \xi, r)$ $(r < t)$ constructed in Theorem D.9.

4.7. Let $(W_t)_{t \in \mathbb{R}_+}$ be a Brownian motion. Consider the population growth model

$$\frac{dN_t}{dt} = (r_t + \alpha W_t) N_t, \qquad (4.59)$$

where N_t is the size of population at time t ($N_0 > 0$ given) and $(r_t + \alpha \cdot W_t)$ is the relative rate of growth at time t. Suppose the process $r_t = r$ is constant.

1. Solve SDE (4.59).
2. Estimate the limit behavior of N_t when $t \to \infty$.
3. Show that if W_t is independent of N_0, then

$$E[N_t] = E[N_0] e^{rt}.$$

An extension model of (4.59) for exponential growth with several independent white-noise sources in the relative growth rate is given as follows. Let $(W_1(t), \ldots, W_n(t))_{t \in \mathbb{R}_+}$ be Brownian motion in $\mathbb{R}^d$, with $\alpha_1, \ldots, \alpha_n$ constants. Then

$$dN_t = \left(r dt + \sum_{k=1}^{n} \alpha_k dW_k(t) \right) N_t, \qquad (4.60)$$

where N_t is, again, the size of population at time t with $N_0 > 0$ given.

4. Solve SDE (4.60).

4.8. Let $(W_t)_{t \in \mathbb{R}_+}$ be a one-dimensional Brownian motion. Show that the process (*Brownian motion on the unit circle*)

$$u_t = (\cos W_t, \sin W_t)$$

is the solution of the SDE (in matrix notation)

$$du_t = -\frac{1}{2} u_t dt + K u_t dW_t, \qquad (4.61)$$

where $K = \begin{bmatrix} 0 & -1 \\ 1 & 0 \end{bmatrix}$.

More generally, show that the process (*Brownian motion on the ellipse*)

$$u_t = (a \cos W_t, b \sin W_t)$$

is a solution of (4.61), where $K = \begin{bmatrix} 0 & -\frac{a}{b} \\ \frac{b}{a} & 0 \end{bmatrix}$.

4.9 (Brownian bridge). For fixed $a, b \in \mathbb{R}$ consider the one-dimensional equation

$$\begin{cases} u(0) = a, \\ du_t = \dfrac{b - u_t}{1 - t} dt - dW_t \qquad (0 \le t < 1). \end{cases}$$

Verify that

$$u_t = a(1 - t) + bt + (1 - t) \int_0^t \frac{dW_s}{1 - s} \qquad (0 \le t < 1)$$

solves the equation and prove that $\lim_{t \to 1} u_t = b$ a.s. The process $(u_t)_{t \in [0,1[}$ is called the *Brownian bridge* (from a to b).

4.10. Solve the following SDEs:

1. $\begin{bmatrix} du_1 \\ du_2 \end{bmatrix} = \begin{bmatrix} 1 \\ 0 \end{bmatrix} dt + \begin{bmatrix} 1 & 0 \\ 0 & u_1 \end{bmatrix} \begin{bmatrix} dW_1 \\ dW_2 \end{bmatrix}$.

2. $du_t = u_t dt + dW_t$. (*Hint:* Multiply both sides by e^{-t} and compare with $d(e^{-t} u_t)$.)

3. $du_t = -u_t dt + e^{-t} dW_t$.

4.11. Consider n-dimensional Brownian motion $\mathbf{W} = (W_1, \ldots, W_n)$ starting at $\mathbf{a} = (a_1, \ldots, a_n) \in \mathbb{R}^n$ $(n \ge 2)$ and assume $|\mathbf{a}| < R$. What is the expected value of the first exit time τ_K of B from the ball

$$K = K_R = \{\mathbf{x} \in \mathbb{R}^n; |\mathbf{x}| < R\}?$$

(*Hint:* Use Dynkin's formula.)

4.12. Find the generators of the following processes.

1. Brownian motion on an ellipse (Problem 4.8).
2. Arithmetic Brownian motion:

$$\begin{cases} u(0) = u_0, \\ du(t) = a\,dt + b\,dW_t. \end{cases}$$

3. Geometric Brownian motion:

$$\begin{cases} u(0) = u_0, \\ du(t) = au(t)dt + bu(t)dW_t. \end{cases}$$

4. (Mean-reverting) Ornstein–Uhlenbeck process:

$$\begin{cases} u(0) = u_0 \\ du(t) = (a - bu(t))dt + c\,dW_t. \end{cases}$$

4.13. Find a process $(u_t)_{t\in\mathbb{R}_+}$ whose generator is the following:

1. $Af(x) = f'(x) + f''(x)$, where $f \in BC(\mathbb{R}) \cap C^2(\mathbb{R})$.
2. $Af(t,x) = \frac{\partial f}{\partial t} + cx\frac{\partial f}{\partial x} + \frac{1}{2}\alpha^2 x^2 \frac{\partial^2 f}{\partial x^2}$, where $f \in BC(\mathbb{R}^2) \cap C^2(\mathbb{R}^2)$ and c, α are constants.

4.14. Let $\triangle$ denote the Laplace operator on $\mathbb{R}^n$, $\phi \in BC(\mathbb{R}^n)$ and $\alpha > 0$. Find a solution $(u_t)_{t\in\mathbb{R}_+}$ of the equation

$$\left(\alpha - \frac{1}{2}\triangle\right)u = \phi \quad \text{in } \mathbb{R}^n.$$

Is the solution unique?

4.15. Consider a linear SDE

$$du(t) = [a(t) + b(t)u(t)]dt + [c(t) + d(t)u(t)]dW(t), \tag{4.62}$$

where the functions a, b, c, d are bounded and measurable. Prove:

1. If $a \equiv c \equiv 0$, then the solution $u(t) = u_0(t)$ is given by

$$u_0(t) = u_0(0)\exp\left\{\int_0^t \left[b(s) - \frac{1}{2}d^2(s)\right]ds + \int_0^t d(s)dW_s\right\}.$$

2. Setting $u(t) = u_0(t)v(t)$, show that $u(t)$ is a solution of (4.62) if and only if

$$v(t) = v(0) + \int_0^t [u_0(s)a(s) - c(s)d(s)]ds + \int_0^t c(s)u_0(s)ds.$$

Thus the solution of (4.62) is $u_0(t)v(t)$ with $u(0) = u_0(0)v(0)$.

4.16. Show that the solution (4.17) of (4.15) is a Gaussian process whenever the initial condition $u(t_0)$ is either deterministic or a Gaussian random variable.

4.17. Consider a diffusion process X associated with an SDE with drift $\mu(x,t)$ and diffusion coefficient $\sigma^2(x,t)$. Show that for any $\theta \in \mathbb{R}$ the process

$$Y_\theta(t) = \exp\left\{\theta X(t) - \theta\int_0^t \mu(X(s),s)ds - \frac{1}{2}\int_0^t \sigma^2(X(s),s)ds\right\}, \quad t \in \mathbb{R}_+,$$

is a martingale.

4.18. Consider a diffusion process X associated with an SDE with drift $\mu(x,t) = \alpha t$ and diffusion coefficient $\sigma^2(x,t) = \beta t$, with $\alpha \geq 0$ and $\beta > 0$. Let T_a be the first passage time to the level $a \in \mathbb{R}$; evaluate

$$E\left[e^{-\lambda T_a^2}\Big| X(0) = 0\right] \text{ for } \lambda > 0.$$

(*Hint:* Use the result of Problem 4.17)

4.19. Let X be a diffusion process associated with an SDE with drift $\mu(x,t) = -\alpha x$ and constant diffusion coefficient $\sigma^2(x,t) = \beta$, with $\alpha \in \mathbb{R}_+^*$ and $\beta \in \mathbb{R}$. Show that the moments $q_r(t) = E[X(t)^r]$, $r = 1, 2, \ldots$ of $X(t)$ satisfy the system of ordinary differential equations

$$\frac{d}{dt} q_r(t) = -\alpha r q_r(t) + \frac{\beta^2 r(r-1)}{2} q_{r-2}(t), \quad r = 1, 2, \ldots,$$

with the assumption $q_r(t) = 0$ for any integer $r \leq -1$.

4.20. Let X be the diffusion process defined in Problem 4.19. Show that the characteristic function of $X(t)$, defined as $\varphi(v; t) = E[\exp\{ivX(t)\}], v \in \mathbb{R}$, satisfies the partial differential equation

$$\frac{\partial}{\partial t} \varphi(v; t) = -\alpha v \frac{\partial}{\partial v} \varphi(v; t) - \frac{1}{2} \beta^2 v^2 \varphi(v; t).$$

5

Stability, Stationarity, Ergodicity

Here we shall consider multidimensional diffusion processes $\{\mathbf{u}(t), t \in I\}$ in $\mathbb{R}^d$, $(d \in \mathbb{N} \setminus \{0\})$ solution on a time interval $I \subset \mathbb{R}_+$ of a d-dimensional system of stochastic differential equations of the form

$$d\mathbf{u}(t) = \mathbf{a}(t, \mathbf{u}(t))dt + b(t, \mathbf{u}(t))d\mathbf{W}(t), \tag{5.1}$$

subject to a suitable initial condition.

We may remind here the existence and uniqueness theorem (see Theorem 4.4 for the one-dimensional case; for a multidimensional SDE, see, e.g., Friedman (2004, p. 98)).

Theorem 5.1 *Suppose that*

1. *the components of the parameter functions* $\mathbf{a}(t, \mathbf{x}) = (a_1(t, \mathbf{x}), \dots, a_d(t, \mathbf{x}))'$ *and* $b(t, \mathbf{x}) = (b_{ij}(t, \mathbf{x}))_{i=1,\dots,d, j=1,\dots,m}$ *are real-valued functions, measurable with respect to* $(t, \mathbf{x}) \in [t_0, T] \times \mathbb{R}^d$.
2. *a real constant C exists such that, for any $t \in [t_0, T]$, and $\mathbf{x} \in \mathbb{R}^d$:*

$$|\mathbf{a}(t, \mathbf{x})| + \sum_{i,j=1}^{d} |\sigma_{ij}(t, \mathbf{x})| \leq C(1 + |\mathbf{x}|);$$

3. *a real constant B exists such that, for any $t \in [t_0, T]$, and $\mathbf{x}, \mathbf{y} \in \mathbb{R}^d$:*

$$|\mathbf{a}(t, \mathbf{x}) - \mathbf{a}(t, \mathbf{y})| + \sum_{i,j=1}^{d} |\sigma_{ij}(t, \mathbf{x}) - \sigma_{ij}(t, \mathbf{y})| \leq B(|\mathbf{x} - \mathbf{y}|).$$

4. $\mathbf{u_0}$ *is a d-dimensional random vector independent of the sigma algebra* $\mathcal{F}_t = \sigma(W_s, t_0 \leq t \leq T)$, *such that* $E[|\mathbf{u_0}|^2] < +\infty$.

© Springer Science+Business Media New York 2015

V. Capasso, D. Bakstein, *An Introduction to Continuous-Time Stochastic Processes*, Modeling and Simulation in Science, Engineering and Technology, DOI 10.1007/978-1-4939-2757-9_5

Then the SDE (5.1) subject to the initial condition

$$\mathbf{u}(t_0) = \mathbf{u_0}, \tag{5.2}$$

admits a unique solution $\{\mathbf{u}(t; t_0, \mathbf{u}_0), t \in [t_0, T]\} \in \mathcal{C}([t_0, T])$.

Moreover, there exists a constant $K > 0$, depending on the constants B, C, and on T, such that

$$E[\sup_{t_0 \leq t \leq T} |\mathbf{u}(t)|^2] \leq K(1 + E[|\mathbf{u}(t_0)|^2]). \tag{5.3}$$

The above theorem can be improved, as anticipated in Remark 4.7, as follows.

Theorem 5.2 *Under the same assumptions of Theorem 5.1, but Assumption 3. therein, substituted by*

$3'.$ *For any $N > 0$ a real constant B_N exists such that, for any $t \in [t_0, T]$, and $\mathbf{x}, \mathbf{y} \in \mathbb{R}^d$, subject to $|\mathbf{x}| \leq N$, $|\mathbf{y}| \leq N$,*

$$|\mathbf{a}(t, \mathbf{x}) - \mathbf{a}(t, \mathbf{y})| + \sum_{i,j=1}^{d} |\sigma_{ij}(t, \mathbf{x}) - \sigma_{ij}(t, \mathbf{y})| \leq B_N(|\mathbf{x} - \mathbf{y}|). \tag{5.4}$$

Then the SDE (5.1) subject to the initial condition

$$\mathbf{u}(t_0) = \mathbf{u_0}, \tag{5.5}$$

admits a unique solution $\{\mathbf{u}(t; t_0, \mathbf{u}_0), t \in [t_0, T]\} \in \mathcal{C}([t_0, T])$.

Moreover, there exists a constant $K > 0$, depending on the constant C, and on T, such that

$$E[\sup_{t_0 \leq t \leq T} |\mathbf{u}(t)|^2] \leq K(1 + E[|\mathbf{u}(t_0)|^2]). \tag{5.6}$$

Remark 5.3. We wish to remark that for the estimate (5.6) only Assumption 2. is required, while Assumption $3'$ is required for uniqueness.

A global (in time) existence result for the solutions of Equation (5.1) can be obtained under more restrictive assumptions on the coefficients, as in the following proposition.

Proposition 5.4 *Suppose that the parameters of system (5.1) are continuous with respect to all their variables, and satisfy, uniformly in the whole $\mathbb{R}_+ \times \mathbb{R}^d$, the following assumptions*

$A_1.$ *a real constant C exists such that, for any $t \in \mathbb{R}_+$, and $\mathbf{x} \in \mathbb{R}^d$:*

$$|\mathbf{a}(t, \mathbf{x})| + \sum_{i,j=1}^{d} |\sigma_{ij}(t, \mathbf{x})| \leq C(1 + |\mathbf{x}|);$$

A_2.   *a real constant B exists such that, for any $t \in \mathbb{R}_+$, and $\mathbf{x}, \mathbf{y} \in \mathbb{R}^d$:*

$$|\mathbf{a}(t, \mathbf{x}) - \mathbf{a}(t, \mathbf{y})| + \sum_{i,j=1}^{d} |\sigma_{ij}(t, \mathbf{x}) - \sigma_{ij}(t, \mathbf{y})| \leq B(|\mathbf{x} - \mathbf{y}|).$$

Then the SDE system (5.1) admits a unique solution for any time $t \geq t_0$, such that $E[|\mathbf{u}(t; t_0, \mathbf{u}_0)|] < +\infty$, for all $t \geq t_0$.

Proof. See, e.g., Ikeda and Watanabe (1989, p. 177), or Has'minskii (1980, p. 83). □

5.1 Time of explosion and regularity

The results in the previous section apply whenever Assumptions 2. and 3′ on the parameters are guaranteed. In general this is not true, so that the solution of the initial value problem for Equation (5.1) may not exist globally, i.e. for all times in a given interval of time, finite or not. A more general existence result may be indeed obtained if we relax Assumption 3′, and admit that the solution may blow up (explode) in a finite time.

So, for the time being we shall only assume that Assumptions 2. and 3′ hold in every cylinder $I \times U_R \subset \mathbb{R}_+ \times \mathbb{R}^d$, where we have denoted by U_R the ball $U_R := \{\mathbf{x} \in \mathbb{R}^d | |\mathbf{x}| \leq R\}$, for $R \in \mathbb{R}^*$.

Given a deterministic point $\mathbf{x}_0 \in \mathbb{R}^d$, we denote by $\mathbf{u}(t; t_0, \mathbf{x}_0)$ the solution of the SDE (5.1) subject to the initial condition $\mathbf{u}(t_0; t_0, \mathbf{x}_0) = \mathbf{x}_0$, at any time $t > t_0$ at which it exists with $E[|\mathbf{u}(t; t_0, \mathbf{x}_0)|^2] < +\infty$.

Let now $n_0 \in \mathbb{N}$, and consider the balls $U_n := \{\mathbf{x} \in \mathbb{R}^d | |\mathbf{x}| \leq n\}$, for any $n \in \mathbb{N}, n > n_0$, and denote by τ_n the first exit time from the ball U_n of the process solution of the SDE (5.1), subject to the deterministic initial condition $\mathbf{x}_0 \in U_n$.

By varying $n \in \mathbb{N}, n > n_0$, we may constitute a sequence of processes $\{\mathbf{u}_n(t; t_0, \mathbf{x}_0), t > t_0\}$, each of which is well defined up to time τ_n. Thus

$$\tau_n := \inf\{t > t_0 | |\mathbf{u}_n(t; t_0, \mathbf{x}_0)| > n\}.$$

Since, for $n_0 < n < m$, it is $U_n \subset U_m$, by uniqueness, it can be shown that any two processes $\mathbf{u}_n(t; t_0, \mathbf{x}_0)$ and $\mathbf{u}_m(t; t_0, \mathbf{x}_0)$ are indistinguishable up to time τ_n.

It is clear that the sequence of stopping times $(\tau_n)_{n \in \mathbb{N}, n > n_0}$ is monotonically increasing, so that it will admit a limit τ, for $n \to \infty$, either finite or infinite. This stopping time τ is called *time of explosion* of the process $\{\mathbf{u}(t; t_0, \mathbf{x}_0), t \geq t_0\}$. It is not difficult to show that the definition of the explosion time given above is independent of the choice of the sequence of bounded domains $(U_n)_{n \in \mathbb{N}, n > n_0}$, provided that $\sup_{\mathbf{x} \in U_n} |\mathbf{x}| \to +\infty$ as $n \to \infty$. By continuity of the solutions of SDEs, we may state that $\mathbf{u}(\tau; t_0, \mathbf{x}_0) = \lim_{n \to \infty} \mathbf{u}(\tau_n; t_0, \mathbf{x}_0)$; thus, $|\mathbf{u}(\tau; t_0, \mathbf{x}_0)| = \lim_{n \to \infty} |\mathbf{u}(\tau_n; t_0, \mathbf{x}_0)| = \lim_{n \to \infty} n = +\infty$.

Definition 5.5. We say that the solution of the SDE (5.1) started at a point $\mathbf{x}_0 \in \mathbb{R}^d$ *explodes* if
$$P(\tau < +\infty | \mathbf{u}(t_0) = \mathbf{x}_0) > 0.$$

Definition 5.6. We say that the SDE system (5.1) is *regular* if, for any initial condition $\mathbf{x}_0 \in \mathbb{R}^d$,
$$P(\tau = +\infty | \mathbf{u}(t_0) = \mathbf{x}_0) = 1.$$

Corollary 5.7 *Under the assumptions of Proposition 5.4 the SDE system (5.1) is regular.*

Interesting conditions for regularity are offered in the following theorem.

We know that the drift vector of the diffusion processes solutions of (5.1) is given by $\mathbf{a}(t, \mathbf{x})$, and the diffusion matrix by

$$\sigma_{ij} = (bb')_{ij} = \sum_{k=1}^{m} b_{ik} b_{jk}, \quad i, j = 1, \ldots, d. \tag{5.7}$$

As usual, in the sequel we shall refer to the operator $\mathcal{L}$ such that, for any function $\phi \in C^{1,2}(\mathbb{R}_+ \times \mathbb{R}^d)$,

$$\mathcal{L}\phi(t, \mathbf{x}) := \frac{\partial \phi(t, \mathbf{x})}{\partial t} + \sum_{i=1}^{d} a_i(t, \mathbf{x}) \frac{\partial \phi(t, \mathbf{x})}{\partial x_i} + \frac{1}{2} \sum_{i,j=1}^{d} \sigma_{ij}(t, \mathbf{x}) \frac{\partial^2 \phi(t, \mathbf{x})}{\partial x_i \partial x_j}, \tag{5.8}$$

with $t \in \mathbb{R}_+, \mathbf{x} \in \mathbb{R}^d$.

In the autonomous case the parameters $\mathbf{a}(\mathbf{x})$, and $b(\mathbf{x})$ are time independent. In this case solutions of

$$d\mathbf{u}(t) = \mathbf{a}(\mathbf{u}(t))dt + b(\mathbf{u}(t))d\mathbf{W}(t) \tag{5.9}$$

are time-homogeneous Markov diffusion processes, with drift $\mathbf{a}(\mathbf{x})$, and diffusion matrix $\sigma(\mathbf{x}) = b(\mathbf{x})b'(\mathbf{x})$. We shall then consider the operator L_0 such that, for any function $\phi \in C^2(\mathbb{R}^d)$,

$$L_0\phi(\mathbf{x}) := \sum_{i=1}^{d} a_i(\mathbf{x}) \frac{\partial \phi(\mathbf{x})}{\partial x_i} + \frac{1}{2} \sum_{i,j=1}^{d} \sigma_{ij}(\mathbf{x}) \frac{\partial^2 \phi(\mathbf{x})}{\partial x_i \partial x_j}, \tag{5.10}$$

with $\mathbf{x} \in \mathbb{R}^d$.

Theorem 5.8 *Let the conditions of the theorem on existence and uniqueness on the parameters of the SDE system (5.1) apply on any cylinder $I \times U_R$ ($U_R := \{\mathbf{x} \in \mathbb{R}^d | |\mathbf{x}| < R\}, R > 0$), and let $v \in C^{1,2}(\mathbb{R}_+ \times \mathbb{R}^d)$ be a nonnegative real-valued function such that, for some constant $C > 0$,*

(i) $\mathcal{L}v(t, \mathbf{x}) \leq Cv(t, \mathbf{x}), \quad t \in \mathbb{R}_+, \mathbf{x} \in \mathbb{R}^d;$

(ii) $\inf_{|\mathbf{x}|>R} v(t, \mathbf{x}) \to +\infty, \quad t \in \mathbb{R}_+, R \to +\infty.$

Then

$$E[v(t, \mathbf{u}(t; t_0, \mathbf{u}_0))] \leq E[v(t_0, \mathbf{u}_0)]e^{C(t-t_0)}, \tag{5.11}$$

provided the expected value on the right-hand side exists, which is guaranteed whenever the initial condition is a deterministic one.

Proof. See Has'minskii (1980, page 84).   □

Under the assumptions of Theorem 5.8, by Dynkin's formula, for a deterministic initial condition $\mathbf{x}_0 \in \mathbb{R}^d$, it implies that

$$P(\tau_n \leq t) \leq \frac{v(t_0, \mathbf{x}_0)}{\inf_{|\mathbf{x}|\geq n, s>t_0} v(s, \mathbf{x})}e^{C(t-t_0)}, \tag{5.12}$$

if τ_n denotes the first exit time of $\mathbf{u}(t; t_0, \mathbf{x}_0)$ from the ball U_n, $n \in \mathbb{N}$, defined above,

$$\tau_n := \inf\{t \in \mathbb{R}_+ || \mathbf{u}(t; t_0, \mathbf{x}_0)| > n\}.$$

Corollary 5.9 *Under the assumptions of Theorem 5.8 the SDE system (5.1) is regular.*

An interesting consequence of Theorem 5.8 is the following corollary assuring the existence of an invariant region for the solution of the SDE system (5.1).

Corollary 5.10 *Let D and $(D_n)_{n\in\mathbb{N}}$ be open sets in $\mathbb{R}^d$ such that (here $\bar{D}_n$ denotes the closure of D_n)*

$$D_n \subset D_{n+1}, \quad \bar{D}_n \subset D, \quad D = \bigcup_n D_n,$$

and suppose $\mathbf{a}$ and b satisfy conditions A_1 and A_2 above on each cylinder $I \times D_n$, for some $t_0 \in \mathbb{R}_+$. Suppose further that a nonnegative function $v \in C^{1,2}([t_0, +\infty[\times D)$ exists which satisfies

(i)   *for some positive constant C*

$$\mathcal{L}v(t, x) \leq Cv(t, x) \tag{5.13}$$

for any $t > t_0$, and $x \in D$;

(ii)

$$\lim_{n\to\infty} \inf_{\substack{t>t_0 \\ x\in D\backslash D_n}} v(t, x) = +\infty.$$

Then, for any initial condition $\mathbf{x}_0$, *such that* $P(\mathbf{x}_0 \in D) = 1$, *the conclusion of Theorem 5.8 holds.*

Moreover $\mathbf{u}(t; t_0, \mathbf{x}_0) \in D$ *almost surely for all* $t > t_0$. *Thus*

$$P(\tau_D = +\infty) = 1,$$

where τ_D *is the first exit time of* $\mathbf{u}(t, t_0, \mathbf{x}_0)$ *from* D *(which means that* D *is an invariant region for the SDE system 5.1).*

Proof. See Has'minskii (1980, p. 86), and Gard (1988, p. 132).   □

The above corollary is of great importance in applications, as one can see in the following one.

5.1.1 Application: A Stochastic Predator-Prey model

In Barra et al. (1978) the following prey-predator model had been considered

$$\begin{cases} du_1 = u_1[a_1 - b_{11}u_1 - b_{12}u_2]dt + u_1k_1(u_1)dW_1, \\ du_2 = u_2[-a_2 + b_{21}u_1 - b_{22}u_2]dt + u_2k_2(u_2)dW_2, \end{cases} \tag{5.14}$$

with all parameters positive, and k_i, $i = 1, 2$ Lipschitzian positive, and bounded functions in the open positive quadrant. Let $\mathbf{x}^* = (x_1^*, x_2^*)$ denote the nontrivial equilibrium of the corresponding deterministic system $(k_i \equiv 0, \ i = 1, 2)$; then by taking the function

$$v(\mathbf{x}) = \sum_{i=1}^{2} x_i^* \left(\frac{x_i}{x_i^*} - \ln \frac{x_i}{x_i^*} - 1 \right), \tag{5.15}$$

the assumptions of Corollary 5.10 are satisfied for D the open positive quadrant; thus, system (5.14) admits a global solution in time and D is an invariant set, as desirable.

5.2 Stability of Equilibria

For the time being consider a temporally homogeneous processes which are solutions of a d−dimensional system of stochastic differential equations of the form (5.9), such that the operator L_0 defined in (5.10) is uniformly elliptic (see Appendix C) in an open bounded subset Ω of $\mathbb{R}^d$, and the elliptic problem

$$\begin{cases} L_0[\phi] = -1 \text{ in } \Omega, \\ \phi = 0 \qquad \text{on } \partial\Omega \end{cases}$$

has a bounded solution $\phi \in C^2(\bar{\Omega})$.

If we denote by $\tau_\mathbf{x}$ the first exit time from Ω of the process solution of the SDE system (5.9), subject to a deterministic initial condition $\mathbf{x} \in \Omega$, by Dynkin's formula we know that

$$E[\tau_{\mathbf{x}}] = \phi(\mathbf{x}).$$

It then follows that $\tau_{\mathbf{x}} < +\infty$ almost surely and thus the trajectory started at $\mathbf{x}$ exits Ω in a finite time with probability 1. Therefore, even though $\mathbf{0} \in \Omega$ might have been an asymptotically stable equilibrium $(\mathbf{a}(\mathbf{0}) = \mathbf{0})$ for the associated deterministic system, the addition of the Wiener noise in this case has made it "unstable" for the stochastic system. It becomes then of great relevance to reexamine the concepts of stability of equilibria for an SDE system corresponding to Equation (5.1).

Anyhow we shall assume that $\mathbf{0}$ is an equilibrium for system (5.1), i.e. we suppose that $\mathbf{a}(t, \mathbf{0}) = \mathbf{0}$, and $b(t, \mathbf{0}) = 0$, for any $t \in \mathbb{R}_+$. We shall also assume that system (5.1) is regular; in particular, we assume that

1. the conditions of the existence and uniqueness theorem are satisfied globally on $[t_0, +\infty)$;
2. $\mathbf{a}$ and b are continuous.

Denote by $\{\mathbf{u}(t; t_0, \mathbf{c}), t \in [t_0, +\infty)\}$, the unique solution of system (5.1) subject to a deterministic initial condition $\mathbf{c} \in \mathbb{R}^d$.

With $\mathbf{c}$ being deterministic, all moments $E[|\mathbf{u}(t; t_0, \mathbf{c})|^k]$, $k > 0$, exist for every $t \geq t_0$.

Let $v \in C^{1,2}([t_0, +\infty) \times \mathbb{R}^d)$ be a real-valued positive definite function and define the process $V(t) := v(t, \mathbf{u}(t, t_0, \mathbf{c}))$. By Itô's formula

$$dV(t) = \mathcal{L}[v](t, \mathbf{u}(t))dt + \sum_{i=1}^{d} \sum_{j=1}^{m} \frac{\partial}{\partial x_i} v(t, \mathbf{u}(t)) b_{ij}(t, \mathbf{u}(t)) dW_j(t). \tag{5.16}$$

If we require that

$$\forall t \geq t_0, \forall \mathbf{x} \in \mathbb{R}^d: \ \mathcal{L}[v](t, \mathbf{x}) \leq 0, \tag{5.17}$$

then

$$E[dV(t)] \leq 0,$$

and hence

$$E[\mathcal{L}[v](t, \mathbf{u}(t))dt] \leq 0. \tag{5.18}$$

Functions $v(t, \mathbf{x})$ that satisfy (5.17) are the stochastic equivalents of Lyapunov functions.

We may observe that, by integrating equation (5.16), we obtain

$$V(t) = v(t_0, \mathbf{c}) + \int_{t_0}^{t} \mathcal{L}[v](r, \mathbf{u}(r)) dr$$

$$+ \int_{t_0}^{t} \sum_{i=1}^{d} \sum_{j=1}^{m} \frac{\partial}{\partial x_i} v(r, \mathbf{u}(r)) b_{ij}(r, \mathbf{u}(r)) dW_j(r),$$

$$V(s) = v(t_0, \mathbf{c}) + \int_{t_0}^{s} \mathcal{L}[v](r, \mathbf{u}(r)) dr$$

$$+ \int_{t_0}^{s} \sum_{i=1}^{d} \sum_{j=1}^{m} \frac{\partial}{\partial x_i} v(r, \mathbf{u}(r)) b_{ij}(r, \mathbf{u}(r)) dW_j(r),$$

and subtracting one from the other gives

$$V(t) - V(s) = \int_{s}^{t} \mathcal{L}[v](r, \mathbf{u}(r)) dr + \int_{s}^{t} \sum_{i=1}^{d} \sum_{j=1}^{m} \frac{\partial}{\partial x_i} v(r, \mathbf{u}(r)) b_{ij}(r, \mathbf{u}(r)) dW_j(r).$$

By putting

$$H(t) = \int_{s}^{t} \sum_{i=1}^{d} \sum_{j=1}^{m} \frac{\partial}{\partial x_i} v(r, \mathbf{u}(r)) b_{ij}(r, \mathbf{u}(r)) dW_j(r),$$

and denoting by $\mathcal{F}_t$ the σ-algebra generated by all Wiener processes up to time t, we obtain

$$E[V(t) - V(s)|\mathcal{F}_s] = E\left[\int_{s}^{t} \mathcal{L}[v](r, \mathbf{u}(r)) dr | \mathcal{F}_s\right] + E[H(t)|\mathcal{F}_s]. \qquad (5.19)$$

By known properties of the Itô integral, we may recognize that $H(t)$ is a zero mean martingale with respect to $\{\mathcal{F}_t, t \in \mathbb{R}_+\}$ Therefore

$$E[H(t)|\mathcal{F}_s] = H(s) = 0.$$

Then (5.19) can be written as

$$E[V(t) - V(s)|\mathcal{F}_s] = E\left[\int_{s}^{t} \mathcal{L}[v](r, \mathbf{u}(r)) dr | \mathcal{F}_s\right],$$

and by (5.17)

$$E[V(t) - V(s)|\mathcal{F}_s] \leq 0.$$

Thus $V(t)$ is a supermartingale with respect to $\{\mathcal{F}_t, t \in \mathbb{R}_+\}$. By the supermartingale inequality

$$\forall [a,b] \subset [t_0, +\infty) :\cdot P\left(\sup_{a \leq t \leq b} v(t, \mathbf{u}(t)) \geq \epsilon\right) \leq \frac{1}{\epsilon} E[v(a, \mathbf{u}(a))]$$

and, for $a = t_0$, $\mathbf{u}(a) = \mathbf{c}$ (constant), $b \to +\infty$ we obtain

$$P\left(\sup_{t_0 \leq t \leq +\infty} v(t, \mathbf{u}(t)) \geq \epsilon\right) \leq \frac{1}{\epsilon} v(t_0, \mathbf{c}) \qquad \forall \epsilon > 0, \mathbf{c} \in \mathbb{R}^m.$$

If we suppose that $\lim_{\mathbf{c} \to \mathbf{0}} v(t_0, \mathbf{c}) = 0$, then

$$\lim_{\mathbf{c} \to \mathbf{0}} P\left(\sup_{t_0 \le t \le +\infty} v(t, \mathbf{u}(t)) \ge \epsilon \right) \le \frac{1}{\epsilon} v(t_0, \mathbf{c}) = 0 \qquad \forall \epsilon > 0, \qquad (5.20)$$

and hence, for all $\epsilon_1 > 0$, there exists a $\delta(\epsilon_1, t_0)$ such that

$$\forall |\mathbf{c}| < \delta : \ P\left(\sup_{t_0 \le t \le +\infty} v(t, \mathbf{u}(t)) \ge \epsilon \right) \le \epsilon_1.$$

If we suppose that
$$|\mathbf{u}(t)| > \epsilon_2 \Rightarrow v(t, \mathbf{u}(t)) > \epsilon,$$

as, for example, if v is the Euclidean norm, then (5.20) can be written as

$$\lim_{\mathbf{c} \to \mathbf{0}} P\left(\sup_{t_0 \le t \le +\infty} |u(t, t_0, \mathbf{c})| \ge \epsilon \right) = 0 \qquad \forall \epsilon > 0.$$

Definition 5.11. The point $\mathbf{0}$ is a *stochastically stable* equilibrium of (5.1) if

$$\lim_{\mathbf{c} \to \mathbf{0}} P\left(\sup_{t_0 \le t \le +\infty} |u(t, t_0, \mathbf{c})| \ge \epsilon \right) = 0 \qquad \forall \epsilon > 0.$$

The point $\mathbf{0}$ is *asymptotically stochastically stable* if

$$\begin{cases} \mathbf{0} \text{ is stochastically stable,} \\ \lim_{\mathbf{c} \to \mathbf{0}} P(\lim_{t \to +\infty} u(t, t_0, \mathbf{c}) = \mathbf{0}) = 1. \end{cases}$$

The point $\mathbf{0}$ is *globally asymptotically stochastically stable* if

$$\begin{cases} \mathbf{0} \text{ is stochastically stable,} \\ P(\lim_{t \to +\infty} u(t, t_0, \mathbf{c}) = \mathbf{0}) = 1 \ \forall \mathbf{c} \in \mathbb{R}^m. \end{cases}$$

Theorem 5.12 *The following two statements can be shown to be true (see also Arnold (1974) and Schuss (1980)):*

1. *If $L[v](t, \mathbf{x}) \le 0$, for all $t \ge t_0$, $\mathbf{x} \in B_h$ (B_h denotes the open ball centered at $\mathbf{0}$, with radius h), then $\mathbf{0}$ is stochastically stable.*
2. *If $v(t, \mathbf{x}) \le \omega(\mathbf{x})$ for all $t \ge t_0$, with positive definite $\omega(\mathbf{x})$ and negative definite $L[v]$, then $\mathbf{0}$ is asymptotically stochastically stable.*

Example 5.13. Consider, for $a, b \in \mathbb{R}$, the one-dimensional linear equation

$$du(t) = au(t)dt + bu(t)dW(t),$$

subject to a given initial condition $u(0) = u_0$. We know that the solution is given by

$$u(t) = u_0 \exp\left\{\left(a - \frac{b^2}{2}\right)t + bW(t)\right\}.$$

By the strong law of large numbers (see Proposition 2.188)

$$\frac{W(t)}{t} \to 0 \text{ a.s.} \qquad \text{for } t \to +\infty,$$

and we have

- $u(t) \to 0$ almost surely, if $a - \frac{b^2}{2} < 0$,
- $u(t) \to +\infty$ almost surely, if $a - \frac{b^2}{2} > 0$.

If $a = \frac{b^2}{2}$, then

$$u(t) = u_0 \exp\{bW(t)\},$$

and therefore

$$P\left(\limsup_{t \to +\infty} u(t) = +\infty\right) = 1.$$

Let us now consider the function $v(x) = |x|^\alpha$ for some $\alpha \in \mathbb{R}_+ - \{0\}$. Then

$$\mathcal{L}[v](x) = \left(a + \frac{1}{2}b^2(\alpha - 1)\right)\alpha|x|^\alpha.$$

It is easily seen that, if $a - \frac{b^2}{2} < 0$, we can choose α such that $0 < \alpha < 1 - \frac{2a}{b^2}$ and obtain a Lyapunov function v with

$$\mathcal{L}[v](x) \le -kv(x)$$

for $k > 0$. This confirms the global asymptotic stability of 0 for the stochastic differential equation.

The result in the preceding example may be extended to the nonlinear case by local linearization techniques (see Gard (1988, p. 139)).

Theorem 5.14 *Consider the scalar stochastic differential equation*

$$du(t) = a(t, u(t))dt + b(t, u(t))dW(t),$$

where, in addition to the existence and uniqueness conditions, the functions a and b are such that two real constants a_0 and b_0 exist so that

$$a(t, x) = a_0 x + \bar{a}(t, x),$$
$$b(t, x) = b_0 x + \bar{b}(t, x),$$

for any $t \in \mathbb{R}_+$ and any $x \in \mathbb{R}$, with $\bar{a}(t, x) = o(x)$ and $\bar{b}(t, x) = o(x)$, uniformly in t. Then, if $a_0 - \frac{b_0^2}{2} < 0$, the equilibrium solution $u^{eq} \equiv 0$ of equation (5.14) is stochastically asymptotically stable.

Proof. Consider again the function

$$v(x) = |x|^\alpha$$

for some $\alpha > 0$. From Itô's formula we obtain

$$\mathcal{L}[v](x)$$
$$= \left(a_0 + \frac{\bar{a}(t,x)}{x} + \frac{1}{2}(\alpha - 1)\left(b_0 + \frac{\bar{b}(t,x)}{x} \right)^2 \right) \alpha |x|^\alpha$$
$$= \left(a_0 - \frac{b_0^2}{2} + \frac{\bar{a}(t,x)}{x} + \frac{1}{2}\alpha b_0^2 + (\alpha - 1)\left(b_0 \frac{\bar{b}(t,x)}{x} + \frac{\bar{b}^2(t,x)}{2x^2} \right) \right) \alpha |x|^\alpha.$$

Choose $\alpha > 0$ and $r > 0$ sufficiently small so that for $x \in\,]-r, 0[\cup]0, r[$ we have

$$\left| \frac{\bar{a}(t,x)}{x} \right| + \frac{1}{2}\alpha b_0^2 + \left| (\alpha - 1)\left(b_0 \frac{\bar{b}(t,x)}{x} + \frac{\bar{b}^2(t,x)}{2x^2} \right) \right| < \left| a_0 - \frac{b_0^2}{2} \right|.$$

We may then claim that a constant $k > 0$ exists such that

$$\mathcal{L}[v](x) \le -kv(x),$$

from which the required result follows.   □

5.3 Stationary distributions

We now consider the autonomous multidimensional case, i.e., a stochastic differential equation in $\mathbb{R}^d$ of the form (5.9) that we report here for convenience

$$d\mathbf{u}(t) = \mathbf{a}(\mathbf{u}(t))dt + b(\mathbf{u}(t))d\mathbf{W}(t). \tag{5.21}$$

The preceding results provide conditions for the asymptotic stability of $\mathbf{0}$ as an equilibrium solution. In particular, we obtain that, for suitable assumptions on the parameters and for a suitable initial condition $\mathbf{c} \in \mathbb{R}^n$, we have

$$\lim_{t \to +\infty} \mathbf{u}(t, 0, \mathbf{c}) = \mathbf{0}, \qquad \text{a.s.}$$

We may notice that almost sure convergence implies convergence in law of $\mathbf{u}(t, 0, \mathbf{c})$ to the degenerate random variable $\mathbf{u}^{eq} \equiv \mathbf{0}$, i.e., the convergence of the transition probability to a degenerate invariant distribution with density $\delta_0(\mathbf{x})$, the standard Dirac delta function:

$$P(t, \mathbf{x}, \mathbf{B}) \to \int_{\mathbf{B}} \delta_0(\mathbf{x})d\mathbf{x} \text{ for any } \mathbf{B} \in \mathcal{B}_{\mathbb{R}^n}.$$

If (5.21) does not have an equilibrium, we may still investigate the possibility that an asymptotically invariant (but not necessarily degenerate) distribution exists for the solution of the stochastic differential equations.

5.3.1 Recurrence and transience

As for Markov chains (see, e.g., Kemeny and Snell (1960), Norris (1998)) the concepts of recurrence and transience of a process are instrumental for analyzing the possible existence of an invariant distribution and the applicability of ergodic theorems.

For simplicity we shall consider only temporally homogeneous processes which are solutions of systems of stochastic differential equations with time independent coefficients as in (5.21).

Consider an open set U in $\mathbb{R}^d$, and let U^c denote its complement.

Definition 5.15. We say that the stochastic process $(\mathbf{u}(t))_{t \in \mathbb{R}_+}$ solution of (5.21) is *recurrent* with respect to U, or $U-$recurrent, if it is regular, and, for any $\mathbf{x} \in U$, and any open subset $V \subset U$,

$$P(\mathbf{u}(\tau_m) \in V \quad \text{for a sequence of finite random times} \quad \tau_m$$
$$\text{increasing to} \quad +\infty | \mathbf{u}(0) = \mathbf{x}) = 1. \tag{5.22}$$

If, in the above, $U = \mathbb{R}^d$, then we just say that the process is recurrent.

Definition 5.16. We say that the stochastic process $(\mathbf{u}(t))_{t \in \mathbb{R}_+}$ solution of (5.21) is *transient* with respect to U, or $U-$transient, if it is regular, and, for any $\mathbf{x} \in U$,
$$P(\lim_{t \to +\infty} |\mathbf{u}(t)| = +\infty | \mathbf{u}(0) = \mathbf{x}) = 1. \tag{5.23}$$

If, in the above, $U = \mathbb{R}^d$, then we just say that the process is transient.

It is well known (see, e.g., Itô and McKean (1965)) that an $n-$dimensional Brownian motion is recurrent if $n \leq 2$, and transient if $n \geq 3$. This fact will follow from the results of this chapter.

In addition to Assumptions A_1, and A_2 introduced at the beginning of this chapter, we now introduce two additional assumptions on the coefficients of the system of stochastic differential equations (5.21) , which will be instrumental in the following analysis.

A_3. The matrix $(\sigma_{ij})_{1 \leq i,j \leq d}$ is positive definite for any $\mathbf{x} \in \mathbb{R}^d$;
A_4.   As $|\mathbf{x}| \to +\infty$

$$\sigma_{ij}(\mathbf{x}) \to \sigma_{ij}^0 \tag{5.24}$$

and

$$\sum_{i=1}^n x_i a_i(x) \to 0, \tag{5.25}$$

where the matrix $(\sigma_{ij}^0)_{1 \leq i,j \leq d}$ admits at least three positive eigenvalues.
A_5.   As $|\mathbf{x}| \to +\infty$

$$\sigma_{ij}(\mathbf{x}) - \sigma_{ij}^0 = o\left(\frac{1}{\ln |\mathbf{x}|}\right), \tag{5.26}$$

$$\sum_{i=1}^{n} |a_i(\mathbf{x})| = o\left(\frac{1}{|\mathbf{x}| \ln |\mathbf{x}|}\right), \tag{5.27}$$

and the matrix $(\sigma_{ij}^0)_{1 \leq i,j \leq d}$ admits precisely two positive eigenvalues.

The following theorems allow the identification of either transience or recurrence of the process $(\mathbf{u}(t)_{t \in \mathbb{R}_+})$, solution of the system of SDEs (5.21) in terms of its coefficients.

Theorem 5.17

1.   If Assumptions A_1, A_2, A_3, A_4 hold, then the solution of (5.21) is transient.

2.   If Assumptions A_1, A_2, A_3, A_5 hold, then the solution of (5.21) is recurrent.

Proof. See Friedman (2004, pp. 197–203).   □

The following theorem relates the recurrence with respect to an open bounded subset of $\mathbb{R}^d$ to the recurrence with respect to the whole $\mathbb{R}^d$.

Theorem 5.18 *Suppose that the diffusion matrix $(\sigma_{ij})_{1 \leq i,j \leq d}$ is nonsingular, i.e. its smallest eigenvalue is bounded away from zero in any bounded domain of $\mathbb{R}^d$. If the process $(\mathbf{u}(t)_{t \in \mathbb{R}_+})$, solution of the system of SDEs (5.21) is recurrent with respect to some open bounded subset $U \subset \mathbb{R}^d$, then $(\mathbf{u}(t)_{t \in \mathbb{R}_+})$, is recurrent with respect to any nonempty open bounded subset of $\mathbb{R}^d$.*

Proof. See Has'minskii (1980, p. 111).   □

We may then claim that, under the assumptions of Theorem 5.18, recurrence with respect to the whole $\mathbb{R}^d$ is a consequence of recurrence with respect to any open bounded subset of $\mathbb{R}^d$.

Equivalent definitions for recurrence and transience can be expressed in terms of exit times.

Given an open bounded subset $U \subset \mathbb{R}^d$, denote by $\tau_{U^c} := \inf\{t > 0 | \mathbf{u}(t) \in U\}$ the first exit time from U^c (hence the first time of visit of U)

Definition 5.19. We say that the stochastic process $(\mathbf{u}(t))_{t \in \mathbb{R}_+}$ solution of (5.21) is *recurrent* with respect to U, or $U-$recurrent, if it is regular, and, for any $\mathbf{x} \in U^c$,

$$P(\tau_{U^c} < +\infty | \mathbf{u}(0) = \mathbf{x}) = 1. \tag{5.28}$$

Definition 5.20. We say that the stochastic process $(\mathbf{u}(t))_{t \in \mathbb{R}_+}$ solution of (5.21) is *transient* with respect to U, or $U-$transient, if it is regular, and, for any $\mathbf{x} \in U^c$,

$$P(\tau_{U^c} < +\infty | \mathbf{u}(0) = \mathbf{x}) < 1. \tag{5.29}$$

The following lemma adds an important class property on the finiteness of the mean time of first visit to any open subset U of $\mathbb{R}^d$.

Lemma 5.21. *Suppose that system (5.21) satisfies the assumptions of Theorem 5.18. If the mean value $E[\tau_{U_0^c}|\mathbf{u}(0) = \mathbf{x_0}]$ is finite for some open bounded subset U_0 of $\mathbb{R}^d$, and some $\mathbf{x_0} \in U_0^c$, then $E[\tau_{U^c}|\mathbf{u}(0) = \mathbf{x}]$ is finite for any open bounded subset U of $\mathbb{R}^d$, and any $\mathbf{x} \in U^c$.*

Proof. See Has'minskii (1980, p. 116).   □

This result leads to the following definition.

Definition 5.22. A homogeneous recurrent Markov process, solution of system (5.21), is said *positive recurrent* if its mean recurrence time for some (and then for any) open bounded subset of $\mathbb{R}^d$ is finite. Otherwise it is said *null recurrent*.

Consider an open bounded subset U of $\mathbb{R}^d$. From Dynkin's Formula (4.55) we know that, if $\phi(\mathbf{x})$ is a finite solution of problem (4.56) at point $\mathbf{x} \in U$, then the mean value $E[\tau_{\mathbf{x}}]$ of the first exit time from U, for a trajectory of (5.21) started at point $\mathbf{x} \in U$, at time 0, is finite.

Actually this result does not relate explicitly to the parameters of the SDE system. Has'minskii (1980, Chapter III) offers various results about conditions on the coefficients of (5.21) that may guarantee the finiteness of $E[\tau_{\mathbf{x}}]$.

A sufficient condition for the finiteness of the mean recurrence time, i.e. the positive recurrence is the following one, based on Lyapunov functions.

Theorem 5.23 *Assume that the Markov process $(\mathbf{u}(t)_{t \in \mathbb{R}_+})$, solution of the system of SDEs (5.21) in $\mathbb{R}^d$ is regular. It is positive recurrent if there exists an open bounded subset $U \subset \mathbb{R}^d$, and a nonnegative real valued function $v \in C^2(U)$ such that, for some constant $C > 0$,*

(i) $L_0 v(\mathbf{x}) \leq C, \quad \mathbf{x} \in U$;
(ii) $\inf_{|\mathbf{x}| > R} v(\mathbf{x}) \to +\infty, \quad \text{as} \quad R \to +\infty$.

Proof. See Has'minskii (1980, p. 99).   □

An interesting corollary of the above theorem is the following one.

Corollary 5.24 *Under the assumptions of the above theorem, suppose that the set U is bounded with respect to one of the coordinates, i.e. there exist an $i \in \{1, \dots, d\}$, and $x_i^{(0)}, x_i^{(1)} \in \mathbb{R}$, such that, for any $\mathbf{x} \in U: \quad x_i^{(0)} \leq x_i \leq x_i^{(1)}$. Suppose further that $0 < \sigma_0 < \sigma_{ii}(\mathbf{x})$, and $a_i(\mathbf{x}) < a_0$ (or $a_i(\mathbf{x}) > a_0$) for any $\mathbf{x} \in U$. Then the random variable τ_U admits a finite mean value, so that the process $(\mathbf{u}(t)_{t \in \mathbb{R}_+})$, solution of the system of SDEs (5.21) is positive recurrent.*

Proof. See Has'minskii (1980, p. 99).   □

An obvious consequence of the above corollary is the following one.

Corollary 5.25 *Under the assumptions of the above theorem, suppose that the set U is bounded, the parameters of the SDE system (5.21) are bounded* in U, the diffusion matrix is nonsingular in U. Then the random variable τ_U admits a finite mean value, so that the process $(\mathbf{u}(t)_{t\in\mathbb{R}_+})$, solution of the system of SDEs (5.21) is positive recurrent.

Finally we report a theorem which guarantees necessary and sufficient conditions for the finiteness of the mean recurrence time.

Theorem 5.26 *Assume that the parameters of the system of SDEs (5.21) satisfy conditions A_1 and A_2 in every compact set of $\mathbb{R}^d$, so that its solution is regular. Assume further that the diffusion matrix satisfies the following nondegeneracy condition; for any $\mathbf{x} \in \mathbb{R}^d$, there exists an $M(\mathbf{x}) > 0$ such that*

$$\sum_{i,j=1}^{d} \sigma_{ij}\xi_i\xi_j \geq M(\mathbf{x})|\xi|^2, \quad \text{for all} \quad \xi \in \mathbb{R}^d. \tag{5.30}$$

A necessary and sufficient condition for the finiteness of the exit time τ_U is the existence of a nonnegative real valued function $v \in C^2(U)$ such that

$$L_0 v(\mathbf{x}) = -1, \quad \mathbf{x} \in U. \tag{5.31}$$

The function $\phi(\mathbf{x}) := E_{\mathbf{x}}[\tau_U]$, $\mathbf{x} \in U$, is then the smallest positive solution of the boundary value problem

$$\begin{cases} L_0[\phi] = -1, \ in \ U, \\ \phi = 0, \qquad on \ \partial U. \end{cases} \tag{5.32}$$

Under the above conditions, we may further state that $E_{\mathbf{x}}[\tau_U] \leq v(\mathbf{x})$.

Proof. See Has'minskii (1980, p. 102). □

Additional criteria for recurrence and existence of invariant measures for multidimensional diffusions can be found in Bhattacharya (1978), and the references therein.

5.3.2 Existence of a stationary distribution

As in Has'minskii (1980, p. 117), we shall make the following assumptions on the Markov process $(\mathbf{u}(t)_{t\in\mathbb{R}_+})$, solution of the system of SDEs (5.21) in $\mathbb{R}^d$.

B.   An open bounded subset U of $\mathbb{R}^d$ exists with a sufficiently regular (with respect to the elliptic operator L_0) boundary Γ such that

B_1.   In U, and some neighborhood thereof, the smallest eigenvalue of the diffusion matrix $(\sigma_{ij}(\mathbf{x}))_{1\leq i,j\leq d}$ is bounded away from zero.

B_2.   For any $\mathbf{x} \in U^c$ the mean exit time $E[\tau_{U^c}|\mathbf{u}(0) = \mathbf{x}]$ is finite, and $\sup_{\mathbf{x}\in K} E[\tau_{U^c}|\mathbf{u}(0) = \mathbf{x}] < +\infty$, for any compact set $K \subset \mathbb{R}^d$.

Remark 5.27. Thanks to Assumption B_1, we may state that we are in the conditions of Theorem 5.18, so that Assumption B_2 guarantees the positive recurrence of the process all over $\mathbb{R}^d$. Notice that Assumption B_1 implies some sort of strong connectivity of $\mathbb{R}^d$ with respect to the process; in the language of finite Markov chains, this would mean that the whole state space constitutes a unique recurrent class. We might then expect that, under Assumptions B, the process admits a unique stationary distribution to which all initial distributions would converge. This is the *leit motiv* of the following results.

Theorem 5.28 *If the Markov process $(\mathbf{u}(t)_{t\in\mathbb{R}_+}$, solution of the system of SDEs (5.21) in $\mathbb{R}^d$ satisfies Assumptions B, then it admits an invariant distribution.*

Proof. See Has'minskii (1980, p. 119).   □

We may remind that an invariant distribution of the Markov process $(\mathbf{u}(t))_{t\in\mathbb{R}_+}$, having a homogenous transition measure $\{p(\mathbf{x},t,A); \mathbf{x} \in \mathbb{R}^d, t \in \mathbb{R}_+, A \in \mathcal{B}_{\mathbb{R}^d}\}$, is a probability measure μ on $\mathcal{B}_{\mathbb{R}^d}$ such that, for any $A \in \mathcal{B}_{\mathbb{R}^d}$,

$$\int_{\mathbb{R}^d} \mu(d\mathbf{x})p(\mathbf{x},t,A) = \mu(A). \tag{5.33}$$

Equivalently, for any real valued integrable function f with respect to the measure μ,

$$\int_{\mathbb{R}^d} \mu(d\mathbf{x})E[f(\mathbf{u}(t))|\mathbf{u}(0) = \mathbf{x}] = \int_{\mathbb{R}^d} \mu(d\mathbf{x})f(\mathbf{x}). \tag{5.34}$$

5.3.3 Ergodic theorems

As a consequence of the results in the above sections, we may state the first ergodic theorem.

Theorem 5.29 *Let the Markov process $(\mathbf{u}(t))_{t\in\mathbb{R}_+}$, solution of the system of SDEs (5.21) in $\mathbb{R}^d$ satisfy Assumptions B, and let μ denote its stationary distribution. Then, for any real valued integrable function f with respect to the measure μ,*

$$\frac{1}{T} \int_0^T dt f(\mathbf{u}(t)) \xrightarrow[T\to\infty]{} \int_{\mathbb{R}^d} \mu(d\mathbf{x})f(\mathbf{x}), \quad P - \text{a.s.} \tag{5.35}$$

Proof. See Has'minskii (1980, p. 121).   □

A trivial consequence of the above theorem is the following corollary, which guarantees the uniqueness of the invariant distribution.

Corollary 5.30 *Under the assumptions of Theorem 5.29 the stationary distribution of the process $(\mathbf{u}(t)_{t\in\mathbb{R}_+})$ is unique.*

Let $\{p(\mathbf{x}, t, A); \mathbf{x} \in \mathbb{R}^d, t \in \mathbb{R}_+, A \in \mathcal{B}_{\mathbb{R}^d}\}$, denote the homogeneous transition probability of the Markov process $(\mathbf{u}(t)_{t \in \mathbb{R}_+})$, solution of the system of SDEs (5.21) in $\mathbb{R}^d$.

Theorem 5.31 *Let the Markov process* $(\mathbf{u}(t)_{t \in \mathbb{R}_+})$, *solution of the system of SDEs (5.21) in* $\mathbb{R}^d$ *satisfy Assumptions B, and let* μ *denote its stationary distribution. Then, for any real valued continuous and bounded function* f,

$$\int_{\mathbb{R}^d} p(\mathbf{x}, t, d\mathbf{y}) f(\mathbf{y}) \underset{t \to \infty}{\longrightarrow} \int_{\mathbb{R}^d} \mu(d\mathbf{y}) f(\mathbf{y}). \tag{5.36}$$

For any continuity set A *of* μ, *i.e. a measurable set* $A \in \mathcal{B}_{\mathbb{R}^d}$, *having a boundary* ∂A *such that* $\mu(\partial A) = 0$,

$$p(\mathbf{x}, t, A) \underset{t \to \infty}{\longrightarrow} \mu(A) \tag{5.37}$$

P*-a.s. with respect to* $\mathbf{x} \in \mathbb{R}^d$.

Proof. See Has'minskii (1980, p. 130). $\qquad\qquad\qquad\qquad\qquad\qquad$ $\square$

In presence of an invariant region, we may apply the following result.

Theorem 5.32 *Given the same assumptions as in Corollary 5.10, suppose further that* $n_0 \in \mathbb{N}$ *and* $M, k \in \mathbb{R}_+ \setminus \{0\}$ *exist, such that*

1. $\sum_{i,j=1}^d \sigma_{ij}(\mathbf{x}) \xi_i \xi_j \geq M |\xi|^2$ *for all* $\mathbf{x} \in \bar{D}_{n_0}, \xi \in \mathbb{R}^d$;
2. $L[v](\mathbf{x}) \leq -k$ *for all* $\mathbf{x} \in D \setminus \bar{D}_{n_0}$.

Then there exists an invariant distribution $\tilde{P}$ *with nowhere-zero density in* D, *such that for any* $\mathbf{B} \in \mathcal{B}_{\mathbb{R}^d}, \mathbf{B} \subset D$:

$$P(t, \mathbf{x}, \mathbf{B}) \to \tilde{P}(\mathbf{B}) \text{ as } t \to +\infty,$$

where $P(t, \mathbf{x}, \mathbf{B})$ *is the transition probability* $P(t, \mathbf{x}, \mathbf{B}) = P(\mathbf{u}(t, \mathbf{x}) \in \mathbf{B})$ *for the solution of the given stochastic differential equation.*

Proof. See Has'minskii (1980, p. 134), and Gard (1988, p.145). $\qquad\qquad$ $\square$

A relevant result concerning the existence of an invariant distribution has been obtained in Veretennikov (1997) (see also Veretennikov (2005), and the references therein).

Theorem 5.33 *Assume that the parameters of the system of SDEs (5.21) satisfy conditions* A_1 *and* A_2 *in every compact set of* $\mathbb{R}^d$, *so that its solution is regular. Assume further that the diffusion matrix is nondegenerate in* $\mathbb{R}^d$. *A nontrivial invariant distribution exists if there exist constants* $M_0 \geq 0$, *and* $r > (d/2) + 1$, *such that*

$$\left(\mathbf{a}(\mathbf{x}), \frac{\mathbf{x}}{|\mathbf{x}|} \right) \leq -\frac{r}{|\mathbf{x}|}, \quad |\mathbf{x}| \geq M_0. \tag{5.38}$$

Application: A Stochastic Food Chain

As a foretaste of the next part on applications of stochastic processes we take an example from Gard (1988, p. 177). Consider the deterministic system, representing a food chain,

$$
\begin{cases}
\dfrac{dz_1}{dt} = z_1[a_1 - b_{11}z_1 - b_{12}z_2], \\[2mm]
\dfrac{dz_2}{dt} = z_2[-a_2 + b_{21}z_1 - b_{22}z_2 - b_{23}z_3], \\[2mm]
\dfrac{dz_3}{dt} = z_3[-a_3 + b_{32}z_2 - b_{33}z_3].
\end{cases}
$$

If we suppose now that the three species' growth rates exhibit independent Wiener noises with scaling parameters $\sigma_i > 0$, $i = 1, 2, 3$, respectively, i.e.

$$
a_i\, dt \to a_i\, dt + \sigma_i\, dW_i, \quad i = 1, 2, 3,
$$

this leads to the following stochastic differential system,

$$
\begin{cases}
du_1 = u_1[a_1 - b_{11}u_1 - b_{12}u_2]dt + u_1\sigma_1 dW_1, \\
du_2 = u_2[-a_2 + b_{21}u_1 - b_{22}u_2 - b_{23}u_3]dt + u_2\sigma_2 dW_2, \\
du_3 = u_3[-a_3 + b_{32}u_2 - b_{33}u_3]dt + u_3\sigma_3 dW_3,
\end{cases}
\tag{5.39}
$$

subject to suitable initial conditions. This system represents a food chain in which the three species' growth rates exhibit independent Wiener noises with scaling parameters $\sigma_i > 0$, $i = 1, 2, 3$, respectively. If we assume that all the parameters a_i and b_{ij} are strictly positive and constant for any $i, j = 1, 2, 3$, it can be shown that, in the absence of noise, the corresponding deterministic system admits, in addition to the trivial one, a unique nontrivial feasible equilibrium $\mathbf{x}^{eq} \in \mathbb{R}^3_+$. This one is globally asymptotically stable in the so-called feasible region $\mathbb{R}^3_+ \setminus \{\mathbf{0}\}$, provided that the parameters satisfy the inequality

$$
a_1 - \left(\frac{b_{11}}{b_{21}}\right) a_2 - \left(\frac{b_{11}b_{22} + b_{12}b_{21}}{b_{21}b_{32}}\right) a_3 > 0.
$$

This result is obtained through the Lyapunov function

$$
v(\mathbf{x}) = \sum_{i=1}^{n} c_i \left(x_i - x_i^{eq} - x_i^{eq} \ln \frac{x_i}{x_i^{eq}} \right),
$$

provided that the $c_i > 0$, $i = 1, 2, 3$, are chosen to satisfy

$$
c_1 b_{12} - c_2 b_{21} = 0 = c_2 b_{23} - c_3 b_{32}.
$$

In fact, if one denotes by B the interaction matrix $(b_{ij})_{1 \le i,j \le 3}$ and $C = \mathrm{diag}(c_1, c_2, c_3)$, the matrix

$$CB + B'C = -2 \begin{pmatrix} c_1 b_{11} & 0 & 0 \\ 0 & c_2 b_{22} & 0 \\ 0 & 0 & c_3 b_{33} \end{pmatrix}$$

is negative definite. The derivative of v along a trajectory of the deterministic system is given by

$$\dot{v}(\mathbf{x}) = \frac{1}{2} (\mathbf{x} - \mathbf{x}^{eq}) \cdot [CB + B'C] (\mathbf{x} - \mathbf{x}^{eq}),$$

which is then negative definite, thus implying the global asymptotic stability of $\mathbf{x}^{eq} \in \mathbb{R}^3_+$.

Returning to the stochastic system, consider the same Lyapunov function as for the deterministic case. By means of Itô's formula we obtain

$$L_0[v](\mathbf{x}) = \frac{1}{2} \left((\mathbf{x} - \mathbf{x}^{eq}) \cdot [CB + B'C] (\mathbf{x} - \mathbf{x}^{eq}) + \sum_{i=1}^{3} c_i \sigma_i^2 x_i^{eq} \right).$$

It can now be shown that, if the σ_i, $i = 1, 2, 3$, satisfy

$$\sum_{i=1}^{3} c_i \sigma_i^2 x_i < 2 \min_i \{c_i b_{ii} x_i^{eq}\},$$

then the ellipsoid

$$(\mathbf{x} - \mathbf{x}^{eq}) \cdot [CB + B'C] (\mathbf{x} - \mathbf{x}^{eq}) + \sum_{i=1}^{3} c_i \sigma_i^2 x_i^{eq} = 0$$

lies entirely in $\mathbb{R}^3_+$. One can then take as D_{n_0} any neighborhood of the ellipsoid such that $\bar{D}_{n_0} \subset \mathbb{R}^3_+$ and the conditions of Theorem 5.32 are met. As a consequence the stochastic system (5.39) admits an invariant distribution with nowhere-zero density in $\mathbb{R}^3_+$.

Notice that this is not a realistic model as far as the parameters are concerned (see, e.g., Mao et al. (2002)) since the parameters affected by the Brownian noise may become negative; still the solution remains positive as required by the model.

An additional interesting application to stochastic population dynamics can be found in Roozen (1987).

5.4 The one-dimensional case

5.4.1 Stationary solutions

In the one-dimensional case Equation (5.9) reduces to

$$du(t) = a(u(t))dt + b(u(t))dW_t. \tag{5.40}$$

By assuming sufficient regularity of the parameters a and b in $\mathbb{R}$, we know that the solution $\{u(t; 0, x_0); t \in \mathbb{R}_+\}$ of (5.40), subject to the initial condition $u(0) = x_0 \in \mathbb{R}$, is regular and admits a probability density $f(x_0; x, t)$ solution of the Fokker-Planck equation (see (4.48))

$$\frac{\partial}{\partial t} f(x_0; x, t) = -\frac{\partial}{\partial x}[a(x)f(x_0; x, t)] + \frac{1}{2}\frac{\partial^2}{\partial x^2}[b^2(x)f(x_0; x, t)], \qquad (5.41)$$

for $x_0 \in \mathbb{R}$, and $t \in \mathbb{R}_+$.

A regular time invariant solution π of Equation (5.41) will then satisfy the following

$$0 = -\frac{d}{dx}[a(x)\pi(x)] + \frac{1}{2}\frac{d^2}{dx^2}[b^2(x)\pi(x)].$$

Given sufficient regularity on the parameters, we may then state that

$$-[a(x)\pi(x)] + \frac{1}{2}\frac{d}{dx}[b^2(x)\pi(x)] = constant. \qquad (5.42)$$

If we further impose that both $\pi(x)$ and $\frac{d}{dx}\pi(x)$ tend to 0, as $x \to \pm\infty$, then the $constant = 0$, and Equation (5.42) becomes

$$-[a(x)\pi(x)] + \frac{1}{2}\frac{d}{dx}[b^2(x)\pi(x)] = 0. \qquad (5.43)$$

By imposing further that $b^2(x) > 0$ and $\pi(x) > 0$ for $x \in \mathbb{R}$, we may easily find a solution of the form

$$\pi(x) = \frac{K}{b^2(x)}\exp\{-\Phi(x)\}, \qquad (5.44)$$

with

$$\Phi(x) = -\int_0^x \frac{2a(y)}{b^2(y)}dy,$$

and normalizing constant

$$K = \left\{\int_{-\infty}^{+\infty} \frac{1}{b^2(x)}\exp\{-\Phi(x)\}dx\right\}^{-1}.$$

Hence the condition for a nontrivial $\pi > 0$ in (5.44) is that

$$0 < K < +\infty.$$

Example 5.34. In the case $a(x) = \mu \in \mathbb{R}$, and $b^2(x) = \sigma^2 \in \mathbb{R}_+^*$, given constants, we have

$$\Phi(x) = -\int_0^x \frac{2\mu}{\sigma^2} dy = -\frac{2\mu}{\sigma^2} x, \quad x \in \mathbb{R},$$

leading to $K = 0$, so that we do not have a density with the required regularities.

This is the case of the standard Brownian motion, for which $a(x) = 0 \in \mathbb{R}$, and $b^2(x) = 1, \in \mathbb{R}$.

Example 5.35. (Ornstein-Uhlenbeck) In the case $a(x) = -kx$, with $k \in \mathbb{R}_+^*$, and $b^2(x) = \sigma^2 \in \mathbb{R}_+^*$, given constants, we have

$$\Phi(x) = \frac{k}{\sigma^2} x^2, \quad x \in \mathbb{R},$$

leading to $K = \frac{2}{\sigma^2} \left(\frac{\pi}{k/\sigma^2} \right)^{-\frac{1}{2}}$, so that

$$\pi(x) = \left(\frac{k}{\pi\sigma^2} \right)^{\frac{1}{2}} \exp\left\{ -\frac{k}{\sigma^2} x^2 \right\}, \quad x \in \mathbb{R},$$

which is a Gaussian density.

More in general, let us consider the time homogenous SDE (5.40) in the state space $E = [\alpha, \beta] \subset \mathbb{R}$.

As above, we denote by $f(x_0; x, t)$ the transition density of Equation (5.40), i.e. the conditional pdf of $u(t)$ at x, given $u(0) = x_0$.

Definition 5.36. The boundary point α is said *accessible* from the interior of E if and only if, for any $\varepsilon > 0$, and any $x_0 \in (\alpha, \beta)$, there exists a time $t > 0$ such that $\int_\alpha^{\alpha+\varepsilon} f(x_0; y, t) dy > 0$. Similarly, an interior point $x_0 \in (\alpha, \beta)$, is said *accessible* from the boundary point α if and only if for any $\varepsilon > 0$, there exists a time $t > 0$ such that $\int_{x_0-\varepsilon}^{x_0+\varepsilon} f(\alpha; y, t) dy > 0$.

Definition 5.37. The boundary point α is said a *regular boundary point* if and only if α is accessible from the interior of E, and any interior point of E is accessible from the boundary point α.

Clearly the same definitions apply to the boundary point β. The following theorem holds.

Theorem 5.38 *Let $(u(t))_{t \in \mathbb{R}_+}$ be the solution of the time homogenous SDE (5.40) in the state space $E = [\alpha, \beta] \subset \mathbb{R}$, and assume that both α and β are regular boundary points. If the normalized solution $g : S \to \mathbb{R}_+$ of the following equation*

$$-[a(x)g(x)] + \frac{1}{2}\frac{d}{dx}[b^2(x)g(x)] = 0,$$

is unique, and it satisfies

$$\lim_{x \to \alpha} [a(x)g(x)] = \lim_{x \to \beta} [a(x)g(x)] = \lim_{x \to \alpha} [b^2(x)g(x)] = \lim_{x \to \beta} [b^2(x)g(x)] = 0,$$

then g is the density of the unique stationary distribution of the process $(u(t))_{t \in \mathbb{R}_+}$.

If the above conditions hold, the stationary density is given by

$$g(x) = \frac{K}{b^2(x)} \exp(-\Phi(x)),$$

with

$$\Phi(x) = -\int_0^x \frac{2a(y)}{b^2(y)} dy,$$

and K is the normalizing constant.

Proof. See, e.g., Tan (2002, p. 318). For a general theory, the interested reader may refer to Skorohod (1989). □

Example 5.39. (*Diffusion approximation of the Wright model of population genetics*) In the Wright model of population genetics presented in Tan (2002, ; pages 279, 320) (see also Ludwig (1974, ; pages 74–77)), given two alleles A and a, the Markov chain $(X_n)_{n \in \mathbb{N}}$ describes the number of A allele in a large diploid population of size N.

The rescaled process $u(t) = \frac{1}{2N} X(t)$, $t \in \mathbb{R}_+$, in absence of selection, is approximated by a diffusion process in the state space $E = [0, 1]$, with drift parameter

$$a(x) = -\gamma_1 x + \gamma_2(1 - x),$$

and diffusion parameter

$$b^2(x) = x(1 - x),$$

where $\gamma_i > 0$, $i = 1, 2$, so that both 0 and 1 are regular boundary points.

Under these conditions, the stationary distribution of $(u(t))_{t \in \mathbb{R}_+}$ exists and its density is the Beta distribution given by

$$g(x) = \frac{1}{B(2\gamma_2, 2\gamma_1)} x^{2\gamma_2 - 1} (1 - x)^{2\gamma_1 - 1}, \quad x \in [0, 1],$$

where $B(\cdot, \cdot)$ is the special function Beta.

Example 5.40. (*Diffusion approximation of a two-stage model of carcinogenesis*) In a two-stage model of carcinogenesis presented in Tan (2002; pages 263, 323), the number of initiated cells $(I_t)_{t \in \mathbb{R}_+}$ is modelled as a birth-and-death process with immigration. If the number of normal stem cells N_0 is very large, the rescaled process $u(t) = \frac{1}{N_0} I(t)$, $t \in \mathbb{R}_+$, is approximated by a diffusion process in the state space $E = [0, +\infty)$, with drift parameter

$$a(x) = -\xi x + \frac{\lambda}{N_0},$$

and diffusion parameter

$$b^2(x) = \frac{\omega}{N_0} x,$$

where $\xi = d - b$, and $\omega = d + b$, with both $b, d > 0$, and $\lambda \geq 0$.

Under these conditions, the stationary distribution of $(u(t))_{t \in \mathbb{R}_+}$ exists only under the condition $d > b$, so that $\xi = d - b > 0$. The invariant density is then given by

$$g(x) = \frac{\gamma_2^{\gamma_1}}{\Gamma(\gamma_1))} x^{\gamma_1 - 1} \exp\{-\gamma_2 x\}, \quad x \in \mathbb{R}_+.$$

Additional interesting examples can be found, e.g., in Cai and Lin (2004).

5.4.2 First passage times

It is worth recalling Dynkin's formula for an autonomous one-dimensional SDE (5.40). Under sufficient regularity on the parameters that guarantee existence and uniqueness of an initial value problem, let $\{u(t; x); t \in \mathbb{R}_+\}$ denote the solution of the SDE (5.40) subject to the initial condition $u(0; x) = x \in \mathbb{R}$. Given $\alpha, \beta \in \mathbb{R}$, with $\alpha < x < \beta$, let τ be a Markov time associated with $\{u(t; x); t \in \mathbb{R}_+\}$. For any real function $\phi \in C^2(\alpha, \beta)$, the following holds

$$\phi(u(\tau, x)) = \phi(x) + \int_0^\tau L_0\phi(u(t', x))dt'$$

$$+ \int_0^\tau b(u(t', x))\frac{d}{dx}\phi(u(t', x))dW_{t'}, \quad (5.45)$$

with

$$L_0\phi(x) := \frac{1}{2}b^2(x)\frac{d^2}{dx^2}\phi(x) + a(x)\frac{d}{dx}\phi(x), \quad x \in (\alpha, \beta). \quad (5.46)$$

By taking expectations on both sides we then obtain the well-known Dynkin's Formula for autonomous one-dimensional SDEs.

$$E[\phi(u(\tau, x))] = \phi(x) + E[\int_0^\tau L_0\phi(u(t', x))dt']. \quad (5.47)$$

Let us now analyze the first passage times for the one-dimensional case. Let $\{u(t; x); t \in \mathbb{R}_+\}$ denote the solution of the SDE (5.40) subject to the initial condition $u(0; x) = x \in \mathbb{R}$. Given $\alpha, \beta \in \mathbb{R}$, with $\alpha < x < \beta$, we shall denote by $\tau_x[\alpha, \beta]$ the first exit time of $u(t; x)$ from the interval (α, β), i.e.

$$\tau_x[\alpha, \beta] := \inf\{t \geq 0 | u(t; x) \notin (\alpha, \beta)\}$$

$$= +\infty \quad \text{if the set on the right hand side is empty.} \quad (5.48)$$

As a consequence of Dynkin's formula (5.47) we have the following theorem.

Theorem 5.41 *If $b(x) > 0$, for any $x \in [\alpha, \beta]$, then the random variable $\tau_x[\alpha, \beta]$ is finite a.s., and $v(x) := E[\tau_x[\alpha, \beta]]$ is the solution of the boundary value problem*

$$\begin{cases} \dfrac{1}{2}b^2(x)\dfrac{d^2}{dx^2}v(x) + a(x)\dfrac{d}{dx}v(x) = -1, & \text{in } (\alpha, \beta), \\ v(\alpha) = v(\beta) = 0 \end{cases}$$

In this case we may obtain an explicit solution (which is left to Exercise 5.2).

Corollary 5.42 *Under the assumptions of Theorem 5.41, let*

$$s(x) := \exp\{-\int_0^x \frac{2a(y)}{b^2(y)}dy\}, \quad x \in (\alpha, \beta); \qquad (5.49)$$

$$m(x) := (b^2(y)\,s(x))^{-1}, \quad x \in (\alpha, \beta); \qquad (5.50)$$

and

$$S(x) := \int_0^x s(y)dy, \quad x \in (\alpha, \beta). \qquad (5.51)$$

Then the solution of problem (5.41) is given by

$$E[\tau_x[\alpha, \beta]] = 2\left[\frac{S(x) - S(\alpha)}{S(\beta) - S(\alpha)} \int_\alpha^\beta dy\, s(y) \int_\alpha^y dz\, m(z) \right.$$

$$\left. - \int_\alpha^x dy\, s(y) \int_\alpha^y dz\, m(z) \right]. \qquad (5.52)$$

The function m, defined in (5.50), is known as the *speed density*, and the function S, defined in (5.65), is known as *scale function* of the process. It is worth noticing that the scale function is strictly monotone increasing.

We wish now to compute the probabilities of first exit from the interval $[\alpha, \beta]$ of the solution $u(t; x)$, when the initial condition is $x \in (\alpha, \beta)$.

Denote by

$$p_\alpha(x) := P(u(t; x) \text{ hits } \alpha \text{ before } \beta), \quad x \in (\alpha, \beta), \qquad (5.53)$$

and by

$$p_\beta(x) := P(u(t; x) \text{ hits } \beta \text{ before } \alpha), \quad x \in (\alpha, \beta). \qquad (5.54)$$

Again thanks to Dynkin's formula we may state the following theorem.

Theorem 5.43 *If $b(x) > 0$, for any $x \in [\alpha, \beta]$, then*

$$p_\alpha(x) = \frac{S(\beta) - S(x)}{S(\beta) - S(\alpha)}, \quad x \in (\alpha, \beta), \qquad (5.55)$$

while

$$p_\beta(x) = \frac{S(x) - S(\alpha)}{S(\beta) - S(\alpha)}, \quad x \in (\alpha, \beta). \tag{5.56}$$

Proof. Thanks to Dynkin's formula, we just need to remind that p_α is the solution of the following boundary value problem

$$\begin{cases} \frac{1}{2}b^2(x)\frac{d^2}{dx^2}u(x) + a(x)\frac{d}{dx}u(x) = 0, & \text{in } (\alpha, \beta), \\ u(\alpha) = 1, \ u(\beta) = 0, \end{cases}$$

while p_β is the solution of the following boundary value problem

$$\begin{cases} \frac{1}{2}b^2(x)\frac{d^2}{dx^2}u(x)] + a(x)\frac{d}{dx}u(x) = 0, & \text{in } (\alpha, \beta), \\ u(\alpha) = 0, \ u(\beta) = 1 \end{cases}$$

(see Exercise 5.3).

$\square$

Remark 5.44. It is of interest the case $a(x) = 0$, for any $x \in (\alpha, \beta)$. In this case the expression (5.55) reduces to

$$p_\alpha(x) = \frac{\beta - x}{\beta - \alpha}, \quad x \in (\alpha, \beta), \tag{5.57}$$

while (5.58) reduces to

$$p_\beta(x) = \frac{x - \alpha}{\beta - \alpha}, \quad x \in (\alpha, \beta). \tag{5.58}$$

5.4.3 Ergodic theorems

With respect to the Ergodic Theorem 5.29, and Theorem 5.28 for the existence of an invariant distribution, more explicit results can be obtained in the one-dimensional case (5.40).

Under the conditions of Theorem 5.43 in the previous section, we know that the probability that the process $u(t; x)$, started at $x \in (\alpha, \beta)$, hits β before α is given by (5.58). So that, if we refer to any point $y \in (\alpha, \beta)$, such that $\alpha < x < y$, and denote by τ_y (resp. τ_α) the hitting time of y (resp. α),

$$P(\tau_y < \tau_\alpha) = \frac{S(x) - S(\alpha)}{S(y) - S(\alpha)}, \quad x \in (\alpha, \beta). \tag{5.59}$$

We may then claim that

$$P(\tau_y < +\infty | u(0) = x) = \lim_{\alpha \to -\infty} \frac{S(x) - S(\alpha)}{S(y) - S(\alpha)}, \quad x \in (\alpha, \beta). \tag{5.60}$$

Thus, if $S(-\infty) = -\infty$, it will be

$$P(\tau_y < +\infty | u(0) = x) = 1, \quad x \in (\alpha, \beta). \tag{5.61}$$

Similarly, for $y \in (\alpha, \beta)$, such that $y < x < \beta$,

$$P(\tau_y < +\infty | u(0) = x) = \lim_{\beta \to +\infty} \frac{S(x) - S(\beta)}{S(y) - S(\beta}, \quad x \in (\alpha, \beta). \tag{5.62}$$

Thus, if $S(+\infty) = +\infty$, it will be

$$P(\tau_y < +\infty | u(0) = x) = 1, \quad x \in (\alpha, \beta). \tag{5.63}$$

On the other hand, it is not difficult to prove that, if one of the values $S(-\infty)$ or $S(+\infty)$ is finite, then for some $y \in (\alpha, \beta)$

$$P(\tau_y < +\infty | u(0) = x) < 1, \quad x \in (\alpha, \beta). \tag{5.64}$$

Hence we may state the following theorem (see also Gihman and Skorohod (1972, p. 115), or Friedman (2004, p. 219)).

Theorem 5.45 *Under the conditions of Theorem 5.43, if $S(-\infty) = -\infty$, and $S(+\infty) = +\infty$, then the solution of (5.40) is recurrent; in all other cases, it is transient.*

Suppose that the parameters of system (5.40) are continuous with respect to their variables, and satisfy, uniformly in the whole $\mathbb{R}$, the assumptions A_1, and A_2 of Section 5.1. Then we know that the SDE (5.40) is regular. Assume further that $b(x) > 0$, for any $x \in \mathbb{R}$.

Under the above assumptions the scale function

$$S(x) := \int_0^x \exp\{-\int_0^x \frac{2a(y)}{b^2(y)} dy\} dy, \quad x \in \mathbb{R}, \tag{5.65}$$

is strictly monotone increasing.

We are under the conditions of Theorem 5.43, and assume further that $S(-\infty) = -\infty$, and $S(+\infty) = +\infty$, so that the solution of the SDE (5.40) is recurrent.

The process $Y(t) := S(u(t)), t \in \mathbb{R}_+$, will be itself a diffusion, solution of the SDE

$$dY(t) = \sigma(Y(t))dW_t, \tag{5.66}$$

with

$$\sigma^2(y) = b^2(S^{-1}(y))[s(S^{-1}(y))]^2 > 0, \quad \text{for any } y \in \mathbb{R}. \tag{5.67}$$

To Equation (5.66) all assumptions made on Equation (5.40) apply. From now on we will refer to the following assumption.

(D) $0 < D := \left(\int_{-\infty}^{+\infty} \frac{dy}{\sigma^2(y)} \right)^{-1} < +\infty.$

It is not difficult to realize that, under Assumption D, the function

$$F(y) := D \int_{-\infty}^{y} \frac{dz}{\sigma^2(z)}, \quad y \in \mathbb{R}, \tag{5.68}$$

is a nontrivial invariant distribution for the SDE (5.66) (see Gihman and Skorohod (1972, p. 138)).

As a consequence of the above, the following ergodic theorems can be proved for Equation (5.66); they can easily be translated to Equation (5.40). For the proofs of both theorems, we refer to Gihman and Skorohod (1972, pp. 141–144).

Theorem 5.46 *If Assumption D holds, for any real measurable function φ such that*

$$\int_{-\infty}^{+\infty} \frac{|\varphi(z)|}{\sigma^2(z)} dz < +\infty, \tag{5.69}$$

the following holds:

$$\lim_{T \to +\infty} \frac{1}{T} \int_0^T \varphi(Y(t; y)) dt = D \int_{-\infty}^{+\infty} \frac{\varphi(z)}{\sigma^2(z)} dz, \quad \text{a.s.,} \tag{5.70}$$

for any initial condition $Y(0) = y \in \mathbb{R}$.

Theorem 5.47 *Under Assumption D, if we denote by*

$$F_t(z; y) := P(Y(t; y) \leq z), \quad z \in \mathbb{R}, \tag{5.71}$$

then

$$\lim_{t \to +\infty} F_t(z; y) = D \int_{-\infty}^{z} \frac{1}{\sigma^2(x)} dx, \quad \text{for any } z \in \mathbb{R}, \tag{5.72}$$

for any initial condition $Y(0) = y \in \mathbb{R}$.

5.5 Exercises and Additions

5.1. Let $u(t)$, $t \in \mathbb{R}_+$, be the solution of the SDE

$$du(t) = a(u(t))dt + \sigma(u(t))dW(t)$$

subject to the initial condition

$$u(0) = u_0 > 0.$$

Provided that $a(0) = \sigma(0) = 0$, show that, for every $\varepsilon > 0$, there exists a $\delta > 0$ such that

$$P_{u_0}\left(\lim_{t\to+\infty} u(t) = 0\right) \geq 1 - \varepsilon$$

whenever $0 < u_0 < \delta$ if and only if

$$\int_0^\delta \exp\left\{\int_0^y \frac{2a(x)}{\sigma^2(x)}\right\} dy < \infty.$$

Further, if $\sigma(x) = \sigma_0 x + o(x)$, and similarly $a(x) = a_0 x + o(x)$, then the stability condition is

$$\frac{a_0}{\sigma_0^2} < \frac{1}{2}.$$

5.2. Let $u(t)$ be the solution of the SDE

$$du(t) = a(u(t))dt + b(u(t))dW(t)$$

subject to an initial condition

$$u(0) = x \in (\alpha, \beta) \subset \mathbb{R}.$$

Show that the mean $\mu_T(x)$ of the first exit time

$$T = \inf\left\{t \geq 0 \mid u(t) \notin (\alpha, \beta)\right\}$$

is the solution of the ordinary differential equation

$$-1 = a(x)\frac{d\mu_T}{dx} + \frac{1}{2}b^2(x)\frac{d^2\mu_T}{dx^2},$$

subject to the boundary conditions

$$\mu_T(\alpha) = \mu_T(\beta) = 0.$$

Find an explicit solution for this problem (see, e.g., Karlin and Taylor (1981, p. 193)).

5.3. Let $u(t)$ be the solution of the SDE

$$du(t) = a(u(t))dt + b(u(t))dW(t),$$

subject to an initial condition

$$u(0) = x \in (\alpha, \beta) \subset \mathbb{R}.$$

Show that the probability of hitting the boundary (for the first time) at α is given by

$$P(u(\tau_x) = \alpha) = 1 - \frac{\int_\alpha^x \Phi(y)dy}{\int_\alpha^\beta \Phi(y)dy},$$

where

$$\Phi(y) = \exp\left\{-2\int_\alpha^y \frac{a(z)}{b^2(z)}dz\right\}.$$

5.4. [Gard 1988, p. 108] Show that the solution of the following SDE, subject to the conditions on parameters $r, \beta, K > 0$ and an initial condition $X(0) = x > 0$

$$dX(t) = rX(t)(K - X(t))dt + \beta X(t)dW(t),$$

is given by

$$X(t) = \frac{\exp\left\{(rK - \frac{1}{2}\beta^2) + \beta W(t)\right\}}{\frac{1}{x} + r\int_0^t \exp\left\{(rK - \frac{1}{2}\beta^2) + \beta W(s)\right\}ds}.$$

5.5. [Friedman 2004, p. 223] Let $u(t)$ be the solution of the SDE

$$du(t) = a(u(t))dt + b(u(t))dW(t)$$

subject to a suitable initial condition $u(0)$, and set

$$I_1(x) = \int_{-\infty}^x dy \exp\left\{-2\int_0^y \frac{a(z)}{b^2(z)}dz\right\},$$

$$I_2(x) = \int_x^{+\infty} dy \exp\left\{-2\int_0^y \frac{a(z)}{b^2(z)}dz\right\}.$$

Prove that, if $I_1(x) < +\infty$, and $I_2(x) < +\infty$, then

$$P\{\lim_{t\to+\infty} u(t) = +\infty\} = P\{\sup_{t>0} u(t) = +\infty\} = E\left[\frac{I_1(u(0))}{I_1(u(0)) + I_2(u(0))}\right],$$

$$P\{\lim_{t\to+\infty} u(t) = -\infty\} = P\{\inf_{t>0} u(t) = -\infty\} = E\left[\frac{I_2(u(0))}{I_1(u(0)) + I_2(u(0))}\right].$$

Part II

Applications of Stochastic Processes

6

Applications to Finance and Insurance

The financial industry is one of the most influential driving forces behind the research into stochastic processes. This is due to the fact that it relies on stochastic models for valuation and risk management. But perhaps more surprisingly, it was also one of the main drivers that led to their initial discovery.

As early as 1900, Louis Bachelier, a young French doctorate researcher, analyzed financial contracts, also referred to as *financial derivatives*, traded on the Paris bourse and in his thesis (Bachelier 1900) attempted to lay down a mathematical foundation for their valuation. He observed that the prices of the underlying assets evolved randomly, and he employed a normal distribution to model them. This was a few years before Einstein (1905), in the context of physics, published a model of, effectively, Brownian motion, later formalized by the work of Wiener, which in turn led to the development of Itô theory in the 1950s and 1960s (Itô and McKean 1965), representing the interface of classical and stochastic mathematics. All these then came to prominence through Robert Merton's (1973) as well as Black and Scholes' (1973) derivation of their partial differential equation and formula for the pricing of financial options contracts. These represented direct applications of the then already known backward Kolmogorov equation (4.36) and Feynman–Kac formula (4.39). Still today they serve as the most widely used basic model of mathematical finance.

Furthermore, in his work, Bachelier concluded that the observed prices of assets traded on the exchange represent equilibria, meaning that one or more buyers and sellers are happy to trade a certain amount at the same time. If the market is efficient and rational, their riskless profit expectations must therefore be zero. The latter represents the economic concept of no-arbitrage, which mathematically is closely connected to martingales. Both

© Springer Science+Business Media New York 2015   313
V. Capasso, D. Bakstein, *An Introduction to Continuous-Time Stochastic Processes*, Modeling and Simulation in Science, Engineering and Technology, DOI 10.1007/978-1-4939-2757-9_6

are fundamental building blocks of all financial modeling involving stochastic processes, as was demonstrated by Harrison and Kreps (1979) and Harrison and Pliska (1981).

Many books on mathematical finance start out by describing discrete-time stochastic models before deriving the continuous-time equivalent. However, in line with all the preceding chapters on the theory of stochastic processes, we will only focus on continuous-time models. Discrete-time models in practice serve, primarily, for numerical solutions of continuous processes but also for an intuitive introduction to the topic. We refer the interested reader to the classics by Wilmott et al. (1993) for the former and Pliska (1997) as well as Cox et al. (1979) for the latter.

In this chapter we commence with the mathematical modeling of the concept of no-arbitrage and then apply it in the context of the original Black–Scholes–Merton model. We employ the latter for the valuation of different types of financial contracts. In the subsequent section we give an overview of different models of interest rates and yield curves, followed by a description of extensions to the Black–Scholes–Merton model like time dependence, jump diffusions, and stochastic volatility. The final section introduces models of insurance and default risk.

6.1 Arbitrage-Free Markets

In economic theory the usual definition of a market is a physical or conceptual place where supply meets demand for goods and services or, more generally, assets. These are exchanged in certain ratios. The latter are typically formulated in terms of a base monetary measuring unit, namely a currency, and called prices. This motivates the following definition of a market for the purpose of (continuous-time) stochastic modeling.

Definition 6.1. A filtered probability space $(\Omega, \mathcal{F}, P, (\mathcal{F}_t)_{t \in [0,T]})$ endowed with adapted stochastic processes $\left(S_t^{(i)}\right)_{t \in [0,T]}$, $i = 0, \ldots, n$, representing *asset prices* in terms of particular currencies, is called a *market*.

Asset prices are usually considered stochastic because they change over time, and unpredictably so, due to a multitude of factors like supply vs. demand or other external shocks.

Remark 6.2. The risky assets $\left(S_t^{(i)}\right)_{t \in [0,T]}$, $i = 1, \ldots, n$, are RCLL stochastic processes, thus their future values are not predictable.

In reality, no asset is entirely safe. Nonetheless, for modeling purposes it is often convenient to consider the concept of a riskless asset.

Remark 6.3. If we define, say, $S_t^{(0)} := B_t$ as a riskless asset, then $(B_t)_{t \in [0,T]}$ is a deterministic, and thus predictable, process.

Furthermore, in a market, it is possible to exchange or trade assets. This is represented by defining holding and portfolio processes.

Definition 6.4. A *holding process* $\mathbf{H}_t = \left(H_t^{(0)}, H_t^{(1)}, \ldots, H_t^{(n)}\right)$, which is adapted and predictable with respect to the filtration $(\mathcal{F}_t)_{t \in [0,T]}$, together with the asset processes $\left(S_t^{(i)}\right)_{t \in [0,T]}$, $i = 0, \ldots, n$, generate the *portfolio process*

$$\Pi_t = \mathbf{H}_t \cdot \left(B_t, S_t^{(1)}, \ldots, S_t^{(n)}\right)',$$

where $(\Pi_t)_{t \in [0,T]}$ is also adapted to $(\mathcal{F}_t)_{t \in [0,T]}$.

As usual, in the equation above, A' denotes the transpose of matrix A, and $v \cdot w$ denotes the scalar product of vectors v and w.

Note that the drivers of the asset price and holding processes are fundamentally different. The former are exogenously driven by information, aggregate supply/demand, and other external factors in the market, whereas the latter are controlled by a particular market participant. In its simplest form, the individual holding process is considered to be sufficiently small such that it has no influence on the asset price processes. The respective underlying random variables also have different dimensions. Each price process S_t^i is stated in currency per unit, whereas each H_t^i represents a dimensionless scalar. It is also often important to distinguish the following two cases.

Definition 6.5. If $T < \infty$, then the market has a *finite horizon*. Otherwise, if $T = +\infty$, then the market is said to have an *infinite horizon*.

So far, the definition of a market and its properties are insufficient to guarantee that the mathematical model is a realistic one in terms of economics. For this purpose, conditions have to be imposed on the various processes that constitute the market:

Proposition 6.6 *A realistic mathematical model of a finite-horizon market has to satisfy the following conditions:*

1. *(Conservation of funds and nonexplosive portfolios). For every* $0 \leq T < +\infty$ *the holding process* $\mathbf{H}_t$ *has to satisfy:*

$$\Pi_T = \Pi_0 + \int_0^T H_t^{(0)} dB_t + \sum_{i=1}^n \int_0^T H_t^{(i)} dS_t^{(i)}, \tag{6.1}$$

along with the nonexplosion condition

$$\left| \int_0^T d\Pi_t \right| < \infty \quad a.s.$$

The conservation-of-funds condition is also called the self-financing *portfolios* property.

2. *(Nonarbitrage).* *A deflated portfolio process $(\Pi_t^*)_{t\in[0,T]}$ with almost surely $\Pi_0^* = 0$ and $\Pi_T^* > 0$ or, equivalently, with almost surely $\Pi_0^* < 0$ and $\Pi_T^* \geq 0$ is inadmissible. Here $\Pi_t^* = \Pi_t/S_t^{(j)}$ for any arbitrary numeraire or deflator asset j.*

3. *(Trading or credit limits).* *Either $(\mathbf{H}_t)_{t\in[0,T]}$ is square-integrable and of bounded variance or $\Pi_t \geq c$ for all t, with $-\infty < c \leq 0$ constant and arbitrary.*

Condition 1 is intuitively obvious as, like the conservation of mass principle in physics, no wealth can vanish, nor can it grow to infinity in a finite horizon. For condition 3 there is a standard example (Exercise 6.4) demonstrating that in continuous time there exist arbitrage opportunities if it is not satisfied. Lastly, condition 2 is also obvious, in the sense that if an investor were able to create riskless wealth above the return of the riskless asset (in economic language: "a free lunch"), it would lead to unlimited profits. Hence the model would be ill posed. Formally, the first fundamental theorem of asset pricing has to be satisfied.

Theorem 6.7 (First fundamental theorem of asset pricing). *If in a particular market there exists an equivalent martingale (probability) measure $Q \sim P$ (Definition A.53) for any arbitrary deflated portfolio process $(\Pi_t^*)_{t\in[0,T]}$, namely,*

$$\Pi_0^* = E_P\left[\Pi_t^* \Lambda_t\right] = E_Q\left[\Pi_t^*\right] \qquad \forall t \in [0, T],$$

where Λ_t is the Radon–Nikodym derivative (Remark 4.36)

$$\frac{dQ}{dP} = \Lambda_t \qquad on \ \mathcal{F}_t,$$

then the market is free of arbitrage opportunities, provided the conditions of Girsanov's Theorem 4.35 are satisfied.

Proof. For a proof in the general continuous-time case, we refer to Delbaen and Schachermeyer (1994). □

We now make the first step into the application of valuing financial options, more generally called contingent claims.

Definition 6.8. A *financial derivative* or *contingent claim* $(V_t(S_t^{(i)}))_{t\in[0,T]}$, $i = 0,\dots,n$ is an $\mathbb{R}$-valued function of the underlying asset processes $(S_t^{(i)})_{t\in[0,T]}$ adapted to the filtered probability space $(\Omega, \mathcal{F}, P, (\mathcal{F}_t)_{t\in[0,T]})$.

Definition 6.9. A deflated contingent claim $(V_t^*(S_t^{(i)}))_{t\in[0,T]}$, $i = 0,\dots,n$ is *attainable* if there exists a holding process $\tilde{\mathbf{H}}_t = (\tilde{H}_t^{(0)}, \tilde{H}_t^{(1)},\dots,\tilde{H}_t^{(n)})$, generating the deflated portfolio process $(\tilde{\Pi}_t^*(\tilde{\mathbf{H}}_t))_{t\in[0,T]}$, such that

$$(V_t^*(S_t^{(i)}))_{t\in[0,T]} = (\tilde{\Pi}_t^*(\tilde{\mathbf{H}}_t))_{t\in[0,T]} \qquad \forall t \in [0, T].$$

Definition 6.10. A market is *complete* if and only if every deflated contingent claim $(V_t^*(S_t^{(i)}))_{t \in [0,T]}$, $i = 0, \ldots, n$ is attainable.

Theorem 6.11 (Second fundamental theorem of asset pricing).
If there exists a unique equivalent martingale (probability) measure $Q \sim P$ for any arbitrary deflated portfolio process $(\Pi_t^*)_{t \in [0,T]}$ in a particular market, then the market is complete.

Proof. For a proof in the general continuous-time case, we refer the reader to Shiryaev and Cherny (2001). $\square$

We attempt to make the significance of the two fundamental theorems more intuitive and thereby demonstrate the duality between the concepts of nonarbitrage and the existence of a martingale measure. Assume a particular portfolio in an arbitrage-free market has value $\tilde{\Pi}_T(\omega)$ for each $\omega \in \mathcal{F}_T$. If another portfolio $\hat{\Pi}_0$ can be created so that a self-financing trading strategy $(\hat{\mathbf{H}}_t)_{t \in [0,T]}$ exists replicating $\tilde{\Pi}_T(\omega)$, namely,

$$\hat{\Pi}_T(\omega) = \hat{\mathbf{H}}_0 \cdot \mathbf{S}_0 + \sum_{i=0}^{n} \int_0^T H_t^{(i)} dS_t^{(i)} \geq \tilde{\Pi}_T(\omega) \qquad \forall \omega \in \mathcal{F}_T,$$

then, necessarily,

$$\hat{\Pi}_t \geq \tilde{\Pi}_t \quad \forall t \in [0,T] \tag{6.2}$$

and, in particular,

$$\hat{\Pi}_0 \geq \tilde{\Pi}_0.$$

Otherwise there exists an arbitrage opportunity by buying the cheaper portfolio and selling the overvalued one. In fact, by this argumentation, the value of $\tilde{\Pi}_0$ has to be the solution of the constrained optimization problem

$$\tilde{\Pi}_0 = \min_{(H_t)_{t \in [0,T]}} \hat{\Pi}_0,$$

subject to the value-conservation condition (6.1) and the (super)replication condition (6.2). Hence if we can find an equivalent measure Q under which

$$E_Q \left[\sum_{i=0}^{n} \int_0^T H_t^{(i)} dS_t^{(i)} \right] \leq 0,$$

then the value of the replicated portfolio has to satisfy

$$\tilde{\Pi}_0 = \max_Q E_Q \left[\tilde{\Pi}_T \right],$$

subject to, again, the value-conservation condition and the (super)martingale condition

$$E_Q\left[\hat{\Pi}_t\right] \geq \hat{\Pi}_0.$$

The latter can be considered the so-called dual formulation of the replication problem. By the second fundamental theorem of asset pricing, if Q is unique, then all inequalities turn to equalities and

$$\tilde{\Pi}_0 = E_Q\left[\tilde{\Pi}_T\right] = \hat{\Pi}_0. \tag{6.3}$$

This result states that the nonarbitrage value of an arbitrary portfolio in an arbitrage-free and complete market is its expectation under the unique equivalent martingale measure.

Here we have implicitly assumed that the values of the portfolios are stated in terms of a numeraire of value 1. Generally, a *numeraire* asset or *deflator* serves as a measure in whose units all other assets are stated. The following theorem states that a particular numeraire can be interchanged with another.

Theorem 6.12 (Numeraire invariance theorem). *A self-financing holding strategy* $(\mathbf{H}_t)_{t\in[0,T]}$ *remains self-financing under a change of almost surely positive numeraire asset, i.e., if*

$$\frac{\Pi_T}{S_T^{(i)}} = \frac{\Pi_0}{S_T^{(i)}} + \int_0^T d\left(\frac{\Pi_t}{S_t^{(i)}}\right),$$

then

$$\frac{\Pi_T}{S_T^{(j)}} = \frac{\Pi_0}{S_T^{(j)}} + \int_0^T d\left(\frac{\Pi_t}{S_t^{(j)}}\right),$$

with $i \neq j$, *provided* $\int_0^T d\Pi_t < \infty$.

Proof. We arbitrarily choose $S_t^{(i)} = 1$ for all $t \in [0,T]$, and for notational simplicity write $S_t^{(j)} \equiv S_t$. Now it suffices to show that if

$$\Pi_T = \Pi_0 + \int_0^T d\Pi_t = \Pi_0 + \int_0^T \mathbf{H}_t \cdot d\mathbf{S}_t, \tag{6.4}$$

then this implies

$$\frac{\Pi_T}{S_T} = \frac{\Pi_0}{S_0} + \int_0^T \mathbf{H}_t \cdot d\left(\frac{\mathbf{S}_t}{S_t}\right). \tag{6.5}$$

Taking the differential and substituting (6.4),

$$d\left(\frac{\Pi_t}{S_t}\right) = \frac{d\Pi_t}{S_t} + \Pi_t d\left(\frac{1}{S_t}\right) + d\Pi_t d\left(\frac{1}{S_t}\right)$$

$$= \mathbf{H}_t \cdot \left(\left(\frac{d\mathbf{S}_t}{S_t}\right) + \mathbf{S}_t d\left(\frac{1}{S_t}\right) + d\mathbf{S}_t d\left(\frac{1}{S_t}\right)\right),$$

after integration gives (6.5).   □

In fact, in an arbitrage-free complete market, for every choice of numeraire there will be a distinct equivalent martingale measure. As we will demonstrate, the change of numeraire may be a convenient valuation technique of portfolios and contingent claims.

6.2 The Standard Black–Scholes Model

The Black–Scholes–Merton market has a particularly simple and intuitive form. It consists of a *riskless account* process $(B_t)_{t\in[0,T]}$, following

$$\frac{dB_t}{B_t} = rdt,$$

with constant *instantaneous riskless interest rate* r, so that

$$B_t = B_0 e^{rt} \qquad \forall t \in [0, T]$$

and typically $B_0 \equiv 1$ normalized. Here r describes the instantaneous time value of money, namely, how much relative wealth can be earned when saved over an infinitesimal instance dt, or, conversely, how it is discounted if received in the future.

Furthermore, there exists a *risky asset* process $(S_t)_{t\in[0,T]}$, following geometric Brownian motion (Example 4.11)

$$\frac{dS_t}{S_t} = \mu dt + \sigma dW_t$$

with a constant drift μ and a constant volatility σ scaling a Wiener process dW_t, resulting in

$$S_t = S_0 \exp\left\{\left(\mu - \frac{1}{2}\sigma^2\right)t + \sigma W_t\right\} \qquad \forall t \in [0, T].$$

Both assets are adapted to the filtered probability space $(\Omega, \mathcal{F}, P, (\mathcal{F}_t)_{t\in[0,T]})$. The market has a finite horizon and is free of arbitrage as well as complete. To demonstrate this we take B_t as the numeraire asset and attempt to find an equivalent measure Q for which the discounted process

$$S_t^* := \frac{S_t}{B_t} \tag{6.6}$$

is a local martingale. Invoking Itô's formula gives

$$dS_t^* = S_t^*((\mu - r)dt + \sigma dW_t), \tag{6.7}$$

which, by Girsanov's Theorem 4.35, shows that

$$W_t^Q = W_t + \frac{\mu - r}{\sigma}t$$

turns (6.6) into a martingale, namely,

$$S_t^* = S_0^* \exp\left\{-\frac{1}{2}\sigma^2 t + \sigma W_t^Q\right\} \qquad \forall t \in [0, T], \tag{6.8}$$

and therefore

$$S_0^* = E_Q[S_t^*]$$

under the equivalent measure Q, given by

$$\frac{dQ}{dP} = \exp\left\{-\frac{\mu - r}{\sigma}W_t - \left(\frac{\mu - r}{\sigma}\right)^2 \frac{t}{2}\right\} \qquad \text{on } (\mathcal{F}_t)_{t \in [0,T]}.$$

Now, by the numeraire invariance theorem, this means that there will be unique martingale measures for all possible deflated portfolios, and hence there is no arbitrage in the Black–Scholes model and it is complete. This now allows us to price arbitrary replicable portfolios and contingent claims with formula (6.3). But going back to the primal replication problem, we can derive the Black–Scholes partial differential equation from the conservation-of-funds condition (6.1). Explicitly, the replication constraints for a particular portfolio

$$V_t := \Pi_t$$

in the Black–Scholes model are

$$V_t = \Pi_0 + \int_0^t H_s^{(S)} dS_s + \int_0^t H_s^{(B)} dB_s,$$
$$= H_t^{(S)} S_t + H_t^{(B)} B_t \tag{6.9}$$

subject to the sufficient nonexplosion condition

$$\int_0^t \left|H_s^{(B)}\right| ds + \int_0^t \left|H_s^{(S)}\right|^2 ds < \infty \qquad \text{a.s.,}$$

and because by definition

$$V_0 = H_0^{(S)} S_0 + H_0^{(B)} B_0 = \Pi_0.$$

Invoking Itô's formula, we obtain

$$dV_t = \frac{\partial V_t}{\partial t} dt + \frac{\partial V_t}{\partial S_t} dS_t + \frac{1}{2}\sigma^2 S_t^2 \frac{\partial^2 V_t}{\partial S_t^2} dt \tag{6.10}$$

on the left-hand side of equation (6.9) and

$$dV_t = H_t^{(S)} dS_t + H_t^{(B)} dB_t \tag{6.11}$$

on the right. If we equate (6.10) and (6.11), as well as choose

$$H_t^{(S)} = \frac{\partial V_t}{\partial S_t} \qquad \text{and} \qquad H_t^{(B)} = \frac{V_t}{B_t} - \frac{\partial V_t}{\partial S_t}\frac{S_t}{B_t},$$

then the *hedging strategy* $(H_t^{(S)}, H_t^{(B)})_{\forall t \in [0,T]}$ remains predictable with respect to $(\mathcal{F}_t)_{t \in [0,T]}$, and is thus risk free. Rearranging the result gives the Black–Scholes equation

$$\mathcal{L}_{BS} V_t := \frac{\partial V_t}{\partial S_t} + \frac{1}{2}\sigma^2 S_t^2 \frac{\partial^2 V_t}{\partial S_t^2} + rS_t \frac{\partial V_t}{\partial S_t} - rV_t = 0. \tag{6.12}$$

First, it is notable that the drift scalar μ under P has canceled out when changing to the measure Q. This is given by the logic that hedging will always be riskless and thus the statistical properties of the process are irrelevant as the random factors cancel out. Second, the partial differential equation is a backward Kolmogorov equation [see (4.36)] with killing rate r. As such we know that we require a suitable terminal condition and should look for a solution given by the Feynman–Kac formula (4.39). In fact, the valuation formula (6.3) provides us with exactly that.

Remark 6.13. Common financial derivatives are *forwards* and *options*. They have a particular time T (expiry) value V_T, also called the *payoff*. The payoff of a forward is

$$V_T^F = S_T - K.$$

So-called *vanilla options* are *calls* and *puts*, whose respective payoffs are

$$V_T^C = \max\{S_T - K, 0\} \qquad \text{(call)}$$

and

$$V_T^P = \max\{K - S_T, 0\} \qquad \text{(put)},$$

where K is a positive constant of the same dimension as S_T, called the *strike* price.

As was demonstrated in Theorems 6.7 and 6.11, in an arbitrage-free and complete market, financial derivatives can be regarded as synthetic portfolios $(\Pi_t)_{t \in [0,T]}$, which provide a certain payoff

$$V_T(\omega) = \Pi_T(\omega) \qquad \forall \omega \in \mathcal{F}_T.$$

Hence, substituting the payoff of a forward into formula (6.3) and employing the normalized riskless asset B_t as numeraire, we obtain

$$V_0^F = E_Q \left[\frac{V_T^F}{B_T} \right] \tag{6.13}$$

$$= E_Q \left[e^{-rT}(S_T - K, 0) \right]$$

$$= e^{-rT} \left(E_Q \left[S_T \right] - K \right). \tag{6.14}$$

Now by (6.7), it becomes obvious that changing to the martingale measure implies setting the drift of the risky asset to r. Hence, using (6.8)

$$E_Q [S_T] = \int_{-\infty}^{\infty} S_T f(S_T) dS_T$$

$$= \int_{-\infty}^{\infty} S_0 e^{(r - \frac{1}{2}\sigma^2)T + \sigma\sqrt{T}x} \varphi(x) dx$$

$$= S_0 e^{rT},$$

where $f(x)$ is the log-normal density of S_T (1.3) and $\varphi(x)$ the standard normal density (1.2), after substitution into (6.14), finally resulting in

$$V_0^F = S_0 - e^{-rT} K. \tag{6.15}$$

The value of a forward (6.15) is not dependent on the volatility σ of S_t in the Black–Scholes–Merton market. A forward is considered to be a linear financial derivative.

Definition 6.14. A contingent claim $(V_t(S_t^{(i)}))_{t \in [0,T]}$, $i = 0, \dots, n$ is called *linear* if it does not depend on the distribution of any $S_t^{(i)}$.

Also note that a forward can be replicated statically. In (6.11) the hedging strategy results in

$$H_t^{(F)} = \frac{\partial V_t^F}{\partial S_t} = 1 \quad \text{and} \quad H_t^{(B)} = -Ke^{-rT},$$

both of which are independent of t. Conversely, we have the payoff of a call option:

$$V_0^C = E_Q \left[e^{-rT} \max \{ S_T - K, 0 \} \right]$$

$$= e^{-rT} \left(E_Q \left[S_T I_{[S_T > K]}(S_T) \right] - K E_Q \left[I_{[S_T > K]}(S_T) \right] \right). \tag{6.16}$$

Similarly to a forward we obtain the integrals

$$E_Q\left[S_T I_{[S_T > K]}(S_T)\right] = \int_K^\infty S_T f(S_T) dS_T \tag{6.17}$$
$$= \int_{-d_2}^\infty S_0 e^{\left(r - \frac{1}{2}\sigma^2\right)T + \sigma\sqrt{T}x} \varphi(x) dx$$
$$= S_0 e^{rT} \Phi(d_1)$$

and

$$E_Q\left[I_{[S_T > K]}(S_T)\right] = Q(S_T > K) = \Phi(d_2), \tag{6.18}$$

where again [see (1.2)] $\varphi(x)$ is the standard normal density and $\Phi(x)$ its cumulative distribution, and

$$d_1 = \frac{\ln\frac{S_0}{K} + \left(r + \frac{1}{2}\sigma^2\right)T}{\sigma\sqrt{T}}, \tag{6.19}$$

as well as $d_2 = d_1 - \sigma\sqrt{T}$ (we leave the interim steps in the derivation as an exercise). Hence the so-called *Black–Scholes formula* for a call option is

$$V_{BS}(S_0) := V_0^C = S_0 \Phi(d_1) - K e^{-rT} \Phi(d_2), \tag{6.20}$$

and similarly, the Black–Scholes put formula is

$$V_0^P = K e^{-rT} \Phi(-d_2) - S_0 \Phi(-d_1). \tag{6.21}$$

In fact, both are related through the so-called *put–call parity*:

$$V_t^F = V_t^C - V_t^P. \tag{6.22}$$

Obviously, options are nonlinear (also called *convex*) financial derivatives because their value generally depends on the distribution of $(S_t)_{\forall t \in [0,T]}$. However, call and put options in particular only depend on the terminal distribution of S_T.

Digital Options and Martingale Probabilities

As was shown for contingent claims that only depend on the terminal distribution of S_T, we can simply substitute their respective payoff kernel V_T into (6.13). A *binary* or *digital* call option has the simple payoff $V_T = I_{[S_T \geq K]}(S_T)$. Hence its value is

$$V_0^D = e^{-rT} E_Q[I_{[S_T \geq K]}(S_T)]$$
$$= e^{-rT} \Phi(d_2),$$

as was already demonstrated in (6.18). In fact, the option has the interpretation

$$V_0^D = e^{-rT} Q(S_T > K), \tag{6.23}$$

meaning it is the probability under the martingale measure of the risky asset value exceeding the strike at expiry T. In fact, if $(V_t^C)_{t \in [0,T]}$ is a call option, then

$$-\frac{\partial V_t^C}{\partial K} = V_t^D, \tag{6.24}$$

i.e., the derivative of a call option with respect to the strike is the negative discounted probability of being in the money at expiry under the risk-neutral martingale measure. Also, from (6.24) it can be directly observed that

$$V_t^D = \lim_{dK \downarrow 0} \frac{V_t^C(K + dK) - V_t^C(K)}{dK},$$

hence the digital option is a linear call-option spread, which is model-independent, if the values of V_t^C are known.

Barrier Options and Exit Times

A common example of derivatives that depend on the entire path of an underlying random variable $(S_t)_{t \in [0,T]}$ are so-called barrier options. In their simplest form they are put or call options with the additional feature that if the underlying random variable hits a particular upper or lower barrier (or both) at any time in $[0, T]$, an event is triggered. One particular example is a so-called down-and-out call option $(D_t)_{t \in [0,T]}$ that becomes worthless when a lower level b is hit. Hence the payoff is

$$D_T = \max\{S_T - K, 0\} I_{[\min_{t \in [0,T]} S_t > b]}. \tag{6.25}$$

Here the time $\tau = \inf\{t \in [0, T] | S_t \leq b\}$ is a stopping time and, more specifically, a first exit time, as defined in Definition 2.91. Also note that τ can be directly inferred from $(S_t)_{t \in [0,T]}$. Thus, inserting the payoff (6.25) into the standard valuation formula (6.13), we need to calculate

$$\begin{aligned}
D_0 &= e^{-rT} E_Q \left[\max\{S_T - K, 0\} I_{[\min_{t \in [0,T]} S_t > b]} \right] \\
&= e^{-rT} \left(E_Q \left[(S_T - K) I_{[\min_{t \in [0,T]} S_t > b \cap S_T > K]} \right] \right) \\
&= e^{-rT} \left(E_Q \left[S_T I_{[\min_{t \in [0,T]} S_t > b \cap S_T > K]} \right] \right. \\
&\qquad \left. - KQ \left(\min_{t \in [0,T]} S_t > b \cap S_T > K \right) \right).
\end{aligned} \tag{6.26}$$

It is not difficult to see that the latter probability can be transformed as

$$Q\left(\min_{t\in[0,T]} W_t^Q > g(b) \cap W_T^Q > g(K)\right)$$

$$= Q\left(W_T^Q < -g(K)\right) - Q\left(\min_{t\in[0,T]} W_t^Q < g(b) \cap W_T^Q > g(K)\right), \quad (6.27)$$

where

$$g(x) = \frac{\ln\frac{x}{S_0} - \left(r - \frac{1}{2}\sigma^2\right)T}{\sigma}.$$

Now using the reflection principle of Lemma 2.180, we see that the last term of (6.27) can be rewritten as

$$Q\left(\left(\tilde{W}_T^Q < g(b) \cup W_T^Q < g(b)\right) \cap W_T^Q > g(K)\right)$$

$$= Q\left(\tilde{W}_T^Q < g(b) \cap W_T^Q > g(K)\right)$$

$$= Q(W_T^Q < 2g(b) - g(K)).$$

Since W_T^Q is a standard Brownian motion under Q, we obviously have the probability law

$$Q\left(W_T^Q < y\right) = \Phi\left(\frac{y}{\sqrt{T}}\right)$$

for any $y \in \mathbb{R}$. Backsubstitution gives the solution of the last term of (6.26). We leave the remaining (rather cumbersome) steps of the derivation to Exercise 6.8. Eventually, the result turns out as

$$D_0 = V_{BS}(S_0) - \left(\frac{S_0}{b}\right)^{1-\frac{2r}{\sigma^2}} V_{BS}\left(\frac{b^2}{S_0}\right)$$

in terms of the Black–Scholes price (6.20).

American Options and Stopping Times

Options, like vanilla calls and puts, that only depend on the terminal distribution of S_T are called *European* options. Conversely, options that can be exercised at any $\tau \in [0,T]$ at the holder's discretion are called *American*. It can be shown through replication nonarbitrage arguments (e.g., Øksendal 1998; Musiela and Rutkowski 1998) that their valuation formula is

$$V_0^* = \sup_{\tau\in[0,T]} E_Q\left[V_\tau^*\right],$$

where τ is a stopping time (Definition 2.44). In general, we are dealing with so-called optimal stopping or free boundary problems, and there are usually no closed-form solutions, because τ, unlike for simple barrier options, cannot be inferred directly from the level of S_t. The American option value in the Black–Scholes model can be posed in terms of a linear complementary problem (e.g., Wilmott et al. 1993). Defining the value of immediate exercise as P_t, we have

$$\mathcal{L}_{BS}V_t \leq 0 \quad \text{and} \quad V_t \geq P_t, \tag{6.28}$$

with

$$\mathcal{L}_{BS}V_t(V_t - P_t) = 0 \quad \text{and} \quad V_T = P_T.$$

Now, if there exists an early exercise region $\mathcal{R} = \{S_\tau | \tau < T\}$, we necessarily have $V_\tau = P_\tau$. If $\mathcal{L}_{BS}V_\tau = \mathcal{L}_{BS}P_\tau > 0$, then this represents a contradiction of (6.28). Therefore, in this case, early exercise can never be optimal, as, for instance, for a call option with payoff $P_\tau = \max\{S_\tau - K, 0\}$, because

$$\mathcal{L}_{BS}\max\{S_\tau - K, 0\} = \frac{1}{2}\sigma^2 K^2 \delta(S_\tau - K) + rKI_{[S_\tau > K]}(S_\tau) \geq 0,$$

where δ represents the Dirac delta. Conversely, if $V(\mathcal{A}) < P(\mathcal{A})$ for some region $\mathcal{A}$, then $\mathcal{A} \subseteq \mathcal{R}$, meaning it would certainly be optimal to exercise within this region and generally within a larger one. As an example, for a put option with $P_\tau = \max\{K - S_\tau, 0\}$, we have that

$$V_0(0) = Ke^{-rT} < P_T(0) = K.$$

Hence, an American put V_t^A has a higher value than a European one V_t^E. In fact, Musiela and Rutkowski (1998) demonstrate that in the Black–Scholes model it can be represented as

$$V_t^A = V_{BS} + E_Q\left[\int_t^T e^{-r(\tau-t)}rKI_{[S_\tau \in \mathcal{R}]}(S_\tau)d\tau\right].$$

Typically, American options are valued employing numerical methods.

6.3 Models of Interest Rates

The Black–Scholes model incorporates the concept of the *time value of money* through the instantaneous continuously compounded riskless short rate r. However, it assumes that this rate is deterministic and even constant throughout time or, in other words, the term structure (of interest rates) is flat and has no volatility. But in reality it is neither. In fact, a significant part of the financial markets is related to debt or, as it is more commonly called, fixed income instruments. The latter, in their simplest form, are future cash flows promised to a beneficiary by a debtor, who may be a government, corporation, individual, etc. The buyer of the debt hopes to pay as little upfront as possible and earn a maximum stream of interest payments; the converse is true of the debtor. These securities can be regarded as derivatives on interest rates. The latter are used as a tool of expressing the discount between the value of money today and money to be received in the future. In reality, this discount tends to be a function of the time to maturity T of the debt,[10] and, moreover, it changes continuously and unpredictably. These concepts can be formalized in a simple discount-bond market.

Definition 6.15. A filtered probability space $(\Omega, \mathcal{F}, P, (\mathcal{F}_t)_{t \in [0,T]})$ endowed with adapted stochastic processes $(B_t^{(i)})_{t \in [0,T_i]}$, $i = 0, \ldots, n$, $T_n \leq T$, with

$$B_{T_i}^{(i)} = 1 \qquad \forall i = 0, \ldots, n,$$

representing *discount-bond prices* is called a *discount-bond market*. The *term structure* of (continuously compounded) *zero rates* $(r(t, T))_{\forall t, T; 0 \leq t \leq T}$ is given by the relationship

$$B_t^{(i)} = e^{-r(t,T_i)(T_i - t)} \quad \forall i.$$

By the fundamental theorems of asset pricing, the discount-bond market is free of arbitrage if there exist equivalent martingale measures for all discount-bond ratios $B_t^{(i)}/B_t^{(j)}$, $i, j \in \{0, \ldots, n\}$. But instead of evolving the discount-bond prices directly, models for fixed income derivatives focus on the dynamics of the underlying interest rates. We will give brief summaries of the main approaches to interest rate modeling.

Short Rate Models

Motivated by the Black–Scholes model, the first stochastic modeling approaches were performed on the concept of the short rate.

Definition 6.16. The instantaneous *short rate*

$$r_t := r(t, t) \qquad \forall t \in [0, T]$$

is connected to the value of a discount bond through

$$B_t^{(i)} = E_Q \left[e^{-\int_t^{T_i} r_s ds} \right] \qquad \forall i = 0, \ldots, n, \qquad (6.29)$$

under the risk-neutral measure Q.

Vasicek (1977) proposed that the short rate follows a Gaussian process

$$dr_t = \mu_r(t, r_t)dt + \sigma_r(t, r_t)dW_t^P$$

under the *physical* or empirical measure P. This then results in a nonarbitrage relationship between the short rate and bond processes of different maturities based on the concept of a market price of risk process $(\lambda_t)_{t \in [0,T]}$.

---

[10] Other very important factors are, for example, the creditworthiness of the debtor or the rate of inflation.

Proposition 6.17. *Let the short rate r_t follow the diffusion process*

$$dr_t = \mu_r(r_t, t)dt + \sigma_r(r_t, t)dW_t^P.$$

Furthermore, assume that the discount bonds $B_t^{(i)}$ with $t \leq T_i$ for all i have interest rates as their sole risky factor and follow the sufficiently regular stochastic processes

$$dB_t^{(i)} = \mu_i(r, t, T_i)dt + \sigma_i(r, t, T_i)dW_t^P \qquad \forall i. \tag{6.30}$$

Then the nonarbitrage bond drifts are given by

$$\mu_i = r_t + \sigma_i \lambda(r_t, t),$$

where $\lambda(r_t, t)$ is the market price of the interest rate risk process.

Proof. Let us define the portfolio process $(\Pi_t)_{t \in [0,T]}$ as

$$\Pi_t = H_t^{(1)} B_t^{(1)} + H_t^{(2)} B_t^{(2)} \tag{6.31}$$

and normalize it by putting $H_t^{(1)} \equiv 1$ and $H_t^{(2)} := H_t$ for all t. The dynamics over a time interval dt are then given by

$$d\Pi_t = dB_t^{(1)} + H_t dB_t^{(2)}. \tag{6.32}$$

Invoking Itô's formula we have

$$\mu_i = \frac{\partial B^{(i)}}{\partial t} + \mu_r \frac{\partial B^{(i)}}{\partial r} + \frac{1}{2}\sigma_r \frac{\partial^2 B^{(i)}}{\partial r^2}$$

for the bond drift and

$$\sigma_i = \sigma_r \frac{\partial B^{(i)}}{\partial r} \tag{6.33}$$

for the bond volatility. Substituting both along with (6.30) into (6.32) after cancelations, we obtain

$$d\Pi_t = (\mu_1 - H_t\mu_2)dt + \left(\sigma_r \frac{\partial B^{(1)}}{\partial r} - H_t\sigma_r \frac{\partial B^{(2)}}{\partial r}\right)dW_t^P.$$

It becomes obvious that when choosing the hedge ratio as

$$H_t = \sigma_r \frac{\partial B^{(1)}}{\partial r} \left(\sigma_r \frac{\partial B^{(2)}}{\partial r}\right)^{-1}, \tag{6.34}$$

the Wiener process dW_t^P, and hence all risk, vanishes so that

$$d\Pi_t = r_t \Pi_t dt, \tag{6.35}$$

meaning that the bond must earn the riskless rate. Now, substituting (6.33), (6.34), and (6.31) into (6.35), after rearrangement, we get the relationship

$$\frac{\mu_1 - r_t B_t^{(1)}}{\sigma_1} = \frac{\mu_2 - r_t B_t^{(2)}}{\sigma_2}.$$

Observing that the two sides do not depend on the opposite index, we can write

$$\frac{\mu_i - r_t B_t^{(i)}}{\sigma_i} = \lambda(r_t, t) \quad \forall i,$$

where $\lambda(r_t, t)$ is an adapted process, independent of T_i.   □

Corollary 6.18 *By changing to the risk-neutral measure Q given by*

$$\frac{dQ}{dP} = \exp\left\{ -\int_0^t \lambda dW_s^P - \int_0^t \frac{\lambda^2}{2} ds \right\} \qquad on \ \mathcal{F}_t,$$

the risk-neutralized short rate process is given by

$$dr_t = (\mu_r - \sigma_r \lambda)dt + \sigma_r dW_t^Q,$$

where

$$W_t^Q = W_t^P + \int_0^t \lambda ds.$$

The reason why λ arises is that the short rate, representing the stochastic variable, contrary to the asset price process S_t in the Black–Scholes model, is not directly tradeable, meaning that a portfolio $H_t r_t$ is meaningless. One cannot buy units of it directly for hedging. In practice, however, λ is rarely calculated explicitly. Instead, in a short rate modeling framework some functional forms of μ_r and σ_r are specified and their parameters *calibrated* to observed market prices. This implies that one is moving from a physical measure P to a risk-neutral measure Q. For that purpose it is useful to choose the short rate processes such that there exists a tractable analytic solution for the bond price. In fact, the Vasicek SDE under the measure Q for the short rate is chosen to be the mean-reverting Ornstein–Uhlenbeck process (Example 4.11)

$$dr_t = (a - br_t)dt + \sigma dW_t^Q,$$

which, by using it in (6.29), leads one to conjecture that the solution of a discount bond maturing at time T, namely with terminal condition $B_T^{(T)} = 1$, is of the form

$$B_t^{(T)} = e^{C(t,T) - D(t,T)r_t},$$

thereby preserving the Markov property of the process. Some cumbersome, yet straightforward, calculations show that

$$D(t,T) = \frac{1}{b}\left(1 - e^{-b(T-t)}\right) \tag{6.36}$$

and

$$C(t,T) = \frac{\sigma^2}{2}\int_t^T (D(s,T))^2 ds - a\int_t^T D(s,T)ds. \tag{6.37}$$

It becomes apparent that the model only provides three parameters to describe the dynamics of a potentially complex term structure. Therefore, another common model is that of Hull and White (1990), also called the extended Vasicek model, which makes all the parameters time-dependent, namely,

$$dr_t = (a_t - b_t r_t)dt + \sigma_t dW_t^Q,$$

thereby allowing a richer description of the yield curve dynamics.

Heath–Jarrow–Morton Approach

As an evolution in interest rate modeling Heath et al. (1992) defined an approach assuming a yield curve to be specified by a continuum of traded bonds and evolved it through *instantaneous forward rates* $f(t,T)$ instead of the short rate. The former are defined through the expression

$$B_t^{(T)} = e^{-\int_t^T f(t,s)ds}, \tag{6.38}$$

and thus

$$f(t,T) = -\frac{\partial \ln B_t^{(T)}}{\partial T}$$

and

$$f(t,t) = r_t. \tag{6.39}$$

In fact, the Heath–Jarrow–Morton approach is very generic, and most other models are just specializations of it. It assumes that forward rates, under the risk-neutral measure Q associated with the riskless account numeraire, follow the SDE

$$df(t,T) = \mu(t,T)dt + \boldsymbol{\sigma}(t,T) \cdot d\mathbf{W}_t^Q, \tag{6.40}$$

where $\boldsymbol{\sigma}(t,T)$ and $d\mathbf{W}_t^Q$ are n-dimensional. In fact, due to nonarbitrage arguments, the drift function $\mu(t,T)$ can be fully specified. Invoking Itô's formula on (6.38), we obtain the relationship

$$\frac{dB_t^{(T)}}{B_t^{(T)}} = \left(r_t - \int_t^T \mu(t,s)ds + \frac{1}{2}\left|\int_t^T \boldsymbol{\sigma}(t,s)ds\right|^2 \right) dt$$
$$- \int_t^T \boldsymbol{\sigma}(t,s) \cdot d\mathbf{W}_t^Q$$

because

$$\int_t^T \frac{\partial f(t,s)}{\partial t}ds = \int_t^T \mu(t,s)ds + \int_t^T \boldsymbol{\sigma}(t,s) \cdot d\mathbf{W}_t^Q,$$

and by noting (6.39) and (6.40) as well as Fubini's Theorem A.42. But now, for the deflated discount bond to be a martingale, the drift has to be r_t. Thus

$$\int_t^T \mu(t,s)ds = \frac{1}{2}\left|\int_t^T \boldsymbol{\sigma}(t,s)ds\right|^2,$$

and so

$$\mu(t,T) = \boldsymbol{\sigma}(t,T) \cdot \int_t^T \boldsymbol{\sigma}(t,s)ds \qquad (6.41)$$

Substituting (6.41) into (6.40), we obtain arbitrage-free processes of a continuum of forward rates, driven by one or more Wiener processes:

$$df(t,T) = \boldsymbol{\sigma}(t,T) \cdot \int_t^T \boldsymbol{\sigma}(t,s)ds + \boldsymbol{\sigma}(t,T) \cdot d\mathbf{W}_t^Q.$$

Unlike for short rate models, no market price of risk appears. This is due to the fact that forward rates are actually tradeable, as the following section will demonstrate.

Brace–Gatarek–Musiela Approach

As a very intuitive yet powerful approach, Brace et al. (1997) and other authors (Miltersen et al. 1997; Jamshidian 1997) in parallel introduced a model of discrete forward rates $\left(F_t^{(i)}\right)_{t\in[0,T]}$, $i = 1,\ldots,n$, that span a yield curve through the discrete discount bonds

$$B_t^{(k)} = \prod_{i=1}^k \left(1 + F_t^{(i)}(T_i - T_{i-1})\right)^{-1}, \qquad 1 \le k \le n. \qquad (6.42)$$

The forward rates are assumed to follow the system of SDEs

$$d\mathbf{F}_t = \boldsymbol{\mu}(t,\mathbf{F}_t)dt + \Sigma(t,\mathbf{F}_t)d\mathbf{W}_t,$$

where Σ is a diagonal matrix containing the respective volatilities and $d\mathbf{W}_t$ is a vector of Wiener processes with correlations

$$E\left[dW_t^{(i)}dW_t^{(j)}\right] = \rho_{ij}dt.$$

In particular, all forward rate processes are considered to be of the log-normal form

$$\frac{dF_t^{(i)}}{F_t^{(i)}} = \mu^{(i)}(t, \mathbf{F}_t)dt + \sigma_t^{(i)}dW_t^{(i)} \qquad \forall i. \tag{6.43}$$

Again, similar to the Heath–Jarrow–Morton model, a martingale nonarbitrage argument determines the drift $\mu^{(i)}$ for each forward rate $F_t^{(i)}$. To see this, we can write (6.42) as a recurrence relation, and after rearrangement we obtain

$$F_t^{(i)}B_t^{(i)} = \frac{B_t^{(i-1)} - B_t^{(i)}}{T_i - T_{i-1}}, \tag{6.44}$$

which states that the left-hand side is equivalent to a portfolio of traded assets and has to be driftless under the martingale measure associated with a numeraire asset. In fact, we have a choice of numeraire asset among all combinations of available bonds (6.42). We arbitrarily choose a bond $B_t^{(N)}$, $1 \le N \le n$, with associated *forward measure* Q_N, and thus

$$E^{Q_N}\left[d\left(F_t^{(i)}\frac{B_t^{(i)}}{B_t^{(N)}}\right)\right] = 0. \tag{6.45}$$

The derivation is left as an exercise, and the end result is

$$\mu_t^{(i)} = \begin{cases} -\sum_{j=i+1}^{N} \frac{(T_{j+1}-T_j)F_t^{(j)}\sigma_t^{(i)}\sigma_t^{(j)}\rho_{ij}}{1+(T_{j+1}-T_j)F_t^{(j)}} & \text{if } i < N, \\ 0 & \text{if } i = N, \\ \sum_{j=N+1}^{n} \frac{(T_{j+1}-T_j)F_t^{(j)}\sigma_t^{(i)}\sigma_t^{(j)}\rho_{ij}}{1+(T_{j+1}-T_j)F_t^{(j)}} & \text{if } i > N. \end{cases} \tag{6.46}$$

This model is particularly appealing as it directly takes real-world observable inputs like forward rates and their volatilities and also discrete compounding/discounting. But the potentially large number of Brownian motions makes the model difficult to handle computationally, as it may require large-scale simulations.

6.4 Extensions and Alternatives to Black–Scholes

In practice Black–Scholes is the most commonly used model, despite its simplicity. Much of the modern research into financial mathematics looks at extensions and alternatives that try to improve on some of its shortcomings.

As already discussed, the introduction of stochastic interest rate processes is a significant step. Another important issue is that the volatility parameter σ is constant across time t and underlying level S_t. However, in reality, put and call options of different strikes K and expiries T are traded on exchanges, and their prices $\hat{V}(T, K)$ are directly observable. This allows one to invert V_{BS} and determine so-called *implied volatilities* $\sigma_{imp}(T, K)$ because the simple Black–Scholes formula for calls (6.20) and puts (6.21) are one-to-one mappings between prices of options for respective T and K to their volatility parameter σ. The implied volatility is then such that

$$V_{BS}(\sigma_{imp}(T, K)) = \hat{V}(T, K).$$

Quoting option prices in terms of their implied volatility makes them directly comparable across T and K in the sense that an option with a higher implied volatility is relatively more expensive than one with a lower-implied volatility.[11]

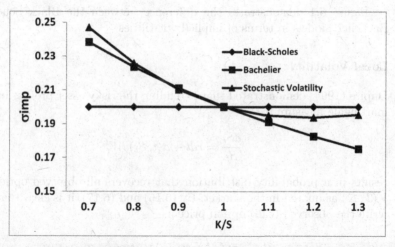

Fig. 6.1. The implied volatility skew generated by the Black–Scholes, Bachelier, and stochastic volatility models

If the Black–Scholes model were an accurate description of the real world, then $\sigma_{imp}(T, K) = \sigma$ constant. But in the real world this is not the case. Usually implied volatilities are dependent on both K and T. Typical shapes of the implied volatility surface across the strike are so-called skews or smiles. If the underlying is regarded as being floored, namely $S_t > 0$, then usually $\sigma_{imp}(K_1, T) > \sigma_{imp}(K_2, T)$, if $K_1 < K_2$, giving a negative correlation between S_t and σ_{imp}. Intuitively this can be explained by the fact that a fixed change

---

[11] By put–call parity (6.22), the implied volatility for puts and calls in an arbitrage-free market has to be identical for all pairs T, K.

in S_t has a relatively larger impact the smaller the level of S_t. One simple way of capturing this negative correlation is to change the process of S_t to a normal model, namely $(S_t)_{t\in[0,T]}$, following arithmetic Brownian motion (Example 4.11)

$$dS_t = \mu dt + \sigma dW_t.$$

This is sometimes referred to as the Bachelier (1900) model. The derivation of the call option formula is straightforward and similar to the Black–Scholes model (do as an exercise), resulting in

$$V_0^C = (S_0 - K)\Phi(d) + S_0\sigma\sqrt{T}\varphi(d),$$

with Φ and φ as in (6.18) and

$$d = \frac{S_0 - K}{S_0\sigma\sqrt{T}}.$$

Figure 6.1 demonstrates the difference between the Black–Scholes and Bachelier models in terms of implied volatilities.

Local Volatility

Dupire (1994) demonstrated that extending the risky asset process under the martingale measure to

$$\frac{dS_t}{S_t} = rdt + \sigma(t, S_t)dW_t^Q \tag{6.47}$$

results in a probability distribution that recovers all observed option prices $\hat{V}(T, K)$ as quoted in the market. By (6.16) and (6.17), it is clear that we can write the observed (call) option price as

$$\hat{V}(T, K) = e^{-rT} \int_K^\infty (S_T - K)f(0, S_0, T, S_T)dS_T.$$

Having differentiated with respect to K we obtained the cumulative distribution as (6.23) and (6.24). Differentiating one more time with respect to K, we obtain the so-called *risk-neutral transition density*

$$f(0, S_0, T, K) = e^{rT}\frac{\partial^2 \hat{V}}{\partial K^2}. \tag{6.48}$$

Now, by Theorem 4.56, $f(0, S_0, T, K)$ has to satisfy the Kolmogorov forward equation

$$\frac{\partial f}{\partial T} = \frac{1}{2}\frac{\partial^2(\sigma^2(T, K)K^2 f)}{\partial K^2} - \frac{\partial(rKf)}{\partial K} \tag{6.49}$$

with initial condition

$$f(S_0, 0, x, 0) = \delta(x - S_0).$$

Substituting (6.48) into (6.49), integrating twice with respect to K (after applying Fubini's Theorem A.42, when changing the order of integration), and noting the boundary condition

$$\lim_{K \to \infty} \frac{\partial \hat{V}}{\partial T} = 0,$$

we obtain

$$\frac{\partial \hat{V}}{\partial T} = \frac{1}{2} \sigma^2(T, K) K^2 \frac{\partial^2 \hat{V}}{\partial K^2} - rK \frac{\partial \hat{V}}{\partial K}.$$

Thus

$$\sigma(T, K) = \sqrt{\frac{\frac{\partial \hat{V}}{\partial T} + rK \frac{\partial \hat{V}}{\partial K}}{\frac{1}{2} K^2 \frac{\partial^2 \hat{V}}{\partial K^2}}},$$

fully specifying the process (6.47). Note that this is a one-factor model with a time- and level-dependent volatility parameter. While it recovers all option prices perfectly at $t = 0$, it may no longer do so at $t > 0$ if S_t has changed, meaning that while the static properties of the model are satisfactory, the dynamic ones may not be. To improve on these, various multifactor modeling approaches exist.

Jump Diffusions

Merton (1976) introduced an extension to the Black–Scholes model that appended the risky asset process by a Poisson process $(N_t)_{t \in [0,T]}$, with $N_0 = 0$ and constant intensity λ, independent of $(W_t)_{t \in [0,T]}$, to allow asset prices to move discontinuously. The compensated risky asset price process now follows

$$\frac{dS_t}{S_{t-}} = (r - \lambda m)dt + \sigma dW_t^Q + J_t dN_t, \tag{6.50}$$

under the risk-neutral equivalent martingale measure, with $(J_t)_{t \in [0,T]}$ an i.i.d. sequence of random variables valued in $[-1, \infty[$ of the form $J_t = J_i I_{[t > \tau_i[}(t)$ with $J_0 = 0$, τ_i an increasing sequence of times, and where

$$E[dN_t] = \lambda dt \qquad \text{and} \qquad E[J_t] = m.$$

Then the solution of (6.50) can be written as

$$S_T = S_0 \exp \left\{ \left(r - \frac{\sigma^2}{2} - \lambda m \right) T + \sigma W_T^Q \right\} \prod_{i=1}^{N_T} (1 + J_i).$$

Defining an option value process by $(V_t(S_t))_{t \in [0,T]}$, we apply Itô's formula, along with its extension, to Poisson processes and assume that jump risk in the market can be diversified [see Merton (1976) and the references therein], so that we can use the chosen risk-neutral measure Q. We obtain

$$[\mathcal{L}_{BS} V_t](S_t) = \lambda \left(m \frac{\partial V_t}{\partial S_t} - E[V_t(J_t S_t) - V_t(S_t)] \right).$$

The solution to this partial differential equation can still be written in the form (6.16). But closed-form expressions of the expectation and probability terms only exist for special cases. Two such cases were identified by Merton (1976), first when $N_t \in \{0, 1\}$ and $J_1 = -1$, i.e., the case where there exists the possibility of a single jump that puts the risky asset into the absorbing state 0. Then the solution for, say, a call option is V_{BS} but with a modified risk-free rate $r + \lambda$. The second case is where $J_t > -1$ and

$$\ln J_t \sim N(\mu, \gamma^2),$$

so that

$$m = e^{\mu + \frac{1}{2}\gamma^2}.$$

Then

$$V_0^C = \sum_{i=0}^{\infty} \frac{e^{-\lambda m T}(\lambda m T)^i}{i!} V_B(\sigma_i, r_i),$$

where the risk-free rate is given by

$$r_i = r + \frac{i}{T} \left(\mu + \frac{\gamma^2}{2} \right) - \lambda(m - 1)$$

and the volatility by

$$\sigma_n = \sqrt{\sigma^2 + \frac{i}{T}\gamma^2}.$$

Another, semiclosed-form, expression exists when J_t are exponentially distributed (Kou 2002), but usually the solution has to be written in terms of Fourier transforms that need to be solved numerically.

A Model for Spot Prices in the Electricity Market

Another application of jump diffusion processes is the market for electricity spot prices. The latter tend to exhibit severe sudden spikes. In Branger et al. (2010) the spot price has been modelled as the sum of a (deterministic) seasonal component plus a jump-diffusion component $X = (X_t)_{t \in \mathbb{R}_+}$, and a spike component $Y = (Y_t)_{t \in \mathbb{R}_+}$. Under suitable assumptions, the dynamics of X and Y, under a risk neutral probability measure Q, are given by

$$dX_t = k\left(-\frac{\lambda_t \sigma}{k} - X_t\right) + \sigma dW_t^Q + J_t^X dN_t^X; \tag{6.51}$$

$$dY_t = -\gamma Y_t + J_t^Y dN_t^X; \tag{6.52}$$

where

(i) $W = (W_t^Q)_{t \in \mathbb{R}_+}$ is a standard Wiener process under Q;

(ii) $N^X = (N_t^X)_{t \in \mathbb{R}_+}$ is a time inhomogeneous Poisson process with deterministic intensity $(h_t^X)_{t \in \mathbb{R}_+}$;

(iii) $N^Y = (N_t^Y)_{t \in \mathbb{R}_+}$ is a time inhomogeneous Poisson process with deterministic intensity $(h_t^Y)_{t \in \mathbb{R}_+}$;

(iv) $J_t^X, t \in \mathbb{R}_+$ is a family of real valued random variables with time dependent pdf $g_t^X, t \in \mathbb{R}_+$;

(v) $J_t^Y, t \in \mathbb{R}_+$ is a family of real valued random variables with time dependent pdf $g_t^Y, t \in \mathbb{R}_+$.

Finally, $\lambda_t dt$ stands for the compensation of the price for taking one unit of diffusion risk due to W_t^Q; it is assumed a time dependency.

It is left as an exercise (Exercise 6.14) to recognize that the final value problem for the prize $V(x, y; t)$ of a European-style option is given by

$$\frac{\partial}{\partial t}V + (\mathcal{L}_D + \mathcal{L}_J - r)V = 0, \tag{6.53}$$

subject to a suitable final value.

The diffusion part leads to the differential operator

$$\mathcal{L}_D v(x, y; t) := \frac{\sigma^2}{2}\frac{\partial^2}{\partial x^2}v(x, y; t) + k\left(-\frac{\lambda_t \sigma}{k} - x\right)\frac{\partial}{\partial x}v(x, y; t), \tag{6.54}$$

while the jump part leads to the integro-differential operator

$$\mathcal{L}_J v(x, y; t) := h_t^X \int_{-\infty}^{+\infty} (v(x + z^X, y; t) - v(x, y; t))g_t^X(z^X)dz^X$$

$$+ h_t^Y \int_{-\infty}^{+\infty} (v(x, y + z^Y; t) - v(x, y; t))g_t^Y(z^Y)dz^Y$$

$$- \gamma y \frac{\partial}{\partial y}v(x, y; t). \tag{6.55}$$

For further details, and for the well-posedness and numerical solution of the above final value problem, see, Branger et al. (2010).

Further extensions of the classical Black-Scholes model including fractional Brownian noise, the reader may refer, e.g., to Dai-Heyde (1996), Zähle (2002) and the references therein. In Shevchenko (2014) SDE models including Wiener noise, fractional Brownian motion, and jumps are discussed together with their relevance in economic and financial models.

Stochastic Volatility

As demonstrated in Fig. 6.1, the volatility skew across K/S need not be a straight line, but it may have a pronounced curvature or so-called *smile*. A model that gives this feature and on top of it adjusts for the dynamic shortcomings of local volatility is *stochastic volatility*, where $(\sigma_t)_{t\in[0,T]}$ is a stochastic process. In a general representation we can write it as a system of SDEs:

$$dS_t = \mu(\sigma_t, S_t, t)dt + f(\sigma_t, S_t, t)dW_t^S,$$
$$dg(\sigma_t) = h(\sigma_t, t)dt + w(\sigma_t, t)dW_t^\sigma,$$
$$< dW_t^S, dW_t^\sigma > = \rho dt.$$

Stochastic volatility models have an additional driving factor and increase the tail density of the distribution of S_t. Two popular specific models are Heston and SABR. We will give a brief outline on how these models are applied. First, the Heston model (Heston 1993) has the specific form

$$dS_t = \mu S_t dt + \sqrt{v_t} S_t dW_t^S,$$
$$dv_t = a(m - v_t)dt + b\sqrt{v_t}dW_t^v,$$
$$< dW_t^S, dW_t^v > = \rho dt,$$

where $v_t = \sigma_t^2$ is a stochastic variance process. If we extend the Black–Scholes market with a traded option $\hat{V}_t$, then the contingent claim-hedging equation (6.9) becomes

$$V_t = H_t^{(B)}B_t + H_t^{(S)}S_t + H_t^{(V)}\hat{V}_t. \tag{6.56}$$

Applying Itô's formula on both sides, choosing appropriate hedges $H_t^{(B)}$, $H_t^{(S)}$, $H_t^{(V)}$, and following a similar argument on the risk-neutral drift conditions as in Proposition 6.17, it can be shown [Exercise 6.15 or see Lewis (2000)] that V_t follows the equation

$$\mathcal{L}_{BS}V_t + a'(m' - v_t)\frac{\partial V_t}{\partial v_t} + \frac{1}{2}b^2 v_t^2 \frac{\partial^2 V_t}{\partial v_t^2} + \rho b v_t S_t \frac{\partial^2 V_t}{\partial S_t \partial v_t} = 0, \tag{6.57}$$

where a' and m' are the risk-neutralized parameters under the Q measure and $\mathcal{L}_{BS}$ is employing $\sqrt{v_t}$ instead of σ. Heston (1993) derives a closed-form solution of (6.57) through Fourier transforms.

A second widely used model is referred to as SABR (Hagan et al. 2002), standing for stochastic alpha, beta, rho. It is specified by the system

$$dS_t = \sigma_t S_t^\beta dW_t^S,$$
$$d\sigma_t = \sigma_t \nu dW_t^\sigma,$$
$$< dW_t^S, dW_t^\sigma > = \rho dt,$$

with $\sigma_0 = \alpha > 0$, $0 \le \beta \le 1$, $\nu \ge 0$, and $-1 \le \rho \le 1$. For simplicity the model is typically written in forward space or, alternatively, without much loss of generality, $r = 0$. Hagan et al. (2002) show through singular perturbation analysis, by considering ν to be small, that the model has an asymptotic expansion solution similar to the Black–Scholes formula

$$V_0^C = S_0 \Phi(d_3) - K\Phi(d_4),$$

with

$$d_{3,4} = \frac{\ln \frac{S_0}{K} \pm \frac{1}{2}\hat{\sigma}^2 T}{\hat{\sigma}\sqrt{T}},$$

where the so-called implied SABR volatility approximation is given by

$$\hat{\sigma} = \frac{\alpha}{(S_0 K)^{\frac{(1-\beta)}{2}}\left[1 + \frac{(1-\beta)^2}{24}\ln^2 \frac{S_0}{K} + \frac{(1-\beta)^4}{1920}\ln^4 \frac{S_0}{K}\right]} \left(\frac{z}{x(z)}\right)$$

$$\left[1 + T\left(\frac{(\alpha - \alpha\beta)^2}{24(S_0 K)^{1-\beta}} + \frac{\rho\beta\nu\alpha}{4(S_0 K)^{\frac{1-\beta}{2}}} + \frac{(2 - 3\rho^2)\nu^2}{24}\right)\right],$$

$$z = \frac{\nu}{\alpha}(S_0 K)^{\frac{(1-\beta)}{2}}\ln \frac{S_0}{K},$$

$$x(z) = \ln \frac{\sqrt{1 - 2\rho z + z^2} + z - \rho}{1 - \rho}.$$

Typically the SABR model is used through its analytical approximation formula as an interpolation scheme between different traded options across T and K. The approximation formula is fairly precise for S_0/K not far from 1 and ν not too large.

6.5 Insurance Risk

Another very important application of stochastic processes is in the field of insurance. These are typically discrete event dynamics in a continuous-time framework. The model often has to give information about the probability and time of default or ruin of an asset or company (see, e.g., Embrechts et al. (1997)).

Ruin Probabilities

A typical one-company insurance portfolio is modeled as follows. The initial value of the portfolio is the so-called *initial reserve* $u \in \mathbb{R}_+^*$. At random times $\sigma_n \in \mathbb{R}_+^*$ (not to be confused with volatilities), a random claim $U_n \in \mathbb{R}^*$ occurs for $n \in \mathbb{N}^*$. During the time interval $]0, t] \subset \mathbb{R}_+^*$ an amount $\Pi_t \in \mathbb{R}_+^*$ of income is collected through *premia*.

The *cumulative claims process* up to time $t > 0$ is then given by

$$X_t = \sum_{k=1}^{\infty} U_k I_{[\sigma_k \leq t]}(t).$$

In this way the value of the portfolio at time t, the so-called *risk reserve*, is given by

$$R_t = u + \Pi_t - X_t.$$

The *claims surplus process* is given by

$$S_t = X_t - \Pi_t.$$

If we assume that premiums are collected at a constant rate $\beta > 0$, then

$$\Pi_t = \beta t, \qquad t > 0.$$

Now, the *time of ruin* $\tau(u)$ of the insurance company is a function of the initial reserve level. It is the first time when the claims surplus process crosses this level, namely,

$$\tau(u) := \min\{t > 0 | R_t < 0\} = \min\{t > 0 | S_t > u\}.$$

Hence, an insurance company is interested in the ruin probabilities; first, the *finite-horizon ruin probability*, which is defined as

$$\psi(u, x) := P(\tau(u) \leq x) \qquad \forall x \geq 0;$$

second, the *probability of ultimate ruin*, defined as

$$\psi(u) := \lim_{x \to +\infty} \psi(u, x) = P(\tau(u) \leq +\infty).$$

It may also be interested in the *survival probability* defined as

$$\bar{\psi}(u) = 1 - \psi(u).$$

It is clear that

$$\psi(u, x) = P\left(\max_{0 \leq t \leq x} S_t > u\right).$$

The preceding model shows that the marked point process $(\sigma_n, U_n)_{n \in \mathbb{N}^*}$ on $(\mathbb{R}_+^* \times \mathbb{R}_+^*)$ plays an important role. As a particular case, we consider the marked Poisson process with *independent marking*, i.e., the case in which $(\sigma_n)_{n \in \mathbb{N}^*}$ is a Poisson process on $\mathbb{R}_+^*$ and $(U_n)_{n \in \mathbb{N}^*}$ is a family of i.i.d. $\mathbb{R}_+^*$-valued random variables, independent of the underlying point process $(\sigma_n)_{n \in \mathbb{N}^*}$. In this case, we have that the interoccurrence times between claims $T_n = \sigma_n - \sigma_{n-1}$ (with $\sigma_0 = 0$) are independent and identically exponentially distributed random variables with a common parameter $\lambda > 0$ (Rolski et al. 1999). In this way the number of claims N_t during $]0, t]$, $t > 0$, i.e., the underlying counting process

$$N_t = \sum_{k=1}^{\infty} I_{[\sigma_k \leq t]}(t),$$

is a Poisson process on $\mathbb{R}_+^*$ with intensity λ. Now, let the claim sizes U_n be i.i.d. with common cumulative distribution function F_U and let $(U_n)_{n \in \mathbb{N}^*}$ be independent of $(N_t)_{t \in \mathbb{R}_+}$. We may notice that in this case the cumulative claim process

$$X_t = \sum_{k=1}^{N_t} U_k = \sum_{k=1}^{\infty} U_k I_{[\sigma_k \leq t]}(t), \qquad t > 0,$$

is a *compound Poisson process*. Clearly, the latter has stationary independent increments and, in fact, is a Lévy process, so that we can state the following theorem.

Theorem 6.19 (Karlin and Taylor 1981, p. 428). *Let $(X_t)_{t \in \mathbb{R}_+^*}$ be a stochastic process having stationary independent increments, and let $X_0 = 0$. $(X_t)_{t \in \mathbb{R}_+^*}$ is then a compound Poisson process if and only if its characteristic function $\phi_{X_t}(z)$ is of the form*

$$\phi_{X_t}(z) = \exp\left\{-\lambda t (1 - \phi(z))\right\}, \qquad z \in \mathbb{R},$$

where $\lambda > 0$ and ϕ is a characteristic function.

With respect to the preceding model, ϕ is the common characteristic function of the claims $U_n, n \in \mathbb{N}^*$. If μ and σ^2 are the mean and the variance of U_1, respectively, we have

$$E[X_t] = \mu \lambda t,$$
$$Var[X_t] = \left(\sigma^2 + \mu^2\right) \lambda t.$$

We may also obtain the cumulative distribution function of X_t through the following argument:

$$P(X_t \leq x) = P\left(\sum_{k=1}^{N_t} U_k \leq x\right)$$

$$= \sum_{n=0}^{\infty} P\left(\sum_{k=1}^{N_t} U_k \leq x \,\middle|\, N_t = n\right) P(N_t = n)$$

$$= \sum_{n=0}^{\infty} \frac{(\lambda t)^n e^{-\lambda t}}{n!} F^{(n)}(x)$$

for $x \geq 0$ (it is zero otherwise), where

$$F^{(n)}(x) = P(U_1 + \cdots + U_n \leq x), \qquad x \geq 0,$$

with

$$F^{(0)}(x) = \begin{cases} 1 & \text{for } x \geq 0, \\ 0 & \text{for } x < 0. \end{cases}$$

In the special case of exponentially distributed claims, with common parameter $\mu > 0$, we have

$$F_U(u) = P(U_1 \leq u) = 1 - e^{-\mu u}, \qquad u \geq 0,$$

so that $U_1 + \cdots + U_n$ follows a gamma distribution with

$$F_U^{(n)}(u) = 1 - \sum_{k=0}^{n-1} \frac{(\mu u)^k e^{-\mu u}}{k!} = \frac{\mu^n}{(n-1)!} \int_0^u e^{-\mu v} v^{n-1} dv$$

for $n \geq 1$, $u \geq 0$. The following theorem holds for exponentially distributed claim sizes.

Theorem 6.20. *Let*

$$F_U(u) = 1 - e^{-\mu u}, \qquad u \geq 0.$$

Then

$$\psi(u, x) = 1 - e^{-\mu u - (1+c)\lambda x} g(\mu u + c\lambda x, \lambda x),$$

where

$$c = \mu \frac{\beta}{\lambda},$$

$$g(z, \theta) = J(\theta z) + \theta J^{(1)}(\theta z) + \int_0^z e^{z-v} J(\theta v) dv - \frac{1}{c} \int_0^{c\theta} e^{c\theta - v} J\left(zc^{-1}v\right) dv,$$

with $\theta > 0$. Here

$$J(x) = \sum_{n=0}^{\infty} \frac{x^n}{n! n!}, \qquad x \geq 0,$$

and $J^{(1)}(x)$ is its first derivative.

Proof. See, e.g., Rolski et al. (1999, p. 196). □

For the general compound Poisson model, we may provide information about the finite-horizon ruin probability $P(\tau(u) \leq x)$ by means of martingale methods. We note again that, in terms of the claims surplus process S_t, we have

$$\tau(u) = \min\{t | S_t > u\}, \qquad u \geq 0,$$

and

$$\psi(u, x) = P(\tau(u) \leq x), \qquad x \geq 0, u \geq 0.$$

The claims surplus process is then given by

$$S_t = \sum_{k=1}^{N_t} U_k - \beta t,$$

where $\lambda > 0$ is the arrival rate, β the premium rate, and F_U the claim size distribution. Let

$$Y_t = \sum_{k=1}^{N_{t-}} U_k, \quad t \geq 0,$$

be the left-continuous version of the cumulative claim size X_t. Based on the notion of reversed martingales (Rolski et al. 1999, p. 434), it can be shown that the process

$$Z_t = X^*_{x-t}, \quad t \in [0, x[, x > 0,$$

with

$$X^*_t = \frac{Y_t}{u + \beta t} + \int_t^x \frac{Y_v}{v} \frac{u}{(u + \beta v)^2} dv, \qquad 0 < t \leq x,$$

for $u \geq 0$ and $x > 0$, is an $\mathcal{F}_t^X$-martingale. Let

$$\tau^0 = \sup\{v | v \geq x, S_v \geq u\},$$

and $\tau^0 = 0$ if $S(v) < u$ for all $v \in [0, x]$. Then $\tau := x - \tau^0$ is a bounded $\mathcal{F}_t^X$-stopping time. As a consequence,

$$E[Z_\tau] = E[Z_0],$$

i.e.,

$$E\left[\frac{Y_x}{u + \beta x} \bigg| Y_x \leq u + \beta x\right] = E\left[\frac{Y_{\tau^0}}{u + \beta \tau^0} + \int_{\tau^0}^x \frac{Y_v}{v} \frac{u}{(u + \beta v)^2} dv \bigg| S_{x-} \leq u\right].$$

On the other hand, we have

$$P(\tau(u) > x) = P\left(S_x \leq u \cap \tau^0 = 0\right) = P(S_x \leq u) - P\left(S_x \leq u \cap \tau^0 > 0\right).$$

Now, since

$$Y_{\tau^0} = u + \beta\tau^0 \qquad \text{for } \tau^0 > 0,$$

we have

$$E\left[\frac{Y_{\tau^0}}{u+\beta\tau^0}\bigg| S_x \le u\right] = E\left[\frac{Y_{\tau^0}}{u+\beta\tau^0}\bigg| S_x \le u \cap \tau^0 > 0\right]$$
$$= P\left(S_x \le u \cap \tau^0 > 0\right).$$

Thus, for $u > 0$, we have the following result.

Theorem 6.21 (Rolski et al. 1999, p. 434). *For all $u \ge 0$ and $x > 0$,*

$$1 - \psi(u,x) = \max\left\{E\left[1 - \frac{Y_x}{u+\beta x}\right], 0\right\} + E\left[\int_{\tau^0}^x \frac{Y_v}{v}\frac{u}{(u+\beta v)^2}dv\bigg| S_x \le u\right].$$

In particular, for $u = 0$,

$$1 - \psi(0,x) = \max\left\{E\left[1 - \frac{Y_x}{\beta x}\right], 0\right\}.$$

A Stopped Risk Reserve Process

Consider the risk reserve process

$$R_t = u + \beta t - \sum_{k=1}^{N_t} U_k.$$

A useful model for stopping the process is to stop R_t at the time of ruin $\tau(u)$ and let it jump to a *cemetery* state. In other words, consider the process

$$X_t = \begin{cases} (1, R_t) & \text{if } t \le \tau(u), \\ (0, R_{\tau(u)}) & \text{if } t > \tau(u). \end{cases}$$

The process $(X_t, t)_{t\in\mathbb{R}_+}$ is a *piecewise deterministic Markov process* as defined in Davis (1984). The infinitesimal generator of $(X_t, t)_{t\in\mathbb{R}_+}$ is given by

$$\mathcal{A}g(y,t)$$
$$= \frac{\partial g}{\partial t}(y,t) + I_{[y\ge 0]}(y)\left(\beta\frac{\partial g}{\partial y}(y,t) + \lambda\left(\int_0^y g(y-v,t)dF_U(v) - g(y,t)\right)\right)$$

for g satisfying sufficient regularity conditions, so that it is in the domain of $\mathcal{A}$ (Rolski et al. 1999, p. 467). If g does not depend explicitly upon time and $g(y) = 0$ for $y < 0$, then the infinitesimal generator reduces to

$$\mathcal{A}g(y) = \beta\frac{dg}{dy}(y) + \lambda\left(\int_0^y g(y-v)dF_U(v) - g(y)\right).$$

The following theorem holds.

Theorem 6.22. *Under the preceding assumptions,*

1. *The only solution $g(y)$ to $\mathcal{A}g(y) = 0$, such that $g(0) > 0$ and $g(y) = 0$, for $y \in]-\infty, 0[$, is the survival function $\bar{\psi}(y) = P(\tau(u) = +\infty)$.*
2. *Let $x > 0$ be fixed and let $g(y,t)$ solve $\mathcal{A}g = 0$ in $(\mathbb{R} \times [0,x])$ with boundary condition $g(y,x) = I_{[y \geq 0]}(y)$. Then $g(y,0) = P(\tau(y) > x)$ for any $y \in \mathbb{R}$, $x \in \mathbb{R}_+^*$.*

6.6 Exercises and Additions

6.1. Let $(\mathcal{F}_n)_{n \in \mathbb{N}}$ and $(\mathcal{G}_n)_{n \in \mathbb{N}}$ be two filtrations on a common probability space $(\Omega, \mathcal{F}, P)$ such that $\mathcal{G}_n \subseteq \mathcal{F}_n \subseteq \mathcal{F}$ for all $n \in \mathbb{N}$; we say that a real-valued discrete-time process $(X_n)_{n \in \mathbb{N}}$ is an $(\mathcal{F}_n, \mathcal{G}_n)$-*martingale* if and only if

- $(X_n)_{n \in \mathbb{N}}$ is an $\mathcal{F}_n$-adapted integrable process;
- For any $n \in \mathbb{N}$, $E[X_{n+1} - X_n | \mathcal{G}_n] = 0$.

A process $C = (C_n)_{n \geq}$ is called $\mathcal{G}_n$-*predictable* if C_n is $\mathcal{G}_{n-1}$-measurable. Given $N \in \mathbb{N}$, we say that a $\mathcal{G}_n$-predictable process C is *totally bounded by time N* if

- $C_n = 0$ almost surely for all $n > N$;
- There exists a $K \in \mathbb{R}_+$ such that $C_n < k$ almost surely for all $n \leq N$.

Let C be a $\mathcal{G}_n$-predictable process, totally bounded by time N. We say that it is a risk-free $\{\mathcal{G}_n\}_N$-strategy if, further,

$$\sum_{i=1}^{N} C_i(X_i - X_{i-1}) \geq 0 \quad \text{a.s.}, \qquad P\left(\sum_{i=1}^{N} C_i(X_i - X_{i-1}) > 0\right) > 0.$$

Show that there exists a risk-free $\{\mathcal{G}_n\}_N$-strategy for $X = (X_n)_{n \in \mathbb{N}}$ if and only if there does not exist an equivalent measure $\tilde{P}$ such that $X_{n \wedge N}$ is a $(\mathcal{F}_n, \mathcal{G}_n)$-martingale under $\tilde{P}$. This is an extension of the first fundamental theorem of asset pricing, Theorem 6.7. See also Dalang et al. (1990) and Aletti and Capasso (2003).

6.2. Given a filtration $(\mathcal{F}_n)_{n \in \mathbb{N}}$ on a probability space $(\Omega, \mathcal{F}, P)$, the filtration $\mathcal{F}_n^m := \mathcal{F}_{n-m}$ is called an m-*delayed filtration*. An $\mathcal{F}_n$-adapted, integrable stochastic process $X = (X_n)_{n \in \mathbb{N}}$ is an m-*martingale* if it is an $(\mathcal{F}_n, \mathcal{F}_n^m)$-martingale (Problem 6.1). Find a real-valued (2-martingale) X where no profit is available during any unit of time, i.e.,

$$\forall i, \qquad P(X_i - X_{i-1} > 0) > 0, \qquad P(X_i - X_{i-1} < 0) > 0,$$

but admits a risk-free $\{\mathcal{F}_n^1\}_3$-strategy C, i.e.,

$$\sum_{i=1}^{3} C_i(X_i - X_{i-1}) \geq 0 \quad \text{a.s.}, \qquad P\left(\sum_{i=1}^{3} C_i(X_i - X_{i-1}) > 0\right) > 0$$

(Aletti and Capasso 2003).

6.3. With reference to Problem 6.2, consider a risk-free $\{\mathcal{F}_n\}_N$-strategy. Show that there exists an $n \in \{1, \ldots, N\}$ such that

$$C_n(X_n - X_{n-1}) \geq 0 \quad \text{a.s.}, \qquad P\left(C_n(X_n - X_{n-1}) > 0\right) > 0,$$

i.e., if no profit is available during any unit of time, then we cannot have a profit up to time N (Aletti and Capasso 2003).

6.4. Consider a Black–Scholes market with $r = \mu = 0$ and $\sigma = 1$. Then a value-conserving strategy $H_t^{(S)} = 1/\sqrt{T}$ yields a portfolio value of $\Pi_t = \int_0^t dW_s/\sqrt{T-s}$. Show that

$$P(\Pi_\tau \geq c, 0 \leq \tau \leq T) = 1,$$

with c an arbitrary constant and τ a stopping time. Hence any amount can be obtained in finite time. It is easy to see that (unlike conditions 1 and 2) condition 3 of Proposition 6.6 is not automatically satisfied (e.g., Duffie 1996).

6.5. For both the Black–Scholes and the Bachelier models calculate the hedge ratio $H_t^{(S)} = \partial V_t^C/\partial S_t$ for a call option. The latter is also called the *delta* of an option. Furthermore, calculate $\partial V_t^C/\partial t$ (*theta*), $\partial V_t^C/\partial \sigma$ (*vega*), and $\partial^2 V_t^C/\partial S_t^2$ (*gamma*). These hedge ratios are called the *Greeks* of an option.

6.6. Show that in a Black–Scholes market, when using the martingale measure Q^* associated with S_t as the numeraire asset, the probability $Q^*(S_T > K) = \Phi(d_1)$.

6.7. For a drifting Wiener process $X_t = W_t + \mu t$, where W_t is P-Brownian motion and its maximum value attained is

$$M_t = \max_{\tau \in [0,t]} X_\tau,$$

apply the reflection principle and Girsanov's theorem to show that

$$P(X_T \leq a \cap M_T \geq b) = e^{2\mu b} P(X_T \geq 2b - a + 2\mu T)$$

for $a \leq b$ and $b \geq 0$. See also Musiela and Rutkowski (1998) or Borodin and Salminen (1996).

6.8. Referring to the barrier option Problem (6.26), show that

$$Q\left(\min_{t \in [0,T]} W_t^Q > g(b) \cap W_T^Q > g(K)\right)$$
$$= \Phi(d_1) - \left(\frac{b}{S_0}\right)^{\frac{2r}{\sigma^2}-1} \Phi\left(\frac{\ln\frac{b^2}{S_0 K} + \left(r - \frac{1}{2}\sigma^2\right)T}{\sigma\sqrt{T}}\right), \qquad (6.58)$$

where d_1 is given by (6.19). From (6.58) obtain the joint density of S_T and its minimum over $[0, T]$, and thus solve

$$E_Q\left[S_T I_{[\min_{t \in [0,T]} S_t > b \cap S_T > K]}\right].$$

6.9. Two American options have an explicit solution:

- (American Digital Call) It pays 1 unit of currency if $S_T > K$ at expiry and can also be exercised early. Show that its value under Black–Scholes is

$$V_0 = \left(\frac{K}{S_0}\right)^{\frac{2r}{\sigma^2}} \Phi\left(\frac{\ln\frac{S_0}{K} - (r + \frac{1}{2}\sigma^2)T}{\sigma\sqrt{T}}\right) + \frac{S_0}{K}\Phi\left(\frac{\ln\frac{S_0}{K} + (r + \frac{1}{2}\sigma^2)T}{\sigma\sqrt{T}}\right)$$

by considering that you need to solve

$$V_0 = E_Q\left[e^{-r\tau}I_{[\tau \le T]}\right],$$

with the first exit time $\tau = \inf\{t|S_t \ge K\}$.
- (American Perpetual Put) It pays $K - S_\tau$ upon exercise at time τ but has no expiration date. Show that its value under Black–Scholes is

$$V_0 = \frac{\sigma^2}{2r}\hat{S}^{1-\frac{2r}{\sigma^2}}S_0^{-\frac{2r}{\sigma^2}},$$

where

$$\hat{S} = \frac{K}{1 + \frac{\sigma^2}{2r}}$$

is the time-homogeneous optimal exercise level of $(S_t)_{t \ge 0}$. Consider that perpetual options have no theta, namely $\partial V_t/\partial t = 0$, thus turning the Black–Scholes partial differential equation (6.12) into an ordinary differential equation, as well as its boundary conditions at $\hat{S}$.

6.10. Why can it be conjectured that the bond equation in the Vasicek model is of the form

$$B_t^{(T)} = e^{C(t,T) - D(t,T)r_t}? \tag{6.59}$$

Derive the results (6.36) and (6.37). [*Hint:* Derive a partial differential equation for $B_t^{(T)}$ using a similar argumentation as for the Black–Scholes equation. Substitute (6.59) and solve.] Note that interest rate models whose discount-bond solution is of this form are called *affine* [see Problem 2.43 and Hunt and Kennedy (2000)].

6.11. In the Brace–Gatarek–Musiela model, derive the nonarbitrage drifts (6.46) of the log-normal forward rates $F_t^{(i)}$. (*Hint:* In (6.45) note that $\frac{B_t^{(i)}}{B_t^{(N)}}$ is a martingale under Q_N. Given this, derive the drift as

$$\mu_i = -\frac{d\left\langle \ln F_t^{(i)}, \ln \frac{B_t^{(i)}}{B_t^{(N)}}\right\rangle}{dt}$$

and solve.)

6.12. A so-called *par swap rate* $S(t, T_s, T_e)$ has to satisfy

$$S(t, T_s, T_e) = \frac{\sum_{i=s+1}^{e} F_t^{(i-1)} B_t^{(i)} (T_i - T_{i-1})}{A_{s,e}}, \tag{6.60}$$

where

$$A_{s,e} = \sum_{i=s+1}^{e} B_t^{(i)} (t_i - t_{i-1})$$

is called an *annuity*. If relationship (6.44) holds and the forward rates are driven by (6.43), then show that the swap rate process can approximately be written as

$$dS(t, T_s, T_e) = \sigma_{S(t,T_s,T_e)} S(t, T_s, T_e) dW_t^{A_{s,e}},$$

where $\sigma_{S(t,T_s,T_e)}$ is deterministic and $dW_t^{A_{s,e}}$ is a Brownian motion under the martingale measure induced by taking $A_{s,e}$ as numeraire. (*Hint:* Assume that the coefficients of all the forward rates $F_t^{(i)}$ in (6.60) are approximately deterministic, invoke Itô's formula, and apply Girsanov's theorem.) Convince yourself that a contingent claim with a swap rate as underlying (a so-called constant maturity swap or CMS payoff) is a nonlinear instrument.

6.13. The *constant elasticity of variance* market (Cox 1996; Boyle and Tian 1999) is a Black–Scholes market where the risky asset follows

$$dS_t = \mu S_t dt + \sigma S_t^{\frac{\alpha}{2}} dW_t$$

for $0 \leq \alpha < 2$. Show that this market has no equivalent risk-neutral measure.

6.14. For the spot price model show that Equation (6.53) holds.

6.15. Find the appropriate hedge ratios $H_t^{(B)}$, $H_t^{(S)}$, $H_t^{(V)}$ in (6.56) that eliminate the Wiener processes dW_t^S and dW_t^S and thus derive (6.57).

7

Applications to Biology and Medicine

7.1 Population Dynamics: Discrete-in-Space–Continuous-in-Time Models

In the chapter on stochastic processes, the Poisson process was introduced as an example of an RCLL nonexplosive counting process. Furthermore, we reviewed a general theory of counting processes as point processes on a real line within the framework of martingale theory and dynamics. Indeed, for these processes, under the usual regularity assumptions, we can invoke the Doob–Meyer decomposition theorem [see (2.135)*ff*] and claim that any nonexplosive RCLL process $(X_t)_{t \in \mathbb{R}_+}$ satisfies a generalized SDE of the form

$$dX_t = dA_t + dM_t, \tag{7.1}$$

subject to a suitable initial condition. Here A is the compensator of the process, modeling the "evolution," and M is a martingale, representing the "noise."

As was mentioned in the sections on counting and marked point processes, a counting process $(N_t)_{t \in \mathbb{R}_+}$ is a random process that counts the occurrence of certain events over time, namely N_t being the number of such events having occurred during the time interval $]0, t]$. We have noticed that a nonexplosive counting process is RCLL with upward jumps of magnitude 1; here, we impose the initial condition $N_0 = 0$, almost surely. Since we are dealing with those counting processes that satisfy the conditions of the local Doob–Meyer decomposition Theorem 2.143, a nondecreasing predictable process $(A_t)_{t \in \mathbb{R}_+}$ (the compensator) exists such that $(N_t - A_t)_{t \in \mathbb{R}_+}$ is a right-continuous local martingale. Further, we assume that the compensator is absolutely continuous with respect to the usual Lebesgue measure on $\mathbb{R}_+$. In this case we say that $(N_t)_{t \in \mathbb{R}_+}$ has a (predictable) intensity $(\lambda_t)_{t \in \mathbb{R}_+}$ such that

© Springer Science+Business Media New York 2015
V. Capasso, D. Bakstein, *An Introduction to Continuous-Time Stochastic Processes*, Modeling and Simulation in Science, Engineering and Technology, DOI 10.1007/978-1-4939-2757-9_7

$$A_t = \int_0^t \lambda_s ds, \qquad \text{for any } t \in \mathbb{R}_+,$$

and SDE (7.1) can be rewritten as

$$dX_t = \lambda_t dt + dM_t.$$

If the process is integrable and λ is left-continuous with right limits (LCRL), one can easily show that

$$\lambda_t = \lim_{\Delta t \to 0+} \frac{1}{\Delta t} E[N_{t+\Delta t} - N_t | \mathcal{F}_{t-}] \qquad \text{a.s.,}$$

and if we further assume the simplicity of the process, we also have

$$\lambda_t = \lim_{\Delta t \to 0+} \frac{1}{\Delta t} P(N_{t+\Delta t} - N_t = 1 | \mathcal{F}_{t-}) \qquad \text{a.s.;}$$

the latter means that $\lambda_t dt$ is the conditional probability of a new *event* during $[t, t+dt)$ given the history of the process during $[0, t)$. It really represents the model of evolution of the counting process, similar to classical deterministic differential equations.

Example 7.1. Let X be a nonnegative real random variable with absolutely continuous probability law having density f, cumulative distribution function F, survival function $S = 1 - F$, and hazard rate function $\alpha(t) = \dfrac{f(t)}{S(t)}$, $t > 0$. Assume

$$\int_0^t \alpha(s)ds = -\ln(1 - F(t)) < +\infty, \qquad \text{for any } t \in \mathbb{R}_+,$$

but

$$\int_0^{+\infty} \alpha(t)dt = +\infty.$$

Define the univariate process N_t by

$$N_t = I_{[X \leq t]}(t)$$

and let $(\mathcal{N}_t)_{t \in \mathbb{R}_+}$ be the filtration the process generates, i.e.,

$$\mathcal{N}_t = \sigma(N_s, s \leq t) = \sigma\left(X \wedge t, I_{[X \leq t]}(t)\right).$$

Define the left-continuous adapted process Y_t by

$$Y_t = I_{[X \geq t]}(t) = 1 - N_{t-}.$$

It can be easily shown [e.g., Andersen et al. (1993)] that N_t admits

$$A_t = \int_0^t Y_s \alpha(s) ds$$

as a compensator and hence N_t has stochastic intensity λ_t defined by

$$\lambda_t = Y_t \alpha(t), \qquad t \in \mathbb{R}_+.$$

In other words,

$$N_t - \int_0^{X \wedge t} \alpha(s) ds$$

is a local martingale. Here $\alpha(t)$ is a deterministic function, while Y_t, clearly, is a predictable process. This is a first example of what is known as a multiplicative intensity model.

Example 7.2. Let X be a random time as in the previous example, and let U be another random time, i.e., a nonnegative real random variable. Consider the random variable $T = X \wedge U$ and define the processes

$$N_t = I_{[T \leq t]} I_{[X \leq U]}(t)$$

and

$$N_t^U = I_{[T \leq t]} I_{[U < X]}(t)$$

and the filtration

$$\mathcal{N}_t = \sigma \left(N_s, N_s^U, s \leq t \right).$$

The hazard rate function α of X is known as the *net hazard rate*; it is given by

$$\alpha(t) = \lim_{h \to 0+} \frac{1}{h} P[X \leq t + h | X > t].$$

On the other hand, the quantity

$$\alpha^+(t) = \lim_{h \to 0+} \frac{1}{h} P[X \leq t + h | X > t, U > t]$$

is known as the *crude hazard rate* whenever the limit exists. In this case,

$$N_t - \int_0^t I_{[T \geq t]} \alpha(s) ds$$

is a local martingale.

Birth-and-Death Processes

A Markov birth-and-death process provides an example of a bivariate counting process. Let $(X_t)_{t \in \mathbb{R}_+}$ be the size of a population subject to a birth rate λ and a death rate μ. Then the infinitesimal transition probabilities are

$$P\left(X_{t+\Delta t} = j \mid X_{t-} = h\right) = \begin{cases} \lambda h \Delta t + o(\Delta t) & \text{if } j = h+1, \\ \mu h \Delta t + o(\Delta t) & \text{if } j = h-1, \\ 1 - (\lambda h + \mu h)\Delta t + o(\Delta t) & \text{if } j = h, \\ o(\Delta t) & \text{otherwise.} \end{cases}$$

Let $N_t^{(1)}$ and $N_t^{(2)}$ be the number of births and deaths, respectively, up to time $t \geq 0$, assuming $N_0^{(1)} = 0$ and $N_0^{(2)} = 0$. Then

$$(\mathbf{N}_t)_{t \in \mathbb{R}_+} = \left(N_t^{(1)}, N_t^{(2)}\right)$$

is a bivariate counting process with intensity process $(\lambda X_{t-}, \mu X_{t-})_{t \in \mathbb{R}_+}$ (Figs. 7.1 and 7.2). This is an example of a formulation of a Markov process with countable state space as a counting process. In particular, we may write an SDE for X_t as follows:

$$dX_t = \lambda X_{t-} dt - \mu X_{t-} dt + dM_t,$$

where M_t is a suitable martingale noise.

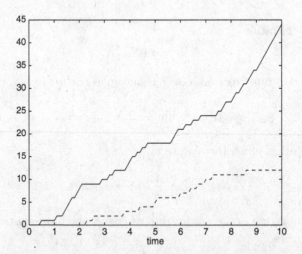

Fig. 7.1. Simulation of a birth-and-death process with birth rate $\lambda = 0.2$, death rate $\mu = 0.05$, initial population $X_0 = 10$, time step $dt = 0.1$, and interval of observation $[0, 10]$. The *continuous line* represents the number of births $N_t^{(1)}$; the *dashed line* represents the number of deaths $N_t^{(2)}$

A Model for Software Reliability

Let N_t denote the number of software failures detected during the time interval $]0, t]$, and suppose that F is the true number of faults existing in the software at time $t = 0$. In the Jelinski–Moranda model (Jelinski and Moranda 1972) it is assumed that N_t is a counting process with intensity

$$\lambda_t = \rho(F - N_{t-}),$$

where ρ is the individual failure rate (Fig. 7.3). This model corresponds to a pure death process in which the total initial population F usually is unknown, as is the rate ρ.

Contagion: The Simple Epidemic Model

Epidemic systems provide models for the transmission of a contagious disease within a population. In the "simple epidemic model" (Bailey 1975; Becker 1989), the total population N is divided into two main classes:

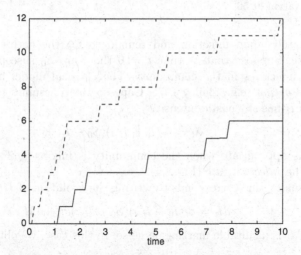

Fig. 7.2. Simulation of a birth-and-death process with birth rate $\lambda = 0.09$, death rate $\mu = 0.2$, initial population $X_0 = 10$, time step $dt = 0.1$, and interval of observation $[0, 10]$. The *continuous line* represents the number of births $N_t^{(1)}$; the *dashed line* represents the number of deaths $N_t^{(2)}$

(S) The class of susceptibles, including those individuals capable of contracting the disease and becoming infectives themselves.

(I) The class of infectives, including those individuals who, having contracted the disease, are capable of transmitting it to susceptibles.

Let I_t denote the number of individuals who have been infected during the time interval $]0, t]$. Assume that individuals become infectious themselves

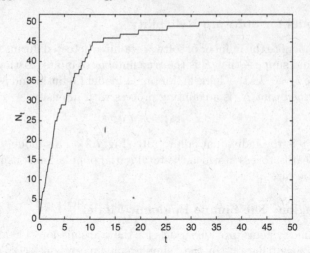

Fig. 7.3. Simulation of a model for software reliability: individual failure rate $\rho = 0.2$, true initial number of faults $F = 50$, time step $dt = 0.1$, and interval of observation $[0, 50]$

immediately upon infection and remain so for the entire duration of the epidemic. Suppose that at time $t = 0$ there are S_0 susceptible individuals and I_0 infectives in the community. The classical model based on the *law of mass action* (e.g., Bailey 1975; Capasso 1993) assumes that the counting process I_t has stochastic intensity

$$\lambda_t = \beta_t(I_0 + I_{t-})(S_0 - I_{t-}),$$

which is appropriate when the community is mixing uniformly. Here β_t is called the *infection rate* (Fig. 7.4).

Formally, this corresponds to writing the evolution of $I(t)$ via the SDE

$$dI_t = \beta_t(I_0 + I_{t-})(S_0 - I_{t-})dt + dM_t,$$

where M_t is a suitable martingale noise. In this case, we obtain

$$\langle M \rangle_t = \int_0^t \lambda_s ds$$

for the variation process $\langle M \rangle_t$, so that

$$M_t^2 - \int_0^t \lambda_s ds$$

is a zero-mean martingale. As a consequence,

$$Var[M_t] = E\left[\int_0^t \lambda_s ds\right] = E[I_t].$$

More general models can be found in Capasso (1990) and the references therein.

Contagion: The General Stochastic Epidemic

For a wide class of epidemic models the total population $(N_t)_{t \in \mathbb{R}_+}$ includes three subclasses. In addition to the classes of susceptibles $(S_t)_{t \in \mathbb{R}_+}$ and infectives $(I_t)_{t \in \mathbb{R}_+}$, already introduced in the simple model, a third class is considered, i.e., (R), the class of *removals*. This comprises those individuals who, having contracted the disease, and thus being already infectives, are no longer in the position of transmitting the disease to other susceptibles because of death, immunization, or isolation. Let us denote the number of removals as $(R_t)_{t \in \mathbb{R}_+}$.

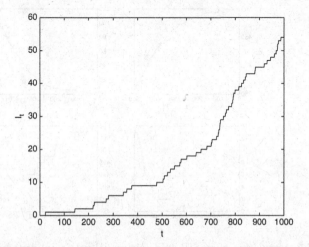

Fig. 7.4. Simulation of a simple epidemic (SI) model: initial number of susceptibles $S_0 = 500$, initial number of infectives $I_0 = 4$, infection rate (constant) $\beta = 5 \times 10^{-6}$, time step $dt = 1$, interval of observation $[0, 1000]$

The process $(S_t, I_t, R_t)_{t \in \mathbb{R}_+}$ is modeled as a multivariate jump Markov process valued in $E' = \mathbb{N}^3$. Actually, if we know the behavior of the total population process N_t, because

$$S_t + I_t + R_t = N_t \qquad \text{for any } t \in \mathbb{R}_+,$$

then we need to provide a model only for the bivariate process $(S_t, I_t)_{t \in \mathbb{R}_+}$, which is now valued in $E = \mathbb{N}^2$. The only nontrivial elements of a resulting intensity matrix Q (Sect. 2.6) are given by

- $q_{(s,i),(s+1,i)} = \alpha$, birth of a susceptible;
- $q_{(s,i),(s-1,i)} = \gamma s$, death of a susceptible;
- $q_{(s,i),(s,i+1)} = \beta$, birth of an infective;
- $q_{(s,i),(s,i-1)} = \delta i$, removal of an infective;
- $q_{(s,i),(s-1,i+1)} = \kappa s i$, infection of a susceptible.

For $\alpha = \beta = \gamma = 0$ we have the so-called *general stochastic epidemic* (e.g., Bailey 1975; Becker 1989). In this case the total population is constant (assume $R_0 = 0$; Fig. 7.5):

$$N_t \equiv N = S_0 + I_0 \qquad \text{for any } t \in \mathbb{R}_+.$$

Contagion: Diffusion of Innovations

When a new product is introduced in a market, its diffusion is due to a process of adoption by individuals who are aware of it. Classical models of

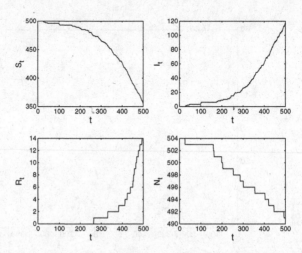

Fig. 7.5. Simulation of an SIR epidemic model with vital dynamics: initial number of susceptibles $S_0 = 500$, initial number of infectives $I_0 = 4$, initial number of removed $R_0 = 0$, birth rate of susceptibles $\alpha = 10^{-4}$, death rate of a susceptible $\gamma = 5 \times 10^{-5}$, birth rate of an infective $\beta = 10^{-5}$, rate of removal of an infective $\delta = 8.5 \times 10^{-4}$, infection rate of a susceptible $k = 1.9 \times 10^{-5}$, time step $dt = 1$, interval of observation $[0, 500]$

innovation diffusion are very similar to epidemic systems, even though in this case rates of adoption (infection) depend upon specific marketing and advertising strategies (Capasso et al. 1994; Mahajan and Wind 1986). In this case the total population N of possible consumers is divided into the following main classes:

(S)   The class of potential adopters, including those individuals capable of adopting the new product, thus themselves becoming adopters.

(A)   The class of adopters, those individuals who, having adopted the new product, are capable of transmitting it to potential adopters.

Let A_t denote the number of individuals who, by time $t \geq 0$, have already adopted a new product that has been put on the market at time $t = 0$.

Suppose that at time $t = 0$ there are S_0 potential adopters and A_0 adopters in the market. In the basic models it is assumed that all consumers are homogeneous with respect to their inclination to adopt the new product. Moreover, all adopters are homogeneous in their ability to persuade others to try new products, and adopters never lose interest but continue to inform those consumers who are not aware of the new product. Under these assumptions the classical model for the adoption rate is again based on the law of mass action (Bartholomew 1976), apart from an additional parameter $\lambda_0(t)$ that describes adoption induced by external actions, independent of the number of adopters, such as advertising, price reduction policy, etc. Then the stochastic intensity for this process is given by

$$\lambda(t) = (\lambda_0(t) + \beta_t A_{t-})(S_0 - A_{t-}),$$

which is appropriate when the community is mixing uniformly. Here β_t is called the *adoption rate* (Fig. 7.6).

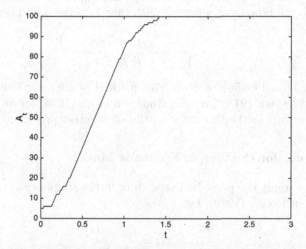

Fig. 7.6. Simulation of the contagion model for diffusion of innovations: external influence $\lambda_0(t) = 5 \times 10^{-4}t$, adoption rate (constant) $\beta = 0.05$, initial potential adopters $S_0 = 100$, initial adopters $A_0 = 5$, time step $dt = 0.01$, interval of observation $[0,3]$

Inference for Multiplicative Intensity Processes

Let

$$dN_t = \alpha_t Y_t dt + dM_t$$

be a stochastic equation for a counting process N_t, where the noise is a zero-mean martingale. Furthermore, let

$$B_s = \frac{J_{s-}}{Y_s} \qquad \text{with} \qquad J_s = I_{[Y_s > 0]}(s).$$

B_t is, like Y_t, a predictable process, so that by the integration theorem,

$$M_t^* = \int_0^t B_s dM_s$$

is itself a zero-mean martingale. Note that

$$M_t^* = \int_0^t B_s dM_s = \int_0^t B_s dN_s - \int_0^t \alpha_s J_{s-} ds,$$

so that

$$E\left[\int_0^t B_s dN_s\right] = E\left[\int_0^t \alpha_s J_{s-} ds\right],$$

i.e., $\int_0^t B_s dN_s$ is an unbiased estimator of $E[\int_0^t \alpha_s J_{s-} ds]$. If α is constant and we stop the process at a time T such that $Y_t > 0, t \in [0, T]$, then

$$\hat{\alpha} = \frac{1}{T} \int_0^T \frac{dN_s}{Y_s}$$

is an unbiased estimator of α. This method of inference is known as Aalen's method (Aalen 1978) [see also Andersen et al. (1993) for an extensive application of this method to the statistics of counting processes].

Inference for the Simple Epidemic Model

We may apply the preceding procedure to the simple epidemic model as discussed in Becker (1989). Let

$$B_s = \frac{I_{[S_s > 0]}(s)}{I_{s-} S_{s-}},$$

and suppose β is constant. Let T be such that $S_t > 0, t \in [0, T]$. Then an unbiased estimator for β would be

$$\hat{\beta} = \frac{1}{T} \int_0^T \frac{dI_s}{S_{s-} I_{s-}} = \frac{1}{T} \frac{1}{S_0 I_0} + \frac{1}{(S_0 - 1)(I_0 - 1)} + \cdots + \frac{1}{(S_T + 1)(I_T + 1)}.$$

The standard error (SE) of $\hat{\beta}$ is

$$\frac{1}{T} \left(\int_0^T B_s^2 dI_s \right)^2.$$

By the central limit theorem for martingales (Rebolledo 1980), we can also deduce that

$$\frac{\hat{\beta} - \beta}{SE(\hat{\beta})}$$

has an asymptotic $N(0,1)$ distribution, which leads to confidence intervals and hypothesis testing on the model in the usual way [see Becker (1989) and the references therein].

Inference for a General Epidemic Model

In Yang (1985) a model was proposed as an extension of the *general epidemic model* presented above. The epidemic process is modeled in terms of a multivariate jump Markov process $(S_t, I_t, R_t)_{t \in \mathbb{R}_+}$, or simply $(S_t, I_t)_{t \in \mathbb{R}_+}$, when the total population is constant, i.e.,

$$N_t := S_t + I_t + R_t = N + 1.$$

In this case, if we further suppose that $S_0 = N$, $I_0 = 1$, $R_0 = 0$, instead of using (S_t, I_t), the epidemic may be described by the number of infected individuals (not including the initial case) $M_1(t)$ and the number of removals $M_2(t) = R_t$ during $]0,t]$, $t \in \mathbb{R}_+^*$. Since we are dealing with a finite total population, the number of infected individuals and the number of removals are bounded, so that

$$E[M_k(t)] \leq N + 1, \qquad k = 1, 2.$$

The processes $M_1(t)$ and $M_2(t)$ are submartingales with respect to the history $(\mathcal{F}_t)_{t \in \mathbb{R}_+}$ of the process, i.e., the filtration generated by all relevant processes. We assume that the two processes admit multiplicative stochastic intensities of the form

$$\Lambda_1(t) = \kappa G_1(t-)(N - M_1(t-)),$$
$$\Lambda_2(t) = \delta(1 + M_1(t-) - M_2(t-)),$$

respectively, where $G_1(t)$ is a known function of infectives in circulation at time t. It models the release of pathogen material by infected individuals. Hence

$$Z_k(t) = M_k(t) - \int_0^t \Lambda_k(s)ds, \qquad k = 1, 2,$$

are orthogonal martingales with respect to $(\mathcal{F}_t)_{t \in \mathbb{R}_+}$. As a consequence, Aalen's unbiased estimators for the infection rate κ and the removal rate δ are given by

$$\hat{\kappa} = \frac{M_1(t)}{B_1(t-)}, \quad \hat{\delta} = \frac{M_1(t)}{B_1(t-)},$$

where

$$B_1(t) = \int_0^t G_1(s)(N - M_1(s))ds,$$

$$B_2(t) = \int_0^t (1 + M_1(s) - M_2(s))ds.$$

Theorem 1.3 in Jacobsen (1982, p. 163) gives conditions for a multivariate martingale sequence to converge to a multivariate normal process. If such conditions are met, then, as $N \to \infty$,

$$\begin{pmatrix} \sqrt{B_1(t)}(\hat{\kappa} - \kappa) \\ \sqrt{B_2(t)}(\hat{\delta} - \delta) \end{pmatrix} \xrightarrow{d} N\left(\begin{pmatrix} 0 \\ 0 \end{pmatrix}, \Gamma \right),$$

where

$$\Gamma = \begin{pmatrix} \kappa & 0 \\ 0 & \delta \end{pmatrix}.$$

In general, it is not easy to verify the conditions of this theorem. They surely hold for the simple epidemic model presented above, where $\delta = 0$. Related results are given in Ethier and Kurtz (1986) and Wang (1977) for a scaled infection rate $\kappa \to \frac{\kappa}{N}$ (see the following section). See also Capasso (1990) for additional models and related inference problems.

7.2 Population Dynamics: Continuous Approximation of Jump Models

A more realistic model than the general stochastic epidemic of the preceding section, which takes into account a rescaling of the force of infection due to the size of the total population, is the following (Capasso (1993)):

$$q_{(s,i),(s-1,i+1)} = \frac{\kappa}{N} si = N\kappa \frac{s}{N} \frac{i}{N}.$$

We may also rewrite

$$q_{(s,i),(s,i-1)} = \delta N \frac{i}{N},$$

so that both transition rates are of the form

$$q_{k,k+l}^{(N)} = N\beta_l \left(\frac{k}{N} \right)$$

for

$$k = (s,i)$$

and;

$$k + l = \begin{cases} (s, i - 1), \\ (s - 1, i + 1). \end{cases}$$

This model is a particular case of the following situation:
Let $E = \mathbb{Z}^d \cup \{\Delta\}$, where Δ is the point at infinity of $\mathbb{Z}^d$, $d \geq 1$. Further, let

$$\beta_l : \mathbb{Z}^d \to \mathbb{R}_+, \qquad l \in \mathbb{Z}^d,$$

$$\sum_{l \in \mathbb{Z}^d} \beta_l(k) < +\infty, \qquad \text{for each} \qquad k \in \mathbb{Z}^d.$$

For f defined on $\mathbb{Z}^d$, and vanishing outside a finite subset of $\mathbb{Z}^d$, let

$$\mathcal{A}f(x) = \begin{cases} \sum_{l \in \mathbb{Z}^d} \beta_l(x)(f(x + l) - f(x)), & x \in \mathbb{Z}^d, \\ 0, & x = \Delta. \end{cases}$$

Let $(Y_l)_{l \in \mathbb{Z}^d}$ be a family of independent standard Poisson processes. Let $X(0) \in \mathbb{Z}^d$ be nonrandom and suppose

$$X(t) = X(0) + \sum_{l \in \mathbb{Z}^d} l Y_l \left(\int_0^t \beta_l(X(s)) ds \right), \qquad t < \tau_\infty, \tag{7.2}$$

$$X(t) = \Delta, \qquad t \geq \tau_\infty, \tag{7.3}$$

where

$$\tau_\infty = \inf \{t | X(t-) = \Delta\}.$$

The following theorem holds (Ethier and Kurtz 1986, p. 327).

Theorem 7.3.

1. *Given $X(0)$, the solution of system (7.2) and (7.3) above is unique.*
2. *If $\mathcal{A}$ is a bounded operator, then X is a solution of the martingale problem for $\mathcal{A}$.*

As a consequence, for our class of models for which

$$q_{k,k+l}^{(N)} = N\beta_l \left(\frac{k}{N} \right), \qquad k \in \mathbb{Z}^d, l \in \mathbb{Z}^d,$$

we have that the corresponding Markov process, which we shall denote by $\hat{X}^{(N)}$, satisfies, for $t < \tau_\infty$:

$$\hat{X}^{(N)}(t) = \hat{X}^{(N)}(0) + \sum_{l \in \mathbb{Z}^d} l Y_l \left(N \int_0^t \beta_l \left(\frac{\hat{X}^{(N)}(s)}{N} \right) ds \right),$$

where the Y_l are independent standard Poisson processes. By setting

$$F(x) = \sum_{l \in \mathbb{Z}^d} l\beta_l(x), \qquad x \in \mathbb{R}^d$$

and

$$X^{(N)} = \frac{1}{N}\hat{X}^{(N)},$$

we have

$$X^{(N)}(t) = X^{(N)}(0) + \sum_{l \in \mathbb{Z}^d} \frac{l}{N}\tilde{Y}_l\left(N\int_0^t \beta_l\left(X^{(N)}(s)\right)ds\right)$$

$$+ \int_0^t F(X^{(N)}(s))ds, \qquad (7.4)$$

where

$$\tilde{Y}_l(u) = Y_l(u) - u$$

is the centered standard Poisson process. The state space for $X^{(N)}$ is

$$E_N = E \cap \left\{\frac{k}{N}, k \in \mathbb{Z}^d\right\}$$

for $E \subset \mathbb{R}^d$. We require that $x \in E_N$ and $\beta(x) > 0$ imply $x + \frac{l}{N} \in E_N$. The generator for $X^{(N)}$ is

$$\mathcal{A}^{(N)}f(x)$$

$$= \sum_{l \in \mathbb{Z}^d} N\beta_l(x)\left(f\left(x + \frac{l}{N}\right) - f(x)\right)$$

$$= \sum_{l \in \mathbb{Z}^d} N\beta_l(x)\left(f\left(x + \frac{l}{N}\right) - f(x) - \frac{l}{N}\nabla f(x)\right) + F(x)\nabla f(x), \qquad x \in E_N.$$

Of interest is the asymptotic behavior of the system for a large value of the scale parameter N.

By the strong law of large numbers, we know that

$$\lim_{N \to \infty} \sup_{u \leq v}\left|\frac{1}{N}\tilde{Y}_l(Nu)\right| = 0, \qquad \text{a.s.},$$

for any $v \geq 0$. As a consequence, the following theorem holds (Ethier and Kurtz 1986, p. 456).

Theorem 7.4. *Suppose that for each compact $K \subset E$*

$$\sum_{l \in \mathbb{Z}^d} |l| \sup_{x \in K} \beta_l(x) < +\infty,$$

and there exists $M_K > 0$ such that

$$|F(x) - F(y)| \le M_K |x - y|, \qquad x, y \in K;$$

suppose $X^{(N)}$ satisfies (7.4) above, with

$$\lim_{N \to \infty} X^{(N)}(0) = x_0 \in \mathbb{R}^d.$$

Then, for every $t \ge 0$,

$$\lim_{N \to \infty} \sup_{s \le t} \left| X^{(N)}(s) - x(s) \right| = 0, \qquad a.s.,$$

where $x(t)$, $t \in \mathbb{R}_+$ is the unique solution of

$$x(t) = x_0 + \int_0^t F(x(s)) ds, \qquad t \ge 0,$$

wherever it exists.

For the application of the preceding theorem to the general stochastic epidemic introduced at the beginning of this section, see Problem 7.9. For a graphical illustration of the foregoing calculations, see Figs. 7.7 and 7.8. Further, interesting examples may also be found in Sect. 6.4 of Tan (2002). For further examples of models for population dynamics described by SDEs, the reader may refer to Mao et al. (2002) and Mao et al. (2005).

7.3 Population Dynamics: Individual-Based Models

The scope of this chapter is to introduce the reader to the modeling of a system of a large, though still finite, population of individuals subject to mutual interaction and random dispersal. These systems may well describe the collective behavior of individuals in herds, swarms, colonies, armies, etc. [examples can be found in Burger et al. (2007), Capasso and Morale (2009), Durrett and Levin (1994), Flierl et al. (1999), Gueron et al. (1996), Okubo (1986), and Skellam (1951)]. Under suitable conditions, the behavior of such systems, in the limit of the number of individuals tending to infinity, may be described in terms of nonlinear reaction-diffusion systems. We may then claim that while SDEs may be utilized for modeling populations at the *microscopic* scale of individuals (Lagrangian approach), partial differential equations provide a *macroscopic* Eulerian description of population densities.

Up to now, Kolmogorov equations like that of Black–Scholes were linear partial differential equations; in this chapter, we derive nonlinear partial differential equations for density-dependent diffusions. This field of research,

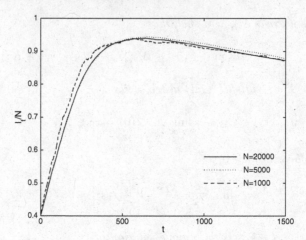

Fig. 7.7. Continuous approximation of a jump model: general stochastic epidemic model with $S_0 = 0.6N$, $I_0 = 0.4N$, $R_0 = 0$, rate of removal of an infective $\delta = 10^{-4}$; infection rate of a susceptible $k = 8 \times 10^{-3}N$; time step $dt = 10^{-2}$; interval of observation $[0, 1500]$. The *three lines* represent the simulated I_t/N as a function of time t for three different values of N

already well established in the general theory of statistical physics (e.g., De Masi and Presutti 1991; Donsker and Varadhan 1989; Méléard 1996), has gained increasing attention since it also provides the framework for the modeling, analysis, and simulation of agent-based models in economics and finance (e.g., Epstein and Axtell 1996).

The Empirical Distribution

We start from the Lagrangian description of a system of $N \in \mathbb{N} \setminus \{0, 1\}$ particles. Suppose the kth particle ($k \in \{1, \ldots, N\}$) is located at $X_N^k(t) \in \mathbb{R}^d$, at time $t \geq 0$. Each $(X_N^k(t))_{t \in \mathbb{R}_+}$ is a stochastic process valued in the state space $(\mathbb{R}^d, \mathcal{B}_{\mathbb{R}^d})$, $d \in \mathbb{N} \setminus \{0\}$, on a common probability space $(\Omega, \mathcal{F}, P)$. An equivalent description of the foregoing system may be given in terms of the (random) Dirac measures $\epsilon_{X_N^k(t)}$ ($k = 1, 2, \ldots, N$) on $\mathcal{B}_{\mathbb{R}^d}$ such that, for any real function $f \in C_0(\mathbb{R}^d)$, we have

$$\int_{\mathbb{R}^d} f(y)\epsilon_{X_N^k(t)}(dy) = f\left(X_N^k(t)\right).$$

As a consequence, information about the collective behavior of the N particles is provided by the so-called *empirical measure*, i.e., the random measure on $\mathbb{R}^d$

$$X_N(t) := \frac{1}{N} \sum_{k=1}^{N} \epsilon_{X_N^k(t)}, \qquad t \in \mathbb{R}_+.$$

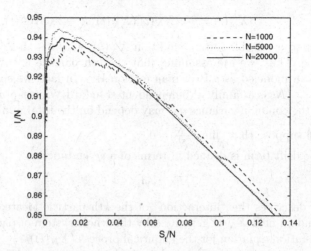

Fig. 7.8. Continuous approximation of a jump model: the same model as in Fig. 7.7 of a general stochastic epidemic model with $S_0 = 0.6N$, $I_0 = 0.4N$, $R_0 = 0$, rate of removal of an infective $\delta = 10^{-4}$, infection rate of a susceptible $k = 8 \times 10^{-3}N$, time step $dt = 10^{-2}$, interval of observation $[0, 1500]$. The *three lines* represent the simulated trajectory $(S_t/N, I_t/N)$ for three different values of N

This measure may be considered as the empirical spatial distribution of the system. It is such that for any $f \in C_0(\mathbb{R}^d)$

$$\int_{\mathbb{R}^d} f(y)[X_N(t)](dy) = \frac{1}{N}\sum_{k=1}^{N} f\left(X_N^k(t)\right).$$

In particular, given a region $B \in \mathcal{B}_{\mathbb{R}^d}$, the quantity

$$[X_N(t)](B) := \frac{1}{N}\mathrm{card}\left\{X_N^k(t) \in B\right\}$$

denotes the relative frequency of individuals, out of N, that at time t stay in B. This is why the measure-valued process

$$X_N : t \in \mathbb{R}_+ \to X_N(t) = \frac{1}{N}\sum_{k=1}^{N} \epsilon_{X_N^k(t)} \in \mathcal{M}_{\mathbb{R}^d}$$

is called the process of *empirical distributions* of the system of N particles.

Evolution Equations

The Lagrangian description of the dynamics of a system of interacting particles is given via a system of SDEs. Suppose that for any $k \in \{1, \dots, N\}$ the process $(X_N^k(t))_{t \in \mathbb{R}_+}$ satisfies the SDE

$$dX_N^k(t) = F_N[X_N(t)](X_N^k(t))dt + \sigma_N dW^k(t), \tag{7.5}$$

subject to a suitable initial condition $X_N^k(0)$, which is an $\mathbb{R}^d$-valued random variable. Thus, we are assuming that the kth particle is subject to random dispersal, modeled as a Brownian motion W^k. In fact, we suppose that W^k, $k = 1, \ldots, N$, is a family of independent standard Wiener processes. Furthermore, the common variance σ_N^2 may depend on the total number of particles; we will suppose that $\lim_{N \to \infty} \frac{\sigma_N^2}{N} = 0$.

The drift term is defined in terms of a given function

$$F_N : \mathcal{M}_{\mathbb{R}^d} \to C(\mathbb{R}^d)$$

and it describes the "interaction" of the kth particle located at $X_N^k(t)$ with the random field $X_N(t)$ generated by the whole system of particles at time t. An evolution equation for the empirical process $(X_N(t))_{t \in \mathbb{R}_+}$ can be obtained thanks to Itô's formula. For each individual particle $k \in \{1, \ldots, N\}$, subject to its SDE, given $f \in C_b^2(\mathbb{R}^d \times \mathbb{R}_+)$, we have

$$
\begin{aligned}
f\left(X_N^k(t), t\right) = {} & f\left(X_N^k(0), 0\right) + \int_0^t F_N[X_N(s)]\left(X_N^k(s)\right) \nabla f\left(X_N^k(s), s\right) ds \\
& + \int_0^t \left[\frac{\partial}{\partial s} f\left(X_N^k(s), s\right) + \frac{\sigma_N^2}{2} \triangle f\left(X_N^k(s), s\right)\right] ds \\
& + \sigma_N \int_0^t \nabla f\left(X_N^k(s), s\right) dW^k(s).
\end{aligned}
\tag{7.6}
$$

Correspondingly, for the empirical process $(X_N(t))_{t \in \mathbb{R}_+}$ we get the following weak formulation of its evolution equation. For any $f \in C_b^{2,1}(\mathbb{R}^d \times \mathbb{R}_+)$ we have

$$
\begin{aligned}
\langle X_N(t), f(\cdot, t) \rangle = {} & \langle X_N(0), f(\cdot, 0) \rangle + \int_0^t \langle X_N(s), F_N[X_N(s)](\cdot) \nabla f(\cdot, s) \rangle ds \\
& + \int_0^t \left\langle X_N(s), \frac{\sigma_N^2}{2} \triangle f(\cdot, s) + \frac{\partial}{\partial s} f(\cdot, s) \right\rangle ds \\
& + \frac{\sigma_N}{N} \int_0^t \sum_k \nabla f\left(X_N^k(s), s\right) dW^k(s).
\end{aligned}
\tag{7.7}
$$

In the previous expressions, we used the notation

$$\langle \mu, f \rangle = \int f(x)\mu(dx) \tag{7.8}$$

for any measure μ on $(\mathbb{R}^d, \mathcal{B}_{\mathbb{R}^d})$ and any (sufficiently smooth) function $f : \mathbb{R}^d \to \mathbb{R}$.

The last term of (7.7) is a martingale with respect to the natural filtration of the process $(X_N(t))_{t \in \mathbb{R}_+}$. Hence we may apply Doob's inequality (Proposition 2.125) such that, for any finite $T > 0$,

$$E\left[\sup_{t\leq T}|M_N(f,t)|^2\right] \leq \frac{4\sigma_N^2\|\nabla f\|_\infty^2 T}{N}.$$

This shows that, for N sufficiently large, the martingale term, which is the only source of stochasticity of the evolution equation for $(X_N(t))_{t\in\mathbb{R}_+}$, tends to zero, for N tending to infinity, since ∇f is bounded in $[0,T]$, and $\frac{\sigma_N^2}{N} \to 0$ for N tending to infinity. Under these conditions we may conjecture that a limiting measure-valued deterministic process $(X_\infty(t))_{t\in\mathbb{R}_+}$ exists whose evolution equation (in weak form) is

$$\langle X_\infty(t), f(\cdot,t)\rangle = \langle X_\infty(0), f(\cdot,0)\rangle + \int_0^t \langle X_\infty(s), F[X_\infty(s)](\cdot)\nabla f(\cdot,s)\rangle\, ds$$

$$+ \int_0^t \left\langle X_\infty(s), \frac{\sigma_\infty^2}{2}\triangle f(\cdot,s) + \frac{\partial}{\partial s}f(\cdot,s)\right\rangle ds$$

for $\sigma_\infty^2 \geq 0$.

Actually, various nontrivial mathematical problems arise in connection with the existence of a limiting measure-valued process $(X_\infty(t))_{t\in\mathbb{R}_+}$. A typical procedure includes the following:

(a) Show the convergence of the stochastic empirical measure process X_N to a deterministic measure process X_∞:

$$X_N \xrightarrow{\mathcal{D}} X_\infty \in C([0,T], \mathcal{M}(\mathbb{R}^d)).$$

(b) Identify the limiting measure process and possibly show its absolute continuity with respect to the usual Lebesgue measure on $\mathbb{R}^d$, i.e., for any $t \in [0,T]$

$$X_\infty = \rho(\cdot,t)\nu^d.$$

(c) Prove existence and uniqueness for the solution of the deterministic density $\rho(x,t)$ of $X_\infty(t)$ satisfying their asymptotic evolution equation.

7.3.1 A Mathematical Detour

In the following subsections we will show how the foregoing procedure has been carried out in particular cases. We start by recalling basic facts regarding the relevant mathematical environment required for carrying out the foregoing procedure.

The Relevant Processes

Within the measurable space $(\mathbb{R}^d, \mathcal{B}_{\mathbb{R}^d})$, consider the following family of stochastic processes on a common probability space:

$$X_N^k(t) \in \mathbb{R}^d, \quad t \in [0,T], \quad 1 \le k \le N,$$

for $N \in \mathbb{N} \setminus \{0\}$; then define the empirical measure associated with the preceding family:

$$X_N(t) = \frac{1}{N} \sum_{k=1}^{N} \epsilon_{X_N^k(t)} \quad \in \mathcal{M}(\mathbb{R}^d).$$

If, for all $1 \le k \le N$, the trajectories of $\{X_N^k(t) \in \mathbb{R}^d, t \in [0,T]\}$ are continuous on $[0,T]$, then

$$X_N := \{X_N(t), t \in [0,T]\} \in C([0,T], \mathcal{M}(\mathbb{R}^d)).$$

The Relevant Metrics

On $\mathcal{M}(\mathbb{R}^d)$ take the *BL (bounded Lipschitz) metric*

$$d_{BL}(\mu, \nu) := \sup_{f \in \mathcal{H}_1} (\langle \mu, f \rangle - \langle \nu, f \rangle) =: \|\mu - \nu\|_1,$$

where

$$\mathcal{H}_1 := \left\{ f \in C_b(\mathbb{R}^d) \mid \|f\|_{Lip} = \sup_{x \in \mathbb{R}^d} |f(x)| + \sup_{x,y \in \mathbb{R}^d, x \ne y} \frac{|f(x) - f(y)|}{|x - y|} \le 1 \right\},$$

and $\langle \mu, f \rangle$ has been defined in (7.8).

Note that on $\mathcal{M}(\mathbb{R}^d)$ BL-convergence is equivalent to weak convergence (Sect. B.1).

Correspondingly, on $C([0,T], \mathcal{M}(\mathbb{R}^d))$, $T > 0$ we shall use the uniform metric with respect to $t \in [0,T]$, so that the distance between $\mu, \nu \in C([0,T], \mathcal{M}(\mathbb{R}^d))$ is given by

$$\sup_{0 \le t \le T} \|\mu(t) - \nu(t)\|_1.$$

On the space of probability measures $\mathcal{M}(C([0,T], \mathcal{M}(\mathbb{R}^d))$ we shall adopt the topology of weak convergence, too.

The Relevant Polish Spaces

Theorem 7.5. $\mathbb{R}^d$, *endowed with the usual Euclidean metric, is a Polish space; hence, $\mathcal{M}(\mathbb{R}^d)$, $C([0,T], \mathcal{M}(\mathbb{R}^d))$, and $\mathcal{M}(C([0,T], \mathcal{M}(\mathbb{R}^d)))$, endowed with the aforementioned metrics, are Polish spaces.*

Recall that in Polish spaces relative compactness and tightness are equivalent.

An important criterion to show relative compactness (tightness) in $C([0,T], \mathcal{M}(\mathbb{R}^d))$ is the following one, derived from Theorem B.91 (Ethier and Kurtz 1986).

Theorem 7.6. *Consider a sequence $(X_N)_{N \in \mathbb{N}}$ of stochastic processes in $C([0,T], \mathcal{M}(\mathbb{R}^d))$, and let $\mathcal{F}_t^N := \sigma\{X_N(s) | s \le t\}$ be the natural filtration associated with $\{X_N(t), t \in [0,T]\}$. Suppose that*

(i) *(Pointwise compactness control) for any real positive ϵ and for any non-negative rational t, a compact $\Gamma_{t,\epsilon}$ exists such that*

$$\inf_N P(X_N(t) \in \Gamma_{t,\epsilon}) > 1 - \epsilon;$$

(ii) *(Small variations during small time intervals) let $\alpha > 0$; for any real $\delta \in (0,1)$, a sequence $(\gamma_N^T(\delta))_{N \in \mathbb{N}}$ of nonnegative real random variables exists such that*

$$\lim_{\delta \to 0} \limsup_{N \to \infty} \mathbb{E}[\gamma_N^T(\delta)] = 0$$

and, for any $t \in [0,T]$,

$$\mathbb{E}[\|X_N(t+\delta) - X_N(t)\|_1^\alpha | \mathcal{F}_t^N] \le \mathbb{E}[\gamma_N^T(\delta) | \mathcal{F}_t^N].$$

Then $(\mathcal{L}(X_N))_{N \in \mathbb{N}}$ is a tight sequence of probability laws.

Within this mathematical setting, procedures (a), (b), and (c) mentioned previously become

(i) Show the relative compactness of the sequence $(\mathcal{L}(X_N))_{N \in \mathbb{N} \setminus \{0\}}$, which corresponds to an existence result for the limit $\mathcal{L}(X)$;

(ii) Show the regularity of the possible limits; we show that the possible limits $\{X(t), t \in [0,T]\}$ are absolutely continuous with respect to the Lebesgue measure for almost all $t \in [0,T]$ $\mathbb{P} -$ a.s.;

(iii) Identify the dynamics of the limit process, i.e., all possible limits are shown to be a solution of a certain deterministic equation that we assume to have a unique solution (this corresponds to the uniqueness of the limit $\mathcal{L}(X)$).

It will be realized that actually items (ii) and (iii) are taken together.

7.3.2 A "Moderate" Repulsion Model

As an example we consider the system [due to Oelschläger (1985)]

$$dX_N^k(t) = -\frac{1}{N} \sum_{m=1, m \neq k}^{N.} \nabla V_N\left(X_N^k(t) - X_N^m(t)\right) dt + dW^k(t), \qquad (7.9)$$

where W^k, $k = 1, \ldots, N$, represent N independent standard Brownian motions valued in $\mathbb{R}^d$ (here all variances are set equal to 1). The kernel V_N is chosen of the form

$$V_N(\mathbf{x}) = \chi_N^d V_1(\chi_N \mathbf{x}), \qquad \mathbf{x} \in \mathbb{R}^d, \tag{7.10}$$

where V_1 is a symmetric probability density with compact support in $\mathbb{R}^d$ and

$$\chi_N = N^{\frac{\beta}{d}}, \qquad \beta \in]0,1[.$$

With respect to the general structure introduced in the preceding subsection on evolution equations, we have assumed that the drift term is given by

$$F_N[X_N(t)]\left(X_N^k(t)\right) = -[\nabla V_N * X_N(t)]\left(X_N^k(t)\right)$$

$$= -\frac{1}{N} \sum_{m=1, m \neq k}^N \nabla V_N\left(X_N^k(t) - X_N^m(t)\right).$$

System (7.9) describes a population of N individuals, subject to random dispersal (Brownian motion) and to repulsion within the range of the kernel V_N. The choice of the scaling (7.10) in terms of the parameter β means that the range of interaction of each individual with the rest of the population is a decreasing function of N (correspondingly, the strength is an increasing function of N). On the other hand, the fact that β is chosen to belong to $]0,1[$ is relevant for the limiting procedure. It is known as *moderate interaction* and allows one to apply suitable convergence results (*laws of large numbers*) (Oelschläger 1985).

For the sake of useful regularity conditions, we assume that

$$V_1 = W_1 * W_1,$$

where W_1 is a symmetric probability density with compact support in $\mathbb{R}^d$, satisfying the condition

$$\int_{\mathbb{R}^d} (1 + |\lambda|^\alpha)|\widetilde{W_1}(\lambda)|^2 d\lambda < \infty \tag{7.11}$$

for some $\alpha > 0$ (here $\widetilde{W_1}$ denotes the Fourier transform of W_1). Henceforth, we also make use of the following notations:

$$W_N(x) = \chi_N^d W_1(\chi_N x),$$
$$h_N(x,t) = (X_N(t) * W_N)(x),$$
$$V_N(x) = \chi_N^d V_1(\chi_N x) = (W_N * W_N)(x),$$
$$g_N(x,t) = (X_N(t) * V_N)(x) = (h_N(\cdot,t) * W_N)(x),$$

so that system (7.9) can be rewritten as

$$dX_N^k(t) = -\nabla g_N(X_N^k(t), t)dt + dW^k(t), \qquad k = 1, \ldots, N.$$

The following theorem holds.

Theorem 7.7. *Let*

$$X_N(t) = \frac{1}{N} \sum_{k=1}^{N} \epsilon_{X_N^k(t)}$$

be the empirical process associated with system (7.9). Assume that

1. *Condition (7.11) holds;*
2. $\beta \in]0, \frac{d}{d+2}]$;
3.
$$\sup_{N \in \mathbb{N}} E\left[\langle X_N(0), \varphi_1 \rangle\right] < \infty, \qquad \varphi_1(x) = (1 + x^2)^{1/2};$$

4.
$$\sup_{N \in \mathbb{N}} E\left[\|h_N(\cdot, 0)\|_2^2\right] < \infty;$$

5.
$$\lim_{N \to \infty} \mathcal{L}\left(X_N(0)\right) = \epsilon_{X_0} \text{ in } \mathcal{M}(\mathcal{M}(\mathbb{R}^d)),$$

where X_0 is a probability measure having a density $p_0 \in C_b^{2+\alpha}(\mathbb{R}^d)$ with respect to the usual Lebesgue measure on $\mathbb{R}^d$.

Then the empirical process X_N converges to a deterministic law X_∞; more precisely,

$$\lim_{N \to \infty} \mathcal{L}(X_N) = \epsilon_{X_\infty} \text{ in } \mathcal{M}(C([0, T], \mathcal{M}(\mathbb{R}^d))),$$

where

$$X_\infty = (X_\infty(t))_{0 \le t \le T} \in C([0, T], \mathcal{M}(\mathbb{R}^d))$$

admits a density

$$p \in C_b^{2+\alpha, 1+\frac{\alpha}{2}}(\mathbb{R}^d \times [0, T]),$$

which satisfies

$$\frac{\partial}{\partial t}p(x, t) = \nabla(p(x, t)\nabla p(x, t)) + \frac{1}{2}\Delta p(x, t), \qquad (7.12)$$
$$p(x, 0) = p_0(x).$$

Equation (7.12) includes nonlinear terms, as in the porous media equation (Oelschläger 1990). This is due to the repulsive interaction between particles, which in the limit produces a density-dependent diffusion. A linear diffusion persists because the variance of the Brownian motions in the individual equations was kept constant. We will see in a second example how it may vanish when the individual variances tend to zero for N tending to infinity. We will not provide a detailed proof of Theorem 7.7, even though we are going to provide a significant outline of it, leaving further details to the referred literature.

By proceeding as in the previous subsection, a straightforward application of Doob's inequality for martingales (Proposition 2.125) justifies the vanishing of the noise term in the following evolution equation for the empirical measure $(X_N(t))_{t \in \mathbb{R}_+}$:

$$\langle X_N(t), f(\cdot, t) \rangle = \langle X_N(0), f(\cdot, 0) \rangle + \int_0^t \langle X_N(s), \nabla g_N(\cdot, s) \nabla f(\cdot, s) \rangle \, ds$$

$$+ \int_0^t \left\langle X_N(s), \frac{\sigma^2}{2} \triangle f(\cdot, s) + \frac{\partial}{\partial s} f(\cdot, s) \right\rangle ds$$

$$+ \frac{\sigma}{N} \int_0^t \sum_k \nabla f \left(X_N^k(s), s \right) dW^k(s),$$

for a given $T > 0$ and any $f \in C_b^{2,1}(\mathbb{R}^d \times [0, T])$. The major difficulty in a rigorous proof of Theorem 7.7 comes from the nonlinear term

$$\varXi_{N,f}(t) = \int_0^t \langle X_N(s), \nabla g_N(\cdot, s) \nabla f(\cdot, s) \rangle \, ds. \tag{7.13}$$

If we rewrite (7.13) in an explicit form, we get

$$\varXi_{N,f}(t) = \int_0^t \frac{1}{N^2} \sum_{k,m=1}^N \nabla V_N \left(X_N^k(s) - X_N^m(s) \right) \nabla f \left(X_N^k(s), s \right) ds.$$

Since, for $\beta > 0$, the kernel $V_N \to \delta_0$, namely the Dirac delta function, this shows that, in the limit, even small changes of the relative position of neighboring particles may have a considerable effect on $\varXi_{N,f}(t)$. But in any case, the regularity assumptions made on the kernel V_N let us state the following lemma, which provides sufficient estimates about g_N and h_N as defined above.

The proof of Theorem 7.7 proceeds by the following steps.

Lemma 7.8. *Under the assumptions of Theorem 7.7,*

(i) The process

$$t \quad \mapsto \quad < X_N(t), \varphi_1 > e^{-Ct}$$

is a supermartingale, for a suitable choice of $C > 0$;

(ii) The process

$$t \quad \mapsto \quad < X_N(t), \varphi_1 > e^{Ct}$$

is a submartingale, for a suitable choice of $C > 0$.

Thanks to Doob's inequalities for martingales, a significant consequence of the foregoing lemma is the following one.

Lemma 7.9. *Given a $T > 0$, for any $\delta > 0$ there exists a compact K_δ in $(\mathcal{M}_P(\mathbb{R}^d), d_{BL})$ such that*

$$\inf_{N \in \mathbb{N}} \mathbb{P}\{X_N(t) \in K_\delta, \ \forall t \in [0, T]\} \geq 1 - \delta.$$

Furthermore, the following can be shown using Itô's formula.

Lemma 7.10. *Given a $T > 0$, for any $\Delta > 0$, there exists a sequence $\{\gamma_N^T(\Delta)\}_{N \in \mathbb{N}}$ of nonnegative random variables such that*

$$\mathbb{E}\left[d_{BL}(X_N(t+\Delta), X_N(t))|\mathcal{F}_t\right] \leq \mathbb{E}\left[\gamma_N^T(\Delta)|\mathcal{F}_t\right] \quad 0 \leq t \leq T - \Delta, N \in \mathbb{N},$$

with

$$\lim_{\Delta \to 0} \limsup_{n \to \infty} \mathbb{E}[\gamma_N^T(\Delta)] = 0.$$

The following proposition is a consequence of the foregoing lemmas, together with Theorem 7.6.

Proposition 7.11 *With X_N as above, the sequence $\mathcal{L}(X_N)$ is relatively compact in the space $\mathcal{M}(C([0, T], \mathcal{M}(\mathbb{R}^d)))$.*

By Proposition 7.11, we may claim that a subsequence of $(\mathcal{L}(X_N))_{N \in \mathbb{N}}$ exists that converges to a probability measure on the space $\mathcal{M}(C([0, T], \mathcal{M}(\mathbb{R}^d)))$. The Skorohod representation Theorem B.71 then assures that a process X_∞^k exists in $C([0, T], \mathcal{M}(\mathbb{R}^d))$ such that

$$\lim_{l \to \infty} X_{N_l} = X_\infty^k, \qquad \text{a.s. with respect to } P.$$

If we can assure the uniqueness of the limit, then all X_∞^k will coincide with some X_∞.

By now, we assume uniqueness, so that we may take $\{N_k\} = \mathbb{N}$; by the Skorohod theorem, we may assert that, corresponding to the possible unique limit law, we can also have an almost sure convergence, i.e.,

$$\lim_{N \to \infty} \sup_{t \leq T} d_{BL}(X_N(t), X_\infty(t)) = 0 \quad \mathbb{P} - \text{a.s.}$$

The following theorems holds (Oelschläger 1985).

Theorem 7.12. *Under the foregoing assumptions, suppose further that $W_1 \in L^2(\mathbb{R}^d)$ is such that, for some $\delta > 0$ and $\alpha_2 > 0$,*

$$\sup_{N \in \mathbb{N}} \mathbb{E}\left[\int_\delta^T \int_{\mathbb{R}^d} (1 + |\lambda|^{\alpha_2})|\widetilde{h}_N(\lambda, t)|^2 d\lambda dt\right] < +\infty.$$

Then the limit measure $X_\infty \in \mathcal{M}_P([0,T] \times \mathbb{R}^d)$ has P-a.s. a density

$$h_\infty \in L^2\left([0,T] \times \mathbb{R}^d\right)$$

with respect to the Lebesgue measure on $[0,T] \times \mathbb{R}^d$, i.e., for any $f \in C_b([0,T] \times \mathbb{R}^d)$

$$\int_0^T \int_{\mathbb{R}^d} f(t,x) X_\infty(dx, dt) = \int_0^T \int_{\mathbb{R}^d} f(t,x) h_\infty(t,x) dx \, dt. \qquad (7.14)$$

Remark 7.13. A priori, the limiting process X_∞ may still be a random process in $C([0,T], \mathcal{M}(\mathbb{R}^d))$. Further, we do not know whether, given a time $t \in [0,T]$, $X_N(t)$ admits a density (possibly deterministic) with respect to the usual Lebesgue measure on $\mathbb{R}^d$. The following analysis leads to an answer to both questions.

The proof of Theorem 7.7 requires further analysis in order to acquire more information about the limit dynamics.

The following result can be shown ($\mathcal{M}_P$ denotes the subspace of probability measures of $\mathcal{M}$).

Theorem 7.14. *Under the hypotheses of Theorem 7.12, let us suppose that a law of large number holds at initial time*

$$\lim_{N \to \infty} \mathcal{L}(X_N(0)) = \delta_{\mu_0} \quad \text{in} \quad \mathcal{M}_P(\mathcal{M}_P(\mathbb{R}^d)),$$

where μ_0 has a density p_0 in $L^2(\mathbb{R}^d) \cap C_b^2(\mathbb{R}^d)$. Then, almost surely, for any $f \in C_b^{1,1}(\mathbb{R}^d, \mathbb{R}_+), 0 \leq t \leq, T,$

$$\langle X_\infty(t), f(\cdot, t) \rangle = \langle \mu_0, f(\cdot, 0) \rangle$$
$$- \frac{1}{2} \int_0^t \langle \nabla h_\infty(\cdot, s), (1 + 2h_\infty(\cdot, s)) \nabla f(\cdot, s) \rangle ds \qquad (7.15)$$
$$+ \int_0^t \langle h_\infty(\cdot, s), \frac{\partial}{\partial s} f(\cdot, s) \rangle ds.$$

So far we have shown that any limit measure $X_\infty \in C([0,T], \mathcal{M}_P(\mathbb{R}^d))$ is a solution of (7.15), with $h_\infty \in L^2\left([0,T] \times \mathbb{R}^d\right)$, satisfying (7.14).

We should prove that for any $t \in [0,T]$, the measure $X_\infty(t)$ is absolutely continuous with respect to the Lebesgue measure, so it admits a density for each $t \in [0,T]$. Thanks to a known result, we can prove that by showing that the Fourier transform of the measure $X_\infty(t)$ is in L^2 for any $t \in [0,T]$; thus, a density exists which belongs to $L^2(\mathbb{R}^d)$, and we prove that it is also L^2 uniformly bounded. Indeed one can show that (Oelschläger 1985)

$$\|p_0\|_2^2 \geq \int_{\mathbb{R}^d} \|\widetilde{X_\infty(\lambda)}\|^2 d\lambda. \qquad (7.16)$$

Thus we may state that for any fixed $t \in [0, T]$ the measure $X_\infty(t)$ has a density with respect to the Lebesgue measure on $\mathbb{R}^d$, and because of (7.14), we also have

$$X_\infty(t) = h_\infty(\cdot, t)\nu^d, \tag{7.17}$$

where ν^d denotes the Lebesgue measure on $\mathbb{R}^d$. Furthermore, again from (7.16) and (7.17), the density is bounded in L^2

$$\|h_\infty(\cdot, t)\|_2 \le \|p_0\|_2.$$

So we may finally state the following theorem.

Theorem 7.15. *Under the hypotheses of Theorem 7.12, let us suppose that a law of large number applies at initial time*

$$\lim_{N \to \infty} \mathcal{L}(X_N(0)) = \delta_{X_0} \quad in \quad \mathcal{M}_P(\mathcal{M}_P(\mathbb{R}^d)),$$

where X_0 has a density p_0 in $L^2(\mathbb{R}^d) \cap C_b^2(\mathbb{R}^d)$. Then, almost surely, the sequence X_N converges in law to a X_∞. For any $t \in [0, T]$ the measure $X_\infty(t)$ has a density $h_\infty(\cdot, t)$ such that for any $f \in C_b^{2,1}(\mathbb{R}^d, \mathbb{R}_+), 0 \le t \le, T,$

$$\langle h_\infty(\cdot, t), f(\cdot, t) \rangle = \langle p_0, f(\cdot, 0) \rangle$$
$$- \frac{1}{2} \int_0^t \langle \nabla h_\infty(\cdot, s), (1 + 2h_\infty(\cdot, s)) \nabla f(\cdot, s) \rangle ds \tag{7.18}$$
$$+ \int_0^t \langle h_\infty(\cdot, s), \frac{\partial}{\partial s} f(\cdot, s) \rangle ds.$$

One can easily see that (7.18) is the weak form of the following partial differential equation:

$$\frac{\partial}{\partial t} \rho(x, t) = \frac{1}{2} \Delta \rho(x, t) + \nabla \cdot (\rho(x, t) \nabla \rho(x))$$
$$\rho(x, 0) = p_0(x), \quad x \in \mathbb{R}^d. \tag{7.19}$$

The uniqueness of the limit h_∞ derives from the uniqueness of the weak solution of the viscous Equation (7.19), in $C_b^{2+\alpha, 1+\alpha/2}(\mathbb{R}^d \times [0, T])$, as it can be achieved via classical arguments (Ladyzenskaja et al. 1968).

We may thus conclude that if we assume that $X_\infty(0)$ admits a deterministic density p_0 at time $t = 0$, then $(X_\infty(t))_{t \in [0,T]}$ satisfies a deterministic evolution equation and is thus itself a deterministic process on $C([0, T], \mathcal{M}(\mathbb{R}^d))$.

From the general theory we know that (7.19) admits a unique solution $p \in C_b^{2+\alpha, 1+\alpha/2}(\mathbb{R}^d \times [0, T])$.

It satisfies itself (7.14), so that we may claim it is a version of the density of the limit measure X_∞, thereby concluding the main theorem.

7.3.3 Ant Colonies

As another example, we consider a model for ant colonies. The latter provide an interesting concept of *aggregation* of individuals. According to a model proposed in Morale et al. (2005) (see also Burger et al. 2007; Capasso and Morale 2009), [based on an earlier model by Grünbaum and Okubo (1994)], in a colony or in an army (in which case the model may be applied to any cross section), ants are assumed to be subject to two conflicting *social forces*: long-range attraction and short-range repulsion. Hence we consider the following basic assumptions (see Figs. 7.9, 7.10, 7.11):

(i) Particles tend to aggregate subject to their interaction within a range of size $R_a > 0$ (finite or not). This corresponds to the assumption that each particle is capable of perceiving the others only within a suitable sensory range; in other words, each particle has a limited knowledge of the spatial distribution of its neighbors.

(ii) Particles are subject to repulsion when they come "too close" to each other.

We may express assumptions (i) and (ii) by introducing in the drift term F_N in (7.5) two additive components (Warburton and Lazarus 1991): F_1, responsible for aggregation, and F_2, for repulsion, such that

$$F_N = F_1 + F_2.$$

Aggregation Term F_1

We introduce a convolution kernel $G_a : \mathbb{R}^d \to \mathbb{R}_+$, having a support confined to a ball centered at $0 \in \mathbb{R}^d$ and radius $R_a \in \bar{\mathbb{R}}_+$ as the range of sensitivity for aggregation, independent of N. A *generalized gradient* operator is obtained as follows. Given a measure μ on $\mathbb{R}^d$, we define the function

$$[\nabla G_a * \mu](x) = \int_{\mathbb{R}^d} \nabla G_a(x - y)\mu(dy), \qquad x \in \mathbb{R}^d,$$

as the classical convolution of the gradient of the kernel G_a with the measure μ. Furthermore, G_a is such that

$$G_a(x) = \hat{G}_a(|x|), \tag{7.20}$$

with $\hat{G}_a$ a decreasing function in $\mathbb{R}_+$. We assume that the aggregation term F_1 depends on such a generalized gradient of $X_N(t)$ at $X_N^k(t)$:

$$F_1[X_N(t)]\left(X_N^k(t)\right) = [\nabla G_a * X_N(t)]\left(X_N^k(t)\right). \tag{7.21}$$

This means that each individual feels this generalized gradient of the measure $X_N(t)$ with respect to the kernel G_a. The positive sign for F_1 and (7.20) expresses a force of attraction of the particle in the direction of increasing concentration of individuals.

We emphasize the great generality provided by this definition of a generalized gradient of a measure μ on $\mathbb{R}^d$. Using particular shapes of G_a, one may include angular ranges of sensitivity, asymmetries, etc. at a finite distance (Gueron et al. 1996).

Repulsion Term F_2

As far as repulsion is concerned, we proceed in a similar way by introducing a convolution kernel $V_N : \mathbb{R}^d \to \mathbb{R}_+$, which determines the range and the strength of influence of neighboring particles. We assume (by anticipating a limiting procedure) that V_N depends on the total number N of interacting particles. Let V_1 be a continuous probability density on $\mathbb{R}^d$ and consider the scaled kernel $V_N(x)$ as defined in (7.10), again with $\beta \in]0, 1[$. It is clear that

$$\lim_{N \to +\infty} V_N = \delta_0,$$

where δ_0 is Dirac's delta function. We define

$$F_2[X_N(t)]\left(X_N^k(t)\right) = -\left(\nabla V_N * X_N(t)\right)\left(X_N^k(t)\right)$$

$$= -\frac{1}{N} \sum_{m=1}^{N} \nabla V_N \left(X_N^k(t) - X_N^m(t)\right). \qquad (7.22)$$

This means that each individual feels the gradient of the population in a small neighborhood. The negative sign for F_2 expresses a drift toward decreasing concentration of individuals. In this case the range of the repulsion kernel decreases to zero as the size N of the population increases to infinity.

Diffusion Term

In this model, randomness may be due to both external sources and "social" reasons. The external sources could, for instance, be unpredictable irregularities of the environment (like obstacles, changeable soils, varying visibility). On the other hand, the innate need of interaction with peers is a social factor. As a consequence, randomness can be modeled by a multidimensional Brownian motion $\mathbf{W}_t$.

The coefficient of $d\mathbf{W}_t$ is a matrix function depending upon the distribution of particles or some environmental parameters. Here, we take into account only the intrinsic stochasticity due to the need of each particle to interact with others. In fact, experiments carried out on ants have shown this need. Hence, simplifying the model, we consider only one Brownian motion

dW_t with the variance of each particle σ_N depending on the total number of particles, not on their distribution. We could interpret this as an approximation of the model by considering all the stochasticities (also those due to the environment) modeled by $\sigma_N dW_t$.

Since σ_N expresses the intrinsic randomness of each individual due to its need for social interaction, it should be decreasing as N increases. Indeed, if the number of particles is large, the mean free path of each particle may reduce down to a limiting value that may eventually be zero:

$$\lim_{N\to\infty} \sigma_N = \sigma_\infty \geq 0. \tag{7.23}$$

Scaling Limits

Let us discuss the two choices for the interaction kernel in the aggregation and repulsion terms, respectively. They anticipate the limiting procedure for N tending to infinity. Here we are focusing on two types of scaling limits, the *McKean–Vlasov limit*, which applies to the long-range aggregation, and the *moderate limit*, which applies to the short-range repulsion. In the previous subsection, we already considered the moderate limit case.

Mathematically the two cases correspond to the choice made on the interaction kernel. In the moderate limit case (e.g., Oelschläger 1985) the kernel is scaled with respect to the total size of the population N via a parameter $\beta \in]0, 1[$. In this case the range of interaction among particles is reduced to zero for N tending to infinity. Thus any particle interacts with many (of order $\frac{N}{\alpha(N)}$) other particles in a small volume (of order $\frac{1}{\alpha(N)}$); if we take $\alpha(N) = N^\beta$, then both $\alpha(N)$ and $\frac{N}{\alpha(N)}$ tend to infinity. In the McKean–Vlasov case (e.g., Méléard 1996) $\beta = 0$, so that the range of interaction is independent of N, and as a consequence any particle interacts with order N other particles.

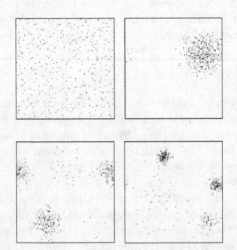

Fig. 7.9. A simulation of the long-range aggregation (7.21) and short-range repulsion (7.22) model for an ant colony with diffusion

This is why in the moderate limit we may speak of *mesoscale*, which lies between the *microscale* for the typical volume occupied by each individual and the *macroscale* applicable to the typical volume occupied by the total population. Obviously, it would be possible also to consider interacting particle systems rescaled by $\beta = 1$. This case is known as the hydrodynamic case, for which we refer the reader to the relevant literature (De Masi and Presutti 1991; Donsker and Varadhan 1989).

The case $\beta > 1$ is less significant in population dynamics. It would mean that the range of interaction decreases much faster than the typical distance between neighboring particles. So most of the time particles do not approach sufficiently close to feel the interaction.

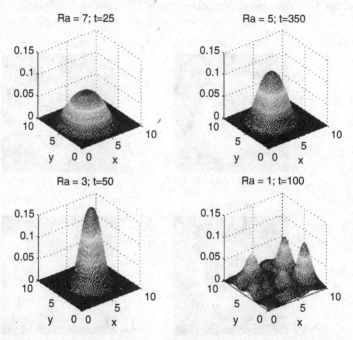

Fig. 7.10. A simulation of the long-range aggregation (7.21) and short-range repulsion (7.22) model for an ant colony with diffusion (smoothed empirical distribution)

Evolution Equations

Again, the fundamental tool for deriving an evolution equation for the empirical measure process is Itô's formula. As in the previous case, the time evolution of any function $f\left(X_N^k(t), t\right)$, $f \in C_b^2(\mathbb{R}^d \times \mathbb{R}_+)$, of the trajectory $\left(X_N^k(t)\right)_{t\in\mathbb{R}_+}$ of the kth individual particle, subject to SDE (7.5), is given by (7.6). By taking into account expressions (7.21) and (7.22) for F_1 and F_2 and (7.8), then from (7.6), we get the following weak formulation of the time evolution of $X_N(t)$ for any $f \in C_b^{2,1}(\mathbb{R}^d \times [0, \infty[)$:

$$\langle X_N(t), f(\cdot, t) \rangle = \langle X_N(0), f(\cdot, 0) \rangle + \int_0^t \langle X_N(s), (X_N(s) * \nabla G_a) \cdot \nabla f(\cdot, s) \rangle \, ds$$

$$- \int_0^t \langle X_N(s), \nabla g_N(\cdot, s) \cdot \nabla f(\cdot, s) \rangle \, ds$$

$$+ \int_0^t \left\langle X_N(s), \frac{\sigma_N^2}{2} \triangle f(\cdot, s) + \frac{\partial}{\partial s} f(\cdot, s) \right\rangle \, ds$$

$$+ \frac{\sigma_N}{N} \int_0^t \sum_k \nabla f\left(X_N^k(s), s\right) \, dW^k(s), \tag{7.24}$$

$$g_N(x, t) = (X_N(t) * V_N)(x).$$

Also for this case we may proceed as in the previous subsection on evolution

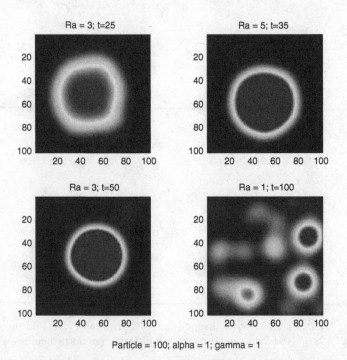

Particle = 100; alpha = 1; gamma = 1

Fig. 7.11. A simulation of the long-range aggregation (7.21) and short-range repulsion (7.22) model for an ant colony with diffusion (two-dimensional projection of smoothed empirical distribution)

equations with the analysis of the last term in (7.24). The process

$$M_N(f, t) = \frac{\sigma_N}{N} \int_0^t \sum_k \nabla f\left(X_N^k(s), s\right) \, dW^k(s), \qquad t \in [0, T],$$

is a martingale with respect to the process's $(X_N(t))_{t \in \mathbb{R}_+}$ natural filtration. By applying Doob's inequality (Proposition 2.125), we obtain

$$E\left[\sup_{t \leq T} |M_N(f,t)|\right]^2 \leq \frac{4\sigma_N^2 \|\nabla f\|_\infty^2 T}{N}.$$

Hence, by assuming that σ_N remains bounded as in (7.23), $M_N(f, \cdot)$ vanishes in the limit $N \to \infty$. This is again the essential reason for the deterministic limiting behavior of the process, since then its evolution equation will no longer be perturbed by Brownian noise.

We will not go into more detail at this point. The procedure is the same as for the previous model. But here we confine ourselves to a formal convergence procedure. Indeed, let us suppose that the empirical process $(X_N(t))_{t \in \mathbb{R}_+}$ tends, as $N \to \infty$, to a deterministic process $(X(t))_{t \in \mathbb{R}_+}$, which for any t is absolutely continuous with respect to the Lebesgue measure on $\mathbb{R}^d$, with density $\rho(x,t)$:

$$\lim_{N \to \infty} \langle X_N(t), f(\cdot, t) \rangle = \langle X(t), f(\cdot, t) \rangle$$

$$= \int f(x,t)\rho(x,t)dx, \qquad t \geq 0.$$

As a formal consequence we get

$$\lim_{N \to \infty} g_N(x,t) = \lim_{N \to \infty} (X_N(t) * V_N)(x) = \rho(x,t),$$

$$\lim_{N \to \infty} \nabla g_N(x,t) = \nabla \rho(x,t),$$

$$\lim_{N \to \infty} (X_N(t) * \nabla G_a)(x) = (X(t) * \nabla G_a(x))$$

$$= \int \nabla G_a(x-y)\rho(y,t)dy.$$

Hence, applying the foregoing limits, from (7.24) we obtain

$$\int_{\mathbb{R}^d} f(x,t)\rho(x,t)dx$$

$$= \int_{\mathbb{R}^d} f(x,0)\rho(x,0)dx$$

$$+ \int_0^t ds \int_{\mathbb{R}^d} dx \left[(\nabla G_a * \rho(\cdot, s))(x) - \nabla \rho(x,s)\right] \cdot \nabla f(x,s)\rho(x,s)$$

$$+ \int_0^t ds \int_{\mathbb{R}^d} dx \left[\frac{\partial}{\partial s} f(x,s)\rho(x,s) + \frac{\sigma_\infty^2}{2} \Delta f(x,s)\rho(x,s)\right], \qquad (7.25)$$

where σ_∞ is defined as in (7.23).

Note that (7.25) is a weak version of the following equation for the spatial density $\rho(x,t)$:

$$\frac{\partial}{\partial t}\rho(x,t) = \frac{\sigma_\infty^2}{2}\Delta\rho(x,t) + \nabla \cdot (\rho(x,t)\nabla\rho(x,t))$$

$$-\nabla \cdot [\rho(x,t)(\nabla G_a * \rho(\cdot,t))(x)], \qquad x \in \mathbb{R}^d, t \geq 0, \quad (7.26)$$

$$\rho(x,0) = \rho_0(x).$$

In the degenerate case, i.e., if (7.23) holds with equality, (7.26) becomes

$$\frac{\partial}{\partial t}\rho(x,t) = \nabla \cdot (\rho(x,t)\nabla\rho(x,t)) - \nabla \cdot [\rho(x,t)(\nabla G_a * \rho(\cdot,t))(x)]. \qquad (7.27)$$

As in the preceding subsection on moderate repulsion, we need to prove the existence and uniqueness of a sufficiently regular solution to (7.27). We refer the reader to Burger et al. (2007) and Nagai and Mimura (1983) as well as to Carrillo (1999) for a general discussion of this topic; for rigorous convergence results in the case $\sigma_\infty > 0$, the reader may refer to Capasso and Morale (2009).

A Law of Large Numbers in Path Space

In this section we supplement our results on the asymptotics of the empirical processes by a law of large numbers in path space. This means that we study the *empirical measures in path space*

$$X_N = \frac{1}{N}\sum_{k=1}^{N}\epsilon_{X_N^k(\cdot)},$$

where $X_N^k(\cdot) = (X_N^k(t))_{0 \leq t \leq T}$ denotes the entire path of the kth particle in the time interval $[0,T]$. The particles move continuously in $\mathbb{R}^d$. Moreover, X_N is a measure on the space $\mathcal{C}([0,T],\mathbb{R}^d)$ of continuous functions from $[0,T]$ to $\mathbb{R}^d$. As in the case of empirical processes, one can prove the convergence of X_N to some limit Y. The proof can be achieved with a few additional arguments from the limit theorem for the empirical processes.

By heuristic considerations in Morale et al. (2005), we get a convergence result for the empirical distribution of the drift $\nabla g_N(\cdot,t)$ of the individual particles

$$\lim_{N\to\infty}\int_0^T \langle X_N(t), |\nabla g_N(\cdot,t) - \nabla\rho(\cdot,t)|\rangle\, dt = 0, \qquad (7.28)$$

$$\lim_{N\to\infty}\int_0^T \langle X_N(t), |X_N(t) * \nabla G_a - \nabla G_a * \rho(\cdot,t)|\rangle\, dt = 0.$$

So (7.28) allows us to replace the drift

$$\nabla g_N(\cdot,t) - X_N(t) * \nabla G_a$$

with the function

$$\nabla\rho(\cdot,t) - \nabla G_a * \rho(\cdot,t)$$

for large N. Hence, for most k, we have $X_k(t) \sim Y(t)$, uniformly in $t \in [0,T]$, where $Y = Y(t)$, $0 \le t \le T$, is the solution of

$$dY(t) = [\nabla G_a * \rho(\cdot,t)(Y(t)) - \nabla\rho(Y(t))]\,dt + \sigma_\infty dW^k(t), \qquad (7.29)$$

with the initial condition, for each $k = 1,\ldots,N$,

$$Y(0) = X_N^k(0). \qquad (7.30)$$

So not only does the density follow the deterministic Equation (7.26), which presents the memory of the fluctuations by means of the term $\frac{\sigma_\infty}{2}\Delta\rho$, but also the stochasticity of the movement of each particle is preserved.

For the degenerate case $\sigma_\infty = 0$, the Brownian motion vanishes as $N \to \infty$. From (7.29) the dynamics of a single particle depends on the density of the whole system. This density is the solution of (7.27), which does not contain any diffusion term. So, not only do the dynamics of a single particle become deterministic, but there is also no memory of the fluctuations present when the number of particles N is finite. The following result confirms these heuristic considerations (Morale et al. 2005).

Theorem 7.16. *For the stochastic system (7.5)–(7.22) make the same assumptions as in Theorem 7.7. Then we obtain*

$$\lim_{N\to\infty} E\left[\frac{1}{N}\sum_{k=1}^{N}\sup_{t\le T}\left|X_N^k(t) - Y(t)\right|\right] = 0,$$

where Y is the solution of (7.29) with the initial solution (7.30) for each $k = 1,\ldots,N$ and ρ is the density of the limit of the empirical processes, i.e., it is the solution of (7.27).

Additional problems of the same kind arising in biology can be found in Champagnat et al. (2006) and Fournier and Méléard (2004).

Long Time Behavior

In this section we investigate the long time behavior of the particle system, for a fixed number N of particles.

Interacting-Diffusing Particles

First of all, let us reconsider our system, with a constant $\sigma \in \mathbb{R}_+^*$,

$$dX_N^k(t) = \left[(\nabla(G - V_N) * X_N)(X_N^k(t))\right]dt$$
$$+\sigma dW^k(t), \qquad k = 1,\ldots,N.$$

It can be shown (Malrieu 2003) that the location of the center of mass $\bar{X}_N$ of N particles,

$$\bar{X}_N(t) = \frac{1}{N}\sum_{k=1}^{N} X_N^k(t),$$

evolves according to the equation

$$d\bar{X}_N(t) = -\frac{1}{N^2}\sum_{k,j=1}^{N} \nabla\left(V_N - G\right)(X_N^k(t) - X_N^j(t))dt + \sigma d\bar{W}(t),$$

where $\bar{W}(t) = \frac{1}{N}\sum_{k=1}^{N} W^k(t)$ is still a Brownian motion; because of the symmetry of kernels V_1 and G, the first term on the right-hand side vanishes, which leads to

$$d\bar{X}_N(t) = \sigma d\bar{W}(t),$$

i.e., the stochastic process $\bar{X}_N$ is a Wiener process. Hence, its law, conditional upon the initial state, is

$$\mathcal{L}\left(\bar{X}_N(t)|\bar{X}_N(0)\right) = \mathcal{L}\left(\bar{X}_N(0), \sigma^2\bar{W}(t)\right) = \mathcal{N}\left(\bar{X}_N(0), \frac{\sigma^2}{N}t\right),$$

with variance $\frac{\sigma^2}{N}t$, which, for any fixed N, increases as t tends to infinity. Consequently we may claim that the probability law of the system does not converge to any nontrivial probability law since otherwise the same would happen for the law of the center of mass.

A Model with a Confining Potential

We then consider a modification to the foregoing system as follows:

$$\begin{aligned} dX_N^k(t) = &\left[\gamma_1 \nabla U(X_N^k(t)) + \gamma_2 \left(\nabla\left(G - V_N\right) * X_N\right)(X_N^k(t))\right]dt \\ &+\sigma dW^k(t), \qquad k = 1,\ldots,N, \end{aligned}$$

where $\gamma_1, \gamma_2 \in \mathbb{R}_+$.

This means that particles are also subject to a force due to the confining potential U. Equations of the type

$$dX_t = -\nabla P(X_t) + \sigma dW_t \tag{7.31}$$

have been thoroughly analyzed in the literature. Under the sufficient condition of strict convexity of the symmetric potential P it has been shown (Malrieu 2003; Carrillo et al. 2003; Markowich and Villani 2000) that system (7.31) does admit a nontrivial invariant distribution.

From a biological point of view a strictly convex confining potential is difficult to explain; it would mean an infinite range of attraction of the force which becomes infinitely strong at the infinite.

A weaker sufficient condition for the existence of a unique invariant measure has been suggested more recently by Veretennikov (1997), Veretennikov (2005), following Has'minskii (1980). This condition states that there exist constants $M_0 \geq 0$ and $r > 0$ such that for $|x| \geq M_0$

$$\left(-\nabla P(x), \frac{x}{|x|}\right) \leq -\frac{r}{|x|}. \tag{7.32}$$

It is easy to prove that without any further condition on the interaction kernels V_N and G, condition (7.32) is satisfied by considering the following condition on U.

There exist constants $M_0 \geq 0$ and $r > 0$ such that

$$\left(\nabla U(x), \frac{x}{|x|}\right) \leq -\frac{r}{|x|}, \quad |x| \geq M_0, \tag{7.33}$$

where $(\cdot, \cdot)$ denotes the usual scalar product in $\mathbb{R}^d$.

We may then apply the results by Veretennikov (2005) and prove the existence of a unique invariant measure for the joint law of the particles locations.

Condition (7.33) means that ∇U may decay to zero as $|x|$ tends to infinity, provided that its tails are sufficiently "fat."

Let $P_N^{x_0}(t)$ denote the joint distribution of N particles at time t, conditional upon a nonrandom initial condition x_0, and let P_S denote the invariant distribution. As far as the convergence of $P_N^{x_0}(t)$ is concerned, for t tending to infinity, as in Veretennikov (2005), one can prove the following result (Capasso and Morale 2009).

Proposition 7.17 *Under the hypotheses of existence and uniqueness and the foregoing assumptions on U, for any k, $0 < k < \tilde{r} - \frac{Nd}{2} - 1$ with $m \in (2k+2, 2\tilde{r} - Nd)$ and $\tilde{r} = \gamma_1 Nr$, there exists a positive constant c such that*

$$\left|P_N^{x_0}(t) - P_N^S\right| \leq c(1 + |x_0|^m)(1 + t)^{-(k+1)},$$

where $\left|P_N^{x_0}(t) - P_N^S\right|$ denotes the total variation distance of the two measures, i.e.,

$$\left|P_N^{x_0}(t) - P_N^S\right| = \sup_{A \in \mathcal{B}_{\mathbb{R}^d}} \left[P_N^{x_0}(t)(A) - P_N^S(A)\right],$$

and x_0 the initial data.

So Proposition 7.17 states a polynomial convergence rate to invariant measure. To improve the rate of convergence, one has to consider more restricted assumptions on U.

Important and interesting results, extending those presented in this chapter, regarding the mean field approximation to a system of N interacting particles whose time evolution is governed by a system of stochastic differential equations, can be found in Bolley (2005) and Bolley et al. (2007).

7.3.4 Price Herding

As an example of herding in economics we present a model for price herding that has been applied to simulate the prices of cars (Capasso et al. 2003). The model is based on the assumption that prices of products of a similar nature and within the same market segment tend to aggregate within a given interaction kernel, which characterizes the segment itself. On the other hand, unpredictable behavior of individual prices may be modeled as a family of mutually independent Brownian motions. Hence we suppose that in a segment of N prices, for any $k \in \{1, \ldots, N\}$ the price $X_N^k(t)$, $t \in \mathbb{R}_+$, satisfies the following system of SDEs:

$$\frac{dX_N^k(t)}{X_N^k(t)} = F_k[\mathbf{X}(t)] \left(X_N^k(t) \right) dt + \sigma_k(\mathbf{X}(t)) dW^k(t).$$

As usual, for a population of prices it is more convenient to consider the evolution of rates. For the force of interaction F_k, which depends upon the vector of all individual prices

$$\mathbf{X}(t) := \left(X_N^1(t), \ldots, X_N^N(t) \right),$$

we assume the following model, similar to the ant colony of the previous subsection:

$$F_k[\mathbf{X}(t)] \left(X_N^k(t) \right) = \frac{1}{N} \sum_{j=1}^{N} \frac{1}{A_{jk}} \left(\frac{I_j(t)}{I_k(t)} \right)^{\beta_{jk}} \nabla K_a \left(X_N^k(t) - X_N^j(t) \right); \quad (7.34)$$

the drift (7.34) includes the following ingredients:

(a) The aggregation kernel

$$K_a(x) = \frac{1}{\sqrt{2\pi a^2}} e^{-\frac{x^2}{2a^2}},$$

$$\nabla K_a(x) = -\frac{x}{a^2} \frac{1}{\sqrt{2\pi a^2}} e^{-\frac{x^2}{2a^2}};$$

(b) The sensitivity coefficient for aggregation

$$\frac{1}{A_{jk}} \left(\frac{I_j(t)}{I_k(t)} \right)^{\beta_{jk}},$$

depending (via the parameters A_{jk} and β_{jk}) on the relative market share $I_j(t)$ of product j with respect to the market share $I_k(t)$ of product k. Clearly, a stronger product will be less sensitive to the prices of competing weaker products;

(c) The coefficient $\frac{1}{N}$ takes into account possible crowding effects, which are also modulated by the coefficients A_{jk}.

As an additional feature a model for inflation may be included in F_k. Given a general rate of inflation $(\alpha_t)_{t\in\mathbb{R}_+}$, F_k may include a term $s_k\alpha_t$ to model via s_k the specific sensitivity of price k. We leave the analysis of the model to the reader, who may refer to Capasso et al. (2003) for details.

Data are shown in Fig. 7.12; parameter estimates are given in Tables 7.1, 7.2, 7.3, 7.4; Fig. 7.13 shows simulated car prices based on such estimates (Bianchi et al. 2005).

7.4 Neurosciences

Stein's Model of Neural Activity

The main component of Stein's model (Stein 1965, 1967) is the depolarization V_t for $t \in \mathbb{R}_+$. A nerve cell is said to be *excited* (or *depolarized*) if $V_t > 0$ and *inhibited* if $V_t < 0$. In the absence of other events, V_t decays according to

$$\frac{dV}{dt} = -\alpha V,$$

where $\alpha = 1/\tau$ is the reciprocal of the nerve membrane time constant $\tau > 0$.

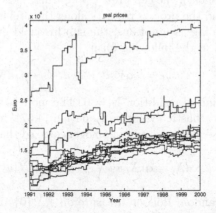

Fig. 7.12. Time series of prices of a segment of cars in Italy during years 1991–2000 (source: *Quattroruote Magazine, Editoriale Domus, Milan, Italy.*)

In the resting state (initial condition), $V_0 = 0$. Afterward, jumps may occur at random times according to independent Poisson processes $(N_t^E)_{t\in\mathbb{R}_+}$ and $(N_t^I)_{t\in\mathbb{R}_+}$, with intensities λ_E and λ_I, respectively, assumed to be strictly positive real constants. If an excitation (a jump) occurs for N^E, at some time $t_0 > 0$, then

$$V_{t_0} - V_{t_0-} = a_E,$$

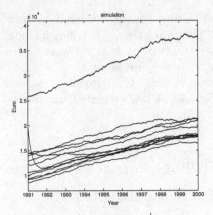

Fig. 7.13. Simulated car prices

whereas if an inhibition (again a jump) occurs for N^I, then

$$V_{t_0} - V_{t_0-} = -a_I,$$

where a_E and a_I are nonnegative real numbers. When V_t attains a given value $\theta > 0$ (the *threshold*), the cell fires. Upon firing, V_t is reset to zero along with N^E and N^I, and the process restarts along the previous model. By collecting all of the foregoing assumptions, the subthreshold evolution equation for V_t may be written in the following form:

$$dV_t = -\alpha V_t dt + a_E dN_t^E - a_I dN_t^I,$$

subject to the initial condition $V_0 = 0$. The model is a particular case of the more general stochastic evolution Equation (4.57); here, the Wiener noise is absent and the equation is time-homogeneous, so that it reduces to the form

$$dX_t = \alpha(X_t)dt + \int_{\mathbb{R}} \gamma(X_t, u)N(dt, du), \qquad (7.35)$$

where N is a random Poisson measure on $\mathbb{R} \setminus \{0\}$ (in (7.35) the integration is over u). In Stein's model $\alpha(x) = -\alpha x$, with $\alpha > 0$ (or simply $\alpha(x) = -x$ if we assume $\alpha = 1$), $\gamma(x, u) = u$, and the Poisson measure N has intensity measure

$$\Lambda((s, t) \times B) = (t - s) \int_B \phi(u)du \qquad \text{for any } s, t \in \mathbb{R}_+, s < t, B \subset \mathcal{B}_{\mathbb{R}}.$$

Here

$$\phi(u) = \lambda_E \delta_0(u - a_E) + \lambda_I \delta_0(u + a_I),$$

where δ_0 denotes the standard Dirac delta distribution. The infinitesimal generator $\mathcal{A}$ of the Markov process $(X_t)_{t \in \mathbb{R}_+}$ in (7.35) is given by

Table 7.1. Estimates for price herding model (7.34) for initial conditions $X_k(0)$ and range of kernel a

Parameter	Method of estimation	Estimate	Std. dev.
$X_1(0)$	ML	1.6209E+00	5.8581E−02
$X_2(0)$	ML	8.4813E−01	6.0740E−03
$X_3(0)$	ML	7.4548E−01	2.3420E−02
$X_4(0)$	ML	1.0189E+00	1.2273E−01
$X_5(0)$	ML	1.4164E+00	1.4417E−01
$X_6(0)$	ML	2.4872E+00	6.2947E−02
$X_7(0)$	ML	1.2084E+00	4.7545E−02
$X_8(0)$	ML	1.0918E+00	4.7569E−02
a	ML	5.0767E+03	6.5267E+02

Table 7.2. Estimates for price herding model (7.34) for parameters A_{ij}

Parameter	Method of estimation	Estimate	Std. dev.
A_{12}	ML	1.0649E−03	3.0865E−02
A_{13}	ML	1.1489E−04	4.1737E−04
A_{14}	ML	1.5779E−03	5.4687E−02
A_{15}	ML	7.6460E−04	1.8381E−02
A_{16}	ML	1.2908E−03	4.0634E−02
A_{17}	ML	1.8114E−03	6.5617E−02
A_{18}	ML	1.5956E−03	5.5572E−02
A_{23}	ML	1.0473E−04	7.2687E−05
A_{24}	ML	1.7397E−04	6.0809E−04
A_{25}	ML	1.7550E−04	5.1100E−04
A_{26}	ML	1.2080E−03	3.7392E−02
A_{27}	ML	9.4809E−04	2.6037E−02
A_{28}	ML	2.7277E−04	2.0135E−03
A_{34}	ML	4.0404E−04	5.5468E−03
A_{35}	ML	1.8136E−04	8.6471E−04
A_{36}	ML	9.5558E−03	4.9764E−01
A_{37}	ML	1.0341E−04	4.4136E−05
A_{38}	ML	7.0953E−04	1.6428E−02
A_{45}	ML	1.0066E−03	2.8485E−02
A_{46}	ML	1.3354E−04	1.3632E−03
A_{47}	ML	2.5239E−04	1.6979E−03
A_{48}	ML	1.1232E−03	3.3652E−02
A_{56}	ML	2.3460E−03	9.2592E−02
A_{57}	ML	1.0143E−03	2.8898E−02
A_{58}	ML	1.1026E−03	3.2724E−02
A_{67}	ML	1.8560E−03	6.8275E−02
A_{68}	ML	2.2820E−03	8.9278E−02
A_{78}	ML	6.4630E−04	1.4003E−02

Table 7.3. Estimates for price herding model (7.34) for parameters β_{ij}

Parameter	Method of estimation	Estimate	Std. dev.
β_{12}	ML	6.8920E−01	5.8447E+00
β_{13}	ML	2.3463E+00	2.7375E+00
β_{14}	ML	7.2454E−01	6.6182E+00
β_{15}	ML	8.4049E−01	6.2349E+00
β_{16}	ML	7.7929E−01	5.6565E+00
β_{17}	ML	6.6793E−01	5.4208E+00
β_{18}	ML	7.6508E−01	5.8422E+00
β_{23}	ML	2.4531E+00	4.5883E−01
β_{24}	ML	1.6924E+00	6.8734E+00
β_{25}	ML	1.6262E+00	5.7128E+00
β_{26}	ML	1.2122E+00	2.1666E+00
β_{27}	ML	7.5140E−01	7.4760E+00
β_{28}	ML	1.3537E+00	6.0109E+00
β_{34}	ML	1.2444E+00	8.1509E+00
β_{35}	ML	1.7544E+00	8.4976E+00
β_{36}	ML	1.0572E+00	8.0208E+00
β_{37}	ML	2.4730E+00	1.9801E−01
β_{38}	ML	1.0674E+00	8.4626E+00
β_{45}	ML	7.5781E−01	6.7267E+00
β_{46}	ML	2.2121E+00	6.9754E+00
β_{47}	ML	1.7360E+00	6.4971E+00
β_{48}	ML	8.1043E−01	6.1451E+00
β_{56}	ML	7.1269E−01	4.5857E+00
β_{57}	ML	7.7251E−01	6.3947E+00
β_{58}	ML	7.0792E−01	6.5014E+00
β_{67}	ML	8.4060E−01	6.8871E+00
β_{68}	ML	8.1190E−01	6.0759E+00
β_{78}	ML	1.0794E+00	8.4994E+00

$$\mathcal{A}f(x) = \alpha(x)\frac{\partial f}{\partial x}(x) + \int_{\mathbb{R}} (f(x + \gamma(x,u)) - f(x))\phi(u)du$$

for any test function f in the domain of $\mathcal{A}$.

The firing problem may be seen as a first passage time through the threshold $\theta > 0$. Let $A =] - \infty, \theta[$. Then the random variable of interest is

$$T_A(x) = \inf\{t \in \mathbb{R}_+ | X_t \in A, X_0 = x \in A\},$$

which is the first exit time from A. If the indicated set is empty, then we set $T_A(x) = +\infty$. The following result holds.

Theorem 7.18.   (Tuckwell 1976; Darling and Siegert 1953). *Let $(X_t)_{t \in \mathbb{R}_+}$ be a Markov process satisfying (7.35), and assume that the existence and uniqueness conditions are fulfilled. Then the distribution function*

$$F_A(x,t) = P(T_A(x) \leq t)$$

satisfies

$$\frac{\partial F_A}{\partial t}(x,t) = \mathcal{A}F_A(\cdot,t)(x), \qquad x \in A, t > 0,$$

subject to the initial condition

$$F_A(x,0) = \begin{cases} 0 & \text{for } x \in A, \\ 1 & \text{for } x \notin A, \end{cases}$$

and boundary condition

$$F_A(x,t) = 1, \qquad x \notin A, x \geq 0.$$

Corollary 7.19 *If the moments*

$$\mu_n(x) = E\left[(T_A(x))^n\right], \qquad n \in \mathbb{N}^*,$$

exist, they satisfy the recursive system of equations

$$\mathcal{A}\mu_n(x) = -n\mu_{n-1}(x), \qquad x \in A, \tag{7.36}$$

subject to the boundary conditions

Table 7.4. Estimates for price herding model (7.34) of s_k and σ_k

Parameter	Method of estimation	Estimate	Std. dev.
s_1	ML	2.0267E−03	2.1858E−04
s_2	ML	5.1134E−03	1.6853E−03
s_3	ML	3.6238E−03	2.5305E−03
s_4	ML	3.6777E−03	2.3698E−03
s_5	ML	1.0644E−04	1.1132E−04
s_6	ML	5.4133E−03	1.2452E−03
s_7	ML	1.0769E−04	1.4414E−04
s_8	ML	2.1597E−03	2.8686E−03
σ_1	MAP	7.0000E−03	2.9073E−06
σ_2	MAP	7.0000E−03	2.9766E−06
σ_3	MAP	7.0000E−03	3.0128E−06
σ_4	MAP	7.0000E−03	2.9799E−06
σ_5	MAP	7.0000E−03	3.0025E−06
σ_6	MAP	7.0000E−03	2.9897E−06
σ_7	MAP	7.0000E−03	2.8795E−06
σ_8	MAP	7.0000E−03	2.9656E−06

$$\mu_n(x) = 0, \qquad x \notin A.$$

The quantity $\mu_0(x)$, $x \in A$, is the probability of X_t exiting from A in a finite time. It satisfies the equation

$$\mathcal{A}\mu_0(x) = 0, \qquad x \in A, \tag{7.37}$$

subject to

$$\mu_0(x) = 1, \qquad x \notin A.$$

The following lemma is due to Gihman and Skorohod (1972, p. 305).

Lemma 7.20. *If there exists a bounded function g on $\mathbb{R}$ such that*

$$\mathcal{A}g(x) \leq -1, \qquad x \in A, \tag{7.38}$$

then $\mu_1 < \infty$ and $P(T_A(x) < +\infty) = 1$.

As a consequence of Lemma 7.20, a neuron in Stein's model fires in a finite time with probability 1 and with finite mean interspike interval. This is due to the fact that the solution of (7.37) is $\mu_0(x) = 1$, $x \in \mathbb{R}$, and this satisfies (7.38). The mean first passage time through θ for an initial value x satisfies, by (7.36),

$$-x\frac{d\mu_1}{dx}(x) + \lambda_E\mu_1(x + a_E) + \lambda_I\mu_1(x - a_I) - (\lambda_E + \lambda_I)\mu_1(x) = -1, \tag{7.39}$$

with $x < \theta$ and boundary condition

$$\mu_1(x) = 0, \qquad \text{for } x \geq \theta.$$

The solution of (7.39) is discussed in Tuckwell (1989), where a diffusion approximation of the original Stein's model of neuron firing is also analyzed.

An optimal control problem for the diffusion approximation of Stein's model was recently analyzed in Lu (2011).

7.5 An SIS Epidemic Model

Let us consider another class of deterministic models for the population dynamics of the spread of infectious diseases which do not induce permanent immunity; this class of models is known as SIS models (see, e.g., Capasso (2008)). In this case the total population of size $N(t)$ at time $t \in \mathbb{R}_+$ includes only two subclasses; the class of susceptibles, of size $S(t)$, $t \in \mathbb{R}_+$, and the class of infectives, of size $I(t)$, $t \in \mathbb{R}_+$, such that $N(t) = S(t) + I(t)$, $t \in \mathbb{R}_+$.

A classical model for such an epidemic system is described in terms of the following system of ordinary differential equations (see, e.g., Hethcote and Yorke (1984))

$$\begin{cases} \dfrac{dS(t)}{dt} = \mu N(t) - \kappa S(t)I(t) + \delta I(t) - \mu S(t) \\ \dfrac{dI(t)}{dt} = \kappa S(t)I(t) - (\mu + \delta)I(t). \end{cases} \tag{7.40}$$

Here μ denotes the death rate of any individual in the population, equal to the birth rate of new susceptibles (any newborn is susceptible); δ^{-1} is the average infection period; and finally κ is the infection rate.

Since, at any time $t \in \mathbb{R}_+$, $N(t) = S(t) + I(t)$, for the model under consideration $N(t) = N(0) \equiv N$ a constant, given by assigning initial conditions at time $t = 0$, i.e. $S(0) = S_0 \in (0, N)$, $I(0) = I_0 \in (0, N)$ such that $S_0 + I_0 = N$. This implies that it is sufficient to analyze only the equation for the infective population

$$\frac{dI(t)}{dt} = \kappa(N - S(t))I(t) - (\mu + \delta)I(t), \qquad (7.41)$$

subject to $I(0) = I_0$.

For this model it is not difficult to show (see, e.g., Hethcote and Yorke (1984)) that the following *threshold theorem* holds.

Theorem 7.21. *Let*

$$R_0^D := \frac{\kappa N}{\mu + \delta}. \qquad (7.42)$$

(i) If $R_0^D \leq 1$, *then, for any initial condition* $I(0) = I_0 \in (0, N)$, $\lim_{t \to +\infty} I(t) = 0$.

(ii) If $R_0^D > 1$, *then, for any initial condition* $I(0) = I_0 \in (0, N)$, $\lim_{t \to +\infty} I(t) = N\left(1 - \frac{1}{R_0^D}\right)$.

This is the reason why the parameter R_0^D, which has an epidemiological meaning, is known as the *threshold parameter* of System (7.40).

In Gray et al. (2011), the authors analyze the consequences on the dynamics of System (7.40) once an environmental perturbation is included on the infection rate κ, of the following form

$$\tilde{\kappa}dt = \kappa dt + \sigma dW(t), \qquad (7.43)$$

where $(W(t))_{t \in \mathbb{R}_+}$ is a standard Wiener process.

Upon such a perturbation, Equation (7.41) becomes

$$dI(t) = I(t)\left([\kappa(N - I(t)) - (\mu + \delta)I(t)]dt + \sigma(N - I(t))dW(t)\right), \qquad (7.44)$$

subject to a deterministic initial condition $I(0) = I_0 \in (0, N)$.

The steps the authors have considered are

1. existence (global in time) of a unique nonnegative solution;
2. threshold theorems for the eventual extinction of the epidemic, or for its persistence;
3. existence of a nontrivial stationary distribution.

7.5.1 Existence of a unique nonnegative solution

Particular attention has to be paid to the fact that a realistic solution of Equation (7.44), started from an $I_0 \in (0, N)$, should stay in $(0, N)$ for all times $t > 0$, almost surely. Indeed the following theorem has been proven.

Theorem 7.22. *For any given deterministic initial value $I(0) = I_0 \in (0, N)$, the stochastic differential equation (7.44) admits a unique positive solution $I(t) \in (0, N)$, for all times $t \in \mathbb{R}_+$, almost surely.*

Proof. The proof is based upon Corollary 5.9 of Chapter 5, by considering the following auxiliary function

$$v(x) := \frac{1}{x} + \frac{1}{N - x}, \quad x \in (0, N). \tag{7.45}$$

In fact it is easy to recognize that

$$v(x) \to +\infty, \quad \text{as} \quad x \downarrow 0^+, \quad \text{or} \quad x \uparrow N^-, \tag{7.46}$$

as required in Assumption (ii) of Theorem 5.8 of Chapter 5.
    Moreover, it is not difficult to show that, for a suitable $C > 0$,

$$L_0 v(x) \leq C v(x), \quad \text{for any} \quad x \in (0, N), \tag{7.47}$$

as requested in Assumption (i) of Theorem 5.8 of Chapter 5   □

7.5.2 Threshold theorem

A new threshold is introduced, smaller than the one for the deterministic case

$$R_0^S := R_0^D - \frac{\sigma^2}{2} \frac{N^2}{\mu + \delta}, \tag{7.48}$$

such that the following theorem holds.

Theorem 7.23.

(i)   If $R_0^S < 1$,   *then, for any deterministic initial condition $I(0) = I_0 \in (0, N)$,*

$$\limsup_{t \to +\infty} \frac{1}{t} \ln I(t) = 0, \quad \text{a.s.};$$

namely $I(t)$ tends to zero exponentially, a.s.

(ii)   If $R_0^S > 1$,   *then, for any deterministic initial condition $I(0) = I_0 \in (0, N)$,*

$$\liminf_{t \to +\infty} I(t) \leq I^*,$$

and

$$\limsup_{t \to +\infty} I(t) \geq I^*,$$

where

$$I^* := \frac{1}{\sigma^2}\left(\sqrt{\kappa^2 - 2\sigma^2(\mu + \delta)} - (\kappa - \sigma^2 N)\right), \qquad (7.49)$$

is the unique root in $(0, N)$ *of*

$$\frac{\sigma^2}{2}(N - I^*)^2 + \kappa I^* - [\kappa N - (\mu + \delta)] = 0.$$

This means that $I(t)$ *will cross the level* I^* *infinitely often, a.s.*

Proof. Let us consider only the case (i). By Itô formula, we have

$$\ln I(t) = \ln I_0 + \int_0^t f(I(s))ds + \int_0^t \sigma(N - I(s))dW(s), \quad t \in \mathbb{R}_+, \quad (7.50)$$

where f is the real-valued function defined by

$$f(x) := -\frac{\sigma^2}{2}(N - x)^2 - \kappa x + \kappa N - (\mu + \delta), \quad x \in (0, N). \qquad (7.51)$$

Under the condition $R_0^S < 1$,

$$f(I(s)) \leq -\frac{\sigma^2}{2}N^2 + \kappa N - (\mu + \delta), \quad x \in (0, N),$$

for any $s \in (0, t)$, so that, for any $t \in \mathbb{R}_+$,

$$\ln I(t) \leq \ln I_0 - \frac{\sigma^2}{2}N^2 + \kappa N - (\mu + \delta) + \int_0^t \sigma(N - I(s))dW(s). \qquad (7.52)$$

This implies

$$\limsup_{t \to +\infty} \frac{1}{t}\ln I(t) \leq -\frac{\sigma^2}{2}N^2 + \kappa N - (\mu + \delta)$$

$$+ \limsup_{t \to +\infty} \frac{1}{t}\int_0^t \sigma(N - I(s))dW(s), \quad \text{a.s.} \qquad (7.53)$$

By the law of large numbers for martingales (see, e.g., Mao (1997, p. 12))

$$\limsup_{t \to +\infty} \frac{1}{t}\int_0^t \sigma(N - I(s))dW(s), \quad \text{a.s.}, \qquad (7.54)$$

from which part (i) of the theorem follows (see Gray et al. (2011) for further details). $\square$

An interesting result concerns the behavior of I^* as a function of σ^2, the variance of the noise added on the infection rate.

Proposition 7.24 *Let $R_0^S > 1$. The value $I^*(\sigma^2)$, as a function of σ^2, for*

$$0 < \sigma^2 < \frac{2(\kappa N - (\mu + \delta))}{N^2},$$

is strictly decreasing, and

$$\lim_{\sigma^2 \downarrow 0^+} I^*(\sigma^2) = N\left(1 - \frac{1}{R_0^D}\right), \tag{7.55}$$

which is the nontrivial equilibrium state of the corresponding SIS deterministic model (7.41).

7.5.3 Stationary distribution

We conclude the analysis, as reported from Gray et al. (2011), with the following theorem.

Theorem 7.25. *If $R_0^S > 1$, then the SDE (7.44) admits a unique stationary distribution, having mean value*

$$m = \frac{2\kappa(R_0^S - 1)(\mu + \delta)}{2\kappa(\kappa - \sigma^2 N) + \sigma^2(\kappa N - (\mu + \delta))}, \tag{7.56}$$

and variance

$$v = \frac{m(\kappa N - (\mu + \delta))}{\kappa} - m^2. \tag{7.57}$$

Proof. The proof is based on Theorem 5.28 of Chapter 5. In order to apply the theorem, we have to consider the following stopping time

$$\tau_{I_0} := \inf\{t \in \mathbb{R}_+ | I(t) \in (a, b)\}, \tag{7.58}$$

and show that, for any $(a, b) \subset (0, N)$, and for any $I_0 \in (a, b)$,

$$E[\tau_{I_0}] < +\infty, \tag{7.59}$$

and moreover, for any $[\alpha, \beta] \subset (0, N)$,

$$\sup_{I_0 \in [\alpha, \beta]} E[\tau_{I_0}] < +\infty. \tag{7.60}$$

See Gray et al. (2011) for a detailed proof.   □

An additional interesting example can be found in Cai and Lin (2004).

7.6 Exercises and Additions

7.1. Consider a birth-and-death process $(X(t))_{t \in \mathbb{R}_+}$ valued in $\mathbb{N}$, as in Sect. 7.1. In integral form the evolution equation for X will be

$$X(t) = X(0) + \alpha \int X(s-)ds + M(t),$$

where $\alpha = \lambda - \mu$ is the survival rate and $M(t)$ is a martingale. Show that

1. $\langle M \rangle(t) = \langle M, M \rangle(t) = (\lambda + \mu) \int_0^t X(s-)ds.$
2. $E[X(t)] = X(0)e^{\alpha t}.$
3. $X(t)e^{-\alpha t}$ is a square-integrable martingale.
4. $Var[X(t)e^{-\alpha t}] = X(0)\dfrac{\lambda + \mu}{\lambda - \mu}(1 - e^{-\alpha t}).$

7.2. (Age-dependent birth-and-death process). An age-dependent population can be divided into two subpopulations, described by two marked counting processes. Given $t > 0$, $U^{(1)}(A_0, t)$ describes those individuals who already existed at time $t = 0$ with ages in $A_0 \in \mathcal{B}_{\mathbb{R}_+}$ and are still alive at time t; and $U^{(2)}(T_0, t)$ describes those individuals who are born during $T_0 \in \mathcal{B}_{\mathbb{R}_+}$, $T_0 \subset [0, t]$ and are still alive at time t. Assume that the age-specific death rate is $\mu(a)$, $a \in \mathbb{R}_+$, and that the birth process $B(T_0), T_0 \in \mathcal{B}_{\mathbb{R}_+}$ admits stochastic intensity

$$\alpha(t_0) = \int_0^{+\infty} \beta(a_0 + t_0)U^{(1)}(da_0, t_0-) + \int_0^{t_0-} \beta(t_0 - \tau)U^{(2)}(d\tau, t_0-),$$

where $\beta(a)$, $a \in \mathbb{R}_+$ is the age-specific fertility rate. Assume now that suitable densities u_0 and b exist on $\mathbb{R}_+$ such that

$$E[U^{(1)}(A_0, 0)] = \int_{A_0} u_0(a)da$$

and

$$E[B(T_0)] = \int_{T_0} b(\tau)d\tau.$$

Show that the following *renewal equation* holds for any $s \in \mathbb{R}_+$:

$$b(s) = \int_0^{+\infty} da\, u_0(a)\, n(s + a)\, \beta(a + s) + \int_0^s d\tau\, \beta(s - \tau)\, n(s - \tau)\, b(\tau),$$

where $n(t) = \exp\left\{ -\int_0^t \mu(\tau)d\tau \right\}$, $t \in \mathbb{R}_+$. The reader may refer to Capasso (1988).

7.3. Let $\bar{E}$ be the closure of an open set $E \subset \mathbb{R}^d$ for $d \geq 1$. Consider a spatially structured birth-and-death process associated with the marked point process defined by the random measure on $\mathbb{R}^d$:

$$\nu(t) = \sum_{i=1}^{I(t)} \varepsilon_{X^i(t)},$$

where $I(t)$, $t \in \mathbb{R}_+$, denotes the number of individuals in the total population at time t, and $X^i(t)$ denotes the random location of the ith individual in $\bar{E}$. Consider the process defined by the following parameters:

1. $\mu : \bar{E} \to \mathbb{R}_+$ is the spatially structured death rate;
2. $\gamma : \bar{E} \to \mathbb{R}_+$ is the spatially structured birth rate;
3. For any $x \in \bar{E}$, $D(x, \cdot) : \mathcal{B}_{\mathbb{R}^d} \to [0,1]$ is a probability measure such that $\int_{\bar{E} \setminus \{x\}} D(x, dz) = 1$; $D(x, A)$ for $x \in \bar{E}$ and $A \in \mathcal{B}_{\mathbb{R}^d}$ represents the probability that an individual born in x will be dispersed in A.

Show that the infinitesimal generator of the process is the operator L defined as follows: for any sufficiently regular test function ϕ

$$L\phi(\nu) = \int_{\bar{E}} \nu(dx) \int_{\mathbb{R}^d} \gamma(x) D(x, dz)[-\phi(\nu) + \phi(\nu + \varepsilon_{x+z})]$$
$$+ \mu(x)[-\phi(\nu) + \phi(\nu - \varepsilon_x)].$$

The reader may refer to Fournier and Méléard (2003) for further analysis.

7.4. Let X be an integer-valued random variable, with probability distribution $p_k = P(X = k)$, $k \in \mathbb{N}$. The probability-generating function of X is defined as

$$g_X(s) = E[s^X] = \sum_{k=0}^{\infty} s^k p_k, \; |s| \leq 1.$$

Consider a homogeneous birth-and-death process $X(t), t \in \mathbb{R}_+$, with birth rate λ, death rate μ, and initial value $X(0) = k_0 > 0$. Show that the probability-generating function $G_X(s; t)$ of $X(t)$ satisfies the partial differential equation

$$\frac{\partial}{\partial t} G_X(s; t) + (1 - s)(\lambda s - \mu) \frac{\partial}{\partial s} G_X(s; t) = 0,$$

subject to the initial condition

$$G_X(s; 0) = s^{k_0}.$$

7.5. Consider now a nonhomogeneous birth-and-death process $X(t), t \in \mathbb{R}_+$, with time-dependent birth rate $\lambda(t)$, death rate $\mu(t)$, and initial value $X(0) = k_0 > 0$. Show that the probability-generating function $G_X(s; t)$ of $X(t)$ satisfies the partial differential equation

$$\frac{\partial}{\partial t} G_X(s; t) + (1 - s)(\lambda(t)s - \mu(t)) \frac{\partial}{\partial s} G_X(s; t) = 0,$$

subject to the initial condition

$$G_X(s; 0) = s^{k_0}.$$

Evaluate the probability of extinction of the population.

The reader may refer to Chiang (1968).

7.6. Consider the general epidemic process as defined in Sect. 7.1, with infection rate $\kappa = 1$ and removal rate δ. Let $G_Z(x, y; t)$ denote the probability-generating function of the random vector $\mathbf{Z}(t) = (S(t), I(t))$, where $S(t)$ denotes the number of susceptibles at time $t \geq 0$ and $I(t)$ denotes the number of infectives at time $t \geq 0$. Assume that $S(0) = s_0$ and $I(0) = i_0$, and let $p(m, n; t) = P(S(t) = m, I(t) = n)$. The joint probability-generating function G will be defined as

$$G_{\mathbf{Z}}(x, y; t) = E[x^{S(t)} y^{I(t)}] = \sum_{m=0}^{s_0} \sum_{n=0}^{s_0 + i_0 - m} p(m, n; t)\, x^m\, y^n.$$

Show that it satisfies the partial differential equation

$$\frac{\partial}{\partial t} G_{\mathbf{Z}}(x, y; t) = y(y - x) \frac{\partial^2}{\partial x \partial y} G_{\mathbf{Z}}(x, y; t) + \delta(1 - y) \frac{\partial}{\partial y} G_{\mathbf{Z}}(x, y; t),$$

subject to the initial condition

$$G_{\mathbf{Z}}(x, y; 0) = x^{s_0} y^{i_0}.$$

7.7. Consider a discrete birth-and-death chain $(Y_n^{(\Delta)})_{n \in \mathbb{N}}$ valued in $S = \{0, \pm\Delta, \pm 2\Delta, \dots\}$, with step size $\Delta > 0$, and denote by $p_{i,j}$ the one-step transition probabilities

$$p_{ij} = P\left(Y_{n+1}^{(\Delta)} = j\Delta \,\middle|\, Y_n^{(\Delta)} = i\Delta\right) \text{ for } i, j \in \mathbb{Z}.$$

Assume that the only nontrivial transition probabilities are

1. $p_{i,i-1} = \gamma_i := \frac{1}{2}\sigma^2 - \frac{1}{2}\mu\Delta$,
2. $p_{i,i+1} = \beta_i := \frac{1}{2}\sigma^2 + \frac{1}{2}\mu\Delta$,
3. $p_{i,i} = 1 - \beta_i - \gamma_i = 1 - \sigma^2$,

where σ^2 and μ are strictly positive real numbers. Note that for Δ sufficiently small, all rates are nonnegative. Consider now the rescaled (in time) process $(Y_{n/\varepsilon}^{(\Delta)})_{n \in \mathbb{N}}$, with $\varepsilon = \Delta^2$; show (formally and possibly rigorously) that the rescaled process weakly converges to a diffusion on $\mathbb{R}$ with drift μ and diffusion coefficient σ^2.

7.8. With reference to the previous problem, show that the same result may be obtained (with suitable modifications) also in the case in which the drift and the diffusion coefficient depend upon the state of the process. For this case show that the probability $\psi(x)$ that the diffusion process reaches c before d, when starting from a point $x \in (c, d) \subset \mathbb{R}$, is given by

$$\psi(x) = \frac{\int_x^d \exp\left\{-\int_c^z \left(2\frac{\mu(y)}{\sigma^2(y)}\right) dy\right\} dz}{\int_c^d \exp\left\{-\int_c^z \left(2\frac{\mu(y)}{\sigma^2(y)}\right) dy\right\} dz}.$$

The reader may refer, e.g., to Bhattacharya and Waymire (1990).

7.9. Consider the general stochastic epidemic with the rescaling proposed at the beginning of Sect. 7.2. Derive the asymptotic ordinary differential system corresponding to Theorem 7.4.

A

Measure and Integration

A.1 Rings and σ-Algebras

Definition A.1. A collection $\mathcal{F}$ of subsets of a set Ω is called a *ring* on Ω if it satisfies the following conditions:

1. $\emptyset \in \mathcal{F}$
2. $A, B \in \mathcal{F} \Rightarrow A \cup B \in \mathcal{F}$
3. $A, B \in \mathcal{F} \Rightarrow A \setminus B \in \mathcal{F}$

Furthermore, $\mathcal{F}$ is called an *algebra* if $\mathcal{F}$ is both a ring and $\Omega \in \mathcal{F}$.

Definition A.2. A ring $\mathcal{F}$ on Ω is called a σ-*ring* if it satisfies the following additional condition:

4. For every countable family $(A_n)_{n \in \mathbb{N}}$ of elements of $\mathcal{F}$: $\bigcup_{n \in \mathbb{N}} A_n \in \mathcal{F}$.

A σ-ring $\mathcal{F}$ on Ω is called a σ-*algebra* if $\Omega \in \mathcal{F}$.

Definition A.3. Every collection $\mathcal{F}$ of elements of a set Ω is called a *semiring* on Ω if it satisfies the following conditions:

1. $\emptyset \in \mathcal{F}$.
2. $A, B \in \mathcal{F} \Rightarrow A \cap B \in \mathcal{F}$.
3. $A, B \in \mathcal{F}, A \subset B \Rightarrow \exists (A_j)_{i \leq j \leq m} \in \mathcal{F}^{\{1,\dots,m\}}$ of disjoint sets such that $B \setminus A = \bigcup_{j=1}^{m} A_j$.

If $\mathcal{F}$ is both a semiring and $\Omega \in \mathcal{F}$, then it is called a *semialgebra*.

Proposition A.4. *A set Ω has the following properties:*

1. If $\mathcal{F}$ is a σ-algebra of subsets of Ω, then it is an algebra.

© Springer Science+Business Media New York 2015
V. Capasso, D. Bakstein, *An Introduction to Continuous-Time*
Stochastic Processes, Modeling and Simulation in Science, Engineering
and Technology, DOI 10.1007/978-1-4939-2757-9

2. If $\mathcal{F}$ is a σ-algebra of subsets of Ω, then
   - $E_1, \ldots, E_n \in \mathcal{F} \Rightarrow \bigcap_{i=1}^{n} E_i \in \mathcal{F}$
   - $E_1, \ldots, E_n, \ldots \in \mathcal{F} \Rightarrow \bigcap_{n=1}^{\infty} E_n \in \mathcal{F}$
   - $B \in \mathcal{F} \Rightarrow \Omega \setminus B \in \mathcal{F}$
3. If $\mathcal{F}$ is a ring on Ω, then it is also a semiring.

Definition A.5. Every pair $(\Omega, \mathcal{F})$ consisting of a set Ω and a σ-ring $\mathcal{F}$ of the subsets of Ω is a *measurable space*. Furthermore, if $\mathcal{F}$ is a σ-algebra, then $(\Omega, \mathcal{F})$ is a *measurable space on which a probability measure can be built*. If $(\Omega, \mathcal{F})$ is a measurable space, then the elements of $\mathcal{F}$ are called $\mathcal{F}$-*measurable* or just *measurable sets*. We will henceforth assume that if a space is measurable, then we can build a probability measure on it.

Example A.6.

1. If $\mathcal{B}$ is a σ-algebra on the set E and $X : \Omega \to E$ a generic mapping, then the set
$$X^{-1}(\mathcal{B}) = \left\{ A \subset \Omega \mid \exists B \in \mathcal{B} \text{ such that } A = X^{-1}(B) \right\}$$
is a σ-algebra on Ω.
2. *Generated σ-algebra.* If $\mathcal{A}$ is a set of the elements of a set Ω, then there exists the smallest σ-algebra of subsets of Ω that contains $\mathcal{A}$. This is the σ-algebra *generated* by $\mathcal{A}$, denoted $\sigma(\mathcal{A})$. If, now, $\mathcal{G}$ is the set of all σ-algebras of subsets of Ω containing $\mathcal{A}$, then it is not empty because it has $\sigma(\Omega)$ among its elements, so that $\sigma(\mathcal{A}) = \bigcap_{\mathcal{C} \in \mathcal{G}} \mathcal{C}$.
3. *Borel σ-algebra.* Let Ω be a topological space. Then the *Borel σ-algebra* on Ω, denoted by $\mathcal{B}_\Omega$, is the σ-algebra generated by the set of all open subsets of Ω. Its elements are called Borelian or Borel-measurable.
4. The set of all left-open, right-closed bounded intervals of $\mathbb{R}$, defined as $(a, b] := \{x \in \mathbb{R} \mid a < x \leq b\}$, for $a, b \in \mathbb{R}$, is a semiring but not a ring.
5. The set of all bounded and unbounded intervals of $\mathbb{R}$ is a semialgebra.
6. If $\mathcal{B}_1$ and $\mathcal{B}_2$ are algebras on Ω_1 and Ω_2, respectively, then the set of rectangles $B_1 \times B_2$, with $B_1 \in \mathcal{B}_1$ and $B_2 \in \mathcal{B}_2$, is a semialgebra.
7. *Product σ-algebra.* Let $(\Omega_i, \mathcal{F}_i)_{1 \leq i \leq n}$ be a family of measurable spaces, and let $\Omega = \prod_{i=1}^{n} \Omega_i$. Defining
$$\mathcal{R} = \left\{ E \subset \Omega \mid \forall i = 1, \ldots, n \ \exists E_i \in \mathcal{F}_i \text{ such that } E = \prod_{i=1}^{n} E_i \right\},$$
then $\mathcal{R}$ is a semialgebra of elements of Ω. The σ-algebra generated by $\mathcal{R}$ is called the *product σ-algebra* of the σ-algebras $(\mathcal{F}_i)_{1 \leq i \leq n}$.

Proposition A.7. *Let $(\Omega_i)_{1 \leq i \leq n}$ be a family of topological spaces with a countable base, and let $\Omega = \prod_{i=1}^{n} \Omega_i$. Then the Borel σ-algebra $\mathcal{B}_\Omega$ is identical to the product σ-algebra of the family of Borel σ-algebras $(\mathcal{B}_{\Omega_i})_{1 \leq i \leq n}$.*

A.2 Measurable Functions and Measure

Definition A.8. Let $(\Omega_1, \mathcal{F}_1)$ and $(\Omega_2, \mathcal{F}_2)$ be two measurable spaces. A function $f : \Omega_1 \to \Omega_2$ is *measurable* if

$$\forall E \in \mathcal{F}_2 \colon f^{-1}(E) \in \mathcal{F}_1.$$

Remark A.9. If $(\Omega, \mathcal{F})$ is not a measurable space, i.e., $\Omega \notin \mathcal{F}$, then there does not exist a measurable mapping from $(\Omega, \mathcal{F})$ to $(\mathbb{R}, \mathcal{B}_\mathbb{R})$ because $\mathbb{R} \in \mathcal{B}_\mathbb{R}$ and $f^{-1}(\mathbb{R}) = \Omega \notin \mathcal{F}$.

Definition A.10. Let $(\Omega, \mathcal{F})$ be a measurable space and $f : \Omega \to \mathbb{R}^n$ a mapping. If f is measurable with respect to the σ-algebras $\mathcal{F}$ and $\mathcal{B}_{\mathbb{R}^n}$, the latter being the Borel σ-algebra on $\mathbb{R}^n$, then f is *Borel-measurable*.

Proposition A.11. *Let $(E_1, \mathcal{B}_1)$ and $(E_2, \mathcal{B}_2)$ be two measurable spaces and $\mathcal{U}$ a set of the elements of E_2, which generates $\mathcal{B}_2$ and $f : E_1 \to E_2$. The necessary and sufficient condition for f to be measurable is $f^{-1}(\mathcal{U}) \subset \mathcal{B}_1$.*

Remark A.12. If a function $f : \mathbb{R}^k \to \mathbb{R}^n$ is continuous, then it is Borel-measurable.

Definition A.13. Let $(\Omega, \mathcal{F})$ be a measurable space. Every Borel-measurable mapping $h : \Omega \to \bar{\mathbb{R}}$ that can only have a finite number of distinct values is called an *elementary function*. Equivalently, a function $h : \Omega \to \bar{\mathbb{R}}$ is elementary if and only if it can be written as the finite sum

$$\sum_{i=1}^{r} x_i I_{E_i},$$

where, for every $i = 1, \ldots, r$, the E_i are disjoint sets of $\mathcal{F}$ and I_{E_i} is the indicator function on E_i.

Theorem A.14 (Approximation of measurable functions through elementary functions). *Let $(\Omega, \mathcal{F})$ be a measurable space and $f : \Omega \to \bar{\mathbb{R}}$ a nonnegative measurable function. There exists a sequence of measurable elementary functions $(s_n)_{n \in \mathbb{N}}$ such that*

1. $0 \le s_1 \le \cdots \le s_n \le \cdots \le f$
2. $\lim_{n \to \infty} s_n = f$

Proposition A.15. *Let $(\Omega, \mathcal{F})$ be a measurable space and $X_n : \Omega \to \mathbb{R}$, $n \in \mathbb{N}$, a sequence of measurable functions converging pointwise to a function $X : \Omega \to \mathbb{R}$; then, X is itself measurable.*

Proposition A.16. *If $f_1, f_2 : \Omega \to \bar{\mathbb{R}}$ are Borel-measurable functions, then so are the functions $f_1 + f_2$, $f_1 - f_2$, $f_1 f_2$, and f_1/f_2, as long as the operations are well defined.*

Lemma A.17. *If $f : (\Omega_1, \mathcal{F}_1) \to (\Omega_2, \mathcal{F}_2)$ and $g : (\Omega_2, \mathcal{F}_2) \to (\Omega_3, \mathcal{F}_3)$ are measurable functions, then so is $g \circ f : (\Omega_1, \mathcal{F}_1) \to (\Omega_3, \mathcal{F}_3)$.*

Proposition A.18. *Let $(\Omega_i, \mathcal{F}_i)_{1 \leq i \leq n}$ be a family of measurable spaces, $\Omega = \prod_{i=1}^{n} \Omega_i$, and $\pi_i : \Omega \to \Omega_i$ for $1 \leq i \leq n$ is the ith projection. Then the product σ-algebra $\bigotimes_{i=1}^{n} \mathcal{F}_i$ of the family of σ-algebras $(\mathcal{F}_i)_{1 \leq i \leq n}$ is the smallest σ-algebra on Ω for which every projection π_i is measurable.*

Proposition A.19. *If $h : (E, \mathcal{B}) \to (\Omega = \prod_{i=1}^{n} \Omega_i, \mathcal{F} = \bigotimes_{i=1}^{n} \mathcal{F}_i)$ is a mapping, then the following statements are equivalent:*

1. *h is measurable.*
2. *For all $i = 1, \ldots n$, $h_i = \pi_i \circ h$ is measurable.*

Proof. $1 \Rightarrow 2$ follows from Proposition A.18 and Lemma A.17. To prove that $2 \Rightarrow 1$, it is sufficient to see that given $\mathcal{R}$, the set of rectangles on Ω, it follows that, for all $B \in \mathcal{R} : h^{-1}(B) \in \mathcal{B}$. Let $B \in \mathcal{R}$. Then for all $i = 1, \ldots, n$, there exists a $B_i \in \mathcal{F}_i$ such that $B = \prod_{i=1}^{n} B_i$. Therefore, by recalling that due to point 2 every h_i is measurable, we have that

$$h^{-1}(B) = h^{-1}\left(\prod_{i=1}^{n} B_i \right) = \bigcap_{i=1}^{n} h_i^{-1}(B_i) \in \mathcal{B}. \qquad \square$$

Corollary A.20. *Let $(\Omega, \mathcal{F})$ be a measurable space and $h : \Omega \to \mathbb{R}^n$ a function. Defining $h_i = \pi_i \circ h : \Omega \to \mathbb{R}$ for $1 \leq i \leq n$, the following two propositions are equivalent:*

1. *h is Borel-measurable.*
2. *For all $i = 1, \ldots, n$, h_i is Borel-measurable.*

Definition A.21. *Let $(\Omega, \mathcal{F})$ be a measurable space. Every function $\mu : \mathcal{F} \to \bar{\mathbb{R}}$ such that*

1. *For all $E \in \mathcal{F}$: $\mu(E) \geq 0$.*
2. *For all $E_1, \ldots, E_n, \ldots \in \mathcal{F}$ such that $E_i \cap E_j = \emptyset$, for $i \neq j$, we have that*

$$\mu\left(\bigcup_{i=1}^{\infty} E_i \right) = \sum_{i=1}^{\infty} \mu(E_i)$$

is a *measure* on $\mathcal{F}$. Moreover, if $(\Omega, \mathcal{F})$ is a measurable space and if

$$\mu(\Omega) = 1,$$

then μ is a *probability measure* or a *probability*. Furthermore, a measure μ is *finite* if

$$\forall A \in \mathcal{F}: \ \mu(A) < +\infty$$

and *σ-finite* if

1. There exists an $(A_n)_{n \in \mathbb{N}} \in \mathcal{F}^\mathbb{N}$ such that $\Omega = \bigcup_{n \in \mathbb{N}} A_n$.
2. For all $n \in \mathbb{N}$: $\mu(A_n) < +\infty$.

Definition A.22. The ordered triple $(\Omega, \mathcal{F}, \mu)$, where Ω denotes a set, $\mathcal{F}$ a σ-ring on Ω, and $\mu : \mathcal{F} \to \bar{\mathbb{R}}$ a measure on $\mathcal{F}$, is a *measure space*. If μ is a probability measure, then $(\Omega, \mathcal{F}, \mu)$ is a *probability space*.[12]

Definition A.23. Let $(\Omega, \mathcal{F}, \mu)$ be a measure space and $\lambda : \mathcal{F} \to \bar{\mathbb{R}}$ a measure on Ω. Then λ is said to be *absolutely continuous* with respect to μ, denoted $\lambda \ll \mu$, if

$$\forall A \in \mathcal{F}: \ \mu(A) = 0 \Rightarrow \lambda(A) = 0.$$

Proposition A.24 (Characterization of a measure). *Let μ be additive on an algebra $\mathcal{F}$ and valued in $\mathbb{R}$ (and not everywhere equal to $+\infty$). The following two statements are equivalent:*

1. *μ is a measure on $\mathcal{F}$.*
2. *For increasing $(A_n)_{n \in \mathbb{N}} \in \mathcal{F}^\mathbb{N}$, where $\bigcup_{n \in \mathbb{N}} A_n \in \mathcal{F}$, we have that*

$$\mu\left(\bigcup_{n \in \mathbb{N}} A_n\right) = \lim_{n \to \infty} \mu(A_n) = \sup_{n \in \mathbb{N}} \mu(A_n).$$

If μ is finite, then 1 and 2 are equivalent to the following statements.

3. *For decreasing $(A_n)_{n \in \mathbb{N}} \in \mathcal{F}^\mathbb{N}$, where $\bigcap_{n \in \mathbb{N}} A_n \in \mathcal{F}$, we have*

$$\mu\left(\bigcap_{n \in \mathbb{N}} A_n\right) = \lim_{n \to \infty} \mu(A_n) = \inf_{n \in \mathbb{N}} \mu(A_n).$$

4. *For decreasing $(A_n)_{n \in \mathbb{N}} \in \mathcal{F}^\mathbb{N}$, where $\bigcap_{n \in \mathbb{N}} A_n = \emptyset$, we have*

$$\lim_{n \to \infty} \mu(A_n) = \inf_{n \in \mathbb{N}} \mu(A_n) = 0.$$

---

[12] Henceforth we will call every measurable space that has a probability measure assigned to it a probability space.

Proposition A.25 (Generalization of a measure). *Let $\mathcal{G}$ be a semiring on E and $\mu : \mathcal{G} \to \mathbb{R}_+$ a function that satisfies the following properties:*

1. *μ is (finitely) additive on $\mathcal{G}$.*
2. *μ is countably additive on $\mathcal{G}$.*
3. *There exists an $(S_n)_{n\in\mathbb{N}} \in \mathcal{G}^{\mathbb{N}}$ such that $E \subset \bigcup_{n\in\mathbb{N}} S_n$.*

Under these assumptions

$$\exists | \, \bar{\mu} : \mathcal{B} \to \bar{\mathbb{R}}_+ \text{ such that } \bar{\mu}|_{\mathcal{G}} = \mu,$$

where $\mathcal{B}$ is the σ-ring generated by $\mathcal{G}$.[13] Moreover, if $\mathcal{G}$ is a semialgebra and $\mu(E) = 1$, then $\bar{\mu}$ is a probability measure.

Proposition A.26. *Let $\mathcal{U}$ be a ring on E and $\mu : \mathcal{U} \to \bar{\mathbb{R}}_+$ (not everywhere equal to $+\infty$) a measure on $\mathcal{U}$. Then, if $\mathcal{B}$ is the σ-ring generated by $\mathcal{U}$,*

$$\exists | \, \bar{\mu} : \mathcal{B} \to \bar{\mathbb{R}}_+ \text{ such that } \bar{\mu}|_{\mathcal{U}} = \mu.$$

Moreover, if μ is a probability measure, then so is $\bar{\mu}$.

Lemma A.27. (Fatou). *Let $(A_n)_{n\in\mathbb{N}} \in \mathcal{F}^{\mathbb{N}}$ be a sequence of random variables and $(\Omega, \mathcal{F}, P)$ a probability space. Then*

$$P(\liminf_n A_n) \leq \liminf_n P(A_n) \leq \limsup_n P(A_n) \leq P(\limsup_n A_n).$$

If $\liminf_n A_n = \limsup_n A_n = A$, then $A_n \to A$.

Corollary A.28. *Under the assumptions of Fatou's Lemma A.27, if $A_n \to A$, then $P(A_n) \to P(A)$.*

A.3 Lebesgue Integration

Let $(\Omega, \mathcal{F})$ be a measurable space. We will denote by $\mathcal{M}(\mathcal{F}, \bar{\mathbb{R}})$ [or, respectively, by $\mathcal{M}(\mathcal{F}, \bar{\mathbb{R}}_+)$] the set of measurable functions on $(\Omega, \mathcal{F})$ and valued in $\bar{\mathbb{R}}$ (or $\bar{\mathbb{R}}_+$).

Proposition A.29. *Let $(\Omega, \mathcal{F})$ be a measurable space and μ a positive measure on $\mathcal{F}$. Then there exists a unique mapping Φ from $\mathcal{M}(\mathcal{F}, \bar{\mathbb{R}}_+)$ to $\bar{\mathbb{R}}_+$, such that:*

1. *For every $\alpha \in \mathbb{R}_+$, $f, g \in \mathcal{M}(\mathcal{F}, \bar{\mathbb{R}}_+)$,*
   $\Phi(\alpha f) = \alpha \Phi(f)$,
   $\Phi(f + g) = \Phi(f) + \Phi(g)$,
   $f \leq g \Rightarrow \Phi(f) \leq \Phi(g)$.

---

[13] $\mathcal{B}$ is identical to the σ-ring generated by the ring generated by $\mathcal{G}$.

2. *For every increasing sequence $(f_n)_{n\in\mathbb{N}}$ of elements of $\mathcal{M}(\mathcal{F}, \bar{\mathbb{R}}_+)$ we have that $\sup_n \Phi(f_n) = \Phi(\sup_n f_n)$ (Beppo–Levi property).*
3. *For every $B \in \mathcal{F}$, $\Phi(I_B) = \mu(B)$.*

Definition A.30. If Φ is the unique functional associated with μ, a measure on the measurable space $(\Omega, \mathcal{F})$, then for every $f \in \mathcal{M}(\mathcal{F}, \bar{\mathbb{R}}_+)$:

$$\Phi(f) = \int^* f(x)d\mu(x) \text{ or } \int^* f(x)\mu(dx) \text{ or } \int^* f(x)d\mu$$

the *upper integral* of μ.

Remark A.31. Let $(\Omega, \mathcal{F})$ be a measurable space, and let Φ be the functional canonically associated with μ measure on $\mathcal{F}$.

1. If $s : \Omega \to \bar{\mathbb{R}}_+$ is an elementary function, and thus $s = \sum_{i=1}^n x_i I_{E_i}$, then

$$\Phi(s) = \int^* s d\mu = \sum_{i=1}^n x_i \mu(E_i).$$

2. If $f \in \mathcal{M}(\mathcal{F}, \bar{\mathbb{R}}_+)$ and defining $\Omega_f = \{s : \Omega \to \bar{\mathbb{R}}_+ | s \text{ elementary}, s \leq f\}$, then Ω_f is nonempty and

$$\Phi(f) = \int^* f d\mu = \sup_{s\in\Omega_f} \int^* s d\mu = \sup_{s\in\Omega_f} \left(\sum_{i=1}^n x_i \mu(E_i) \right).$$

3. If $f \in \mathcal{M}(\mathcal{F}, \bar{\mathbb{R}}_+)$ and $B \in \mathcal{F}$, then by definition

$$\int_B^* f d\mu = \int^* I_B \cdot f d\mu.$$

Definition A.32. Let $(\Omega, \mathcal{F})$ be a measurable space and μ a positive measure on $\mathcal{F}$. An $\mathcal{F}$-measurable function f is *μ-integrable* if

$$\int^* f^+ d\mu < +\infty \text{ and } \int^* f^- d\mu < +\infty,$$

where f^+ and f^- denote the positive and negative parts of f, respectively. The real number

$$\int^* f^+ d\mu - \int^* f^- d\mu$$

is therefore the *Lebesgue integral* of f with respect to μ, denoted by

$$\int f d\mu \text{ or } \int f(x)d\mu(x) \text{ or } \int f(x)\mu(dx).$$

Proposition A.33. *Let $(\Omega, \mathcal{F})$ be a measurable space endowed with measure μ and $f \in \mathcal{M}(\mathcal{F}, \bar{\mathbb{R}}_+)$. Then*

1. $\int^* f d\mu = 0 \Leftrightarrow f = 0 \, a.s.$ with respect to μ.
2. For every $A \in \mathcal{F}, \mu(A) = 0$, we have

$$\int_A^* f d\mu = 0.$$

3. For every $g \in \mathcal{M}(\mathcal{F}, \bar{\mathbb{R}}_+)$ such that $f = g$, a.s. with respect to μ, we have

$$\int^* f d\mu = \int^* g d\mu.$$

Theorem A.34 (Monotone convergence). *Let $(\Omega, \mathcal{F})$ be a measurable space endowed with measure μ, $(f_n)_{n \in \mathbb{N}}$ an increasing sequence of elements of $\mathcal{M}(\mathcal{F}, \bar{\mathbb{R}}_+)$, and $f : \Omega \to \bar{\mathbb{R}}_+$ such that*

$$\forall \omega \in \Omega : \; f(\omega) = \lim_{n \to \infty} f_n(\omega) = \sup_{n \in \mathbb{N}} f_n(\omega).$$

Then $f \in \mathcal{M}(\mathcal{F}, \bar{\mathbb{R}}_+)$ and

$$\int^* f d\mu = \lim_{n \to \infty} \int^* f_n d\mu.$$

Theorem A.35 (Lebesgue's dominated convergence). *Let $(\Omega, \mathcal{F})$ be a measurable space endowed with measure μ, $(f_n)_{n \in \mathbb{N}}$ a sequence of μ-integrable functions defined on Ω, and $g : \Omega \to \bar{\mathbb{R}}_+$ a μ-integrable function such that $|f_n| \leq g$ for all $n \in \mathbb{N}$. If we suppose that $\lim_{n \to \infty} f_n = f$ exists almost surely in Ω, then f is μ-integrable and we have*

$$\int f d\mu = \lim_{n \to \infty} \int f_n d\mu.$$

Lemma A.36. (Fatou). *Let $f_n \in \mathcal{M}(\mathcal{F}, \bar{\mathbb{R}}_+)$. Then*

$$\liminf_n \int^* f_n d\mu \geq \int^* \liminf_n f_n d\mu.$$

Theorem A.37 (Fatou–Lebesgue).

1. Let $|f_n| \leq g \in \mathcal{L}^1$. Then

$$\limsup_n \int f_n d\mu \leq \int \limsup_n f_n d\mu.$$

2. Let $|f_n| \leq g \in \mathcal{L}^1$. Then

$$\liminf_n \int f_n d\mu \geq \int \liminf_n f_n d\mu.$$

3. Let $|f_n| \leq g$ and $f = \lim_n f_n$, almost surely with respect to μ. Then

$$\lim_n \int f_n d\mu = \int f d\mu.$$

Definition A.38. Let $(\Omega, \mathcal{F})$ be a measurable space endowed with measure μ, and let $(E, \mathcal{B})$ be an additional measurable space; let $h : (\Omega, \mathcal{F}) \to (E, \mathcal{B})$ be a measurable function. The mapping $\mu_h : \mathcal{B} \to \bar{\mathbb{R}}_+$, such that $\mu_h(B) = \mu(h^{-1}(B))$ for all $B \in \mathcal{B}$, is a measure on E, called the *induced* or *image measure* of μ via h, and denoted $h(\mu)$.

Proposition A.39. *Given the assumptions of Definition A.38, the function* $g : (E, \mathcal{B}) \to (\mathbb{R}, \mathcal{B}_\mathbb{R})$ *is integrable with respect to* μ_h *if and only if* $g \circ h$ *is integrable with respect to* μ *and*

$$\int g \circ h \, d\mu = \int g \, d\mu_h.$$

Theorem A.40 (Product measure). *Let* $(\Omega_1, \mathcal{F}_1)$ *and* $(\Omega_2, \mathcal{F}_2)$ *be measurable spaces, and let the former be endowed with σ-finite measure μ_1 on $\mathcal{F}_1$. Further suppose that for all $\omega_1 \in \Omega_1$ a measure $\mu(\omega_1, \cdot)$ is assigned on $\mathcal{F}_2$, and that, for all $B \in \mathcal{F}_2$, $\mu(\cdot, B) : \Omega_1 \to \mathbb{R}$ is a Borel-measurable function. If $\mu(\omega_1, \cdot)$ is uniformly σ-finite, then there exists a sequence $(B_n)_{n \in \mathbb{N}} \in \mathcal{F}_2^\mathbb{N}$ such that $\Omega_2 = \bigcup_{n=1}^\infty B_n$ and, for all $n \in \mathbb{N}$, there exists a $K_n \in \mathbb{R}$ such that $\mu(\omega_1, B_n) \leq K_n$ for all $\omega_1 \in \Omega_1$. Then there exists a unique measure μ on the product σ-algebra $\mathcal{F} = \mathcal{F}_1 \otimes \mathcal{F}_2$ such that*

$$\forall A \in \mathcal{F}_1, B \in \mathcal{F}_2 : \qquad \mu(A \times B) = \int_A \mu(\omega_1, B) \mu_1(d\omega_1)$$

and

$$\forall F \in \mathcal{F} : \qquad \mu(F) = \int_{\Omega_1} \mu(\omega_1, F(\omega_1)) \mu_1(d\omega_1).$$

Definition A.41. Let $(\Omega_1, \mathcal{F}_1)$ and $(\Omega_2, \mathcal{F}_2)$ be two measurable spaces endowed with σ-finite measures μ_1 and μ_2 on $\mathcal{F}_1$ and $\mathcal{F}_2$, respectively. Defining $\Omega = \Omega_1 \times \Omega_2$ and $\mathcal{F} = \mathcal{F}_1 \otimes \mathcal{F}_2$, the function $\mu : \mathcal{F} \to \bar{\mathbb{R}}$ with

$$\forall F \in \mathcal{F} : \qquad \mu(F) = \int_{\Omega_1} \mu_2(F(\omega_1)) d\mu_1(\omega_1) = \int_{\Omega_2} \mu_1(F(\omega_2)) d\mu_2(\omega_2)$$

is the unique measure on $\mathcal{F}$ with

$$\forall A \in \mathcal{F}_1, B \in \mathcal{F}_2: \qquad \mu(A \times B) = \mu_1(A) \times \mu_2(B).$$

Moreover, μ is σ-finite on $\mathcal{F}$ as well as a probability measure if μ_1 and μ_2 are as well. The measure μ is the *product measure* of μ_1 and μ_2, denoted by $\mu_1 \otimes \mu_2$.

Theorem A.42 (Fubini).   *Given the assumptions of Definition A.41, let $f : (\Omega, \mathcal{F}) \to (\mathbb{R}, \mathcal{B}_\mathbb{R})$ be a Borel-measurable function such that $\int_\Omega f d\mu$ exists. Then*

$$\int_\Omega f d\mu = \int_{\Omega_1} \int_{\Omega_2} f d\mu_2 d\mu_1 = \int_{\Omega_2} \int_{\Omega_1} f d\mu_1 d\mu_2.$$

Proposition A.43. *Let $(\Omega_i, \mathcal{F}_i)_{1 \leq i \leq n}$ be a family of measurable spaces. Further, let $\mu_1 : \mathcal{F}_1 \to \bar{\mathbb{R}}$ be a σ-finite measure, and let*

$$\forall(\omega_1, \ldots, \omega_j) \in \Omega_1 \times \cdots \times \Omega_j: \qquad \mu(\omega_1, \ldots, \omega_j, \cdot) : \mathcal{F}_{j+1} \to \bar{\mathbb{R}}$$

be a measure on $\mathcal{F}_{j+1}$, $1 \leq j \leq n-1$. If $\mu(\omega_1, \ldots, \omega_j, \cdot)$ is uniformly σ-finite and for every $c \in \mathcal{F}_{j+1}$

$$\mu(\ldots, c) : (\Omega_1 \times \cdots \times \Omega_j, \mathcal{F}_1 \otimes \cdots \otimes \mathcal{F}_j) \to (\bar{\mathbb{R}}, \mathcal{B}_{\bar{\mathbb{R}}})$$

such that

$$\forall(\omega_1, \ldots, \omega_j) \in \Omega_1 \times \cdots \times \Omega_j: \qquad \mu(\ldots, c)(\omega_1, \ldots, \omega_j) = \mu(\omega_1, \ldots, \omega_j, c)$$

is measurable, then, defining $\Omega = \Omega_1 \times \cdots \times \Omega_n$ and $\mathcal{F} = \mathcal{F}_1 \otimes \cdots \otimes \mathcal{F}_n$:

1. *There exists a unique measure $\mu : \mathcal{F} \to \bar{\mathbb{R}}$ such that for every measurable rectangle $A_1 \times \cdots \times A_n \in \mathcal{F}$:*

$$\mu(A_1 \times \cdots \times A_n)$$
$$= \int_{A_1} \mu_1(d\omega_1) \int_{A_2} \mu(\omega_1, d\omega_2) \cdots \int_{A_n} \mu(\omega_1, \ldots, \omega_{n-1}, d\omega_n).$$

   μ is σ-finite on $\mathcal{F}$ and a probability whenever μ_1 and all $\mu(\omega_1, \ldots, \omega_j, \cdot)$ are probability measures;

2. *If $f : (\Omega, \mathcal{F}) \to (\bar{\mathbb{R}}, \mathcal{B}_{\bar{\mathbb{R}}})$ is measurable and nonnegative, then*

$$\int_\Omega f d\mu$$
$$= \int_{\Omega_1} \mu_1(d\omega_1) \int_{\Omega_2} \mu(\omega_1, d\omega_2) \cdots \int_{\Omega_n} f(\omega_1, \ldots, \omega_n) \mu(\omega_1, \ldots, \omega_{n-1}, d\omega_n).$$

Proposition A.44.

1. *Given the assumptions and the notation of Proposition A.43, if we assume that $f = I_F$, then for every $F \in \mathcal{F}$:*

$$\mu(F)$$
$$= \int_{\Omega_1} \mu_1(d\omega_1) \int_{\Omega_2} \mu(\omega_1, d\omega_2) \cdots \int_{\Omega_n} I_F(\omega_1, \ldots, \omega_n) \mu(\omega_1, \ldots, \omega_{n-1}, d\omega_n).$$

2. *For all $j = 1, \ldots, n-1$, let $\mu_{j+1} = \mu(\omega_1, \ldots, \omega_j, \cdot)$. Then there exists a unique measure μ on $\mathcal{F}$ such that for every rectangle $A_1 \times \cdots \times A_n \in \mathcal{F}$ we have*

$$\mu(A_1 \times \cdots \times A_n) = \mu_1(A_1) \cdots \mu_n(A_n).$$

If $f : (\Omega, \mathcal{F}) \to (\bar{\mathbb{R}}, \mathcal{B}_{\bar{\mathbb{R}}})$ is measurable and positive, or else if $\int_\Omega f d\mu$ exists, then

$$\int_\Omega f d\mu = \int_{\Omega_1} d\mu_1 \cdots \int_{\Omega_n} f d\mu_n,$$

and the order of integration is arbitrary. The measure μ is the product measure of $\mu_1, \ldots, \mu_n$ and is denoted by $\mu_1 \otimes \cdots \otimes \mu_n$.

Definition A.45. Let $(v_i)_{1 \leq i \leq n}$ be a family of measures defined on $\mathcal{B}_{\mathbb{R}}$, and

$$v^{(n)} = v_1 \otimes \cdots \otimes v_n$$

their product measure on $\mathcal{B}_{\mathbb{R}^n}$. The *convolution product* of $v_1, \ldots, v_n$, denoted by $v_1 * \cdots * v_n$, is the induced measure of $v^{(n)}$ on $\mathcal{B}_{\mathbb{R}}$ via the function $f : (x_1, \ldots, x_n) \in \mathbb{R}^n \mapsto \sum_{i=1}^n x_i \in \mathbb{R}$.

Proposition A.46. *Let v_1 and v_2 be measures on $\mathcal{B}_{\mathbb{R}}$. Then for every $B \in \mathcal{B}_{\mathbb{R}}$ we have*

$$v_1 * v_2(B) = \int_B d(v_1 * v_2) = \int_{\mathbb{R}} I_B(z) d(v_1 * v_2) = \int \int I_B(x_1 + x_2) d(v_1 \otimes v_2).$$

A.4 Lebesgue–Stieltjes Measure and Distributions

Definition A.47. Let $\mu : \mathcal{B}_{\mathbb{R}} \to \bar{\mathbb{R}}$ be a measure. It then represents a *Lebesgue–Stieltjes* measure if for every interval I we have that $\mu(I) < +\infty$.

Definition A.48. Every function $F : \mathbb{R} \to \mathbb{R}$ that is right-continuous and increasing is a *(generalized) distribution function* on $\mathbb{R}$.

It is in fact possible to establish a one-to-one relationship between the set of Lebesgue–Stieltjes measures and the set of distribution functions in the sense that to every Lebesgue–Stieltjes measure can be assigned a distribution function and vice versa.

Proposition A.49. *Let μ be a Lebesgue–Stieltjes measure on $\mathcal{B}_{\mathbb{R}}$ and the function $F : \mathbb{R} \to \mathbb{R}$ defined, apart from a constant, as*

$$F(b) - F(a) = \mu(]a, b]) \qquad \forall a, b \in \mathbb{R}, a < b.$$

Then F is a distribution function, in particular the one assigned to μ.

Conversely, the following holds.

Proposition A.50. *Let F be a distribution function, and let μ be defined on bounded intervals of $\mathbb{R}$ by*

$$\mu(]a, b]) = F(b) - F(a) \qquad \forall a, b \in \mathbb{R}, a < b.$$

There exists a unique extension of μ that is a Lebesgue–Stieltjes measure on $\mathcal{B}_{\mathbb{R}}$. This measure is the Lebesgue–Stieltjes measure canonically associated with F.

Definition A.51. Every measure $\mu : \mathcal{B}_{\mathbb{R}^n} \to \bar{\mathbb{R}}$ that for every bounded interval I of $\mathbb{R}^n$ has $\mu(I) < +\infty$ is a Lebesgue–Stieltjes measure on $\mathbb{R}^n$.

Definition A.52. Let $f : \mathbb{R} \to \mathbb{R}$ be of constant value 1, and we consider the function $F : \mathbb{R} \to \mathbb{R}$ with

$$F(x) - F(0) = \int_0^x f(t)dt \qquad \forall x > 0,$$

$$F(0) - F(x) = \int_x^0 f(t)dt \qquad \forall x < 0,$$

where $F(0)$ is fixed and arbitrary. This function F is a distribution function, and its associated Lebesgue–Stieltjes measure is called a *Lebesgue measure* on $\mathbb{R}$. It is such that

$$\mu(]a, b]) = b - a, \qquad \forall a, b \in \mathbb{R}, a < b.$$

Definition A.53. Let $(\Omega, \mathcal{F}, \mu)$ be a space with σ-finite measure μ, and consider another measure $\lambda : \mathcal{F} \to \bar{\mathbb{R}}_+$. λ is said to be defined through its *density* with respect to μ if there exists a Borel-measurable function $g : \Omega \to \bar{\mathbb{R}}_+$ with

$$\lambda(A) = \int_A g d\mu \qquad \forall A \in \mathcal{F}.$$

This function g is the density of λ with respect to μ. In this case λ is absolutely continuous with respect to μ ($\lambda \ll \mu$). If μ is a Lebesgue measure on $\mathbb{R}$, then g is the density of μ. A measure ν is called μ-*singular* if there exists $N \in \mathcal{F}$

such that $\mu(N) = 0$ and $\nu(N \setminus \mathcal{F}) = 0$. Conversely, if also $\mu(N) = 0$ whenever $\nu(N) = 0$, then the two measures are *equivalent* (denoted $\lambda \sim \mu$).

Theorem A.54 (Radon–Nikodym). *Let $(\Omega, \mathcal{F})$ be a measurable space, μ a σ-finite measure on $\mathcal{F}$, and λ an absolutely continuous measure with respect to μ. Then λ is endowed with density with respect to μ. Hence there exists a Borel-measurable function $g : \Omega \to \bar{\mathbb{R}}_+$ such that*

$$\lambda(A) = \int_A g \, d\mu, \qquad A \in \mathcal{B}.$$

A necessary and sufficient condition for g to be μ-integrable is that λ is bounded. Moreover, if $h : \Omega \to \bar{\mathbb{R}}_+$ is another density of λ, then $g = h$, almost surely with respect to μ.

Theorem A.55 (Lebesgue–Nikodym). *Let ν and μ be a measure and a σ-finite measure on $(E, \mathcal{B})$, respectively. There exists a $\mathcal{B}$-measurable function $f : E \to \bar{\mathbb{R}}_+$ and a μ-singular measure ν' on $(E, \mathcal{B})$ such that*

$$\nu(B) = \int_B f \, d\mu + \nu'(B) \qquad \forall B \in \mathcal{B}.$$

Furthermore,

1. *ν' is unique.*
2. *If $h : E \to \bar{\mathbb{R}}_+$ is a $\mathcal{B}$-measurable function with*

$$\nu(B) = \int_B h \, d\mu + \nu'(B) \qquad \forall B \in \mathcal{B},$$

then $f = h$ almost surely with respect to μ.

Definition A.56. A function $F : \mathbb{R} \to \mathbb{R}$ is *absolutely continuous* if, for all $\epsilon > 0$, there exists a $\delta > 0$ such that for all $]a_i, b_i[\subset \mathbb{R}$ for $1 \le i \le n$ with $]a_i, b_i[\cap]a_j, b_j[= \emptyset$, $i \ne j$,

$$b_i - a_i < \delta \Rightarrow \sum_{i=1}^{n} |F(b_i) - F(a_i)| < \epsilon.$$

Proposition A.57. *Let F be a distribution function. Then the following two propositions are equivalent:*

1. *F is absolutely continuous.*
2. *The Lebesgue measure canonically associated with F is absolutely continuous.*

Proposition A.58. *Let* $f : [a, b] \to \mathbb{R}$ *be a mapping. The following two statements are equivalent:*

1. f *is absolutely continuous.*
2. *There exists a Borel-measurable function* $g : [a, b] \to \mathbb{R}$ *that is integrable with respect to the Lebesgue measure and*

$$f(x) - f(a) = \int_a^x g(t)dt \qquad \forall x \in [a, b].$$

This function g *is the density of* f.

Proposition A.59. *If* $f : [a, b] \to \mathbb{R}$ *is absolutely continuous, then*

1. f *is differentiable almost everywhere in* $[a, b]$.
2. f', *the first derivative of* f, *is integrable in* $[a, b]$, *and we have that*

$$f(x) - f(a) = \int_a^x f'(t)dt.$$

Theorem A.60 (Fundamental theorem of calculus). *If* $f : [a, b] \to \mathbb{R}$ *is integrable in* $[a, b]$ *and*

$$F(x) = \int_a^x f(t)dt \qquad \forall x \in [a, b],$$

then

1. F *is absolutely continuous in* $[a, b]$.
2. $F' = f$ *almost everywhere in* $[a, b]$.

Conversely, if we consider a function $F : [a, b] \to \mathbb{R}$ *that satisfies points 1 and 2, then*

$$\int_a^b f(x)dx = F(b) - F(a).$$

Proposition A.61. *If* $f : [a, b] \to \mathbb{R}$ *is differentiable in* $[a, b]$ *and has integrable derivatives, then*

1. f *is absolutely continuous in* $[a, b]$.
2. $f(x) = \int_a^x f'(t)dt.$

Definition A.62. Let $(\Omega, \mathcal{F}, \mu)$ be a measure space, and $p > 0$. The set of Borel-measurable functions defined on Ω, such that $\int_\Omega |f|^p d\mu < +\infty$, is a vector space on $\mathbb{R}$; it is denoted with the symbols $\mathcal{L}^p(\mu)$ or $\mathcal{L}^p(\Omega, \mathcal{F}, \mu)$. Its elements are called integrable functions, to the exponent p. In particular, elements of $\mathcal{L}^2(\mu)$ are said to be square-integrable functions. Finally, $\mathcal{L}^1(\mu)$ coincides with the space of functions integrable with respect to μ.

A.5 Radon Measures

Consider a complete metric space E endowed with its Borel σ-algebra $\mathcal{B}_E$.

Definition A.63. A σ-finite measure μ on $\mathcal{B}_E$ is called

(i)   *locally finite* if, for any point $x \in E$, there exists an open neighborhood U of x such that $\mu(U) < +\infty$.

(ii) *inner regular* if

$$\mu(A) = \sup\left\{\mu(K) \,|\, K \quad \text{compact}, \quad K \subset A\right\} \qquad \forall A \in \mathcal{B}_E.$$

(iii) *outer regular* if

$$\mu(A) = \sup\left\{\mu(U) \,|\, U \quad \text{open}, \quad A \subset U\right\} \qquad \forall A \in \mathcal{B}_E.$$

(iv) *regular* if it is both inner and outer regular.

(v)   a *Radon measure* if it is an inner regular and locally finite measure.

Proposition A.64. *The usual Lebesgue measure on $\mathbb{R}^d$ is a regular Radon measure. However, not all σ-finite measures on $\mathbb{R}^d$ are regular.*

Proof. See, e.g., Klenke (2008, p. 247). $\qquad\qquad\square$

Proposition A.65. *If μ is a Radon measure on a locally compact and complete metric space E endowed with its Borel σ-algebra, then*

$$\mu(K) < +\infty, \qquad \forall K \quad \text{compact subset of } E.$$

$$\left|\int_E f d\mu\right| < +\infty$$

for any real-valued continuous function f with compact support.

Proof. See, e.g., Karr (1991, p. 411). $\qquad\qquad\square$

Let us now stick to a locally compact and complete metric space E endowed with its Borel σ-algebra $\mathcal{B}_E$.

Definition A.66. A Radon measure μ on $\mathcal{B}_E$ is

(i) A *point* or *(counting)* measure if $\mu(A) \in \mathbb{N}$, for any $A \in \mathcal{B}_E$.

(ii) A *simple point* measure if μ is a point measure and $\mu(\{x\}) \leq 1$ for any $x \in E$.

(iii) A *diffuse* measure if $\mu(\{x\}) = 0$ for any $x \in E$.

The fundamental point measure is the Dirac measure ϵ_x associated with a point $x \in E$; it is defined by

$$\epsilon_x(A) = \begin{cases} 1, & \text{if } x \in A, \\ 0, & \text{if } x \notin A. \end{cases}$$

A point $x \in E$ is called an *atom* if $\mu(\{x\}) > 0$.

Proposition A.67. *A Radon measure μ on a locally compact and complete metric space E endowed with its Borel σ-algebra has an at most countable set of atoms. It can be decomposed as*

$$\mu = \mu_d + \sum_{i=1}^{K} a_i \epsilon_{x_i},$$

where μ_d is a diffuse measure, $K \in \mathbb{N} \cup \{\infty\}$, $a_i \in \mathbb{R}_+^*$, $x_i \in E$. The decomposition is unique up to reordering.

Proof. See, e.g., Karr (1991, p. 412). □

A Radon measure is *purely atomic* if its diffuse component is zero.

Remark A.68. A purely atomic measure is a point measure if and only if $a_i \in \mathbb{N}$ for each i, and in this case the family $\{x_i, \ i = 1, \ldots, K\}$ can have no accumulation points in E.

A.6 Stochastic Stieltjes Integration

Suppose $(\Omega, \mathcal{F}, P)$ is a given probability space with $(X_t)_{t \in \mathbb{R}_+}$ a measurable stochastic process whose sample paths $(X_t(\omega))_{t \in \mathbb{R}_+}$ are of locally bounded variation for any $\omega \in \Omega$. Now let $(H_s)_{s \in \mathbb{R}_+}$ be a measurable process whose sample paths are locally bounded for any $\omega \in \Omega$. Then the process $H \bullet X$ defined by

$$(H \bullet X)_t(\omega) = \int_0^t H(s, \omega) dX_s(\omega), \qquad \omega \in \Omega, t \in \mathbb{R}_+$$

is called the *stochastic Stieltjes integral* of H with respect to X. Clearly, $((H \bullet X)_t)_{t \in \mathbb{R}_+}$ is itself a stochastic process.

If we assume further that X is progressively measurable and H is $\mathcal{F}_t$-predictable with respect to the σ-algebra generated by X, then $H \bullet X$ is progressively measurable. In particular, if $N = \sum_{n \in \mathbb{N}^*} \epsilon_{\tau_n}$ is a point process on $\mathbb{R}_+$, then for any nonnegative process H on $\mathbb{R}_+$, the stochastic integral $H \bullet N$ exists and is given by

$$(H \bullet N)_t = \sum_{n \in \mathbb{N}^*} I_{[\tau_n \leq t]}(t) H(\tau_n).$$

Theorem A.69. *Let M be a martingale of locally integrable variation, i.e., such that*

$$E\left[\int_0^t d|M_s|\right] < \infty \qquad \text{for any } t > 0,$$

and let C be a predictable process satisfying

$$E\left[\int_0^t |C_s|d|M_s|\right] < \infty \qquad \textit{for any } t > 0.$$

Then the stochastic integral $C \bullet M$ is a martingale.

B

Convergence of Probability Measures on Metric Spaces

B.1 Metric Spaces

For more details on the following and further results, refer to Loève (1963); Dieudonné (1960), and Aubin (1977).

Definition B.1. Consider a set R. A *distance (metric)* on R is a mapping $\rho : R \times R \to \mathbb{R}_+$ that satisfies the following properties.

D1. For any $x, y \in R$, $\rho(x, y) = 0 \Leftrightarrow x = y$.
D2. For any $x, y \in R$, $\rho(x, y) = \rho(y, x)$.
D3. For any $x, y, z \in R$, $\rho(x, z) \leq \rho(x, y) + \rho(y, z)$ (triangle inequality).

Definition B.2. A *metric space* is a set R endowed with a metric ρ; we shall write (R, ρ). Elements of a metric space will be called *points*.

Definition B.3. Given a metric space (R, ρ), a point $a \in R$, and a real number $r > 0$, the *open ball* (or the *closed ball*) of center a and radius r is the set $B(a, r) := \{x \in R | \rho(a, x) < r\}$ (or $B'(a, r) := \{x \in R | \rho(a, x) \leq r\}$).

Definition B.4. In a metric space (R, ρ), an *open set* is any subset A of R such that for any $x \in A$ there exists an $r > 0$ such that $B(a, r) \subset A$.

The empty set is open, and so is the entire space R.

Proposition B.5. *The union of any family of open sets is an open set. The intersection of a finite family of open sets is an open set.*

Definition B.6. The family $\mathcal{T}$ of all open sets in a metric space is called its *topology*. In this respect the couple $(R, \mathcal{T})$ is a *topological space*.

© Springer Science+Business Media New York 2015
V. Capasso, D. Bakstein, *An Introduction to Continuous-Time Stochastic Processes*, Modeling and Simulation in Science, Engineering and Technology, DOI 10.1007/978-1-4939-2757-9

Definition B.7. The *interior* of a set A is the largest open subset of A.

Definition B.8. In a metric space (R, ρ), a *closed set* is any subset of R that is the complement of an open set.

The empty set is closed, and so is the entire space R.

Proposition B.9. *The intersection of any family of closed sets is a closed set. The union of a finite family of closed sets is a closed set.*

Definition B.10. In a metric space (R, ρ), the *closure* of a set A is the smallest subset of R containing A. It is denoted by $\bar{A}$. Any element of the closure of A is called a *point of closure* of A.

Proposition B.11. *A closed set is the intersection of a decreasing sequence of open sets. An open set is the union of an increasing sequence of closed sets.*

Definition B.12. A topological space is called a *Hausdorff topological space* if it satisfies the following property:

(HT) For any two distinct points x and y there exist two disjoint open sets A and B such that $x \in A$ and $y \in B$.

Proposition B.13. *A metric space is a Hausdorff topological space.*

Definition B.14. In a metric space (R, ρ), the *boundary* of a set A is the set $\partial A = \bar{A} \cap (R \setminus A)$. Here $R \setminus A$ is the complement of A.

Definition B.15. Given two metric spaces (R, ρ) and (R', ρ'), a function $f : R \to R'$ is *continuous* if for any open set A' in (R', ρ'), the set $f^{-1}(A')$ is an open set in (R, ρ).

Definition B.16. Two metric spaces (R, ρ) and (R', ρ') are said to be *homeomorphic* if a function $f : R \to R'$ exists satisfying the following two properties:

1. f is a bijection (an invertible function).
2. f is bicontinuous, i.e., both f and its inverse f^{-1} are continuous.

The function f above is called a *homeomorphism*.

Definition B.17. Given two distances ρ and ρ' on the same set R, we say that they are *equivalent distances* if the identity $i_R : x \in R \mapsto x \in R$ is a homeomorphism between the metric spaces (R, ρ) and (R', ρ').

Remark B.18. We may remark here that the notions of open set, closed set, closure, boundary, and continuous function are *topological notions*. They depend only on the topology induced by the metric. The topological properties of a metric space are invariant with respect to a homeomorphism.

Definition B.19. Given a subset A of a metric space (R, ρ), its *diameter* is given by $\delta(A) = \sup_{x \in A, y \in A} d(x, y)$. A is *bounded* if its diameter is finite.

Definition B.20. Given two metric spaces (R, ρ) and (R', ρ'), a function $f : R \to R'$ is *uniformly continuous* if for any $\epsilon > 0$ a $\delta > 0$ exists such that $x, y \in R$, $\rho(x, y) < \delta$ implies $\rho'(f(x), f(y)) < \epsilon$.

Proposition B.21. *A uniformly continuous function is continuous. (The converse is not true in general.)*

Remark B.22. The notions of diameter of a set and of uniform continuity of a function are *metric notions*.

Definition B.23. Let A, B be two subsets of a metric space R. A is said to be *dense* in B if $B \subseteq \bar{A}$. A is said to be *everywhere dense* in R if $\bar{A} = R$.

Definition B.24. A metric space R is said to be *separable* if it contains an everywhere dense countable subset.

Here are some examples of separable spaces with their corresponding everywhere dense countable subsets.

- The space $\mathbb{R}$ of real numbers with distance function $\rho(x, y) = |x - y|$, with the set $\mathbb{Q}$.
- The space $\mathbb{R}^n$ of ordered n-tuples of real numbers $x = (x_1, x_2, \ldots, x_n)$ with distance function $\rho(x, y) = \left\{ \sum_{k=1}^{n} (y_k - x_k)^2 \right\}^{\frac{1}{2}}$, with the set of all vectors with rational coordinates.
- The space $\mathbb{R}_0^n$ of ordered n-tuples of real numbers $x = (x_1, x_2, \ldots, x_n)$ with distance function $\rho_0(x, y) = \max \{ |y_k - x_k|; 1 \leq k \leq n \}$ with the set of all vectors with rational coordinates.
- $C^2([a, b])$, the totality of all continuous functions on the segment $[a, b]$ with distance function $\rho(x, y) = \int_a^b [x(t) - y(t)]^2 dt$ with the set of all polynomials with rational coefficients.

Definition B.25. A family $\{G_\alpha\}$ of open sets in metric space R is called a *basis* of R if every open set in R can be represented as the union of a (finite or infinite) number of sets belonging to this family.

Definition B.26. R is said to be a space with countable basis if there is at least one basis in R consisting of a countable number of elements.

Theorem B.27. *A necessary and sufficient condition for R to be a space with countable basis is that there exists in R an everywhere dense countable set.*

Corollary B.28. *A metric space R is separable if and only if it has a countable basis.*

Definition B.29. A *covering* of a set is a family of sets whose union contains the set. If the number of elements of the family is countable, then we have a *countable covering*. If the sets of the family are open, then we have an open *covering*.

Theorem B.30. *If R is a separable space, then we can select a countable covering from each of its open coverings.*

Theorem B.31. *Every separable metric space R is homeomorphic to a subset of $\mathbb{R}^{\infty}$.*

Definition B.32. In a metric space (R, ρ), a sequence $(x_n)_{n \in \mathbb{N}}$ is any function from $\mathbb{N}$ to R.

Definition B.33. We say that a sequence $(x_n)_{n \in \mathbb{N}}$ admits a *limit* $b \in R$ (is convergent to b) if b is such that for any open set V, with $x \in V$, there exists an $n_V \in \mathbb{N}$ such that for any $n > n_V$ we have $x_n \in V$. We write $\lim_{n \to \infty} x_n = b$.

Definition B.34. A subsequence of a sequence $(x_n)_{n \in \mathbb{N}}$ is any sequence $k \in \mathbb{N} \mapsto x_{n_k} \in R$ such that $(n_k)_{k \in \mathbb{N}}$ is strictly increasing.

Proposition B.35. *If $\lim_{n \to \infty} x_n = b$, then $\lim_{k \to \infty} x_{n_k} = b$ for any subsequence of $(x_n)_{n \in \mathbb{N}}$.*

Definition B.36. b is called a *cluster point* of a sequence $(x_n)_{n \in \mathbb{N}}$ if a subsequence exists having b as a limit.

Proposition B.37. *Given a subset A of a metric space (R, ρ), for any $a \in \bar{A}$ there exists a sequence of elements of A converging to a.*

Proposition B.38. *If x is the limit of a sequence $(x_n)_{n \in \mathbb{N}}$, then x is the unique cluster point of $(x_n)_{n \in \mathbb{N}}$. Conversely, $(x_n)_{n \in \mathbb{N}}$ may have a unique cluster point x, and still this does not imply that x is the limit of $(x_n)_{n \in \mathbb{N}}$ (see Aubin 1977, p. 67 for a counterexample).*

Definition B.39. In a metric space (R, ρ), a *Cauchy sequence* is a sequence $(x_n)_{n \in \mathbb{N}}$ such that for any $\epsilon > 0$ an integer $n_0 \in \mathbb{N}$ exists such that $m, n \in \mathbb{N}$, $m, n > n_0$ implies $\rho(x_m, x_n) < \epsilon$.

Proposition B.40. *In a metric space, any convergent sequence is a Cauchy sequence. The converse is not true in general.*

Proposition B.41. *In a metric space, if a Cauchy sequence $(x_n)_{n \in \mathbb{N}}$ has a cluster point x, then x is the limit of $(x_n)_{n \in \mathbb{N}}$.*

Definition B.42. A metric space R is called *complete* if any Cauchy sequence in R is convergent to a point of R.

Definition B.43. A subspace of a metric space (R, ρ) is any nonempty subset F of R endowed with the restriction of ρ to $F \times F$.

Proposition B.44. *If a subspace of a metric space R is complete, then it is closed in R. In a complete metric space, any closed subspace is complete.*

Definition B.45. A metric space R is said to be *compact* if any arbitrary open covering $\{O_\alpha\}$ of the space R contains a finite subcovering.

Definition B.46. A metric space R is called *precompact* if, for all $\epsilon > 0$, there is a finite covering of R by sets of diameter $< \epsilon$.

Remark B.47. The notion of compactness is a topological one, whereas the notion of precompactness is a metric one.

Theorem B.48. *For a metric space R, the following three conditions are equivalent:*

1. *R is compact.*
2. *Any infinite sequence in R has at least a limit point.*
3. *R is precompact and complete.*

Proposition B.49. *Every precompact metric space is separable.*

Proposition B.50. *In a compact metric space, any sequence that has only one cluster value a converges to a.*

Proposition B.51. *Any continuous mapping of a compact metric space into another metric space is uniformly continuous.*

Definition B.52. A *compact set* (or *precompact set*) in a metric space R is any subset of R that is compact (or precompact) as a subspace of R.

Proposition B.53. *Any precompact set is bounded.*

Proposition B.54. *Any compact set in a metric space is closed. In a compact metric space, any closed subset is compact.*

Proposition B.55. *Any compact set in a metric space is complete.*

Definition B.56. A set M in a metric space R is said to be *relatively compact* if $M = \bar{M}$.

Theorem B.57. *A relatively compact set is precompact. In a complete metric space, a precompact set is relatively compact.*

Proposition B.58. *A necessary and sufficient condition that a subset M of a metric space R be relatively compact is that every sequence of points of M has a cluster point in R.*

Definition B.59. A metric space R is said to be *locally compact* if for every point $x \in R$ there exists a compact neighborhood of x in R.

Theorem B.60. *Let R be a locally compact metric space. The following properties are equivalent:*

1. *There exists an increasing sequence (U_n) of open relatively compact sets in R such that $\bar{U}_n \subset U_{n+1}$ for every n, and $R = \cup_n U_n$.*
2. *R is the countable union of compact subsets.*
3. *R is separable.*

Convergence of Probability Measures

Let now (S, ρ) be a separable metric space endowed with the σ-algebra $\mathcal{S}$ of Borel subsets generated by the topology induced by ρ. As usual, given a probability space $(\Omega, \mathcal{F}, P)$, an S-valued random variable X is an $\mathcal{F} - \mathcal{S}$-measurable function $X : (\Omega, \mathcal{F}) \to (S, \mathcal{S})$.

Definition B.61. A sequence $(X_n)_{n \in \mathbb{N}}$ of random variables, with values in the common measurable space $(S, \mathcal{S})$, converges almost surely to the random variable X (notation $X_n \overset{\text{a.s.}}{\to} X$) if for almost all $\omega \in \Omega$, $X_n(\omega)$ converges to $X(\omega)$ with respect to the metric ρ.

In a metric space, in the foregoing definition only the elements of $(X_n)_{n \in \mathbb{N}}$ are required to be measurable, i.e., random variables, since in any case the limit function will automatically be itself measurable, i.e., a random variable (e.g., Dudley 2005, p. 125). We further remark that, since (S, ρ) is a separable metric space, for any two S-valued random variables X and Y, the distance $\rho(X, Y)$ is a real-valued random variable, so that the following definition makes sense.

Definition B.62. A sequence $(X_n)_{n \in \mathbb{N}}$ of random variables with values in the common measurable space $(S, \mathcal{S})$ converges (*in probability*) to the random variable X (notation $X_n \overset{P}{\to} X$) if for any $\varepsilon > 0$,

$$P(\rho(X_n, X) > \varepsilon) \to 0,$$

as $n \to \infty$.

Theorem B.63. *For random variables valued in a separable metric space, almost sure convergence implies convergence in probability.*

The converse of this theorem does not hold in general, though the following theorem holds.

Theorem B.64. *For random variables $(X_n)_{n\in\mathbb{N}}$ and X, valued in a separable metric space, $X_n \xrightarrow{P} X$ if and only if for every subsequence of $(X_n)_{n\in\mathbb{N}}$ there exists a subsubsequence that converges to X a.s.*

Proof. See, e.g., Dudley (2005, p. 288). □

Within the foregoing framework, let $\mathcal{L}^0(\Omega, \mathcal{F}, S, \mathcal{S})$ or simply $\mathcal{L}^0(S, \mathcal{S})$ denote the set of all $\mathcal{F} - \mathcal{S}$-measurable functions (i.e., S-valued random variables); we will then denote by $L^0(S, \mathcal{S})$ the set of equivalence classes of elements of $\mathcal{L}^0(S, \mathcal{S})$ with respect to the usual P-a.s. equality. Given two elements $X, Y \in \mathcal{L}^0(S, \mathcal{S})$, define

$$\alpha(X, Y) := \inf\{\varepsilon \geq 0 \mid P(\rho(X, Y) > \varepsilon) \leq \varepsilon\}.$$

Theorem B.65. *On $L^0(S, \mathcal{S})$, α is a metric that metrizes convergence in probability, so that for random variables $(X_n)_{n\in\mathbb{N}}$ and X, valued in the separable metric space S, $X_n \xrightarrow{P} X$ if and only if $\alpha(X_n, X) \to 0$.*

Proof. See, e.g., Dudley (2005, p. 289). □

The metric α is called the *Ky Fan* metric.

Theorem B.66. *If (S, ρ) is a complete separable metric space, then $L^0(S, \mathcal{S})$, endowed with the Ky Fan metric α, is complete.*

Proof. See, e.g., Dudley (2005, p. 290). □

Let (S, ρ) be a metric space endowed with its Borel σ-algebra $\mathcal{S}$ as above. Let $P, P_1, P_2, \ldots$ be probability measures on $(S, \mathcal{S})$, and let $C_b(S)$ be the class of all continuous bounded real-valued functions on S.

Definition B.67. A sequence of probability measures $(P_n)_{n\in\mathbb{N}}$ on $(S, \mathcal{S})$ converges weakly to a probability measure P (notation $P_n \xrightarrow{W} P$) if

$$\int_S f dP_n \to \int_S f dP$$

for every function $f \in C_b(S)$.

Proposition B.68. *If (S, ρ) is a metric space, then P and Q are two probability laws on $\mathcal{S}$, and, for any $f \in C_b(S)$, $\int_S f dP = \int_S f dQ$, then $P = Q$.*

An important consequence of the previous proposition is uniqueness of the weak limit of a sequence of probability laws.

Definition B.69. A sequence $(X_n)_{n \in \mathbb{N}}$ of random variables with values in a common measurable space $(S, \mathcal{S})$ converges in distribution to the random variable X (notation $X_n \overset{\mathcal{D}}{\to} X$) if the probability laws P_n of the X_n converge weakly to the probability law P of X:

$$P_n \overset{\mathcal{W}}{\to} P.$$

If we denote by $\mathcal{L}(X)$ the probability law of a random variable X, then the foregoing convergence can be equivalently written as

$$\mathcal{L}(X_n) \overset{\mathcal{W}}{\to} \mathcal{L}(X).$$

Proposition B.70. *If (S, ρ) is a separable metric space, for random variables $(X_n)_{n \in \mathbb{N}}$ and X, valued in S,*

$$X_n \overset{P}{\to} X \Rightarrow X_n \overset{\mathcal{D}}{\to} X.$$

Recall that if for some $x \in S$, $\mathcal{L}(X) = \epsilon_x$, i.e., X is a degenerate random variable, then

$$X_n \overset{P}{\to} X \Longleftrightarrow X_n \overset{\mathcal{D}}{\to} X.$$

Theorem B.71 (Skorohod representation theorem). *Consider a sequence $(P_n)_{n \in \mathbb{N}}$ of probability measures and a probability measure P on a separable metric space $(S, \mathcal{S})$ such that $P_n \xrightarrow[n \to \infty]{\mathcal{W}} P$. Then there exists a sequence of S-valued random variables $(Y_n)_{n \in \mathbb{N}}$ and a random variable Y defined on a common (suitably extended) probability space such that Y_n has probability law P_n, Y has probability law P, and*

$$Y_n \xrightarrow[n \to \infty]{\text{a.s.}} Y.$$

Proof. See, e.g., Billingsley (1968).   □

Consider sequences of random variables $(X_n)_{n \in \mathbb{N}}$ and $(Y_n)_{n \in \mathbb{N}}$ valued in a metric separable space (S, ρ) having a common domain; it makes sense to speak of the distance $\rho(X_n, Y_n)$, i.e., the function with value $\rho(X_n(\omega), Y_n(\omega))$ at ω. Since S is separable, $\rho(X_n, Y_n)$ is a random variable (Billingsley 1968, p. 225), and we have the following theorem.

Theorem B.72. *If $X_n \overset{\mathcal{D}}{\to} X$ and $\rho(X_n, Y_n) \overset{P}{\to} 0$, then $Y_n \overset{\mathcal{D}}{\to} X$.*

Let h be a measurable mapping of the metric space S into another metric space S'. If P is a probability measure on $(S, \mathcal{S})$, then we denote by $h(P)$ the probability measure induced by h on $(S', \mathcal{S}')$, defined by $h(P)(A) = P(h^{-1}(A))$ for any $A \in \mathcal{S}'$. Let D_h be the set of discontinuities of h.

Theorem B.73. *If* $P_n \overset{\mathcal{W}}{\to} P$ *and* $P(D_h) = 0$, *then* $h(P_n) \overset{\mathcal{W}}{\to} h(P)$.

For a random element X of S, $h(X)$ is a random element of S' (since h is measurable), and we have the following corollary.

Corollary B.74. *If* $X_n \overset{D}{\to} X$ *and* $P(X \in D_h) = 0$, *then* $h(X_n) \overset{D}{\to} h(X)$.

We recall now one of the most frequently used results in analysis.

Theorem B.75. (Helly). *For every sequence* $(F_n)_{n \in \mathbb{N}}$ *of distribution functions there exists a subsequence* $(F_{n_k})_{k \in \mathbb{N}}$ *and a nondecreasing, right-continuous function* F *(a generalized distribution function) such that* $0 \le F \le 1$ *and* $\lim_k F_{n_k}(x) = F(x)$ *at continuity points x of F.*

Definition B.76. *A set* A *in* S *such that* $P(\partial A) = 0$ *is called a* P-*continuity set.*

Theorem B.77 (Portmanteau theorem). *Let* $(P_n)_{n \in \mathbb{N}}$ *and* P *be probability measures on a metric space* (S, ρ) *endowed with its Borel σ-algebra. These five conditions are equivalent:*

1. $P_n \overset{\mathcal{W}}{\to} P$.
2. $\lim_n \int f \, dP_n = \int f \, dP$ *for all bounded, uniformly continuous real functions f.*
3. $\limsup_n P_n(F) \le P(F)$ *for all closed F.*
4. $\liminf_n P_n(G) \ge P(G)$ *for all open G.*
5. $\lim_n P_n(A) = P(A)$ *for all P-continuity sets A.*

Consider a metric space (S, ρ). Given a bounded real-valued function f on S, we may consider its Lipschitz seminorm defined as

$$\|f\|_L := \sup_{x \ne y} \frac{|f(x) - f(y)|}{\rho(x, y)}$$

and its supremum norm $\|f\|_\infty := \sup_x |f(x)|$. Let

$$\|f\|_{BL} := \|f\|_L + \|f\|_\infty,$$

and consider the set $BL(S, \rho)$ of all bounded real-valued Lipschitz functions on S, i.e.,

$$BL(S, \rho) := \{f : S \to \mathbb{R} | \ \|f\|_{BL} < \infty\}.$$

Theorem B.78. *Let* (S, ρ) *be a metric space.*

1. $BL(S, \rho)$ *is a vector space.*
2. $\|\cdot\|_{BL}$ *is a norm.*
3. $(BL(S, \rho), \|\cdot\|_{BL})$ *is a Banach space.*

For any two probability laws P and Q on the Borel σ-algebra of (S, ρ) we may define

$$\beta(P,Q) := \sup\left\{ |\int f dP - \int f dQ| \mid \|f\|_{BL} \leq 1 \right\}.$$

Theorem B.79. *Let (S, ρ) be a metric space endowed with its Borel σ-algebra $\mathcal{S}$. β is a metric on the set of all probability laws on $\mathcal{S}$.*

Now on a metric space (S, ρ) consider any subset $A \subset S$ and for any $\varepsilon > 0$ let

$$A^\varepsilon := \{ y \in S \mid \rho(x, y) < \varepsilon \text{ for some } x \in A \}.$$

For any two probability laws P and Q on the Borel σ-algebra $\mathcal{S}$ we may define

$$\gamma(P,Q) := \inf \{ \varepsilon > 0 \mid P(A) \leq Q(A^\varepsilon) + \varepsilon, \text{ for all } A \in \mathcal{S} \}.$$

Theorem B.80. *Let (S, ρ) be a metric space endowed with its Borel σ-algebra $\mathcal{S}$. γ is a metric on the set of all probability laws on $\mathcal{S}$.*

The metric γ is known as the *Prohorov metric*, or sometimes the *Lévy–Prohorov metric*.

Theorem B.81. *Let (S, ρ) be a separable metric space endowed with its Borel σ-algebra $\mathcal{S}$; consider a sequence $(P_n)_{n \in \mathbb{N}}$ and a P probability measure on $\mathcal{S}$. These four statements are equivalent.*

(a) $P_n \overset{W}{\to} P$
(b) $\lim_n \int f dP_n = \int f dP$ *for all functions* $f \in BL(S, \rho)$
(c) $\lim_n \beta(P_n, P) = 0$
(d) $\lim_n \gamma(P_n, P) = 0$

Proof. See, e.g., Dudley (2005, p. 395). □

The fact that convergence in probability implies convergence in law can be expressed in terms of the Prohorov and the Ky Fan metrics as follows.

Theorem B.82. *Let (S, ρ) be a separable metric space endowed with its Borel σ-algebra $\mathcal{S}$, and let X, Y be two S-valued random variables defined on the same probability space. Then*

$$\gamma(\mathcal{L}(X), \mathcal{L}(Y)) \leq \alpha(X, Y).$$

For an interesting account about metrics on probability measures and the relationships among them, we recommend the reader to refer to Gibbs and Su (2002).

Convergence of Empirical Measures

Consider a metric space (S, ρ) endowed with its Borel σ-algebra $\mathcal{S}$, and let $(X_n)_{n \in \mathbb{N}^*}$ be a sequence of i.i.d. S-valued random variables defined on the same probability space $(\Omega, \mathcal{F}, P)$. The sequence $(P_n)_{n \in \mathbb{N}^*}$ of *empirical measures* associated with $(X_n)_{n \in \mathbb{N}^*}$ is defined by

$$P_n(B)(\omega) := \frac{1}{n} \sum_{j=1}^{n} \epsilon_{X_j(\omega)}(B), \qquad B \in \mathcal{S}, \quad \omega \in \Omega,$$

where ϵ_x is the usual Dirac measure associated with a point $x \in S$.

The following theorem is a generalization of the Glivenko–Cantelli theorem, also known as the Fundamental Theorem of Statistics.

Theorem B.83 (Varadarajan). *Let (S, ρ) be a separable metric space endowed with its Borel σ-algebra $\mathcal{S}$; let $(X_n)_{n \in \mathbb{N}^*}$ be a sequence of i.i.d. S-valued random variables defined on the same probability space $(\Omega, \mathcal{F}, P)$; and let P_X denote their common probability law on $\mathcal{S}$. Then the sequence of empirical measures $(P_n)_{n \in \mathbb{N}^*}$ associated with $(X_n)_{n \in \mathbb{N}^*}$ converges to P_X almost surely, i.e.,*

$$P(\{\omega \in \Omega \mid P_n(\cdot)(\omega) \to P_X\}) = 1.$$

Proof. See, e.g., Dudley (2005, p. 399). □

On the set of probability measures on $(S, \mathcal{S})$, we may refer to the topology of weak convergence.

Definition B.84. Let Π be a family of probability measures on $(S, \mathcal{S})$. Π is said to be *relatively compact* if every sequence of elements of Π contains a weakly convergent subsequence, i.e., for every sequence $(P_n)_{n \in \mathbb{N}}$ in Π there exists a subsequence $(P_{n_k})_{k \in \mathbb{N}}$ and a probability measure P [defined on $(S, \mathcal{S})$, but not necessarily an element of Π] such that $P_{n_k} \overset{w}{\to} P$.

Theorem B.85. *Let $(P_n)_{n \in \mathbb{N}}$ be a relatively compact sequence of probability measures and P an additional probability measure on $(S, \mathcal{S})$. Then the following propositions are equivalent:*

(a) $P_n \overset{w}{\to} P$.
(b) *All weakly converging subsequences of $(P_n)_{n \in \mathbb{N}}$ weakly converge to P.*

Definition B.86. A family Π of probability measures on the general metric space $(S, \mathcal{S})$ is said to be *tight* if, for all $\epsilon > 0$, there exists a compact set K_ε such that

$$P(K_\varepsilon) > 1 - \epsilon \qquad \forall P \in \Pi.$$

B.2 Prohorov's Theorem

Prohorov's theorem gives, under suitable hypotheses, equivalence among relative compactness and tightness of families of probability measures.

Theorem B.87 (Prohorov). *Let Π be a family of probability measures on the measurable space $(S, \mathcal{S})$. Then*

1. *If Π is tight, then it is relatively compact.*
2. *Suppose S is separable and complete; if Π is relatively compact, then it is tight.*

Proof. See, e.g., Billingsley (1968).   □

Corollary B.88. *Let (S, ρ) be a Polish space endowed with its Borel σ-algebra $\mathcal{S}$; then, the metric space of all probability measures on $\mathcal{S}$ is complete with either metric β or γ.*

Proof. See, e.g., Dudley (2005, p. 405).   □

B.3 Donsker's Theorem

Weak Convergence and Tightness in $C([0,1])$

Consider a probability measure P on $(\mathbb{R}^\infty, \mathcal{B}_{\mathbb{R}^\infty})$, and let π_k be the projection from $\mathbb{R}^\infty$ to $\mathbb{R}^k$, defined by $\pi_{i_1,\ldots,i_k}(x) = (x_{i_1},\ldots,x_{i_k})$. The functions $\pi_k(P) : \mathbb{R}^k \to [0,1]$ are called *finite-dimensional distributions* corresponding to P. It is possible to show that probability measures on $(\mathbb{R}^\infty, \mathcal{B}_{\mathbb{R}^\infty})$ converge weakly if and only if all the corresponding finite-dimensional distributions converge weakly. Let $C := C([0,1])$ be the space of continuous functions on $[0,1]$ with uniform topology, i.e., the topology obtained by defining the distance between two points $x, y \in C$ as $\rho(x,y) = \sup_t |x(t) - y(t)|$. We shall denote with $(C, \mathcal{C})$ the space C with the topology induced by this metric ρ. For $t_1,\ldots,t_k$ in $[0,1]$, let $\pi_{t_1\ldots t_k}$ be the mapping that carries point x of C to point $(x(t_1),\ldots,x(t_k))$ of $\mathbb{R}^k$. The finite-dimensional distributions of a probability measure P on $(C, \mathcal{C})$ are defined as the measures $\pi_{t_1\ldots t_k}(P)$. Since these projections are continuous, the weak convergence of probability measures on $(C, \mathcal{C})$ implies the weak convergence of the corresponding finite-dimensional distributions, but the converse fails (perhaps in the presence of singular measures), i.e., weak convergence of finite-dimensional distributions of a sequence of probability measures on C is not a sufficient condition for weak convergence of the sequence itself in C. One can prove (e.g., Billingsley 1968) that an additional condition is needed, i.e., relative compactness of the sequence. Since C is a Polish space, i.e., a separable and complete metric space, by Prohorov's theorem we have the following result.

Theorem B.89. *Let $(P_n)_{n \in \mathbb{N}}$ and P be probability measures on $(C, \mathcal{C})$. If the sequence of the finite-dimensional distributions of P_n, $n \in \mathbb{N}$ converge weakly to those of P, and if $(P_n)_{n \in \mathbb{N}}$ is tight, then $P_n \overset{W}{\to} P$.*

To use this theorem we provide here some characterization of tightness. Given a $\delta \in]0, 1]$, a δ-continuity modulus of an element x of C is defined by

$$w_x(\delta) = w(x, \delta) = \sup_{|s-t|<\delta} |x(s) - x(t)|, \qquad 0 < \delta \leq 1.$$

Let $(P_n)_{n \in \mathbb{N}}$ be a sequence of probability measures on $(C, \mathcal{C})$.

Theorem B.90. *The sequence $(P_n)_{n \in \mathbb{N}}$ is tight if and only if these two conditions hold:*

1. *For each positive η there exists an a_η such that*

$$P_n(x||x(0)| > a_\eta) \leq \eta, \qquad n \geq 1.$$

2. *For each positive ϵ and η there exists a δ, with $0 < \delta < 1$, and an integer n_0 such that*

$$P_n(x|w_x(\delta) \geq \epsilon) \leq \eta, \quad n \geq n_0.$$

The following theorem gives a sufficient condition for compactness.

Theorem B.91. *If the following two conditions are satisfied:*

1. *For each positive η, there exists an a such that*

$$P_n(x||x(0)| > a) \leq \eta \qquad n \geq 1.$$

2. *For each positive ϵ and η, there exists a δ, with $0 < \delta < 1$, and an integer n_0 such that*

$$\frac{1}{\delta} P_n \left(x \left| \sup_{t \leq s \leq t+\delta} |x(s) - x(t)| \geq \epsilon \right. \right) \leq \eta, \qquad n \geq n_0,$$

for all $t \in [0, 1]$, then the sequence $(P_n)_{n \in \mathbb{N}}$ is tight.

Let X be a mapping from $(\Omega, \mathcal{F}, P)$ into $(C, \mathcal{C})$. For all $\omega \in \Omega$, $X(\omega)$ is an element of C, i.e., a continuous function on $[0, 1]$, whose value at t we denote by $X(t, \omega)$. For fixed t, let $X(t)$ denote the real function on Ω with value $X(t, \omega)$ at ω. Then $X(t)$ is the projection $\pi_t X$. Similarly, let $(X(t_1), X(t_2), \ldots, X(t_k))$ denote the mapping from Ω into $\mathbb{R}^k$ with values $(X(t_1, \omega), X(t_2, \omega), \ldots, X(t_k, \omega))$ at ω. If each $X(t)$ is a random variable, X is said to be a random function. Suppose now that $(X_n)_{n \in \mathbb{N}}$ is a sequence of random functions. According to Theorem B.90, $(X_n)_{n \in \mathbb{N}}$ is tight if and only if the sequence $(X_n(0))_{n \in \mathbb{N}}$ is tight, and for any positive real numbers ϵ and η there exists δ, $(0 < \delta < 1)$ and an integer n_0 such that

$$P\left(w_{X_n}(\delta) \geq \epsilon\right) \leq \eta, \qquad n \geq n_0.$$

This condition states that the random functions X_n, $n \in \mathbb{N}$, do not oscillate too much. Theorem B.91 can be restated in the same way: $(X_n)_{n\in\mathbb{N}}$ is tight if $(X_n(0))_{n\in\mathbb{N}}$ is tight, and if for any positive ϵ and η there exists a δ, $0 < \delta < 1$, and an integer n_0 such that

$$\frac{1}{\delta}P\left(\sup_{t \leq s \leq t+\delta} |X_n(s) - X_n(t)| \geq \epsilon\right) \leq \eta$$

for $n \geq n_0$ and $0 \leq t \leq 1$.

Donsker's Theorem

Let $(\xi_n)_{n\in\mathbb{N}\setminus\{0\}}$ be a sequence of i.i.d. random variables on $(\Omega, \mathcal{F}, P)$ with mean 0 and variance σ^2. We define the sequence of partial sums $S_n = \xi_1 + \cdots + \xi_n$, $n \in \mathbb{N}$, with $S_0 = 0$. Let us construct the sequence of random variables $(X_n)_{n\in\mathbb{N}}$ from the sequence $(S_n)_{n\in\mathbb{N}}$ by means of rescaling and linear interpolation, as follows:

$$X_n\left(\frac{i}{n}, \omega\right) = \frac{1}{\sigma\sqrt{n}} S_i(\omega) \qquad \text{for} \qquad \frac{i}{n} \in [0, 1[;$$

$$\frac{X_n(t) - X_n\left(\frac{i-1}{n}\right)}{X_n\left(\frac{i}{n}\right) - X_n\left(\frac{i-1}{n}\right)} - \frac{t - \frac{i-1}{n}}{\frac{1}{n}} = 0 \qquad \text{for} \qquad t \in \left[\frac{(i-1)}{n}, \frac{i}{n}\right]. \qquad \text{(B.1)}$$

With a little algebra, we obtain

$$X_n(t) = X_n\left(\frac{i-1}{n}\right) + \frac{t - \frac{i-1}{n}}{\frac{1}{n}}\left(X_n\left(\frac{i}{n}\right) - X_n\left(\frac{i-1}{n}\right)\right)$$

$$= \frac{t - \frac{i-1}{n}}{\frac{1}{n}} X_n\left(\frac{i}{n}\right) + \left(\frac{\frac{i}{n} - t}{\frac{1}{n}}\right) X_n\left(\frac{i-1}{n}\right)$$

$$= \frac{1}{\sigma\sqrt{n}} S_{i-1}(\omega) \frac{\frac{i}{n} - t}{\frac{1}{n}} + \frac{t - \frac{(i-1)}{n}}{\frac{1}{n}} \frac{1}{\sigma\sqrt{n}} S_i(\omega)$$

$$= \frac{1}{\sigma\sqrt{n}} S_{i-1}(\omega) \left(\frac{\frac{i}{n} - t}{\frac{1}{n}} + \frac{t - \frac{i}{n} + \frac{1}{n}}{\frac{1}{n}}\right) + \frac{1}{\sigma\sqrt{n}} \frac{t - \frac{(i-1)}{n}}{\frac{1}{n}} \xi_i(\omega)$$

$$= \frac{1}{\sigma\sqrt{n}} S_{i-1}(\omega) + n\left(t - \frac{i-1}{n}\right) \frac{1}{\sigma\sqrt{n}} \xi_i(\omega).$$

Since $i - 1 = [nt]$, if $t \in [\frac{(i-1)}{n}, \frac{i}{n}]$, we may rewrite (B.1) as follows:

$$X_n(t, \omega) = \frac{1}{\sigma\sqrt{n}} S_{[nt]}(\omega) + (nt - [nt]) \frac{1}{\sigma\sqrt{n}} \xi_{[nt]+1}(\omega). \qquad \text{(B.2)}$$

For any fixed ω, $X_n(\cdot, \omega)$ is a piecewise linear function whose pieces' amplitude decreases as n increases. Since the ξ_i and hence the S_i are random variables it follows by (B.2) that $X_n(t)$ is a random variable for each t. Therefore, the X_n are random functions. The following theorem provides a sufficient condition for $(X_n)_{n \in \mathbb{N}}$ to be a tight sequence.

Theorem B.92. *Suppose X_n, $n \in \mathbb{N}$ is defined by (B.2). The sequence $(X_n)_{n \in \mathbb{N}}$ is tight if for each positive ϵ there exists a λ, with $\lambda > 1$, and an integer n_0 such that, if $n \geq n_0$, then*

$$P\left(\max_{i \leq n} |S_{k+i} - S_k| \geq \lambda \sigma \sqrt{n}\right) \leq \frac{\epsilon}{\lambda^2} \qquad \text{(B.3)}$$

holds for all k.

If the sequence $(\xi_n)_{n \in \mathbb{N} \setminus \{0\}}$ is made of i.i.d. random variables, then condition (B.3) reduces to

$$P\left(\max_{i \leq n} |S_i| \geq \lambda \sigma \sqrt{n}\right) \leq \frac{\epsilon}{\lambda^2}. \qquad \text{(B.4)}$$

Let us denote by P_W the probability measure of the Wiener process as defined in Definition 2.167 and whose existence is a consequence of Theorem 2.64. We will refer here to its restriction to $t \in [0, 1]$, so that its trajectories are almost sure elements of $C([0, 1])$.

Theorem B.93 (Donsker). *Let $(\xi_n)_{n \in \mathbb{N} \setminus \{0\}}$ be a sequence of i.i.d. random variables defined on $(\Omega, \mathcal{F}, P)$ with mean 0 and finite, positive variance σ^2:*

$$E[\xi_n] = 0, \qquad E[\xi_n^2] = \sigma^2.$$

Let $S_n = \xi_1 + \xi_2 + \cdots + \xi_n$, $n \in \mathbb{N}$. Then the random functions

$$X_n(t, \omega) = \frac{1}{\sigma \sqrt{n}} S_{[nt]}(\omega) + (nt - [nt]) \frac{1}{\sigma \sqrt{n}} \xi_{[nt]+1}(\omega)$$

satisfy $X_n \xrightarrow{\mathcal{D}} W$.

Proof. We wish to apply Theorem B.89; we first show that the sequence of the finite-dimensional distributions of X_n, $n \in \mathbb{N}$ converge to those of W. Consider first a single time point s; we need to prove that

$$X_n(s) \xrightarrow{\mathcal{W}} W_s.$$

Since

$$\left| X_n(s) - \frac{1}{\sigma \sqrt{n}} S_{[ns]} \right| = (ns - [ns]) \left| \frac{1}{\sigma \sqrt{n}} \xi_{[ns]+1} \right|$$

and since, by Chebyshev's inequality,

$$P\left(\left|\frac{1}{\sigma\sqrt{n}}\xi_{[ns]+1}\right| \geq \epsilon\right) \leq \frac{E\left[\left|\frac{1}{\sigma\sqrt{n}}\xi_{[ns]+1}\right|^2\right]}{\epsilon^2}.$$

$$= \frac{1}{\sigma^2 n\epsilon^2}E\left[\xi_{[ns]+1}^2\right] =$$

$$= \frac{1}{n\epsilon^2} \to 0, \qquad n \to \infty,$$

we obtain

$$\left|X_n(s) - \frac{1}{\sigma\sqrt{n}}S_{[ns]}\right| \xrightarrow{P} 0.$$

Since $\lim_{n\to\infty} \frac{[ns]}{ns} = 1$, by the Central Limit Theorem for i.i.d. variables

$$\frac{1}{\sigma\sqrt{ns}}\sum_{k=1}^{[ns]} \xi_k \xrightarrow{\mathcal{D}} N(0,1),$$

so that

$$\frac{1}{\sigma\sqrt{n}}S_{[ns]} \xrightarrow{\mathcal{D}} W_s.$$

Therefore, by Theorem B.72, $X_n(s) \xrightarrow{\mathcal{D}} W_s$. Consider now two time points s and t with $s < t$. We must prove

$$(X_n(s), X_n(t)) \xrightarrow{\mathcal{D}} (W_s, W_t).$$

Since

$$\left|X_n(t) - \frac{1}{\sigma\sqrt{n}}S_{[nt]}\right| \xrightarrow{P} 0 \qquad \text{and} \qquad \left|X_n(s) - \frac{1}{\sigma\sqrt{n}}S_{[ns]}\right| \xrightarrow{P} 0$$

by Chebyshev's inequality, so that

$$\left\|(X_n(s), X_n(t)) - \left(\frac{1}{\sigma\sqrt{n}}S_{[ns]}, \frac{1}{\sigma\sqrt{n}}S_{[nt]}\right)\right\|_{\mathbb{R}^2} \xrightarrow{P} 0,$$

and by Theorem B.72, it is sufficient to prove that

$$\frac{1}{\sigma\sqrt{n}}\left(S_{[ns]}, S_{[nt]}\right) \xrightarrow{\mathcal{D}} (W_s, W_t).$$

By Corollary B.74 of Theorem B.73 this is equivalent to proving

$$\frac{1}{\sigma\sqrt{n}}\left(S_{[ns]}, S_{[nt]} - S_{[ns]}\right) \xrightarrow{\mathcal{D}} (W_s, W_t - W_s).$$

For independence of the random variables ξ_i, $i = 1, 2, \ldots, n$, the random variables $S_{[ns]}$ and $S_{[nt]} - S_{[ns]}$ are independent, so that

$$\lim_{n \to \infty} E\left[e^{\frac{iu}{\sigma\sqrt{n}}\sum_{j=1}^{[ns]}\xi_j + \frac{iv}{\sigma\sqrt{n}}\sum_{j=[ns]+1}^{[nt]}\xi_j}\right]$$

$$= \lim_{n \to \infty} E\left[e^{\frac{iu}{\sigma\sqrt{n}}\sum_{j=1}^{[ns]}\xi_j}\right] \cdot \lim_{n \to \infty} E\left[e^{\frac{iv}{\sigma\sqrt{n}}\sum_{j=[ns]+1}^{[nt]}\xi_j}\right]. \qquad (B.5)$$

Since $\lim_{n \to \infty} \frac{[ns]}{ns} = 1$, by the Lindeberg Theorem 1.190

$$\frac{1}{\sigma\sqrt{n}}S_{[ns]} \overset{D}{\to} N(0, s),$$

and for the same reason

$$\frac{1}{\sigma\sqrt{n}}(S_{[nt]} - S_{[ns]}) \overset{D}{\to} N(0, t - s),$$

so that

$$\lim_{n \to \infty} E\left[e^{\frac{iu}{\sigma\sqrt{n}}S_{[ns]}}\right] = e^{-\frac{u^2 s}{2}}$$

and

$$\lim_{n \to \infty} E\left[e^{\frac{iv}{\sigma\sqrt{n}}S_{[nt]} - S_{[ns]}}\right] = e^{-\frac{u^2 s}{2}}.$$

Substitution of these two last equations into (B.5) gives

$$\frac{1}{\sigma\sqrt{n}}\left(S_{[ns]}, S_{[nt]} - S_{[ns]}\right) \overset{D}{\to} (W_s, W_t - W_s),$$

and consequently

$$(X_n(s), X_n(t)) \overset{D}{\to} (W_s, W_t).$$

A set of three or more time points can be treated in the same way, and hence the finite-dimensional distributions converge properly. To prove tightness we apply Theorem B.92; under the assumptions of the present theorem, it can be shown (Billingsley 1968, p. 69) that

$$P\left(\max_{i \leq n}|S_i| \geq \lambda\sqrt{n}\sigma\right) \leq 2P\left(|S_n| \geq (\lambda - \sqrt{2})\sqrt{n}\sigma\right).$$

For $\frac{\lambda}{2} > \sqrt{2}$ we have

$$P\left(\max_{i \leq n}|S_i| \geq \lambda\sqrt{n}\sigma\right) \leq 2P\left(|S_n| \geq \frac{\lambda}{2}\sqrt{n}\sigma\right).$$

By the Central Limit Theorem,

$$P\left(|S_n| \geq \frac{1}{2}\lambda\sigma\sqrt{n}\right) \to P\left(|N| \geq \frac{1}{2}\lambda\right) < \frac{8}{\lambda^3}E\left[|N|^3\right],$$

where the last inequality follows by Chebyshev's inequality, and $N \sim N(0,1)$. Therefore, if ϵ is positive, there exists a λ such that

$$\limsup_{n \to \infty} P\left(\max_{i \leq n} |S_i| \geq \lambda \sigma \sqrt{n}\right) < \frac{\epsilon}{\lambda^2},$$

and then, by Theorem B.92, the sequence of the distribution functions of $(X_n)_{n \in \mathbb{N}}$ is tight.   □

An Application of Donsker's Theorem

Donsker's theorem has the following qualitative interpretation: $X_n \overset{\mathcal{D}}{\to} W$ implies that, if τ is small, then a particle subject to independent displacements $\xi_1, \xi_2, \ldots$ at successive times $\tau_1, \tau_2, \ldots$ appears to follow approximately a Brownian motion.

More important than this qualitative interpretation is the use of Donsker's theorem to prove limit theorems for various functions of the partial sums S_n. Using Donsker's theorem it is possible to use the relation $X_n \overset{\mathcal{D}}{\to} W$ to derive the limiting distribution of $\max_{i \leq n} S_i$.

Since $h(x) = \sup_t x(t)$ is a continuous function on C, $X_n \overset{\mathcal{D}}{\to} W$ implies, by Corollary B.74, that

$$\sup_{0 \leq t \leq 1} X_n(t) \overset{\mathcal{D}}{\to} \sup_{0 \leq t \leq 1} W_t.$$

The obvious relation

$$\sup_{0 \leq t \leq 1} X_n(t) = \max_{i \leq n} \frac{1}{\sigma \sqrt{n}} S_i$$

implies

$$\frac{1}{\sigma \sqrt{n}} \max_{i \leq n} S_i \overset{\mathcal{D}}{\to} \sup_{0 \leq t \leq 1} W_t. \tag{B.6}$$

Thus, under the hypotheses of Donsker's theorem, if we knew the distribution of $\sup_t W_t$, we would have the limiting distribution of $\max_{i \leq n} S_i$. The technique we shall use to obtain the distribution of $\sup_t W_t$ is to compute the limit distribution of $\max_{i \leq n} S_i$ in a simple special case and then, using $h(X_n) \overset{\mathcal{D}}{\to} h(W)$, where h is continuous on C or continuous except at points forming a set of Wiener measure 0, we obtain the distribution of $\sup_t W_t$ in the general case.

Suppose that $S_0, S_1, \ldots$ are the random variables for a symmetric random walk starting from the origin; this is equivalent to supposing that ξ_n are independent and satisfy

$$P(\xi_n = 1) = P(\xi_n = -1) = \frac{1}{2}. \tag{B.7}$$

Let us show that if a is a nonnegative integer, then

$$P\left(\max_{0\le i\le n} S_i \ge a\right) = 2P(S_n > a) + P(S_n = a). \tag{B.8}$$

If $a = 0$, then the previous relation is obvious; in fact, since $S_0 = 0$,

$$P\left(\max_{0\le i\le n} S_i \ge 0\right) = 1$$

and obviously, by symmetry of S_n

$$2P(S_n > 0) + P(S_n = 0) = P(S_n > 0) + P(S_n < 0) + P(S_n = 0) = 1.$$

Suppose now that $a > 0$ and put $M_i = \max_{0\le j\le i} S_j$. Since

$$\{S_n = a\} \subset \{M_n \ge a\}$$

and

$$\{S_n > a\} \subset \{M_n \ge a\},$$

we have

$$P(M_n \ge a) - P(S_n = a) = P(M_n \ge a, S_n < a) + P(M_n \ge a, S_n > a)$$

and

$$P(M_n \ge a, S_n > a) = P(S_n > a).$$

Hence we have to show that

$$P(M_n \ge a, S_n < a) = P(M_n \ge a, S_n > a). \tag{B.9}$$

Because of (B.7), all 2^n possible paths $(S_1, S_2, \ldots, S_n)$ have the same probability 2^{-n}. Therefore, (B.9) will follow if we show that the number of paths contributing to the left-hand event is the same as the number of paths contributing to the right-hand event. To show this, it suffices to find a one-to-one correspondence between the paths contributing to the right-hand event and the paths contributing to the left-hand event.

Given a path $(S_1, S_2, \ldots, S_n)$ contributing to the left-hand event in (B.9), match it with the path obtained by reflecting through a all the partial sums after the first one that achieves the height a. Since the correspondence is one-to-one, (B.9) follows. This argument is an example of the reflection principle. See also Lemma 2.180.

Let α be an arbitrary nonnegative number, and let $a_n = -[-\alpha n^{\frac{1}{2}}]$. By (B.9), we have

$$P\left(\max_{i\le n} \frac{1}{\sqrt{n}} S_i \ge a_n\right) = 2P(S_n > a_n) + P(S_n = a_n).$$

Since S_i can assume only integer values and since a_n is the smallest integer greater than or equal to $\alpha n^{\frac{1}{2}}$,

$$P\left(\max_{i \le n} \frac{1}{\sqrt{n}} S_i \ge \alpha\right) = 2P(S_n > a_n) + P(S_n = a_n).$$

By the central limit theorem,

$$P(S_n \ge a_n) \to P(N \ge \alpha),$$

where $N \sim N(0,1)$ and $\sigma^2 = 1$ by (B.7).

Since in the symmetric binomial distribution $S_n \to 0$ almost surely, the term $P(S_n = a_n)$ is negligible. Thus

$$P\left(\max_{i \le n} \frac{1}{\sqrt{n}} S_i \ge \alpha\right) \to 2P\left(N \ge \alpha\right), \qquad \alpha \ge 0. \tag{B.10}$$

By (B.10), (B.6), and (B.7), we conclude that

$$P\left(\sup_{0 \le t \le 1} W_t \le \alpha\right) = \frac{2}{\sqrt{2\pi}} \int_0^\alpha e^{-\frac{1}{2}u^2} du, \qquad \alpha \ge 0. \tag{B.11}$$

If we drop assumption (B.7) and suppose that the random variables ξ_n are i.i.d. and satisfy the hypothesis of Donsker's theorem, then (B.6) holds and from (B.11) we obtain

$$P\left(\frac{1}{\sigma\sqrt{n}} \max_{i \le n} S_i \le \alpha\right) \to \frac{2}{\sqrt{2\pi}} \int_0^\alpha e^{-\frac{1}{2}u^2} du, \qquad \alpha \ge 0.$$

Thus we have derived the limiting distribution of $\max_{i \le n} S_i$ by Lindeberg's theorem. Therefore, if the ξ_n are i.i.d. with $E[\xi_n] = 0$ and $E[\xi_n^2] = \sigma^2$, then the limit distribution of $h(X_n)$ does not depend on any further properties of the ξ_n. For this reason, Donsker's theorem is often called an invariance principle.

C

Diffusion Approximation of a Langevin System

The following result has gained interest in recent literature of Biophysics and Biochemistry (see, e.g., Schuss (2013)). It concerns the possibility to obtain a spatial diffusion approximation for the Langevin system of stochastic equations on $\mathbb{R}$, describing the evolution of the position $X(t)$ and the velocity $V(t)$ of a particle subject to a potential $\Phi(x)$, during time $t \in \mathbb{R}_+$,

$$\begin{cases} dX(t) = V(t)dt, \\ dV(t) = -[\gamma V(t) + \Phi'(X(t))]dt + \sqrt{2\varepsilon\gamma}dW_t. \end{cases} \tag{C.1}$$

Here we have taken a constant friction parameter γ, while ε derives from the Stokes-Einstein relation

$$\varepsilon = \frac{k_B T}{m}, \tag{C.2}$$

in terms of the temperature T, the mass m of the particle, and the Boltzmann constant k_B; $(W_t)_{t\in\mathbb{R}_+}$ denotes a standard Wiener process.

The solution process $(X(t))_{t\in\mathbb{R}_+}$ of the only position is not a Markov process, but the couple $(X(t), V(t))_{t\in\mathbb{R}_+}$ is a diffusion Markov process, whose joint probability density $p(x, v; t)$ is the solution of the following Fokker-Planck equation, for $x \in \mathbb{R}$, $v \in \mathbb{R}$, $t \in \mathbb{R}_+$, also known as Kramers Equation (see, e.g., Risken (1989, p. 87)),

$$\frac{\partial p}{\partial t}(x, v; t) = -\frac{\partial}{\partial x}vp(x, v; t) + \frac{\partial}{\partial v}[\gamma v + \Phi'(x)]p(x, v; t) + \gamma\varepsilon\frac{\partial^2}{\partial v^2}p(x, v; t), \tag{C.3}$$

subject to suitable initial conditions.

In the case of a large friction parameter γ, the above problem can be approximated by a Markov diffusion process involving the sole position $(X(t))_{t\in\mathbb{R}_+}$, by reducing system (C.1) to the following scalar SDE

© Springer Science+Business Media New York 2015

V. Capasso, D. Bakstein, *An Introduction to Continuous-Time Stochastic Processes*, Modeling and Simulation in Science, Engineering and Technology, DOI 10.1007/978-1-4939-2757-9

$$dX(t) = -\frac{1}{\gamma}\Phi'(X(t))dt + \sqrt{\frac{2\varepsilon}{\gamma}}dW_t. \tag{C.4}$$

We start with the following heuristic derivation (Gardiner (2004, p. 197)).

For large γ, we may assume that the second equation in (C.1) reaches very soon a "quasi" stationary value $\overline{V}$ for the velocity, which then satisfies

$$-[\gamma\overline{V} + \Phi'(X(t))]dt + \sqrt{2\varepsilon\gamma}dW_t = 0; \tag{C.5}$$

from which we may then get

$$\overline{V}dt = -\frac{1}{\gamma}\Phi'(X(t))dt + \sqrt{2\varepsilon\gamma}dW_t; \tag{C.6}$$

by substituting this "quasi" stationary value $\overline{V}$ for the velocity in the first equation of (C.1), we obtain Equation (C.4) for the position $X(t)$, which is then known as the high friction or Smoluchowski-Kramers approximation of the Langevin system (C.1).

Recently an extended and an updated treatment based on a rigorous probabilistic approach has been developed in Hottovy et al. (2014).

Here we present the following theorem, based on previous results by Papanicolau (1977), and recently reproposed in Schuss (2010, p. 265), which has been proven by usual asymptotic methods (see also Kervorkian and Cole (1996, p. 138)).

Theorem C.1 *If the initial value problem*

$$\begin{cases} \dfrac{\partial P^0(x,t)}{\partial t} = \dfrac{\partial}{\partial x}\left\{\dfrac{1}{\gamma}\left[\varepsilon\dfrac{\partial P^0(x,t)}{\partial x} + \Phi'(x)P^0(x,t)\right]\right\}, \\ P^0(x,0) = \varphi(x) \end{cases} \tag{C.7}$$

admits a unique solution for $\varphi \in L^1(\mathbb{R}) \cap L^\infty(\mathbb{R})$, such that

$$\varepsilon\frac{\partial P^0(x,t)}{\partial x} + \Phi'(x)P^0(x,t)$$

is bounded for all $t > t_0$ (for some $t_0 \in \mathbb{R}_+$), then, for all $(x,v,t) \in \mathbb{R}\times\mathbb{R}\times\mathbb{R}_+$ such that

$$\gamma \gg |v|\left|\frac{\partial}{\partial x}\ln P^0(x,t) + \frac{1}{\varepsilon}\Phi'(x)\right|,$$

the probability density function $p(x,v;t)$, solution of (C.3), admits the asymptotic expansion

$$p(x,v;t) \sim \frac{e^{-v^2/2\varepsilon}}{\sqrt{2\pi\varepsilon}}\left\{P^0(x,t) - \frac{v}{\gamma}\left[\frac{\partial P^0(x,t)}{\partial x} + \frac{1}{\varepsilon}\Phi'(x)P^0(x,t)\right] + O\left[\frac{1}{\gamma^2}\right]\right\} \tag{C.8}$$

Proof. Here we will only outline the proof, as from Schuss (2010); we encourage the reader to refer to the cited literature for more details.

Under the time scaling $t = \gamma s$, the Fokker-Planck Equation (C.3) can be rewritten in the form

$$0 = -\frac{1}{\gamma}\frac{\partial}{\partial s}p(x, v; s) - v\frac{\partial}{\partial x}p(x, v; s) + \frac{\partial}{\partial v}[\gamma v + \Phi'(x)]p(x, v; s) + \gamma\varepsilon\frac{\partial^2}{\partial v^2}p(x, v; s).$$

(C.9)

By introducing the operators

$$\mathcal{L}_0 p(x, v; s) = \frac{\partial}{\partial v}\left[v + \varepsilon\frac{\partial}{\partial v}\right]p(x, v; s);$$

(C.10)

$$\mathcal{L}_1 p(x, v; s) = -v\frac{\partial}{\partial x}p(x, v; s) + \Phi'(x)\frac{\partial}{\partial v}p(x, v; s);$$

(C.11)

$$\mathcal{L}_2 p(x, v; s) = -\frac{\partial}{\partial s}p(x, v; s),$$

(C.12)

we may rewrite Equation (C.9) as follows

$$[\gamma\mathcal{L}_0 + \mathcal{L}_1 + \frac{1}{\gamma}\mathcal{L}_2]p(x, v; s) = 0.$$

(C.13)

By expanding $p(x, v; s)$ in terms of the small parameter $\frac{1}{\gamma}$

$$p(x, v; s) = p^0(x, v; s) + \frac{1}{\gamma}p^1(x, v; s) + \left[\frac{1}{\gamma^2}\right]p^2(x, v; s) + \cdots,$$

(C.14)

so that, by comparing terms of the same order in $\frac{1}{\gamma}$ in (C.13), we obtain the following hierarchy of equations

$$\mathcal{L}_0 p^0(x, v; s) = 0;$$

(C.15)

$$\mathcal{L}_0 p^1(x, v; s) = -\mathcal{L}_1 p^0(x, v; s);$$

(C.16)

$$\mathcal{L}_0 p^2(x, v; s) = -\mathcal{L}_1 p^1(x, v; s) - \mathcal{L}_2 p^0(x, v; s).$$

(C.17)

Take Equation (C.15)

$$\frac{\partial}{\partial v}\left[v + \varepsilon\frac{\partial}{\partial v}\right]p^0(x, v; s) = 0,$$

(C.18)

subject to the integrability condition

$$\int\int p^0(x, v; s)dxdv = 1.$$

(C.19)

Under the required regularity assumptions on $p^0(x, v; s)$, we may then claim that

$$\left[v + \varepsilon \frac{\partial}{\partial v}\right] p^0(x, v; s) = const. \tag{C.20}$$

If we impose that both $p^0(x, v; s)$ and $\frac{\partial}{\partial v} p^0(x, v; s)$ tend to 0 as $|v| \to +\infty$ (coherently with the integrability condition (C.19)), we obtain $const = 0$, so that Equation (C.20) reduces to

$$\varepsilon \frac{\partial}{\partial v} p^0(x, v; s) = -v p^0(x, v; s), \tag{C.21}$$

which admits solutions of the form

$$p^0(x, v; s) = \frac{e^{-v^2/2\varepsilon}}{\sqrt{2\pi\varepsilon}} P^0(x, s), \tag{C.22}$$

where $P^0(x, s)$ is a function to be determined, satisfying the integrability condition

$$\int P^0(x, s) dx < +\infty. \tag{C.23}$$

Since we have shown that Equation (C.15) admits a nontrivial solution, by the Fredholm alternative theorems (Riesz and Sz–Nagy (1956, p. 161), Michajlov (1978, p. 85)), the solvability condition for next equation (C.16) is (Schuss (2010, p. 230), Gardiner (2004, p. 195))

$$\int \mathcal{L}_1 p^0(x, v; s) dv = 0. \tag{C.24}$$

It is then shown that an integrable solution of Equation (C.16) is given by

$$p^1(x, v; s) = \frac{e^{-v^2/2\varepsilon}}{\sqrt{2\pi\varepsilon}} \left\{ -\frac{v}{\gamma} \left[\frac{\partial P^0(x, s)}{\partial x} + \frac{1}{\varepsilon} \Phi'(x) P^0(x, s) \right] \right\}. \tag{C.25}$$

Once it is shown that Equation (C.16) admits a nontrivial solution, the solvability condition for Equation (C.17) becomes

$$\int (\mathcal{L}_1 p^1(x, v; s) + \mathcal{L}_2 p^0(x, v; s)) dv = 0. \tag{C.26}$$

The above eventually implies that $P^0(x, s)$ is an integrable solution of the following PDE

$$\frac{\partial}{\partial s} P^0(x, s) = \frac{\partial}{\partial x} \left[\varepsilon \frac{\partial}{\partial x} P^0(x, s) + \Phi'(x) P^0(x, s) \right]. \tag{C.27}$$

By taking into account expansion (C.14) and returning to the original time variable $t = \gamma s$, we obtain the approximation (C.25). $\square$

We may like to observe that Equation (C.14) is now the Fokker-Planck equation for the pdf of the solution of the SDE (C.4).

D

Elliptic and Parabolic Equations

We recall here basic facts about the existence and uniqueness of elliptic and parabolic equations; for further details, the interested reader may refer to Friedman (1963, 1964).

D.1 Elliptic Equations

Consider an open bounded $\Omega \subset \mathbb{R}^n$, (for $n \geq 1$). We are given a_{ij}, b_i, and c, $i, j = 1, \ldots, n$, real-valued functions defined on Ω. Consider the partial differential operator

$$M \equiv \frac{1}{2} \sum_{i,j=1}^{n} a_{ij}(x) \frac{\partial^2}{\partial x_i \partial x_j} + \sum_{i=1}^{n} b_i(x) \frac{\partial}{\partial x_i} + c(x). \tag{D.1}$$

The operator M is said to be *elliptic at a point* $x_0 \in \Omega$ if the matrix $(a_{i,j}(x_0))_{i,j=1,\ldots,n}$ is positive-definite, i.e., for any real vector $\xi \neq 0$, $\sum_{i,j=1}^{n} a_{ij}(x_0)\xi_i \xi_j > 0$.

If there is a positive constant μ such that

$$\sum_{i,j=1}^{n} a_{ij}(x)\xi_i \xi_j \geq \mu \mid \xi \mid^2$$

for all $x \in \Omega$, and all $\xi \in \mathbb{R}^n$, then M is said to be *uniformly elliptic* in Ω.

Definition D.1. A *barrier* for M at a point $y \in \partial\Omega$ is a continuous non-negative function w_y defined on $\overline{\Omega}$ that vanishes only at the point y and such that $M[w_y](x) \leq -1$, for any $x \in \Omega$.

© Springer Science+Business Media New York 2015
V. Capasso, D. Bakstein, *An Introduction to Continuous-Time Stochastic Processes*, Modeling and Simulation in Science, Engineering and Technology, DOI 10.1007/978-1-4939-2757-9

Proposition D.2. *Let $y \in \partial\Omega$. If there exists a closed ball K such that $K \cap \Omega = \emptyset$, and $K \cap \bar{\Omega} = \{y\}$, then y has a barrier for M.*

The First Boundary Value or Dirichlet Problem

Given a real-valued function f defined on Ω and a real-valued function ϕ defined on $\partial\Omega$, the Dirichlet problem consists of finding a solution u of the system

$$\begin{cases} M[u](x) = f(x) & \text{in} \quad \Omega, \\ u(x) \quad\;\; = \phi(x) & \text{in} \quad \partial\Omega. \end{cases} \tag{D.2}$$

Theorem D.3. *Assume that M is uniformly elliptic in Ω, that $c(x) \leq 0$, and that a_{ij}, b_i, $(i, j = 1, \ldots, n)$, c, f are uniformly Hölder continuous with exponent α in $\bar{\Omega}$. If every point of $\partial\Omega$ has a barrier, and ϕ is continuous on $\partial\Omega$, then there exists a unique $u \in C^2(\Omega) \cap C^0(\bar{\Omega})$ solution of the Dirichlet problem (D.2).*

Proof. See, e.g., Friedman (1963, 1964).   □

D.2 The Cauchy Problem and Fundamental Solutions for Parabolic Equations

Let

$$L_0 \equiv \frac{1}{2} \sum_{i,j=1}^{n} a_{ij}(x,t) \frac{\partial^2}{\partial x_i \partial x_j} + \sum_{i=1}^{n} b_i(x,t) \frac{\partial}{\partial x_i} + c(x,t) \tag{D.3}$$

be an elliptic operator in $\mathbb{R}^n$, for all $t \in [0, T]$, and let $f : \mathbb{R}^n \times [0, T] \to \mathbb{R}$, $\phi : \mathbb{R} \to \mathbb{R}$ be two appropriately assigned functions.

The Cauchy problem consists in finding a solution $u(x, t)$ of

$$\begin{cases} L[u] \equiv L_0[u] - u_t = f(x,t) & \text{in } \mathbb{R}^n \times]0, T], \\ u(x, 0) = \phi(x) & \text{in } \mathbb{R}^n. \end{cases} \tag{D.4}$$

The solution is understood to be a continuous function defined for $(x, t) \in \mathbb{R}^n \times [0, T]$, with its derivatives $u_{x_i}, u_{x_i x_j}, u_t$ continuous in $\mathbb{R}^n \times]0, T]$.

Theorem D.4. *Let the matrix $(a_{ij}(x,t))_{i,j=1,\ldots,n}$ be a nonnegative definite real matrix, and let*

$$| a_{ij}(x,t) | \leq C, \quad |b_i(x,t)| \leq C(|x|+1), \quad c(x,t) \leq C(|x|^2+1), \tag{D.5}$$

for a suitable constant C. If $L[u] \leq 0$ in $\mathbb{R}^n \times]0, T]$, and if $u(x,t) \geq -B\exp\{\beta|x|^2\}$ in $\mathbb{R}^n \times [0, T]$ (for some B, β positive constants), and if $u(x, 0) \geq 0$ in $\mathbb{R}^n$, then $u(x, t) \geq 0$ in $\mathbb{R}^n \times [0, T]$.

Proof. See, e.g., Friedman (2004, p. 139). □

Corollary D.5. *Let the matrix* $(a_{ij}(x,t))_{i,j=1,...,n}$ *be a nonnegative definite real matrix, and let (D.5) hold. Then there exists at most one solution of the Cauchy problem (D.4) satisfying*

$$|u(x,t)| \leq -B \exp\{\beta|x|^2\}$$

in $\mathbb{R}^n \times [0,T]$ *(for some* B,β *positive constants).*

The next theorem, and consequent corollary, considers different growth conditions on the coefficients of the operator L_0.

Theorem D.6. *Let the matrix* $(a_{ij}(x,t))_{i,j=1,...,n}$ *be a nonnegative definite real matrix, and let*

$$|a_{ij}(x,t)| \leq C(|x|^2 + 1), \quad |b_i(x,t)| \leq C(|x| + 1), \quad c(x,t) \leq C, \qquad (D.6)$$

where C *is a constant. If* $L[u] \leq 0$ *in* $\mathbb{R}^n \times]0,T]$, $u(x,t) \geq -N(|x|^q + 1)$ *in* $\mathbb{R}^n \times [0,T]$ *(where* N, q *are positive constants), and* $u(x,0) \geq 0$ *in* $\mathbb{R}^n$, *then* $u(x,t) \geq 0$ *in* $\mathbb{R}^n \times [0,T]$.

Proof. See, e.g., Friedman (2004, p. 140). □

Corollary D.7. *Let the matrix* $(a_{ij}(x,t))_{i,j=1,...,n}$ *be a nonnegative definite real matrix, and let conditions (D.6) be satisfied; then, there exists at most one solution* u *of the Cauchy problem with*

$$|u(x,t)| \leq N(1 + |x|^q),$$

where N, q *are positive constants.*

Later the following conditions will be required.

(A_1) There exists a $\mu > 0$ such that $\sum_{i,j=1}^{n} a_{ij}(x,t)\xi_i\xi_j \geq \mu\xi^2$ for all $(x,t) \in \mathbb{R}^n \times [0,T]$.

(A_2) The coefficients of L_0 are bounded continuous functions in $\mathbb{R}^n \times [0,T]$, and the coefficients $a_{ij}(x,t)$ are continuous in t, uniformly with respect to $(x,t) \in \mathbb{R}^n \times [0,T]$.

(A_3) The coefficients of L_0 are Hölder continuous functions (with exponent α) in x, uniformly with respect to the variables (x,t) in compacts of $\mathbb{R}^n \times [0,T]$, and the coefficients $a_{ij}(x,t)$ are Hölder continuous (with exponent α) in x, uniformly with respect to $(x,t) \in \mathbb{R}^n \times [0,T]$.

Definition D.8. A *fundamental solution of the parabolic operator* $L_0 - \frac{\partial}{\partial t}$ in $\mathbb{R}^n \times [0,T]$ is a function $\Gamma(x,t;\xi,r)$, defined, for all $(x,t) \in \mathbb{R}^n \times [0,T]$ and all $(\xi,t) \in \mathbb{R}^n \times [0,T]$, $t > r$, such that, for all ϕ with compact support,[14] the function

---

[14] The support of a function $f : \mathbb{R}^n \to \mathbb{R}$ is the closure of the set $\{x \in \mathbb{R}^n | f(x) \neq 0\}$.

$$u(x,t) = \int_{\mathbb{R}^n} \Gamma(x,t;\xi,r)\phi(\xi)d\xi$$

satisfies

(i) $L[u](x,t) - u_t(x,t) = 0$ for $x \in \mathbb{R}^n$, $r < t \leq T$
(ii) $u(x,t) \to \phi(x)$ as $t \downarrow r$, for $x \in \mathbb{R}^n$

Theorem D.9. *If conditions (A_1), (A_2), and (A_3) hold, then there exists a fundamental solution $\Gamma(x,t;\xi,r)$, for $L_0 - \frac{\partial}{\partial t}$, satisfying the inequalities*

$$|D_x^m \Gamma(x,t;\xi,r)| \leq c_1(t-r)^{-\frac{m+n}{2}} \exp\left\{-c_2 \frac{|x-\xi|^2}{t-r}\right\}, \qquad m = 0,1,$$

where c_1 and c_2 are positive constants. The functions $D_x^m \Gamma$, $m = 0,1,2,$, and $D_t \Gamma$ are continuous in $(x,t;\xi,r) \in \mathbb{R}^n \times [0,T] \times \mathbb{R}^n \times [0,T]$, $t > r$, and $L_0[\Gamma] - \Gamma_t = 0$, as a function of (x,t).

Finally, for any bounded continuous function ϕ we have

$$\int_{\mathbb{R}^n} \Gamma(x,t;\xi,r)\phi(x)dx \to \phi(\xi) \text{ for } t \downarrow r.$$

Proof. See, e.g., Friedman (2004, p. 141). □

Theorem D.10. *Let (A_1), (A_2), (A_3) hold, let $f(x,t)$ be a continuous function in $\mathbb{R}^n \times [0,T]$, Hölder continuous in x, uniformly with respect to (x,t) in compacts of $\mathbb{R}^n \times [0,T]$, and let ϕ be a continuous function in $\mathbb{R}^n$. Moreover, assume that*

$$|f(x,t)| \leq Ae^{a_1|x|^2} \quad \text{in } \mathbb{R}^n \times [0,T],$$
$$|\phi(x)| \leq Ae^{a_1|x|^2} \quad \text{in } \mathbb{R}^n,$$

where A, a_1 are positive constants. Then there exists a solution of the Cauchy problem (D.4) in $0 \leq t \leq T^*$, where $T^* = \min\left\{T, \frac{\bar{c}}{a_1}\right\}$ and $\bar{c}$ is a constant, which depends only on the coefficients of L_0, and

$$|u(x,t)| \leq A'e^{a_1'|x|^2} \quad \text{in } \mathbb{R}^n \times [0,T^*],$$

with positive constants A', a_1'.

The solution is given by

$$u(x,t) = \int_{\mathbb{R}^n} \Gamma(x,t;\xi,0)\phi(\xi)d\xi - \int_0^t \int_{\mathbb{R}^n} \Gamma(x,t;\xi,r)f(\xi,r)d\xi dr.$$

The adjoint operator L^* of $L = L_0 - \frac{\partial}{\partial t}$ is given by

$$L^*[v] = L_0^*[v] + \frac{\partial v}{\partial t},$$

$$L_0^*[v](x,t) = \frac{1}{2} \sum_{i,j=1}^n \frac{\partial^2}{\partial x_i \partial x_j}(a_{ij}(x,t)v(x,t)) - \sum_{i=1}^n \frac{\partial}{\partial x_i}(b_i(x,t)v(x,t)) + c(x,t),$$

by assuming that all quoted derivatives of the coefficients exist and are bounded functions.

Definition D.11. A *fundamental solution of the operator* $L_0^* + \frac{\partial}{\partial t}$ in $\mathbb{R}^n \times [0,T]$ is a function $\Gamma^*(x,t;\xi,r)$, defined, for all $(x,t) \in \mathbb{R}^n \times [0,T]$ and all $(\xi,r) \in \mathbb{R}^n \times [0,T]$, $t < r$, such that, for all g continuous with compact support, the function

$$v(x,t) = \int_{\mathbb{R}^n} \Gamma^*(x,t;\xi,r)g(\xi)d\xi$$

satisfies

1. $L^*[v] + v_t = 0$ for $x \in \mathbb{R}^n$, $0 \leq t \leq r$
2. $v(x,t) \to g(x)$ as $t \uparrow r$, for $x \in \mathbb{R}^n$

We consider the following additional condition.

(A_4) The functions $a_{ij}, \frac{\partial a_{ij}}{\partial x_i}, \frac{\partial^2 a_{ij}}{\partial x_i \partial x_j}, b_i, \frac{\partial b_i}{\partial x_i}, c$ are bounded and the coefficients of L_0^* satisfy conditions (A_2) and (A_3).

Theorem D.12. *If* (A_1)–(A_4) *are satisfied, then there exists a fundamental solution* $\Gamma^*(x,t;\xi,r)$ *of* $L_0^* + \frac{\partial}{\partial t}$*; it is such that*

$$\Gamma(x,t;\xi,r) = \Gamma^*(\xi,r;x,t), \qquad t > r.$$

Proof. See, e.g., Friedman (2004, p. 143). □

E

Semigroups of Linear Operators

In this appendix we will report the main results concerning the structure of contraction semigroups of linear operators on Banach spaces, as they are closely related to evolution semigroups of Markov processes. For the present treatment, we refer to the now classic books by Lamperti (1977), Pazy (1983), and Belleni-Morante and McBride (1998).

Throughout this appendix, E will denote a Banach space.

Definition E.1. A one-parameter family $(T_t)_{t \in \mathbb{R}_+}$ of linear operators on E is a strongly continuous semigroup of bounded linear operators or, simply, a C_0 *semigroup* if

(i) $T_0 = I$ (the identity operator)
(ii) $T_{s+t} = T_s T_t$, for all $s, t \in \mathbb{R}_+$
(iii) $\lim_{t \to 0+} \| T_t x - x \| = 0$, for all $x \in E$

Theorem E.2. *Let* $(T_t)_{t \in \mathbb{R}_+}$ *be a* C_0 *semigroup. There exist constants* $\omega \geq 0$ *and* $M \geq 1$ *such that*

$$\| T_t \| \leq M e^{\omega t}, \quad \text{for} \quad t \in \mathbb{R}_+. \tag{E.1}$$

Corollary E.3. *If* $(T_t)_{t \in \mathbb{R}_+}$ *is a* C_0 *semigroup, then, for any* $x \in E$, *the map* $t \in \mathbb{R}_+ \mapsto T_t x \in E$ *is a continuous function.*

Definition E.4. Let $(T_t)_{t \in \mathbb{R}_+}$ be a semigroup of bounded linear operators. The linear operator $\mathcal{A}$ defined by

$$\mathcal{D}_{\mathcal{A}} = \left\{ x \in E \mid \lim_{t \to 0+} \frac{T_t x - x}{t} \quad \text{exists} \right\}$$

© Springer Science+Business Media New York 2015
V. Capasso, D. Bakstein, *An Introduction to Continuous-Time Stochastic Processes*, Modeling and Simulation in Science, Engineering and Technology, DOI 10.1007/978-1-4939-2757-9

$$Ax = \lim_{t \to 0+} \frac{T_t x - x}{t}, \quad \text{for} \quad x \in \mathcal{D}_A,$$

where the limit is taken in the topology of the norm of E.

Theorem E.5. *Let* $(T_t)_{t \in \mathbb{R}_+}$ *be a* C_0 *semigroup, and let* $\mathcal{A}$ *be its infinitesimal generator. Then*

(a) *For* $x \in E$,

$$\lim_{h \to 0} \frac{1}{h} \int_t^{t+h} T_s x \, ds = T_t x.$$

(b) *For* $x \in E$, $\int_0^t T_s x ds \in \mathcal{D}_A$, *and*

$$A\left(\int_0^t T_s x \, ds \right) = T_t x - x.$$

(c) *For* $x \in \mathcal{D}_A$, $T_t x \in \mathcal{D}_A$, *and*

$$\frac{d}{dt} T_t x = A T_t x = T_t A x$$

(the derivative is taken in the topology of the norm of E*).*
(d) *For* $x \in \mathcal{D}_A$,

$$T_t x - T_s x = \int_s^t T_\tau A x \, d\tau = \int_s^t A T_\tau x \, d\tau.$$

Corollary E.6. *If* $\mathcal{A}$ *is the infinitesimal generator of a* C_0 *semigroup, then its domain* $\mathcal{D}_A$ *is dense in* E.

Corollary E.7. *Let* $(T_t)_{t \in \mathbb{R}_+}$, $(S_t)_{t \in \mathbb{R}_+}$ *be* C_0 *semigroups with infinitesimal generators* $\mathcal{A}$, *and* $\mathcal{B}$, *respectively. If* $\mathcal{A} = \mathcal{B}$, *then* $T_t = S_t$, *for* $t \in \mathbb{R}_+$.

Definition E.8. *Let* $(T_t)_{t \in \mathbb{R}_+}$ *be a* C_0 *semigroup. If in* (E.1) $\omega = 0$, *we say that* $(T_t)_{t \in \mathbb{R}_+}$ *is uniformly bounded; if, moreover,* $M = 1$, *we say that* $(T_t)_{t \in \mathbb{R}_+}$ *is a* C_0 *semigroup of contractions.*

The resolvent set $\rho(A)$ of a linear operator A on E (bounded or not) is the set of all complex numbers λ for which the operator $\lambda I - A$ is invertible, and its inverse is a bounded operator on E. The family

$$\left\{ R(\lambda : A) = (\lambda I - A)^{-1}, \quad \lambda \in \rho(A) \right\}$$

is called the *resolvent* of A.

Definition E.9. A linear operator $A : \mathcal{D}_A \subset E \to E$ is *closed* if and only if for any sequence $(x_n)_{n \in \mathbb{N}} \subset \mathcal{D}_A$ such that $u_n \to u$ and $A u_n \to v$ in E we have that $u \in \mathcal{D}_A$ and $v = Au$.

Theorem E.10. *Let* $(T_t)_{t \in \mathbb{R}_+}$ *be a* C_0 *semigroup of contractions.* $\mathcal{A}$ *is its infinitesimal generator if and only if*

(i) $\mathcal{A}$ is a closed linear operator, and $\overline{\mathcal{D}_{\mathcal{A}}} = E$.

(ii) The resolvent set $\rho(\mathcal{A})$ of $\mathcal{A}$ contains $\mathbb{R}_+^*$, and for any $\lambda > 0$,

$$\|R(\lambda : \mathcal{A})\| \leq \frac{1}{\lambda}. \tag{E.2}$$

Further, for any $\lambda > 0$ and any $x \in E$,

$$R(\lambda : \mathcal{A})x = \int_0^{+\infty} e^{-\lambda t} T_t x \, dt.$$

For any $\lambda > 0$, and any $x \in E$, $R(\lambda : \mathcal{A})x \in \mathcal{D}_{\mathcal{A}}$.

Proof. See, e.g., Pazy (1983, p. 8).   □

Note that, since the map $t \to T_t x$ is continuous and uniformly bounded, the integral exists as an improper Riemann integral and defines indeed a bounded linear operator satisfying (E.2).

Theorem E.11. *Let* $(T_t)_{t \in \mathbb{R}_+}$ *be a* C_0 *semigroup of contractions, and let* $\mathcal{A}$ *be its infinitesimal generator. Then, for any* $t \in \mathbb{R}_+$ *and any* $x \in E$,

$$T_t x = \lim_{n \to \infty} \left(I - \frac{t}{n} \mathcal{A} \right)^{-n} x = \lim_{n \to \infty} \left[\frac{n}{t} R\left(\frac{n}{t} : \mathcal{A} \right) \right]^n x.$$

The foregoing theorem induces the notation $T_t = e^{t\mathcal{A}}$.

Finally, based on all the foregoing treatment we may further notice that if $x \in \mathcal{D}_{\mathcal{A}}$, then we know that $T_t x \in \mathcal{D}_{\mathcal{A}}$, for any $t \in \mathbb{R}_+$, and it is the unique solution of the initial value problem

$$\frac{d}{dt} u(t) = \mathcal{A} u(t), \quad t > 0,$$

subject to the initial condition

$$u(0) = x.$$

F

Stability of Ordinary Differential Equations

We consider the system of ordinary differential equations

$$\begin{cases} \frac{d}{dt}\mathbf{u}(t) = \mathbf{f}(t, \mathbf{u}(t)), \, t > t_0, \\ \mathbf{u}(t_0) = \mathbf{c} \end{cases} \tag{F.1}$$

in $\mathbb{R}^d$ and we suppose that, for all $\mathbf{c} \in \mathbb{R}^d$, there exists a unique general solution $\mathbf{u}(t, t_0, \mathbf{c})$ in $[t_0, +\infty[$. We further suppose that $\mathbf{f}$ is continuous in $[t_0, +\infty[\times\mathbb{R}^d$ and that $\mathbf{0}$ is the equilibrium solution of $\mathbf{f}$. Thus $\mathbf{f}(t, \mathbf{0}) = \mathbf{0}$ for all $t \geq t_0$.

Definition F.1. The equilibrium solution $\mathbf{0}$ is *stable* if, for all $\epsilon > 0$:

$$\exists \delta = \delta(\epsilon, t_0) > 0 \text{ such that } \forall \mathbf{c} \in \mathbb{R}^d, |\mathbf{c}| < \delta \Rightarrow \sup_{t_0 \leq t \leq +\infty} |\mathbf{u}(t, t_0, \mathbf{c})| < \epsilon. \tag{F.2}$$

If condition (F.2) is not verified, then the equilibrium solution is *unstable*. The position of the equilibrium is said to be *asymptotically stable* if it is stable and *attractive*, namely, if along with (F.2), it can also be verified that

$$\lim_{t \to +\infty} \mathbf{u}(t, t_0, \mathbf{c}) = \mathbf{0} \qquad \forall \mathbf{c} \in \mathbb{R}^d, |\mathbf{c}| < \delta \text{ (chosen suitably)}.$$

Remark F.2. There may be attraction without stability.

Remark F.3. If $\mathbf{x}^* \in \mathbb{R}^d$ is the equilibrium solution of $\mathbf{f}$, then the position $\mathbf{y}(t) = \mathbf{u}(t) - \mathbf{x}^*$ tends toward $\mathbf{0}$.

Definition F.4. We consider the ball $B_h \equiv \bar{B}_h(0) = \{\mathbf{x} \in \mathbb{R}^d | \|\mathbf{x}\| \leq h\}, h > 0$, which contains the origin. The continuous function $v : B_h \to \mathbb{R}_+$ is *positive-definite* (in the Lyapunov sense) if

© Springer Science+Business Media New York 2015
V. Capasso, D. Bakstein, *An Introduction to Continuous-Time Stochastic Processes*, Modeling and Simulation in Science, Engineering and Technology, DOI 10.1007/978-1-4939-2757-9

$$\begin{cases} v(\mathbf{0}) = 0, \\ v(\mathbf{x}) > 0 \ \forall \mathbf{x} \in B_h \setminus \{\mathbf{0}\}. \end{cases}$$

The continuous function $v : [t_0, +\infty[\times B_h \to \mathbb{R}_+$ is *positive-definite* if

$$\begin{cases} v(t, \mathbf{0}) = 0 & \forall t \in [t_0, +\infty[, \\ \exists \omega : B_h \to \mathbb{R}_+ \text{ positive-definite such that } v(t, \mathbf{x}) \geq \omega(\mathbf{x}) \ \forall t \in [t_0, +\infty[. \end{cases}$$

v is negative-definite if $-v$ is positive-definite.

Now let $v : [t_0, +\infty[\times B_h \to \mathbb{R}_+$ be a positive-definite function endowed with continuous first partial derivatives with respect to t and x_i, $i = 1, \ldots, d$. We consider the function

$$V(t) = v(t, \mathbf{u}(t, t_0, \mathbf{c})) : [t_0, +\infty[\to \mathbb{R}_+,$$

where $\mathbf{u}(t, t_0, \mathbf{c})$ is the solution of (F.1). V is differentiable with respect to t, and we have

$$\frac{d}{dt} V(t) = \frac{\partial v}{\partial t} + \sum_{i=1}^{d} \frac{\partial v}{\partial x_i} \frac{du_i}{dt}.$$

But $\frac{du_i}{dt} = f_i(t, \mathbf{u}(t, t_0, \mathbf{c}))$, therefore

$$\dot{v} \equiv \frac{d}{dt} V(t) = \frac{\partial v}{\partial t} + \sum_{i=1}^{d} \frac{\partial v}{\partial x_i} f_i(t, \mathbf{u}(t, t_0, \mathbf{c})),$$

and this is the derivative of v with respect to time "along the trajectory" of the system. If $\frac{d}{dt} V(t) \leq 0$ for all $t \in (t_0, +\infty[$, then $\mathbf{u}(t, t_0, \mathbf{c})$ does not increase the value v, which measures by how much $\mathbf{u}$ moves away from $\mathbf{0}$. Through this observation, the required stability of the Lyapunov criterion for the stability of $\mathbf{0}$ has been formulated.

Definition F.5. Let $v : [t_0, +\infty[\times B_h \to \mathbb{R}_+$ be a positive-definite function. v is said to be a *Lyapunov function for the system (F.1) relative to the equilibrium position $\mathbf{0}$* if

1. v is endowed with first partial derivatives with respect to t and $x_i, i = 1, \ldots, d$.
2. For all $t \in]t_0, +\infty[$: $\dot{v}(t) \leq 0$ for all $\mathbf{c} \in B_h$.

Theorem F.6 (Lyapunov).

1. *If there exists $v(t, \mathbf{x})$ a Lyapunov function for system (F.1) relative to the equilibrium position $\mathbf{0}$, then $\mathbf{0}$ is stable.*
2. *If, moreover, the Lyapunov function $v(t, \mathbf{x})$ is such that, for all $t \in [t_0, +\infty[$: $v(t, \mathbf{x}) \leq \omega(\mathbf{x})$ with $\mathbf{u}$ being a positive definite function and $\dot{v}$ negative-definite along the trajectory, then $\mathbf{0}$ is asymptotically stable.*

Example F.7. We consider the autonomous linear system

$$\begin{cases} \frac{d}{dt}\mathbf{u}(t) = A\mathbf{u}(t), \ t > t_0, \\ \mathbf{u}(t_0) = \mathbf{c}, \end{cases}$$

where A is a matrix that does not depend on time. A matrix P is said to be positive definite if, for all $\mathbf{x} \in \mathbb{R}^d, \mathbf{x} \neq \mathbf{0} : \mathbf{x}'P\mathbf{x} > 0$. Considering the function $v(\mathbf{x}) = \mathbf{x}'P\mathbf{x}$, we have

$$\dot{v} = \frac{d}{dt}v(\mathbf{u}(t)) = \sum_{i=1}^{d} \frac{\partial v}{\partial x_i}(A\mathbf{u}(t))_i = \mathbf{u}'(t)PA\mathbf{u}(t) + \mathbf{u}'(t)A'P\mathbf{u}(t).$$

Therefore, if P is such that $PA + A'P = -Q$, with Q being positive-definite, then $\dot{v} = -\mathbf{u}'Q\mathbf{u} < 0$ and, by 2 of Lyapunov's theorem, $\mathbf{0}$ is asymptotically stable.

References

Aalen, O.: Nonparametric inference for a family of counting processes. Ann. Stat. **6**, 701–726 (1978)

Aletti, G., Capasso, V.: Profitability in a multiple strategy market. Decis. Econ. Finance **26**, 145–152 (2003)

Andersen, P.K., Borgan, Ø., Gill, R.D., Keiding, N.: Statistical Models Based on Counting Processes. Springer, Heidelberg (1993)

Anderson, W.J.: Continuous-Time Markov Chains: An Application-Oriented Approach. Springer, New York (1991)

Applebaum, D.: Levy Processes and Stochastic Calculus. Cambridge University Press, Cambridge (2004)

Arnold, L.: Stochastic Differential Equations: Theory and Applications. Wiley, New York (1974)

Ash, R.B.: Real Analysis and Probability. Academic, London (1972)

Ash, R.B., Gardner M.F.: Topics in Stochastic Processes. Academic, London (1975)

Aubin, J.-P.: Applied Abstract Analysis. Wiley, New York (1977)

Bachelier, L.: Théorie de la spéculation. Ann. Sci. École Norm. Sup. **17**, 21–86 (1900)

Bailey, N.T.J.: The Mathematical Theory of Infectious Diseases. Griffin, London (1975)

Baldi, P.: Equazioni differenziali stocastiche. UMI, Bologna (1984)

Barra, M., Del Grosso, G., Gerardi, A., Koch, G., Marchetti, F.: Some basic properties of stochastic population models. Lecture Notes in Biomathematics, vol. 32, pp. 155–164. Springer, Heidelberg (1978)

Bartholomew, D.J.: Continuous time diffusion models with random duration of interest. J. Math. Sociol. **4**, 187–199 (1976)

Bauer, H.: Probability Theory and Elements of Measure Theory. Academic, London (1981)

© Springer Science+Business Media New York 2015   457
V. Capasso, D. Bakstein, *An Introduction to Continuous-Time Stochastic Processes*, Modeling and Simulation in Science, Engineering and Technology, DOI 10.1007/978-1-4939-2757-9

Becker, N.: Analysis of Infectious Disease Data. Chapman & Hall, London (1989)

Belleni-Morante, A., McBride, A.C.: Applied Nonlinear Semigroups. Wiley, Chichester (1998)

Bertoin, J.: Lévy Processes. Cambridge University Press, Cambridge (1996)

Bhattacharya, R.N.: Criteria for recurrence and existence of invariant measures for multidimensional diffusions. Ann. Prob. **6**, 541–553 (1978)

Bhattacharya, R.N., Waymire, E.C.: Stochastic Processes with Applications. Wiley, New York (1990)

Biagini, F, Hu, Y., ksendal, B., Zhang, T.: Stochastic Calculus for Fractional Brownian Motion and Applications. Springer, London (2008)

Bianchi, A., Capasso, V., Morale, D.: Estimation and prediction of a nonlinear model for price herding. In: C. Provasi, (Ed.) Complex Models and Intensive Computational Methods for Estimation and Prediction, pp. 365–370. CLUEP, Padova (2005)

Billingsley, P.: Convergence of Probability Measures. Wiley, New York (1968)

Billingsley, P.: Probability and Measure. Wiley, New York (1986)

Black, F., Scholes, M.: The pricing of options and corporate liabilities. J. Pol. Econ. **81**, 637–654 (1973)

Bohr, H.: Almost Periodic Functions. Chelsea, New York (1947)

Bolley, F.: Quantitative concentration inequalities on sample path space for mean field interaction. arXiv:math/0511752v1 [math.PR], 30 Nov 2005

Bolley, F., Guillin, A., Villani, C.: Quantitative concentration inequalities for empirical measures on non-compact spaces. Prob. Theory Relat. Fields **137**, 541–593 (2007)

Borodin, A., Salminen, P.: Handbook of Brownian Motion: Facts and Formulae. Birkhä user, Boston (1996)

Bosq, D.: Linear Processes in Function Spaces. Theory and Applications. Springer, New York (2000)

Boyle, P., Tian, Y.: Pricing lookback and barrier options under the CEV process. J. Fin. Quant. Anal. **34**, 241–264 (1999)

Brace, A., Gatarek, D., Musiela, M.: The market model of interest rate dynamics. Math. Fin. **7**, 127–154 (1997)

Branger, N., Reichmann, O., Wobben, M.: Pricing electricity derivatives on an hourly basis. J. Energ. Market **3**, 51–89 (2010)

Breiman, L.: Probability. Addison-Wesley, Reading, MA (1968)

Bremaud, P.: Point Processes and Queues: Martingale Dynamics. Springer, Heidelberg (1981)

Burger, M., Capasso, V., Morale, D.: On an aggregation model with long and short range interaction. Nonlinear Anal. Real World Appl. **3**, 939–958 (2007)

Cai, G.Q., Lin, Y.K.: Stochastic analysis of the Lotka-Volterra model for ecosystems. Phys. Rev. E **70**, 041910 (2004)

Capasso, V.: A counting process approach for stochastic age-dependent population dynamics. In: Ricciardi, L.M. (Ed.) Biomathematics and Related Computational Problems, pp. 255–269. Kluwer, Dordrecht (1988)

Capasso, V.: A counting process approach for age-dependent epidemic systems. In: Gabriel, J.P. et al., (Eds.) Stochastic Processes in Epidemic Theory. Lecture Notes in Biomathematics, vol. 86, pp. 118–128. Springer, Heidelberg (1990)

Capasso, V.: Mathematical Structures of Epidemic Systems. Springer, Heidelberg (1993). Second corrected printing (2008)

Capasso, V., Di Liddo, A., Maddalena, L.: Asymptotic behaviour of a nonlinear model for the geographical diffusion of innovations. Dyn. Syst. Appl. **3**, 207–220 (1994)

Capasso, V., Morale, D.: Asymptotic behavior of a system of stochastic particles subject to nonlocal interactions. Stoch. Anal. Appl. **27**, 574–603 (2009)

Capasso, V., Morale, D., Sioli, F.: An agent-based model for "price herding", applied to the automobile market. MIRIAM reports, Milan (2003)

Carrillo, J.: Entropy solutions for nonlinear degenerate problems. Arch. Rat. Mech. Anal. **147**, 269–361 (1999)

Carrillo, J.A., McCann, R.J., Villani, C.: Kinetic equilibration rates for granular nedia and related equations: entropy dissipation and mass transportation estimates. Rev. Mat. Iberoam. **19**, 971–1018 (2003)

Champagnat, N., Ferriére, R., Méléard, S.: Unifying evolutionary dynamics: From individula stochastic processes to macroscopic models. Theor. Pop. Biol. **69**, 297–321 (2006)

Chan, K.C., Karolyi, G.A., Longstaff, F.A., Sanders, A.B.: An empirical comparison of alternative models of the short-term interest rate. J. Fin. **47**, 1209–1227 (1992)

Chiang, C.L.: Introduction to Stochastic Processes in Biostatistics. Wiley, New York (1968)

Chow, Y.S., Teicher, H.: Probability Theory: Independence, Interchangeability, Martingales. Springer, New York (1988)

Chung, K.L.: A Course in Probability Theory, 2nd edn. Academic, New York (1974)

Courant, R., Hilbert, D.: Methods of Mathematical Physics, vol. I. Wiley-Interscience, New York (1966)

Cox, J.C.: The constant elasticity of variance option pricing model. J. Portfolio Manage. **22**, 15–17 (1996)

Cox, J.C., Ross, S.A., Rubinstein, M.: Option pricing: A simplified approach. J. Fin. Econ. **7**, 229–263 (1979)

Çynlar, E.: Introduction to Stochastic Processes. Prentice Hall, Englewood Cliffs, NJ (1975)

Dai, W., Heyde, C.C.: Itô's formula with respect to fractional Brownian motion and its application. J. Appl. Math. Stoch. Anal. **9**, 439–448 (1996)

Dalang, R.C., Morton, A., Willinger, W.: Equivalent martingale measures and non-arbitrage in stochastic securities market models. Stochast. Stochast. Rep. **29**, 185–201 (1990)

Daley, D., Vere-Jones, D.: An Introduction to the Theory of Point Processes. Springer, Berlin (1988)

Daley, D., Vere-Jones, D.: An Introduction to the Theory of Point Processes. Volume II: General Theory and Structure. Springer, Heidelberg (2008)

Darling, D.A.D., Siegert, A.J.F.: The first passage time problem for a continuum Markov process. Ann. Math. Stat. **24**, 624–639 (1953)

Davis, M.H.A.: Piecewise-deterministic Markov processes: A general class of non-diffusion stochastic models. J. R. Stat. Soc. Ser. B **46**, 353–388 (1984)

Delbaen, F., Schachermeyer, W.: A general version of the fundamental theorem of asset pricing. Math. Annal. **300**, 463–520 (1994)

De Masi, A., Presutti, E.: Mathematical Methods for Hydrodynamical Limits. Springer, Heidelberg (1991)

Dembo, A., Zeitouni, O.: Large Deviations Techniques and Applications. Corrected printing. Springer, Heidelberg (2010)

Devijver, P.A., Kittler, J.: Pattern Recognition. A Statistical Approach. Prentice-Hall, Englewood Cliffs, NJ (1982)

Dieudonné, J.: Foundations of Modern Analysis. Academic, New York (1960)

Di Nunno, G., Øksendal, B., Proske, F.: Malliavin Calculus for Levy Processes with Applications to Finance. Springer, Berlin/Heidelberg (2009)

Donsker, M.D., Varadhan, S.R.S.: Large deviations from a hydrodynamical scaling limit. Comm. Pure Appl. Math. **42**, 243–270 (1989)

Doob, J.L.: Stochastic Processes. Wiley, New York (1953)

Dudley, R.M.: Real Analysis and Probability. Cambridge University Press, Cambridge (2005)

Duffie, D.: Dynamic Asset Pricing Theory. Princeton University Press, Princeton, NJ (1996)

Dupire, B.: Pricing with a smile. RISK **1**, 18–20 (1994)

Durrett, R., Levin, S.A.: The importance of being discrete (and spatial). Theor. Pop. Biol. **46**, 363–394 (1994)

Dynkin, E.B.: Markov Processes, vols. 1–2. Springer, Berlin (1965)

Einstein, A.: Über die von der molekularkinetischen Theorie der Wärme geforderten Bewegung von in ruhenden Flüssigkeiten suspendierten Teilchen. Annal. Phys. **17**, 549–560 (1905)

Embrechts, P., Klüppelberg, C., Mikosch, T.: Modelling Extreme Events for Insurance and Finance. Springer, Berlin (1997)

Epstein, J., Axtell, R.: Growing Artificial Societies–Social Sciences from the Bottom Up. Brookings Institution Press and MIT Press, Cambridge, MA (1996)

Ethier, S.N., Kurtz, T.G.: Markov Processes, Characterization and Convergence. Wiley, New York (1986)

Feller, W.: An Introduction to Probability Theory and Its Applications. Wiley, New York (1971)

Flierl, G., Grünbaum, D., Levin, S.A., Olson, D.: From individuals to aggregations: The interplay between behavior and physics. J. Theor. Biol. **196**, 397–454 (1999)

Fournier, N., Méléard, S.: A microscopic probabilistic description of a locally regulated population and macroscopic approximations. Ann. Appl. Prob. **14**, 1880–1919 (2004)

Franke, J., Härdle, W.K., Hafner, C.M.: Statistics of Financial Markets: An Introduction, 2nd edn. Springer, Heidelberg (2011)

Friedman, A.: Partial Differential Equations. Krieger, New York (1963)

Friedman, A.: Partial Differential Equations of Parabolic Type. Prentice-Hall, London (1964)

Friedman, A.: Stochastic Differential Equations and Applications. Academic, London (1975). Two volumes bounded as one, Dover, Mineola, NY (2004)

Fristedt, B., Gray, L.: A Modern Approach to Probability Theory. Birkhäuser, Boston (1997)

Gard, T.C.: Introduction to Stochastic Differential Equations. Marcel Dekker, New York (1988)

Gardiner, C.W.: Handbook of Stochastic Methods for Physics, Chemistry and the Natural Sciences, 3rd edn. Springer, Berlin (2004)

Ghanen, R.G., Spanos, P.D.: Stochastic Finite Elements. A Spectral Approach. Revised Edition. Dover Publication, Mineola, NY (2003)

Gibbs, A.L., Su, F.E.: On choosing and bounding probability metrics. Int. Stat. Rev. **70**, 419–435 (2002)

Gihman, I.I., Skorohod, A.V.: Stochastic Differential Equations. Springer, Berlin (1972)

Gihman, I.I., Skorohod, A.V.: The Theory of Random Processes. Springer, Berlin (1974)

Gnedenko, B.V.: The Theory of Probability. Chelsea, New York (1963)

Gray, A., Greenhalgh, D., Hu, L., Mao, X., Pan, J.: A stochastic differential equation SIS epidemic model. SIAM J. Appl. Math. **71**, 876–902 (2011)

Grigoriu, M.: Stochastic Calculus: Applications to Science and Engineering. Birkhäuser, Boston (2002)

Grünbaum, D., Okubo, A.: Modelling social animal aggregations. In: Levin, S.A. (Ed.) Frontiers of Theoretical Biology. Lectures Notes in Biomathematics, vol. 100, pp. 296–325. Springer, New York (1994)

Gueron, S., Levin, S.A., Rubenstein, D.I.: The dynamics of herds: From individuals to aggregations. J. Theor. Biol. **182**, 85–98 (1996)

Hagan, P.S., Kumar, D., Lesniewski, A., Woodward, D.E.: Managing smile risk. Wilmott Mag. **1**, 84–102 (2002)

Harrison, J.M., Kreps, D.M.: Martingales and arbitrage in multiperiod securities markets. J. Econ. Theory **20**, 381–408 (1979)

Harrison, J.M., Pliska, S.R.: Martingales and stochastic integrals in the theory of continuous trading. Stochast. Process. Appl. **11**, 215–260 (1981)

Has'minskii, R.Z.: Stochastic Stability of Differential Equations. Sijthoff & Noordhoff, The Netherlands (1980)

Heath, D., Jarrow, R., Morton, A.: Bond pricing and the term structure of interest rates: A new methodology for contingent claims valuation. Econometrica **1**, 77–105 (1992)

Heston, S.L.: A closed-form solution for options with stochastic volatility with applications to bond and currency options. Rev. Fin. Stud. **6**, 327–343 (1993)

Hethcote, H.W., Yorke, J.A.: Gonorrhea Transmission and Control. Lecture Notes Biomath, vol. 56. Springer, Heidelberg (1984)

Hottovy, S., McDaniel, A., Volpe, G., Wehr, J.: The Smoluchowski-Kramers limit of stochastic differential equations with arbitrary state-dependent friction. Commun. Math. Phys. DOI:10.1007/s00220-014-2233-4

Hunt, P.J., Kennedy, J.E.: Financial Derivatives in Theory and Practice. Wiley, New York (2000)

Hull, J., White, A.: Pricing interest rate derivative securities. Rev. Fin. Stud. **4**, 573–592 (1990)

Ikeda, N., Watanabe, S.: Stochastic Differential Equations and Diffusion Processes. North-Holland, Kodansha (1989)

Itô, K., McKean, H.P.: Diffusion Processes and Their Sample Paths. Springer, Berlin (1965)

Jacobsen, M.: Statistical Analysis of Counting Processes. Springer, Heidelberg (1982)

Jacod, J., Protter, P.: Probability Essentials. Springer, Heidelberg (2000)

Jacod, J., Shiryaev, A.N.: Limit Theorems for Stochastic Processes. Springer Lecture Notes in Mathematics. Springer, Berlin (1987)

Jacod, J., Shiryaev, A.N.: Limit Theorems for Stochastic Processes. Springer, Berlin (2003)

Jamshidian, F.: LIBOR and swap market models and measures. Fin. Stochast. **1**, 43–67 (1997)

Jeanblanc, M., Yor, M., Chesney, M.: Mathematical Methods for Financial Markets. Springer, London (2009)

Jelinski, Z., Moranda, P.: Software reliability research. In: Statistical Computer Performance Evaluation, pp. 466–484. Academic, New York (1972)

Kallenberg, O.: Foundations of Modern Probability. Springer, Berlin (1997)

Karatzas, I., Shreve, S.E.: Brownian Motion and Stochastic Calculus. Springer, New York (1991)

Karlin, S., Taylor, H.M.: A First Course in Stochastic Processes. Academic, New York (1975)

Karlin, S., Taylor H.M.: A Second Course in Stochastic Processes. Academic, New York (1981)

Karr, A.F.: Point Processes and Their Statistical Inference. Marcel Dekker, New York (1986)

Karr, A.F.: Point Processes and Their Statistical Inference. Second Edition Revised and expanded. Marcel Dekker, New York (1991)

Kemeny, J.G., Snell, J.L.: Finite Markov Chains. Van Nostrand, Princeton (1960)

Kervorkian, J., Cole, J.D.: Multiple Scale and Singular Perturbation Methods. Springer, New York (1996)

Khinchin, A.I.: Mathematical Foundations of Information Theory. Dover, New York (1957)

Klenke, A.: Probability Theory. Springer, Heidelberg (2008)

Klenke, A.: Probability Theory. A Comprehensive Course, 2nd edn. Springer, Heidelberg (2014)

Kloeden, P.E., Platen, E.: Numerical Solution of Stochastic Differential Equations. Springer, Heidelberg (1999)

Kolmogorov, A.N.: Foundations of the Theory of Probability. Chelsea, New York (1956)

Kolmogorov, A.N., Fomin, E.S.V.: Elements of Theory of Functions and Functional Analysis. Groylock, Moscow (1961)

Kou, S.: A jump-diffusion model for option pricing. Manage. Sci. **48**, 1086–1101 (2002)

Kyprianou, A.E.: Fluctuations of Lévy Processes with Applications, 2nd edn. Springer, Heidelberg (2014)

Ladyzenskaja, O.A., Solonnikov, V.A., Ural'ceva, N.N.: Linear and Quasilinear Equations of Parabolic Type. AMS, Providence, RI (1968)

Lamperti, J.: Stochastic Processes: A Survey of the Mathematical Theory. Springer, New York (1977)

Lapeyre, B., Pardoux, E., Sentis, R.: Introduction to Monte-Carlo Methods for Transport and Diffusion Equations. Oxford University Press, Oxford (2003)

Last, G., Brandt, A.: Marked Point Processes on the Real Line: The Dynamic Approach. Springer, Heidelberg (1995)

Lewis, A.L.: Option Valuation Under Stochastic Volatility. Finance Press, Newport Beach (2000)

Lipster, R., Shiryaev, A.N.: Statistics of Random Processes, I: General Theory. Springer, Heidelberg (1977)

Lipster, R., Shiryaev, A.N.: Statistics of Random Processes, II: Applications, 2nd edn. Springer, Heidelberg (2010)

Loève, M.: Probability Theory. Van Nostrand-Reinhold, Princeton, NJ (1963)

Lu, L.: Optimal control of input rates of Stein's models. Math. Med. Biol. **28**, 31–46 (2011)

Ludwig, D.: Stochastic Population Theories. Lecture Notes in Biomathematics, vol. 3. Springer, Heidelberg (1974)

Lukacs, E.: Characteristic Functions. Griffin, London (1970)

Mahajan, V., Wind, Y.: Innovation Diffusion Models of New Product Acceptance. Ballinger, Cambridge, MA (1986)

Malrieu, F.: Convergence to equilibrium for granular media equations and their Euler scheme. Ann. Appl. Prob. **13**, 540–560 (2003)

Mandelbrot, K., van Ness, J.: Fractional Brownian motions, fractional noises and applications. SIAM Rev. **10**, 422–437 (1968)

Mao, X.: Stochastic Differential Equations and Applications. Horwood, Chichester (1997)

Mao, X., Marion, G., Renshaw, E.: Environmental Brownian noise suppresses explosions in population dynamics. Stoch. Proc. Appl. **97**, 95–110 (2002)

Mao, X., et al.: Stochastic differential delay equations of population dynamics. J. Math. Anal. Appl. **304**, 296–320 (2005)

Markowich, P.A., Villani, C.: On the trend to equilibrium for the Fokker-Planck equation: An interplay between physics and functional analysis. Math. Contemp. **19**, 1–31 (2000)

Medvegyev, P.: Stochastic Integration Theory. Oxford University Press, Oxford (2007)

Méléard, S.: Asymptotic behaviour of some interacting particle systems: McKean–Vlasov and Boltzmann models. In: Talay, D., Tubaro, L. (Eds.) Probabilistic Models for Nonlinear Partial Differential Equations. Lecture Notes in Mathematics, vol. 1627, pp. 42–95. CIME Subseries. Springer, Heidelberg (1996)

Mercer, J.: Functions of positive and negative type and their connection with the theory of integral equations. Philos. Trans. Roy. Soc. Lond. **209**, 415–446 (1909)

Merton, R.C.: Theory of rational option pricing. Bell J. Econ. Manage. Sci. **4**, 141–183 (1973)

Merton, R.C.: Option pricing when underlying stock returns are discontinuous. J. Fin. Econ. **3**, 125–144 (1976)

Métivier, M.: Notions fondamentales de la théorie des probabilités. Dunod, Paris (1968)

Meyer, P.A.: Probabilités et Potentiel. Ilermann, Paris (1966)

Michajlov, V.P.: Partial Differential Equations. MIR Publishers, Moscow (1978)

Mikosch, T.: Non-Life Insurance Mathematics, 2nd edn. Springer, Berlin/Heidelberg (2009)

Miltersen, K.R., Sandmann, K., Sondermann, D.: Closed form solutions for term structure derivatives with log-normal interest rates. J. Fin. **52**, 409–430 (1997)

Mishura, Y.: Stochastic Calculus for Fractional Brownian Motion and Related Processes. Springer, Berlin (2008)

Morale, D., Capasso, V., Oelschläger, K.: An interacting particle system modelling aggregation behaviour: From individuals to populations. J. Math. Biol. **50**, 49–66 (2005)

Musiela, M., Rutkowski, M.: Martingale Methods in Financial Modelling. Springer, Berlin (1998)

Nagai, T., Mimura, M.: Some nonlinear degenerate diffusion equations related to population dynamics. J. Math. Soc. Jpn. **35**, 539–561 (1983)

Neveu, J.: Mathematical Foundations of the Calculus of Probability. Holden-Day, San Francisco (1965)

Norris, J.R.: Markov Chains. Cambridge University Press, Cambridge (1998)

Nowman, K.B.: Gaussian estimation of single-factor continuous time models of the term structure of interest rate. J. Fin. **52**, 1695–1706 (1997)

Nualart, D.: Stochastic calculus with respect to fractional Brownian motion. Ann. Fac. Sci. Toulouse-Mathématiques **15**, 63–77 (2006)

Oelschläger, K.: A law of large numbers for moderately interacting diffusion processes. Z. Wahrsch. verw. Geb. **69**, 279–322 (1985)

Oelschläger, K.: Large systems of interacting particles and the porous medium equation. J. Differ. Equ. **88**, 294–346 (1990)

Øksendal, B.: Stochastic Differential Equations. Springer, Berlin (1998)

Okubo, A.: Dynamical aspects of animal grouping: Swarms, school, flocks and herds. Adv. BioPhys. **22**, 1–94 (1986)

Papanicolau, G.C.: Introduction to the asymptotic analysis of stochastic equations. In: DiPrima, R.C. (Ed.) Modern Modeling of Continuum Phenomena. Lectures in Applied Mathematics, vol. 16, pp. 109–147. American Mathematical Society, Providence, RI (1977)

Parzen, E.: Statistical inference on time series by Hilbert space methods, Technical Report no. 23, Statistics Department, Stanford University, Stanford, CA, Jan 1959

Parzen, E.: Stochastic Processes. Holden-Day, San Francisco (1962)

Pascucci, A.: Calcolo Stocastico per la Finanza. Springer Italia, Milano (2008)

Pazy, A.: Semigroups of Linear Operators and Applications to Partial Differential Equations. Springer, New York (1983)

Pliska, S.R.: Introduction to Mathematical Finance: Discrete-Time Models. Blackwell, Oxford (1997)

Protter, P.: Stochastic Integration and Differential Equations. Springer, Berlin (1990). Second Edition 2004

Ramsay, J.O., Silverman, B.W.: Functional Data Analysis, 2nd edn. Springer, Berlin (1990). Second Edition 2004

Rebolledo, R.: Central limit theorems for local martingales. Z. Wahrsch. verw. Geb. **51**, 269–286 (1980)

Renardy, M., Rogers, R.C.: An Introduction to Partial Differential Equations, 2nd edn. Springer, New York (2005)

Revuz, D., Yor, M.: Continuous Martingales and Brownian Motion. Springer, Heidelberg (1991)

Riesz, F., Szőkefalvi-Nagy, B.: Functional Analyis. Blackie & Sons, London and Glasgow (1956)

Risken, H.: The Fokker–Planck Equation. Methods of Solution and Applications, 2nd edn. Springer, Heidelberg (1989)

Robert, P.: Stochastic Networks and Queues. Springer, Heidelberg (2003)

Rogers, L.C.G., Williams, D.: Diffusions, Markov Processes and Martingales, vol. 1. Wiley, New York (1994)

Rolski, T., Schmidli, H., Schmidt, V., Teugels, J.: Stochastic Processes for Insurance and Finance. Wiley, New York (1999)

Roozen, H.: Equilibrium and extinction in stochastic population dynamics. Bull. Math. Biol. **49**, 671–696 (1987)

Samorodnitsky, G., Taqqu, M.S.: Stable Non-Gaussian Random Processes. Chapman & Hall/CRC Press, Boca Ration, FL (1994)

Sato, K.I.: Lévy Processes and Infinitely Divisible Distributions. Cambridge University Press, Cambridge (1999)

Schuss, Z.: Theory and Applications of Stochastic Differential Equations. Wiley. New York (1980)

Schuss, Z.: Theory and Applications of Stochastic Processes: An Analytical Approach. Springer, New York (2010)

Schuss, Z.: Brownian Dynamics at Boundaries and Interfaces, in Physics, Chemistry, and Biology. Springer, New York (2013)

Shevchenko, G.: Mixed fractional stochastic differential equations with jumps. Stochastics **86**, 203–217 (2014)

Shiryaev, A.N.: Probability. Springer, New York (1995)

Shiryaev, A.N., Cherny, A.S.: Vector stochastic integrals and the fundamental theorems of asset pricing. Tr. Mat. Inst. Steklova **237**, 12–56 (2002)

Skellam, J.G.: Random dispersal in theoretical populations. Biometrika **38**, 196–218 (1951)

Skorohod, A.V.: Studies in the Theory of Random Processes. Dover, New York (1982)

Skorohod, A.V.: Asymptotic Methods in the Theory of Stochastic Differential Equations. AMS, Providence, RI (1989)

Sobczyk, K.: Stochastic Differential Equations: With Applications to Physics and Engineering. Kluwer, Dordrecht (1991)

Stein, R.B.: A theoretical analysis of neuronal variability. Biophys. J. **5**, 173–194 (1965)

Stein, R.B.: Some models of neuronal variability. Biophys. J. **7**, 37–68 (1967)

Taira, K.: Diffusion Processes and Partial Differential Equations. Academic, New York (1988)

Tan, W.Y.: Stochastic Models with Applications to Genetics, Cancers, AIDS and Other Biomedical Systems. World Scientific, Singapore (2002)

Tucker, H.G.: A Graduate Course in Probability. Academic Press, New York (1967)

Tuckwell, H.C.: On the first exit time problem for temporarily homogeneous Markov process. J. Appl. Prob. **13**, 39–48 (1976)

Tuckwell, H.C.: Stochastic Processes in the Neurosciences. SIAM, Philadelphia (1989)

Vasicek, O.: An equilibrium characterisation of the term structure. J. Fin. Econ. **5**, 177–188 (1977)

Ventcel', A.D.: A Course in the Theory of Stochastic Processes. Nauka, Moscow (1975) (in Russian). Second Edition 1996

Veretennikov, A.Y.: On polynomial mixing bounds for stochastic differential equations. Stoch. Proc. Appl. **70**, 115–127 (1997)

Veretennikov, A.Y.: On subexponential mixing rate for Markov processes. Theory Prob. Appl. **49**, 110–122 (2005)

Wang, F.J.S.: Gaussian approximation of some closed stochastic epidemic models. J. Appl. Prob. **14**, 221–231 (1977)

Warburton, K., Lazarus, J.: Tendency-distance models of social cohesion in animal groups. J. Theor. Biol. **150**, 473–488 (1991)

Wax, N.: Selected Papers on Noise and Stochastic Processes. Dover, New York (1954)

Williams, D.: Probability with Martingales. Cambridge University Press, Cambridge (1991)

Williger, W., Taqqu, M., Teverovsky, V.: Stock market process and long range dependence. Finance Stoch. **3**, 1–13 (1999)

Wilmott, P., Dewynne, J.N., Howison, S.D.: Option Pricing: Mathematical Models and Computation. Oxford Financial Press, Oxford (1993)

Wu, F., Mao, X., Chen, K.: A highly sensitive mean-reverting process in finance and the Euler–Maruyama approximations. J. Math. Anal. Appl. **348**, 540–554 (2008)

Yang, G.: Stochastic epidemics as point processes. In: Capasso, V., Grosso, E., Paveri-Fontana, S.L. (Eds.) Mathematics in Biology and Medicine. Lectures Notes in Biomathematics, vol. 57, pp. 135–144. Springer, Heidelberg (1985)

Zähle, M.: Long range dependence, no arbitrage and the Black-Scholes formula. Stochast. Dynam. **2**, 265–280 (2002)

Nomenclature

"Increasing" is used with the same meaning as "nondecreasing"; "decreasing" is used with the same meaning as "nonincreasing." In strict cases, "strictly increasing/strictly decreasing" is used.

:=	Equal by definition
≡	Coincide
∅	Empty set
$\int^*$	Integral of a nonnegative measurable function, finite or not
∇	Gradient
⊗	Product of σ-algebras or product of measures
∂A	Boundary of a set A
□	End of a proof
$\xrightarrow[n]{a.s.}$	Almost sure convergence
$\xrightarrow[n]{d}$	Convergence in distribution
$\xrightarrow[n]{L^p}$	Convergence in mean of order p
$\xrightarrow[n]{P}$ or $P-\lim$	Convergence in probability
$\xrightarrow[n]{W}$	Weak convergence
$(E, \mathcal{B}_E)$	Measurable space with E a set and $\mathcal{B}_E$ a σ-algebra of parts of E
$(\Omega, \mathcal{F}, P)$	Probability space with Ω a set, $\mathcal{F}$ a σ-algebra of parts of Ω, and P a probability measure on $\mathcal{F}$

© Springer Science+Business Media New York 2015   469
V. Capasso, D. Bakstein, *An Introduction to Continuous-Time Stochastic Processes*, Modeling and Simulation in Science, Engineering and Technology, DOI 10.1007/978-1-4939-2757-9

$[a, b]$	Closed interval of extremes a and b
$[a, b[$ or $[a, b)$	Semiopen interval closed at extreme a and open at extreme b
$]a, b]$ or $(a, b]$	Semiopen interval open at extreme a and closed at extreme b
$]a, b[$ or (a, b)	Open interval of extremes a and b
$|A|$ or $\sharp(A)$	Cardinal number (number of elements) of a finite set A
$|a|$	Absolute value of a number a; or modulus of a complex number a
$\|x\|$	Norm of a point x
$\langle M, N \rangle$	Predictable covariation of martingales M and N
$\langle M \rangle, \langle M, M \rangle$	Predictable variation of martingale M
$\langle f, g \rangle$	Scalar product of two elements f and g in a Hilbert space
$\bar{A}$	Closure of a set A depending on context
A'	Transpose of matrix A
$A \setminus B$	Set of elements of A that do not belong to B
$B(x, r)$ or $B_r(x)$	Open ball centered at x and having radius r
$\bar{C}$	Complement of set C depending on context
$C(A)$	Set of continuous functions from A to $\mathbb{R}$
$C(A, B)$	Set of continuous functions from A to B
$C^k(A)$	Set of functions from A to $\mathbb{R}$ with continuous derivatives up to order k
$C^{k+\alpha}(A)$	Set of functions from A to $\mathbb{R}$ whose kth derivatives are Lipschitz continuous with exponent α
$C_0(A)$	Set of continuous functions on A with compact support
$C_b(A)$ or $BC(A)$	Set of bounded continuous functions on A
$Cov[X, Y]$	Covariance of two random variables X and Y
$E[Y|\mathcal{F}]$	Conditional expectation of random variable Y with respect to σ-algebra $\mathcal{F}$
$E[\cdot]$	Expected value with respect to an underlying probability law clearly identifiable from context
$E_P[\cdot]$	Expected value with respect to probability law P
$E_x[\cdot]$	Expected value conditional upon a given initial state x in a stochastic process
F_X	Cumulative distribution function of a random variable X
$H \bullet X$	Stochastic Stieltjes integral of process H with respect to stochastic process X
I_A	Indicator function associated with a set A, i.e., $I_A(x) = 1$, if $x \in A$, otherwise $I_A(x) = 0$

$L^p(P)$	Set of equivalence classes of a.e. equal integrable functions with respect to measure P
$N(\mu, \sigma^2)$	Normal (Gaussian) random variable with mean μ and variance σ^2
$O(\Delta)$	Of the same order as Δ
$o(\delta)$	Of higher order with respect to δ
P-a.s.	Almost surely with respect to measure P
$P(A\|B)$	Conditional probability of event A with respect to event B
$P * Q$	Convolution of measures P and Q
$P \ll Q$	Measure P is absolutely continuous with respect to measure Q
$P \sim Q$	Measure P is equivalent to measure Q
P_X	Probability law of a random variable X
P_x	Probability law conditional upon a given initial state x in a stochastic process
$Var[X]$	Variance of a random variable X
W_t	Standard Brownian motion, Wiener process
$X \sim P$	Random variable X has probability law P
$a \wedge b$	Minimum of two numbers
$a \vee b$	Maximum of two numbers
a.e.	Almost everywhere
a.s.	Almost surely
$\exp\{x\}$	Exponential function e^x
$f * g$	Convolution of functions f and g
$f \circ X$ or $f(X)$	A function f composed with a function X
$f^{-1}(B)$	Preimage of set B by function f
f^-, f^+	Negative (positive) part of f, i.e., $f^- = \max\{-f, 0\}$ ($f^+ = \max\{f, 0\}$)
$f\|_A$	Restriction of a function f to set A
$\lim_{s\downarrow t}$	Limit for s decreasing while tending to t
$\lim_{s\uparrow t}$	Limit for s increasing while tending to t
$\mathrm{sgn}\{x\}$	Sign function; 1 if $x > 0$, 0 if $x = 0$, -1 if $x < 0$
$\mathbb{C}$	Complex plane
$\mathbb{N}$	Set of natural nonnegative integers
$\mathbb{N}^*$	Set of natural (strictly) positive integers
$\mathbb{Q}$	Set of rational numbers

$\mathbb{R}^n$	n-dimensional Euclidean space
$\bar{\mathbb{R}}$	Extended set of real numbers, i.e., $\mathbb{R} \cup \{-\infty, +\infty\}$
$\mathbb{R}_+$	Set of positive (nonnegative) real numbers
$\mathbb{R}_+^*$	Set of (strictly) positive real numbers
$\mathbb{Z}$	Set of all integers
$\mathcal{A}$	Infinitesimal generator of a semigroup
$\mathcal{B}_{\mathbb{R}^n}$	σ-algebra of Borel sets on $\mathbb{R}^n$
$\mathcal{B}_E$	σ-algebra of Borel sets generated by the topology of E
$\mathcal{D}_A$	Domain of definition of an operator A
$\mathcal{F}_t$ or $\mathcal{F}_t^X$	History of a process $(X_t)_{t \in \mathbb{R}_+}$ up to time t, i.e., σ-algebra generated by $\{X_s, s \le t\}$
$\mathcal{F}_{t-}$	σ-algebra generated by $\sigma(X_s, s < t)$
$\mathcal{F}_{t+}$	$\bigcap_{s>t} \mathcal{F}_t$
$\mathcal{F}_X$	σ-algebra generated by random variable X
$\mathcal{L}(X)$	Probability law of X
$\mathcal{L}^p(P)$	Set of integrable functions with respect to measure P
$\mathcal{M}(\mathcal{F}, \bar{\mathbb{R}}_+)$	Set of all $\mathcal{F}$-measurable functions with values in $\bar{\mathbb{R}}_+$
$\mathcal{M}(E)$	Set of all measures on E
$\mathfrak{P}(\Omega)$	Set of all parts of a set Ω
Δ	Laplace operator
Φ	Cumulative distribution function of a standard normal probability law
Ω	Underlying sample space
δ_{ij}	Kronecker delta, i.e., $= 1$ for $i = j$, $= 0$ for $i \ne j$
δ_x	Dirac delta function localized at x
ϵ_x	Dirac delta measure localized at x
$\sigma(\mathcal{R})$	σ-algebra generated by family of events $\mathcal{R}$
ω	Element of underlying sample space

Index

absolutely continuous, 405, 413
absorbing state, 113
adapted, 113, 145
affine
  rate model, 347
algebra, 401
  σ-, 401
  Borel, 15, 402
  generated, 9, 402
  product, 15, 77, 402
  semi, 401, 402
  smallest, 78, 402, 404
  tail, 48
annuity, 348
arithmetic Brownian motion, 277
asset
  riskless, 314
  risky, 314
attainable, 316
attractive, 453
autonomous, 253

Bachelier model, 334
ball, 419
  closed, 419
  open, 419
barrier, 443
basis, 421
bicontinuous, 420
bijection, 420

binomial
  distribution, 14
  variable, 22, 29
Black–Scholes
  equation, 321
  formula, 323
  model, 319
Borel
  σ-algebra, 402
  –Cantelli lemma, 48
  algebra, 15
  measurable, 402, 403, 409
boundary, 420
boundary point
  accessible, 301
  regular, 301
bounded, 421
bounded Lipschitz norm, 427
Brownian bridge, 88, 141, 277
Brownian motion, 129
  absorbed , 142
  arithmetic, 242, 277
  first passage time, 182
  fractional, 221
  geometric, 242, 277
  Hölder continuity, 140
  Lévy characterization, 134
  recurrence , 182
  reflected, 142

© Springer Science+Business Media New York 2015

V. Capasso, D. Bakstein, *An Introduction to Continuous-Time Stochastic Processes*, Modeling and Simulation in Science, Engineering and Technology, DOI 10.1007/978-1-4939-2757-9

Printed in the United States
By Bookmasters